THE MAGNETOSPHERES OF THE EARTH AND JUPITER

ASTROPHYSICS AND
SPACE SCIENCE LIBRARY

A SERIES OF BOOKS ON THE RECENT DEVELOPMENTS

OF SPACE SCIENCE AND OF GENERAL GEOPHYSICS AND ASTROPHYSICS

PUBLISHED IN CONNECTION WITH THE JOURNAL

SPACE SCIENCE REVIEWS

VOLUME 52
PROCEEDINGS

THE MAGNETOSPHERES
OF THE EARTH AND JUPITER

PROCEEDINGS OF THE NEIL BRICE MEMORIAL SYMPOSIUM,
HELD IN FRASCATI, MAY 28–JUNE 1, 1974

Jointly sponsored by the
LABORATORIO PER IL PLASMA NELLO SPAZIO
OF THE CONSIGLIO NAZIONALE DELLE RICERCHE, ITALY,
AND BY THE SOCIETÀ ITALIANA DI FISICA

Edited by

V. FORMISANO

Laboratorio Plasma Spazio, C.N.R., Italy

D. REIDEL PUBLISHING COMPANY

DORDRECHT-HOLLAND / BOSTON-U.S.A.

Library of Congress Cataloging in Publication Data

Neil Brice Memorial Symposium, Frascati, 1974.
      The magnetospheres of the earth and Jupiter.

      (Astrophysics and space science library ; v. 52)
      Bibliography: p.
      Includes index.
      1. Magnetosphere—Congresses.   2. Jupiter (Planet)—
Congresses.   I. Formisano, V., ed.   II. Laboratorio per il plasma nello
spazio.   III. Società italiana di fisica.   IV. Title.   V. Series.
QC809.M35N45   1974   538'.766   75-4587
ISBN-13:978-94-010-1791-6   e-ISBN-13:978-94-010-1789-3
DOI: 10.1007/978-94-010-1789-3

Published by D. Reidel Publishing Company,
P.O. Box 17, Dordrecht-Holland

Sold and distributed in the U.S.A., Canada, and Mexico
by D. Reidel Publishing Company, Inc.
306 Dartmouth Street, Boston,
Mass. 02116, U.S.A.

TABLE OF CONTENTS

PREFACE

The Symposium 'The Magnetospheres of the Earth and Jupiter' (May 28th–June 1st, 1974 in Frascati) was organized by the 'Laboratorio Plasma Spazio' of the CNR, on the occasion of its moving to the Frascati area.

The main theoretical topic was to be covered by N. Brice, but he died on January 31st, in a plane crash at Pago Pago (Samoa). It seemed appropriate to all of us to honor Neil Brice by renaming the meeting: 'Neil Brice Memorial Symposium'.

The Symposium covered problems of magnetospheric dynamics, of both the Earth and Jupiter, with emphasis on the comparison between the two objects.

The collaboration of American scientists participating in the recent NASA planetary missions has made it possible to have new important scientific results presented to and discussed by the scientific community.

Of the many people who helped to make the meeting such a success, very special thanks goes to Prof. C. Kennel of U.C.L.A. whose contribution to the organization of this symposium has been extremely important.

V. FORMISANO

LIST OF PARTICIPANTS

Ashour-Abdallah, M., GRI/CNET, 3 Ave. de la République, Issy les Moulineaux, France

Amata, E., ESRO ESTEC, Cosmic Rays Div., Domeinweg, Noordwijk, Holland

Auer, R. D., Max-Planck-Inst. für Extraterrestrial Physiks 8046 Garching-München, Germany

Berkey, F. T., The Auroral Obs., P.O. Box 953, 900L Tromsø, Norway

Biermann, L., Max-Planck-Inst. für Physik und Astrophysik, 8 München 40, Fohringer Ring 6, Germany

Bird, M., Inst. für Astrophysik, Univ. Bonn, Auf dem Hügel 71, 53 Bonn, Germany

Boischot, A., Observatoire de Meudon, 92190 Meudon, Dasop, France

Burger, J. J., Cosmic Ray Group, Kamerlingh Onnes Lab., Rijksuniversiteit, Nieuwsteeg 18, Leiden, Holland

Burton, R., GRI/CNET, 3 Ave. de la République, Issy les Moulineaux, France

Cerisier, J. C., Groupe de Recherches Ionos., 4 Ave. de Neptune, 94100 St. Maure, France

Colombo, G., Ist. Mecc. Appl., Università di Padova, Italy

Conforto, A. M., Università di Roma, Ist. Fisica, Roma, Italy

Coroniti, F., U.C.L.A., Los Angeles, Calif. 90024, U.S.A.

Cosmovici, G., Fisica Univ., Lecce, Italy

D'Anna, E., Fisica Univ., Lecce, Italy

Daigne, G., Observ. de Meudon, Dasop 92, Meudon, France

Duchossois, ESRO ESTEC, Domeinweg, Noordwijk, Holland

Dungey, J. W., Imperial Coll., London SW7, England

Eviatar, A., Tel Aviv Univ., Ramat Aviv, Tel Aviv

Fahr, H. J., Inst. für Astrophysik, Universität Bonn, Auf dem Hügel 71, Bonn 53, Germany

Fillius, R. W., Univ. of Calif., San Diego, Calif. U.S.A.

Fredricks, R. W., T.R.W., R1/1070, 1 Space Park Redondo Beach, Calif. 90278, U.S.A.

Freeman, J. W., Rice Univ., Dept. Space Science, Houston, Tex. 77001, U.S.A.

Gerard, E., Observ. de Meudon, Dasop 92190 Meudon, France

Goertz, C. K., State University, Iowa City, U.S.A.

Greenstadt, E. W., T.R.W., 1 Space Park, Redondo Beach, 90278 Calif. U.S.A.

Gregori, G. P., IFA, C.N.R., Roma

Gruenwaldt, H., M.P.I., 8046 Garching-München, Germany

Gurnett, D. A., State University, Iowa City, Iowa 52242, Dept. of Physics, U.S.A.

Haerendel, G., Max-Planck-Inst., 8046 Garching-München, Germany

Harvey, C. C., Obs. de Paris, 92190 Meudon, France
Haskell, G. P., ESRO HQ, 114 Ave. de C. de Gaulle Paris, Neuilly, France
Hide, R., Meteor Office, Met 021, London Road, Bracknell, Berkshire RG 122SZ, England
Hultqvist, B., Kiruna Geoph. Institute, 98101 Kiruna, Sweden
Hurley, K., Centre d'Étude Spatiale, B.P. 4057, 31029 Toulouse, Cedex, France
Intriligator, D. S., Physics Dept., University of Southern Calif. University Park, Los Angeles, Calif. 90007, U.S.A.
Kennel, C. F., UCLA, Dept. of Physics, Los Angeles, Calif. 90024, U.S.A.
Jentsch, V., Max-Planck-Inst., Postfach 40, 3411 Lindau-Harz, Germany
Lay, G., University of Bonn
Le Blanc, J., Obs. de Meudon, Dasop 92190 Meudon, France
Lecacheux, A., Obs. de Meudon, Dasop 92190 Meudon, France
Leggieri, G., Fisica Univ., Lecce, Italy
Luzemann, B., Inst. für Geoph. der T.U., Mendelssohnstr. 1, 33 Braunschweig, Germany
Martelli, G., Univ. of Sussex, Physics Bldg, Falmer Brighton, England
Nakano, G. H., Lockheed Palo Alto, Research Lab. 5210/205, 3251 Hanover Street, Palo Alto, Calif. 94304, U.S.A.
Nigel, A. H., Sheffield Univ., Space Physics Dept., England
Pacault, R., ESRO HQ, 114 Ave. de C. de Gaulle, Neuilly, Paris, France
Perona, G., Istituto di Elettronica, Politecnico di Torino G. Duca degli Abbruzzi 24 Torino, Italy
Prakash, A., Cornell Univ., Radiophysic C., 412 Space Science Bldg. Ithaca, N.Y. 14850, U.S.A.
Randall, B. A., Iowa University, Iowa City, Iowa, U.S.A.
Rasool, M., NASA HQ, Washington, D.C. 20564, U.S.A.
Rosenbauer, H., Max-Planck-Inst., Fohringer Ring 6, 8 München 40, Germany
Roth, M., Inst. d'Aeronom, 3 Ave. Circulaire, 1180 Bruxelles, Belge
Rothwell, P., Univ. of Southampton, Physics Department, Southampton, England
Russell, C. T., Space Science Centre, GG74, Univ. of California Los Angeles, Calif. 90024, U.S.A.
Saint Marc, A., CESR, BP 4057, 31029 Toulouse, Cedex, France
Scarf, F. L., T.R.W., R1/1071, 1 Space Park, Redondo Beach, 90278 Calif., U.S.A.
Schindler, K., Buhr Univ. Theor. Physik, Lehrstuhl IV, Universitätstr. 150, Gebaude NB 7/56, Pstfach 2148, 463 Bochum, Germany
Schuyer, S., ESTEC, ESRO, Domeinweg, Noordwijk, Holland
Shawhan, S., State Univ., Iowa City, U.S.A.
Simpson, J. A., Fermi Inst., Univ. of Chicago, 933 East 56th Street, Chicago 606 37, Ill., U.S.A.
Smith, E., J.P.L., 4800 Oak Grove Drive Pasadena, Calif. 91103, U.S.A.
Solomon, J., GRI/CNET, 3 Ave. de la République, Issy les Moulineaux
Tatnall, A., Sheffield Univ., Space Phys. Dept., Sheffield, England

Thiel, J., Groupe de Rech., 4 Ave. de Neptune, 94 St. Maure des Fosses, France
Thomas, B. T., Imperial College, London SW 7, England
Trulsen, J., Univ. of Troms, P.O. Box 953, 9000 Troms, Norway
Vasyliunas, V., Mass Inst. Tech., 77 Mass. Ave. Cambridge, Mass., U.S.A.
Vesecky, J. F., Astronomy Dept., Univ. of Leicester, Leicester LE1, 7RH England
Vulpetti, G., Ist. Meccanica, Univ. di Roma, Facoltà di Ingegneria, Via Eudossiana
    18, 00184 Roma, Italy
Wehrlin, N., GRI/CNET, 3 Ave. de la République, 92131 Issy les Moulineaux, France
Wenzel, K. P., ESTEC, ESRO, Domeinweg, Noordwijk, Holland
Williams, D. J., N.O.A.A. R43, Dept. of Commerce, Boulder, Colo. 80302, U.S.A.
Witt, N., Inst. für Astrophysik, Univ. Bonn, Auf dem Hügel 71, 53 Bonn, Germany
Wolfe, J. H., Ames, Muffet Field, Stanford, Calif., U.S.A.

and from the Laboratorio Plasma Spazio: Bavassano, B., Bavassano, M. B., Egidi, A.,
    Formisano, V., Mariani, F., Mastrantonio, G., Moreno, G., Pizzella, G., and
    Signorini, C.

PART I

THE MAGNETOSPHERE OF THE EARTH

THE UPSTREAM ESCAPE OF ENERGIZED
SOLAR WIND PROTONS FROM THE BOW SHOCK

EUGENE W. GREENSTADT

Space Sciences Dept., TRW Systems Group, Redondo Beach, Calif. 90278, U.S.A.

1. Introduction

Backstreaming 2–7 keV protons coming from the Earth's bow shock have been identified directly (Asbridge *et al.*, 1968; Scarf *et al.*, 1970), while still higher energy particles have been found upstream from interplanetary shocks (Armstrong *et al.*, 1970)
and from Jupiter's bow shock (Simpson *et al.*, 1974). In addition, it has been concluded that the longest-period hydromagnetic precursors could not have propagated
upstream from the Earth's bow shock, but must have been generated upstream by
some other agency, presumably reflected particles, and swept downstream with the
solar wind (Greenstadt *et al.*, 1970; Fairfield, 1969). Accordingly, it has also been
shown that protons reflected from the bow shock should be accelerated by the interplanetary electric field, seen in the shock frame, to energies comparable to those observed by plasma experiments (Sonnerup, 1969). These energies correspond to particles traveling along **B** at velocities comparable to the rate at which locally-excited
hydromagnetic precursor waves appear to progress upstream (Greenstadt *et al.*,
1970). Recently, however, there have been some systematic observations of backstreaming protons at the Earth's bow shock with parallel velocity components and
total energies much too high to be associated with the usual long-period upstream
waves or to be produced by Sonnerup's simple reflection process (Lin *et al.*, 1974),
and these protons (30–100 keV) were indeed attributed to some unknown acceleration mechanism in the upstream region.

The observations of Lin *et al.* involved protons of high pitch angle, and, although
their reasons for favoring an upstream acceleration were quite different, it may seem
intuitive that high pitch angle particles would have difficulty escaping the shock,
especially at large field-normal angles. Such an inference would superficially support
the notion of energization outside the bow shock. It seems worthwhile therefore to
examine the extent to which the geometry of individual particle motion alone might
select among reflected particles those that can escape upstream and those that cannot.

In the following paragraphs, the geometry of escape is described and some simple
numerical examples are worked out for a few special cases. It is found that protons
with rather high energies and pitch angles can escape the shock at only marginally
quasi-parallel field orientations (i.e., $\theta_{nB} \approx 50°$), even if they have quite moderate
speeds parallel to **B**.

2. The Geometry of Escaping Particles

For simplicity, we place ourselves exclusively in the ecliptic plane so that we have

V. Formisano (ed.), The Magnetospheres of the Earth and Jupiter, 3–10. All Rights Reserved
Copyright © 1975 by D. Reidel Publishing Company, Dordrecht-Holland

an observation point on the ecliptic outside the shock. Solar wind velocity and field vectors $\mathbf{V}_{SW}$ and $\mathbf{B}_{SW}$ are assumed in the ecliptic, and we deal with the shock locally only as a plane whose unit normal lies in the ecliptic. We work only with dimensionless ratios, so that reflected particle velocities are proportional to the solar wind velocity, and all velocities and energies are divided by V_{SW}. Thus we adopt the convention that the projection $u_\parallel$ of a reflected particle's velocity on $\mathbf{B}_{SW}$ is a product of some scalar p and V_{SW}, i.e., $u_\parallel = p V_{SW}$. Thermal velocities are neglected. The particle velocity $u_\perp$ perpendicular to $\mathbf{B}_{SW}$ is treated the same: $u_\perp \equiv P V_{SW}$. Note that $u_\parallel$ is the guiding center velocity *in the plasma frame* and the particle has pitch angle $\alpha = \arctan u_\perp/u_\parallel = \arctan P/p$, also in the plasma (solar wind) frame.

The sketch of Figure 1 defines one set of quantities used here. The curve represents the ecliptic intersection of the bow shock; $\mathbf{n}$ is the local normal at a point located at angle θ_{XR} from the Sun-Earth line (X-axis). The shock is taken to be symmetric about the X axis and is given by $Y^2 = 0.331\,[(X - 75.25)^2 - 3686]$. This is a symmetrized version of Fairfield's (1971) average shock used in an earlier paper (Greenstadt, 1972).

Additional terms are defined in Figure 2. The ecliptic plane contains $\mathbf{X}$, $\mathbf{Y}$, $\mathbf{B}$, $\mathbf{B}_\perp$, and $\mathbf{n}$, and $\mathbf{Z}$ is the usual ecliptic pole. The solar wind impacts the shock at velocity $\mathbf{V}_{SW}$ along $-\mathbf{X}$. We are interested in a proton whose trajectory, given by the vector $\mathbf{S}$ from the origin, follows a spiral along $\mathbf{B}$ away from the shock, as depicted. Its guiding center has speed $u_\parallel = p V_{SW}$, and its Larmor radius is $a_c = u_\perp/\omega_c = P V_{SW}/\omega_c$. We treat the unshocked plasma as if at rest at zero temperature, so the shock, moving upward along $\mathbf{n}$ at speed $V_{SW} \cos\theta_{Xn}$, encounters protons at rest, some of which are picked up by the shock, accelerated, and emitted at phase ϕ and time $t = 0$. These protons spiral up $\mathbf{B}_{SW}$ with the shock in pursuit. Phase angle ϕ is defined as 0 when $\mathbf{u}_\perp$ is parallel to $\hat{\mathbf{B}}_\perp$, i.e., when the reflected proton escapes the shock at $\mathbf{S}(X, Y, Z) = (0, 0, a_c)$.

We are interested in those protons that are not overtaken by the shock after they begin their corkscrew journey away from it. These are the particles that will be detected far upstream, possibly in association with upstream waves. The condition for 'free escape' is $\mathbf{S} \cdot \mathbf{n} > V_{SW} t \cos\theta_{Xn}$ for all $t > 0$. Note that a proton is most vulnerable

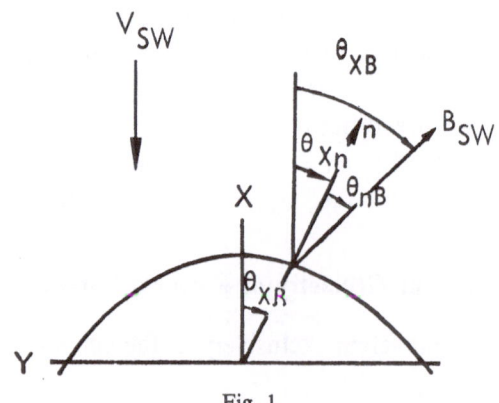

Fig. 1.

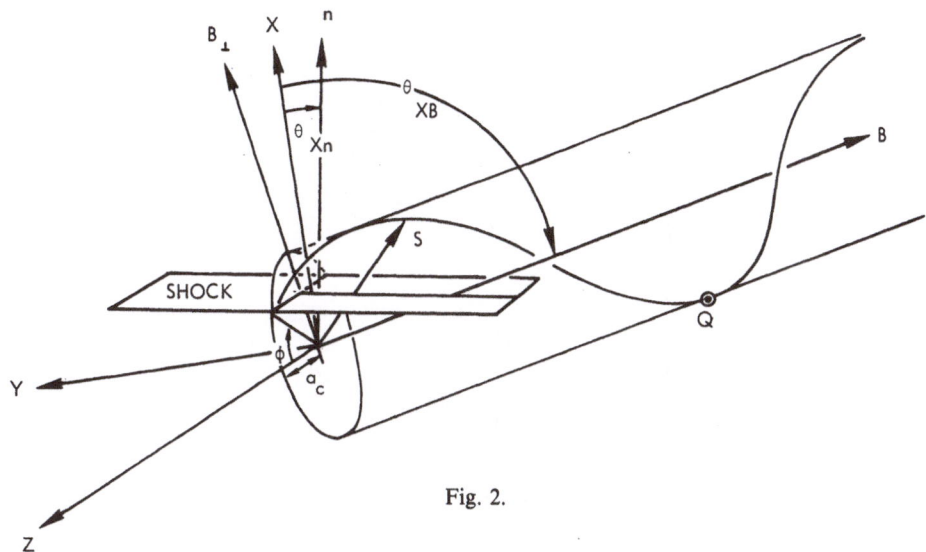

Fig. 2.

to recapture when it circles around to the 'bottom' of its spiral the first time ($\phi +$ $\omega_c t \approx 3\pi/2$), shown as point Q in Figure 2. Only some ratios P/p will permit free escape, and phase ϕ at $t=0$ strongly influences the acceptable range of P/p. If $S_n \equiv \mathbf{S} \cdot \mathbf{n}$, the condition previously stated can be written $S_n(t) - S_n(0) > V_{SW}t \cos\theta_{Xn}$, which, when expanded, yields the inequality:

$$P \sin\theta_{nB} \frac{\sin(\omega_c t + \phi) - \sin\phi}{\omega_c t} + (p \cos\theta_{nB} - \cos\theta_{Xn}) > 0. \qquad (1)$$

Combinations of p, P, and ϕ which satisfy this relation for all t define the free escape particles. Actually, the above expression places a maximum limit on $P/p > 0$ for each value of t, but since the inequality must be satisfied for all t, there is at least such maximum limit for any given ϕ.

3. Numerical Examples

3.1. THE OBSERVED CASE $p = 1.6$

To illustrate the foregoing algebraic result, we recall the Vela-Explorer dual satellite example of 1966. This was the only instance in which the upstream wave progression rate was measured at a specific location on the shock (Greenstadt *et al.*, 1970). In that case, the rate was found to be 1.6 V_{SW}, and the waves were attributed to the probable, but unobserved, reflection of solar wind protons. We assume here for purposes of example, that there actually were reflected particles with $p = 1.6$. In the present notation, that case is represented by $\theta_{nB} = 37°$, $\theta_{Xn} = 21°.5$, and $\theta_{XB} = 58°.5$. Figure 3 shows a vector velocity diagram on a polar plot of P vs ϕ for these parameters. The length of each arrow indicates the greatest relative velocity $P = u/V_{SW}$ a proton may have to escape the shock if it emerges at $t=0$ at a phase position between $\phi = 0$ and 90°, represented by the tail end of one of the arrows. In the figure, we are looking

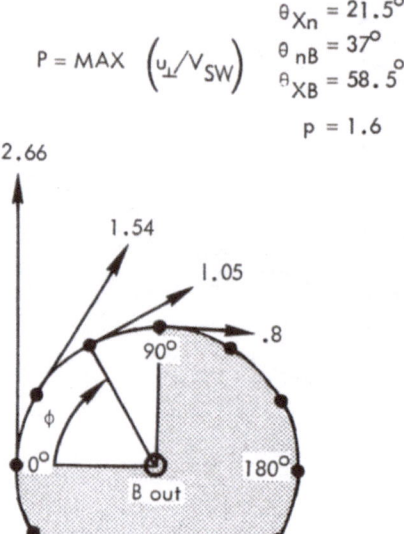

$$\theta_{Xn} = 21.5^\circ$$
$$P = MAX \left(u_\perp / V_{SW}\right) \qquad \theta_{nB} = 37^\circ$$
$$\theta_{XB} = 58.5^\circ$$
$$p = 1.6$$

Fig. 3.

backward along $\mathbf{B}_{SW}$ at the projection of the proton's Larmor circle on a plane perpendicular to $\mathbf{B}_{SW}$; arrows are placed at 30° phase increments.

The computation summarized in Figure 3 allows a range of possible maximum P for the selected quadrant of escape, by reasoning that a stationary proton, initially captured, according to assumption, by the shock at relative normal speed $V_{SW} \cos \theta_{Xn}$, will enter the shock layer at $\phi = 180^\circ$ and emerge after one-half to one cyclotron orbit at $0^\circ \lesssim \phi \lesssim 90^\circ$. Subject to this argument, P would take on values up to about 2.7.

To translate velocity of reflection into detectable particle energy, we note that the energy ratio of a reflected proton of finite pitch angle measured by a directionally-sensitive detector, and its direction of arrival, will depend on the phase ϕ_D of the spiralling particle at its instant of detection and on the angle θ_{XB} the interplanetary field makes with the solar wind flow (along X):

$$E_r/E_{SW} = P^2 + p^2 + 1 - 2(p \cos \theta_{XB} + P \cos \phi_D \sin \theta_{XB}).$$

If ϕ_D is $\pi/2$ when the proton is at the sunward extreme of its Larmor spiral, the highest value of E_r/E_{SW} is achieved when $\phi_D = \pi$, and the lowest when $\phi_D = 0$. If we allow p to vary, Figure 4 shows the behavior of the maximum and minimum of E_r/E_{SW} vs p for $\phi_D = \pi$ and 0, respectively, when P takes on its maximum values of 2.66 and 0.8 at $\phi = 0^\circ$ and 90°. The dashed curve ($P=0$) shows E_r/E_{SW} at zero pitch angle, for comparison. Obviously, a reflected proton of finite P can have measured energy appreciably higher than that permitted when $P=0$. In fact, at $p=1.6$ there is no difficulty in providing backstreaming protons of energy 6 keV or more, such as those recorded

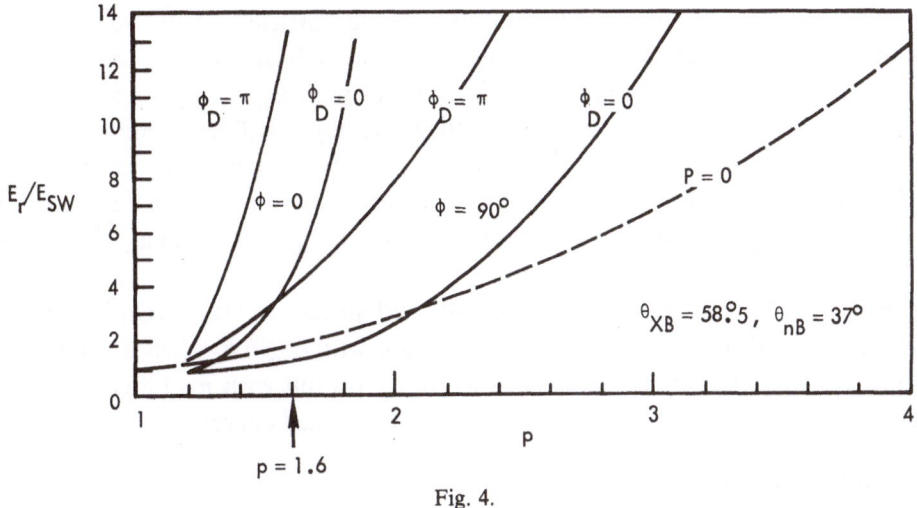

Fig. 4.

by Vela (Asbridge *et al.*, 1968) for $\phi_D = \pi$, $\phi = 0°$. One may interpolate visually to appreciate that the same is true for a range of $\phi_D < \pi$, $\phi > 0°$ as well.

3.2. THE SUBSOLAR POINT

Another specialized case of more general interest is illustrated in Figure 5. Here, the curves represent the maximal detectable energy ratio E_r/E_{SW} (at $\phi_D = \pi$) of protons reflected from the subsolar point of the shock ($\theta_{XR} = \theta_{Xn} = 0$) and traveling along $\mathbf{B}_{SW}$ at the forward edge of the upstream particle (= wave?) region. Exit phases $\phi = 0$ and 90°, with three possible cutoff angles θ_{nB} for each phase, are shown. To clarify the interpretation of Figure 5 by specific example, suppose a satellite-borne proton detector is located in the ecliptic upstream from the bow shock, westward and forward of the subsolar point, and the interplanetary field, which has been perpendicular to

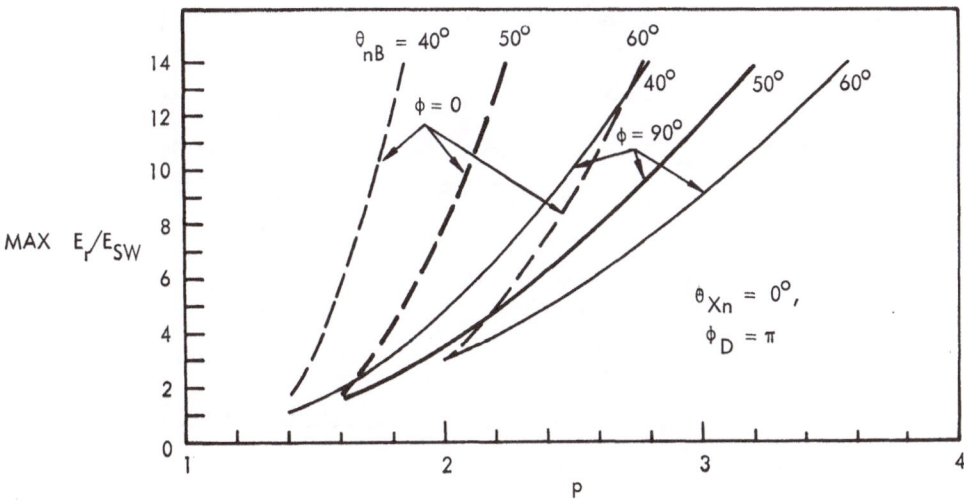

Fig. 5.

the solar wind flow thereby cutting off all reflected particles, rotates suddenly to a stream angle of 50°, connecting the satellite to the subsolar point. Then reflected protons barely emerging from the shock at $\phi = 0°$ with $u_\parallel / V_{sw} = p = 1.6$, after completing one-half of a cyclotron rotation in the shock layer, will be permitted to arrive at the satellite with total energies up to $E_r = 1.6\ E_{sw}$ (middle dashed curve). Alternatively, imagine the satellite moving antisunward in the same upstream region and first encountering the edge of the precursor zone when $\theta_{nB} = 50°$; at that point protons will be detected with $u_\parallel = 1.6\ V_{sw}$ and $E_r / E_{sw} \lesssim 1.6$.

The permissible escape energy rises rapidly with phase, for we see that for a stream angle of 40°, particles with $p = 1.6$, $\phi = 0$ can escape with $E_r \lesssim 6\ E_{sw}$ (left dashed curve). Allowable escape energy also increases with p: a proton with $p = 1.8$ can leave the shock at $\theta_{nB} = 50°$, $\phi = 0°$, with $E_r \lesssim 4\ E_{sw}$ (middle dashed curve).

3.3. HIGH ENERGY (30–70 keV) PROTONS

We turn finally to high energy particles, where p is appreciably higher than 1.6, and we refer to the insert at the upper left of Figure 6. Imagine a proton detector in the ecliptic upstream on the morning side of the shock at the forward edge of the precursor region (circled point), and suppose that the field angle θ_{nB} corresponding to that boundary of the forward region is 40° to 50° at the subsolar point (where the shock normal is parallel to the X axis). Then, for escape angles $\phi = 0°$ and 90° at the subsolar point, our formulas for P and E_r / E_{sw} give the maximal energy ratios vs p shown in the curves in the main part of Figure 6. The figure states, for example, that a proton can leave the subsolar shock at $\phi = 0°$, travel along B_{sw} at 40° to the normal, with parallel component (guiding center velocity) $u_\parallel = 3\ V_{sw}$, and escape upstream with energy as high as $E_r = 100\ E_{sw}$. A bulk velocity of the solar wind corresponding

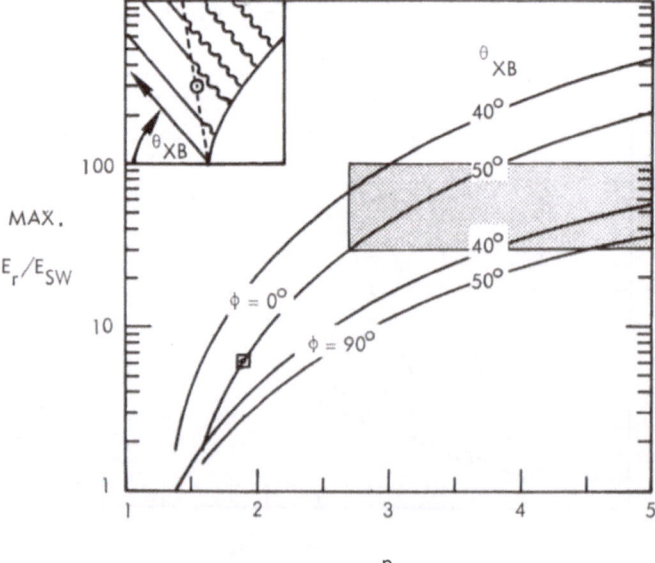

Fig. 6.

to 1 keV would imply $E_r = 100$ keV. Such an example would provide the high energy particles found by Lin *et al.* (1974), if supplied by the shock, without invoking any acceleration enroute. The shaded region of Figure 6 denotes the width of the 30 to 100 keV energy channel of the Lin *et al.* experiment, for a 1 keV solar wind. We see that, for $p = 5$, which was at the extreme of the distribution Lin *et al.* found, even a proton barely escaping at $\phi = 90°$ with the field at 50° to the normal could have total energy high enough to be recorded in their 30–100 keV channel. The shapes and ranges of the curves suggest that a preference for escape angle of intermediate $\phi \approx 45°$ could easily explain both the consistency with which the 30–100 keV channel was occupied for moderate p and the apparent absence of protons above 100 keV even at high p.

4. Discussion

The geometry of escape and the numerical examples described above demonstrate that protons can leave the bow shock and travel upstream with almost arbitrary energy, given only the appropriate p, θ_{Xn}, θ_{nB}, and ϕ, and can satisfy observations with very reasonable selection of values for these parameters. Although the calculations regarding high energy protons cannot be used as evidence that such particles are produced at the bow shock, they do show that such protons, if produced, can escape upstream with the characteristics already observed. However, the maximal energy ratio yielded at the subsolar point by Sonnerup's (1969) formula for protons energized by the interplanetary electric field is approximately 6.7 at $\theta_{nB} = 50°$ (setting his $\delta = \frac{1}{2}$, $\mu = \gamma = 0$). In Figure 6, this would correspond, for $\phi = 90°$, to $p = 1.9$ (square point) and, incidentally, to $P \approx 1.9$. Clearly, if the interplanetary electric field is all there is to work with and p is about as small at the subsolar point as it is on the midmorning flank, where $p \approx 1.6$, the protons of Lin *et al.* cannot be explained without invoking some upstream energization process, as those authors do.

But consider the following: By using inequality (1), we have treated only *perfect* escape. Yet, it seems reasonable that some particles will encounter the shock two or more times, accelerating each time and compounding their total energy until it reaches a high level. A proton of 30 keV has observed velocity of only about 5.5 V_{sw} for a 1 keV solar wind. This does not seem impossible to achieve by *multiple* reflection when a double reflection may multiply the original relative velocity by, say, a factor of 2.56 ($= 1.6^2$), especially remembering the character of quasi-parallel shocks ($\theta_{nB} \lesssim \lesssim 50°$) with their large amplitude pulsations and irregular boundaries. The question of whether energization by multiple reflection in quasi-parallel turbulent waves should be designated as a shock process or an upstream process may thus be only semantic. It is this author's provisional belief that most if not all of the acceleration responsible for the high energies detected by Lin *et al.* occurs close to the nominal shock, although some may be technically 'upstream'. The only apparent difficulty is providing the proper ratio of $P/p \approx \sqrt{5.5} = 2.3$ for such reflected protons.

The provision of adequate P/p by the physics of shock reflection, the introduction of finite temperature, and the representation of three-dimensional reflection in the curved bow shock are left for future analysis. We close by noting that it seems in-tuitive that for a given θ_{nB}, the larger the shock radius of curvature, i.e., the less convex it is locally, the more likely a particle will undergo multiple reflection before free escape upstream. Higher energies should therefore be expected for particles up-stream from interplanetary shocks, and from Jupiter's bow shock than from the Earth's. Such particles have been observed (Armstrong *et al.*, 1970; Simpson *et al.*, 1974), and the acceleration of protons to relativistic energies by multiple reflection in interplanetary shocks has been developed theoretically by Sarris and Van Allen (1974).

Acknowledgement

The material presented in this report was funded by the National Aeronautics and Space Administration under Contract NASW-2398.

References

Armstrong, T. P., Krimigis, S. M., and Behannon, K. W.: 1970, *J. Geophys. Res.* **75**, 5980. 1970.

Asbridge, J. R., Bame, S. J., and Strong, I. B.: 1968, *J. Geophys. Res.* **73**, 5777. 1968.

Fairfield, D. H.: 1969, *J. Geophys. Res.* **74**, 3541.

Fairfield, D. H.: 1971, *J. Geophys. Res.* **76**, 6700.

Greenstadt, E. W.: 1972, *J. Geophys. Res.* **77**, 5467.

Greenstadt, E. W., Green, I. M., Inouye, G. T., Colburn, D. S., Binsack, J. H., and Lyon, E. F.: 1970, *Cosmic Electrodyn.* **1**, 279.

Lin, R. P., Meng, C. I., and Anderson, K. A.: 1974, *J. Geophys. Res.* **79**, 489.

Sarris, E. T. and Van Allen, J. A.: 1974, Univ. of Iowa Report 74-4, January 1974.

Scarf, F. L., Fredricks, R. W., Frank, L. A., Russell, C. T., Coleman, P. J., Jr., and Neugebauer, M.: 1970, *J. Geophys. Res.* **75**, 7316.

Simpson, J. A., Hamilton, D., Lentz, G., McKibben, R. B., Mogro-Campero, A., Perkins, M., Pyle, K. R., and Tuzzolino, A. J.: 1974, *Science* **183**, 306.

Sonnerup, B. U. O.: 1969, *J. Geophys. Res.* **74**, 1301.

MOTIONS OF THE EARTH'S BOW SHOCK

R. D. AUER

*Max-Planck-Institut für Physik und Astrophysik, Institut für extraterrestrische Physik,
8046 Garching, F.R.G.*

1. Introduction

Many aspects of the interaction of the solar wind with the geomagnetic field can be successfully treated by macroscopic fluid equations. This fact can be attributed to the organizing influence of the frozen in magnetic field. The details of the turbulent dissipation processes in the collisionfree shock are, of course, only amenable to a kinetic theory.

However, the gross characteristics of the bow wave (e.g. the discontinuous change of parameters across the shock) can be derived from the model equations of a two-fluid magnetized plasma. The magnetohydrodynamic description in terms of a single fluid has been applied to the impulsive nonlinear interaction of the bow shock with discontinuity surfaces in the solar wind. Attention will be given in the following to the interaction with directional discontinuities (i.e. changes in magnetic field direction only). Moreover, as long as the position and the shape of the shock are discussed in a global manner, ordinary gasdynamic theory will give sufficiently accurate results.

2. Quasistationary Equilibrium Position of the Bow Shock

The simple gasdynamic model for the magnetosphere bow shock system has evolved from supersonic blunt-body theory (Hayes and Probstein, 1959; Spreiter and Jones, 1963). The magnetopause (see Figure 1) presents an obstacle to the supersonic solar wind flow. The subsolar distance D is given in units of the Earth radius R_E by

$$D = \left(\frac{f^2}{k} \frac{B_0^2}{2\pi\varrho v^2} \right)^{1/6} \tag{1}$$

B_0 is the surface strength of the Earth's dipole field at the equator, and v and ϱ denote the solar wind bulk velocity and mass density, respectively. The parameter k characterizes the momentum transfer to the magnetosphere, while f relates the magnetic field just inside the magnetopause to the Earth's undistorted dipole field at that point. Equation (1) states the equilibrium of the geomagnetic field pressure and the dynamic pressure of the impinging solar wind. From a large amount of satellite data Fairfield (1971) has determined a value

$$f^2/k = 1.4. \tag{2}$$

Different theoretical models yield values for f and k which may give a ratio f^2/k differing considerably from the experimental value of Equation (2). The standoff

V. Formisano (ed.), The Magnetospheres of the Earth and Jupiter, 11–17. All Rights Reserved

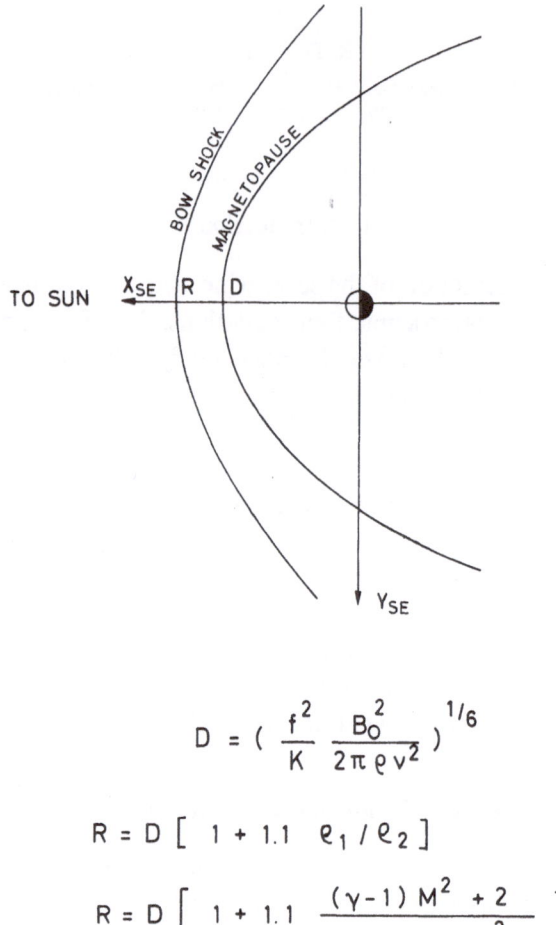

$$D = \left(\frac{f^2}{K} \frac{B_0^2}{2\pi\rho v^2} \right)^{1/6}$$

$$R = D \left[1 + 1.1 \ \rho_1 / \rho_2 \right]$$

$$R = D \left[1 + 1.1 \frac{(\gamma-1)M^2 + 2}{(\gamma+1)M^2} \right]$$

Fig. 1. Intersection lines of bow shock and magnetopause with the ecliptic plane. The relation between subsolar distances from the Earth corresponds to the simple gasdynamic model.

distance R of the accompanying bow wave is primarily determined by mass flow considerations. R varies linearly with the inverse density jump ρ_1/ρ_2 across the bow shock on the stagnation line (Seiff, 1962). Extensive numerical calculations (Spreiter *et al.*, 1966) have led to the formula

$$R = D(1 + 1.1 \ \rho_1/\rho_2) \qquad (3)$$

or inserting the gasdynamic relation for the density jump across the shock

$$R = D \left(1 + 1.1 \frac{(\gamma-1)M^2 + 2}{(\gamma+1)M^2} \right). \qquad (4)$$

From a great body of satellite observations (Fairfield, 1971) a specific heat ratio $\gamma \approx \frac{5}{3}$, corresponding to a classical monatomic gas, was found to be appropriate in this formula (Völk and Auer, 1974). This result is confirmed by HEOS-1 data (Formisano *et al.*, 1973) when Equation (2) has been taken into account (Auer, 1974).

To summarize, the gasdynamic model provides a rather useful means to calculate the standoff distance of the bow shock which is essentially determined by the solar wind dynamic pressure. According to Fairfield (1971) the shock position can be predicted in this way to better than 1 R_E for about 80% of the time. This result shows that intricate plasma characteristics do not significantly affect the global properties of the solar wind flow around the magnetosphere.

However, a γ-value corresponding to 3 degrees of freedom should have been expected from the outset if one recalls the dissipative processes occuring in the shock. In a high Mach number collisionfree shock only turbulent dissipation will be effective enough to stabilize the shock. Thus, a strong coupling of all translational degrees of freedom should result.

3. Motions of the Bow Shock

The equilibrium position of the bow shock, as calculated from the gasdynamic formula, is determined by the solar wind conditions. Average shock velocities of the order of 10 km s^{-1} have been estimated. Such speeds can be readily attributed to quasistationary perturbations caused by Alfvén waves or fast hydromagnetic waves (Völk and Auer, 1974). This process can lead also to exceptionally high shock velocities in very low Mach number situations, such as have been observed, for instance, by Greenstadt *et al.* (1972).

Another class of processes which will initiate rapid bow shock motions is the impulsive interaction with discontinuity surfaces in the solar wind. The case of interplanetary shocks hitting the Earth's bow wave has been discussed by Shen and Dryer (1972). For the interaction with the much more frequent tangential discontinuities, a simple model has been treated by Völk and Auer (1974). Both cases can lead to high shock velocities.

Experimental distinction between tangential and rotational discontinuities in the solar wind is not always simple because of the incomplete determination of plasma parameters especially for anisotropic plasmas (Hudson, 1970). In order to explore the influence of magnetic field direction, the impulsive interaction of the bow shock with directional discontinuities (i.e. discontinuities in magnetic field direction only) will now be discussed (Auer, 1974). We shall restrict ourselves to a special case, namely that the field direction changes by 90° in the sense that the shock geometry is changed from exactly parallel to exactly perpendicular, or vice versa. In this model rather simple jump equations arise and no additional angular parameters are involved. More general calculations have been carried out by Neubauer (1973). The interaction of the bow shock with a directional discontinuity is shown schematically in Figure 2 where the pressure profile is drawn. Starting from a parallel bow shock, the interaction with such an idealized directional discontinuity will lead to a modified perpendicular bow shock travelling away from the Earth and a rather weak transmitted parallel shock. The directional discontinuity, and in addition a tangential discontinuity, will be located between the two shock waves. In the case of an originally per-

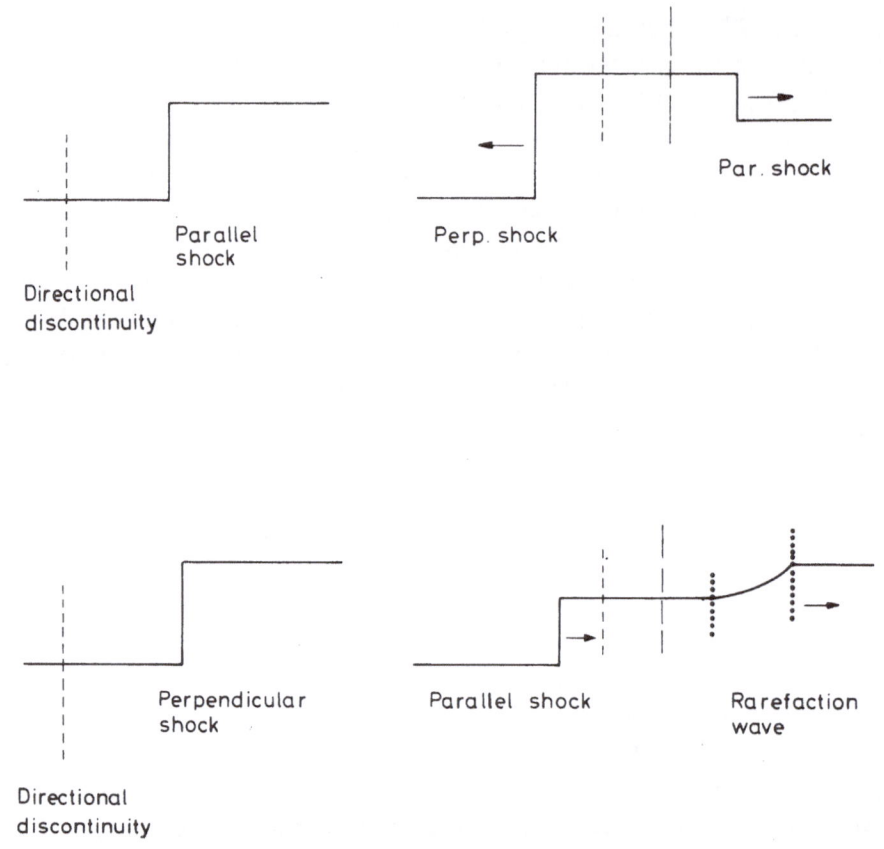

Figure 2.   Schematic presentation of pressure profile and discontinuity splitting in the interaction process. The simultaneous occurence of a pure directional and a pure tangential discontinuity after the interaction is due to the idealized model involved. In general for oblique geometries only one discontinuity with mixed properties will appear (see Neubauer, 1973).

pendicular bow shock, the interaction will lead to a modified, parallel bow shock travelling towards the Earth and a very weak transmitted rarefaction wave. Between both waves again the directional discontinuity and an additional tangential discontinuity are to be found. In Figure 3 the numerical results are presented. The ratio of induced velocity v of the bow shock to upstream magnetosonic speed c_1 is plotted as a function of the original magnetosonic Mach number M_1 with three different values of the plasma β (ratio of thermal to magnetic pressure). The upper part of the Figure corresponds to a transition from a parallel to a perpendicular bow shock accompanied by outward motion, the lower part corresponds to the opposite case where the shock moves inwards. For average solar wind conditions a shock velocity of about 10 km s^{-1} will arise. However, at the flanks of the shock, where the Mach number is smaller, velocities of 20–30 km s^{-1} may be induced.

Thus, the impulsive interaction with a directional discontinuity will induce a local motion of the bow shock which is directed away from the Earth when the modified bow shock has a perpendicular geometry after the interaction. The induced motion

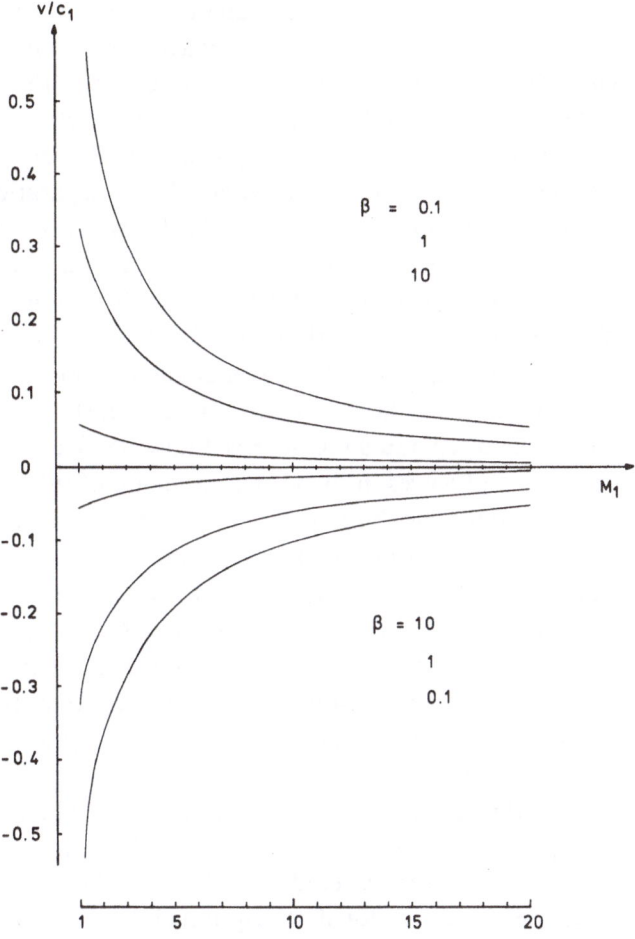

Fig. 3. Numerical results for the induced velocity of the bow shock (see text).

will be directed inwards if the modified shock has a parallel geometry after the interaction. Starting with a velocity of 10 km s^{-1} and assuming an undisturbed travel time of about 10 min, a displacement of about one Earth radius of the shock from the equilibrium position can arise. Such a time interval will elapse before the shock realizes that it is bound to the Earth, because this is the time taken by hydromagnetic waves to carry the information to the magnetopause and back again. Let us assume that such an impulsive interaction occurs at least once, but not too often, during the time interval a satellite takes to traverse the vicinity of the average bow shock location. Then, from this model, the bow wave should in general be found beyond the equilibrium position when seen as a perpendicular shock. It should be found within the equilibrium distance when seen as a parallel shock. This prediction seems to be supported in a statistical way by magnetic field data from the HEOS-2 satellite; this data has been kindly provided by Dr P. C. Hedgecock, Imperial College, London. Only outbound passes have been taken into account, in order to obtain a rather uniform sample of shock crossings. These passes all occured at solar ecliptic latitudes larger than 74°. The data has not been normalized to average dynamic wind pressure and average Mach number. Therefore, a considerable scatter in the shock position is displayed. To order these data points, they have been distributed into two classes, the first class consisting of shock crossings with predominantly perpendicular structure, the second class containing strongly oblique shocks. As separation criterion we have employed the Greenstadt (1972) binary index. In this way it was found that the average shock position, as determined by the HEOS-2 outbound passes, is about 1 R_E closer to the Earth when the shock exhibits a strongly oblique structure compared with the other case of predominantly perpendicular shocks. Exactly this behaviour would have been expected from the model discussed above. Further support is given to this result by HEOS-1 data which has been published by Formisano *et al.* (1973). These authors had separated the observed bow shock crossings according to the existence or absence of waves in the upstream region of the bow shock. This feature is strictly coupled to the shock geometry via Greenstadt's binary index. Thus, effectively their criterion for ordering the data is identical to the criterion used in this work. Formisano *et al.* (1973), however, drew another conclusion. Using the gas-dynamic standoff formula, they computed different values for the specific heat ratio γ for the two separate classes of bow shock positions. Working backwards from their published results we find that again oblique shocks were concentrated about 1 R_E closer to the Earth than were perpendicular ones. Thus our result is independently confirmed by the HEOS-1 observations.

We feel, however, that this spread in bow shock locations should not be attributed to different static equilibrium configurations as described by the standoff formula for different γ values (Formisano *et al.*, 1973). Firstly, it has been stated above that the gasdynamic model yields a good global description of the solar wind flow around the magnetosphere. A γ value close to $\frac{5}{3}$, consistent with turbulent processes in the shock transition which affect all translational degrees of freedom, has been found suitable from the observations (Fairfield, 1971; Völk and Auer, 1973; Auer, 1974). Moreover, interpreting the density jump in Equation (3) purely gasdynamically or even incorporating the magnetic field, will alter the γ value for average conditions by not more than 4%. The values calculated by Formisano *et al.* (1973) for oblique and perpendicular shocks, respectively, differ by a far greater amount than this. Secondly, the existence or absence of upstream waves is no static feature but highly time-dependent and will change with position on the shock surface. For these reasons, an interpretation in terms of a local process, like the impulsive interaction with discontinuity surfaces, which will cause temporary displacements from the quasistationary equilibrium position, seems to be more appropriate.

Finally, it should be noted, that the observed spread in bow shock positions and the correlation with shock geometry is a statistical result. It is highly desirable to investigate single events of impulsive interactions with discontinuity surfaces in the solar wind. However, at least two satellites sampling simultaneously near the bow shock are necessary to identify processes of this kind.

References

Auer, R. D.: 1974, submitted to *J. Geophys. Res.*

Fairfield, D. H.: 1971, *J. Geophys. Res.* **76**, 6700.

Formisano, V., Hedgecock, P. C., Moreno, G., Palmiotto, F., and Chao, J. K.: 1973, *J. Geophys. Res.* **78**, 3731.

Greenstadt, E. W.: 1972, *J. Geophys. Res.* **77**, 5467.

Greenstadt, E. W., Hedgecock, P. C., and Russell, C. T.: 1972, *J. Geophys. Res.* **77**, 1116.

Hayes, W. D. and Probstein, R. F.: 1959, *Hypersonic Flow Theory*, Academic Press, New York.

Hudson, P. D.: 1970, *Planetary Space Sci.* **18**, 1611.

Neubauer, F. M.: 1973, internal report GAMMA 23, Technische Universität Braunschweig.

Seiff, A.: 1962, NASA SP-24, p. 19.

Shen, W. W. and Dryer, M.: 1972, *J. Geophys. Res.* **77**, 4627.

Spreiter, J. R. and Jones, W. P.: 1963, *J. Geophys. Res.* **68**, 3555.

Spreiter, J. R., Summers, A. L., and Alksne, A. Y.: 1966, *Planetary Space Sci.* **14**, 223.

Völk, H. J. and Auer, R. D.: 1974, *J. Geophys. Res.* **79**, 40. (See also *EOS Trans. AGU* **53**, 1108, 1972.)

ON THE EARTH'S BOW SHOCK MOTION AND SPEED

V. FORMISANO and G. MASTRANTONIO

Laboratorio Plasma Spazio, C.N.R., Frascati, Italy

1. Introduction

The Earth's bow shock has been shown to have often a speed rather high (50–150 km s^{-1}) on the basis of different methods of investigation (Formisano *et al.*, 1971; Greenstadt *et al.*, 1972; Formisano *et al.*, 1973; Guha *et al.*, 1972). The causes of the shock motion and speed are not well understood, the resonant oscillator proposed by Smit (1968) being not suitable for the Earth's magnetosphere. Shen and Dryer (1972) studied the interactions of the Earth's bow shock with interplanetary shocks as these are causes of bow shock motions; the frequency of interplanetary shocks is rather low (1 per month) so this kind of interaction cannot be invoked to explain the bow shock motion and speed.

Völk and Auer (1973) have recently considered the interaction of Tangential Discontinuities (TD) and Alfvén waves with the bow shock in order to explain the shock motion and speed. The conclusion of these authors was that Alfvén waves can cause very slow shock motions (up to 10 km s^{-1}) while a TD hitting the bow shock can generate much higher speeds if the density ratio across the TD is high enough.

Völk and Auer (1973) however believed that large density ratios across TD were rarely observed, therefore found it difficult to explain the average value (75–85 km s^{-1}) found by Formisano *et al.* (1973).

It is our intention, here, to verify experimentally with the HEOS-1 plasma data the importance of TD for the bow shock motion and speed.

2. Observations

According to the results of the study of Völk and Auer (1973) given in Table I a TD with $\Delta N/N > 50\%$ would generate a shock motion with speed $\gtrsim 50$ km s^{-1}. Speeds larger than 50 km s^{-1} may give an average bow shock speed of 75–85 km s^{-1}.

We have therefore searched in the HEOS-1 plasma data for large TD defined by

$$\left| \frac{N_2 - N_1}{N_1} \right| > 50\% \quad \text{or} \quad \left| \frac{V_2 - V_1}{V_1} \right| > 10\%, \tag{1}$$

where N_1, N_2, V_1, V_2 were contiguous measurements taken over 6-min intervals (see Bonetti *et al.*, 1969).

As a very large number of these discontinuities do exist, we have counted how many of them were observed per day. A separate study is under development to understand the properties of such discontinuities. Indeed at the moment we cannot exclude the possibility that we are dealing with magnetosonic waves, because, although they are

V. Formisano (ed.), The Magnetospheres of the Earth and Jupiter, 19–24. All Rights Reserved

TABLE I

Shock speed as function of density ratio

N_2/N_1	1/5	1/4	1/3	1/2	1	2	3	4
$V_s(\text{km s}^{-1})$	−154	−131	−101	−62	0	57	87	108

not rotational or contact discontinuities (5–10% only may be shocks), the total internal pressure is not balanced in 60–70% of the cases. Figure 1 shows the histogram of the number of discontinuities per day found in the solar wind plasma data from December 1968 to April 1969 and from September 1969 to April 1970.

On average 12.6 discontinuities defined by Equation (1) are observed per day in the solar wind. In other words on average every two hours is observed in the solar wind a TD capable of generating a shock speed larger than or of the order of 50 km s^{-1}. As an eccentric satellite travels in 2 hours a length equal to the amplitude of the bow shock motion, the probability of crossing the bow shock while it is in motion because of one of these discontinuities is very high.

Furthermore the data show that on average every 26 hours a TD is observed with a density ratio larger than 3. These discontinuities should generate, according to the Völk and Auer process, shock speeds larger than 100 km s^{-1}.

It appears that there are enough strong discontinuities in the solar wind to explain the measured shock speeds with the Völk and Auer mechanism.

We have therefore tried to study in detail some particular case in order to check quantitatively this theory.

On day 83–84, 1969, many bow shock crossings were observed by the satellites HEOS 1, OGO 5 and Explorer 33.

Due to the position of the satellites (see Figure 2) an almost continuous monitoring

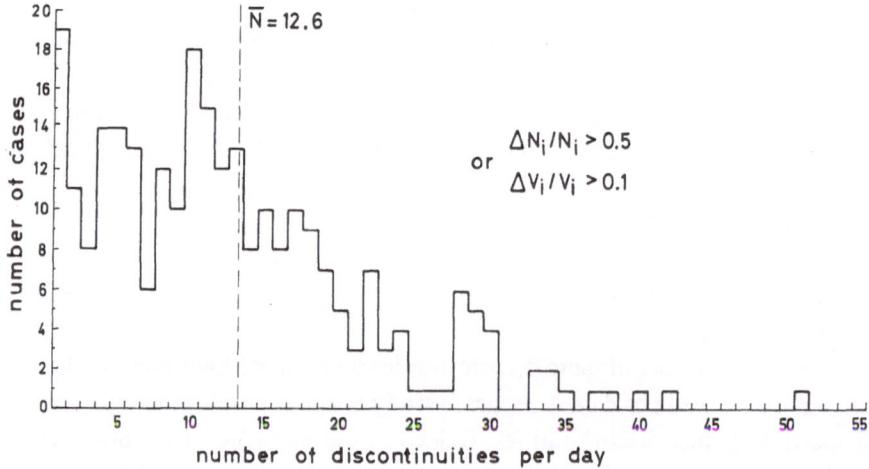

Fig. 1.   Histogram of the number of large discontinuities per day. On average there is a discontinuity every $\simeq$ 2 h.

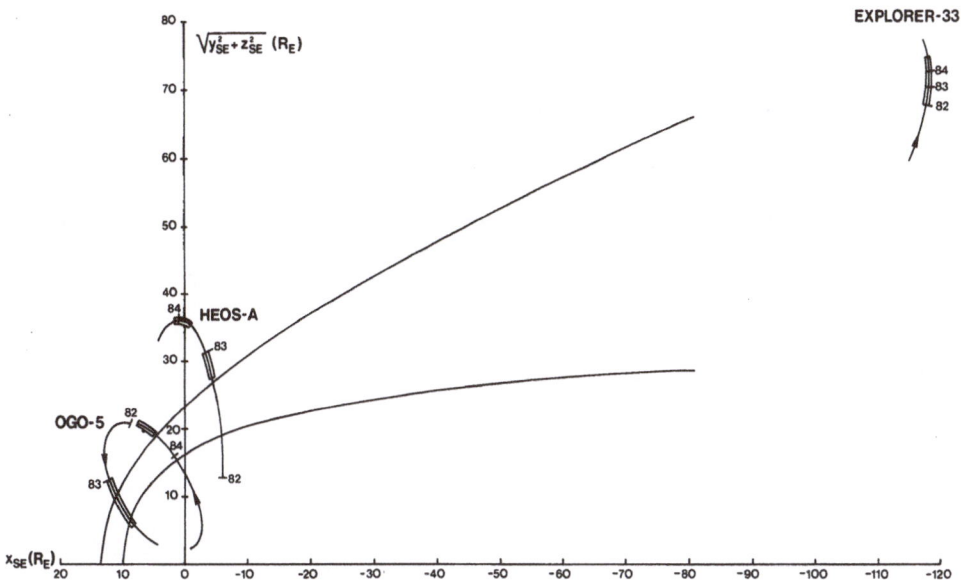

Fig. 2. HEOS-1, OGO-5 and Explorer-33 orbits for day 82, 83, 84, 1969 discussed in the text.

of the shock motion was possible. HEOS 1 crossed the bow shock 31 times in 36 h, OGO 5 24 times in 15 h and Explorer 33, 39 times in ≈ 52 h.

On average, therefore there was almost one shock crossing every hour, the number of discontinuities satisfying Equation (1) observed by HEOS 1 in the solar wind being indeed 35 in ≈ 35 h (see Figure 3).

Although statistically there is a good agreement between the frequency of TD in the solar wind and the motion of the bow shock, no one-to-one correlation is evident, from Figure 3, between TD and shock crossings of the satellites.

It is instructive to study in more detail one example in which a large TD, observed by Explorer 33 and Explorer 35 was able to move the shock from HEOS 1 to OGO 5 at a very large speed. This is the case of the bow shock crossing observed by HEOS 1 on day 43, 1969, 1628 UT. Magnetic field and plasma data from HEOS 1 have been shown by Formisano *et al.*, 1971; Figure 4 of this reference shows also the magnetic field intensity observed by Explorer 35 (always in the solar wind). OGO-5 magnetic field data have been shown by Greenstadt *et al.*, 1972. OGO 5 also observed the bow shock at ≈ 1628 UT and the measured shock speed was $V_s \gtrsim 117$ km s^{-1}; Formisano *et al.*, 1973 obtained for this shock crossing a speed of $V_s \simeq 80\text{--}100$ km s^{-1}.

Völk and Auer, 1973, studying this event, found that "from the OGO-5 magneto-meter data, no TD nor a transmitted shock could be found in the magnetosheath data for the time in question". However the Explorer 35 magnetic field data in Figure 4 of Formisano *et al.* (1971) show, just before the HEOS-1 shock crossing, a $1\text{--}2\,\gamma$ decrease. We have therefore looked in the Explorer 33 plasma and magnetic field data, for a discontinuity with a large increase of the number density N_p. Explorer 33 is located $\approx 87\ R_E$ radially away from HEOS 1 and OGO 5, so a delay time of 23 min

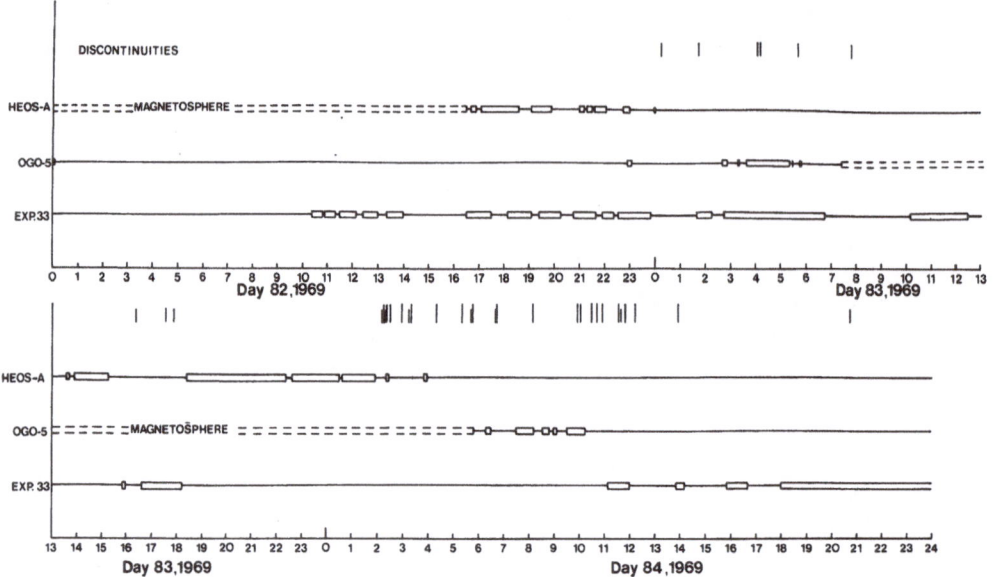

Fig. 3. Bow shock crossing of HEOS 1, OGO 5 and Explorer 33 for day 82, 83, 84, 1969. Magnetosheath periods are indicated by thick segments. Vertical segments on top of each strip indicate large discontinuities observed by HEOS 1 in the solar wind.

should be expected. At Explorer 33 a large increase of the proton number density (from ~ 1 to 2.7 P cm^{-3}) is observed between 1648 and 1654 UT with a magnetic field behavior similar to that observed by Explorer 35 just before the HEOS-1 and OGO-5 shock crossing. We think therefore that a TD compressed the bow shock generating the observed high speed. The density ratio across the TD was 2.7 so a speed of 82 km s^{-1} is predicted by the Völk and Auer model. The observed shock speed seems somewhat higher (100–117 km s^{-1}). While the OGO-5 data do not show any transmitted shock the HEOS-1 magnetic field shows a discontinuity (from 16 to 20 γ) practically coincident with the Explorer 35 discontinuity, that we may interpret as a transmitted shock. As this discontinuity is not observed by OGO 5, it is possible that the life-time of the transmitted shock is very short, because, perhaps, of its low Mach number and of the properties of the magnetosheath plasma in which the shock has to travel.

It should be noted, indeed, that the OGO-5 magnetic field data show a sharp gradient where we would expect to observe the transmitted shock.

3. Discussion and Conclusion

From the analysis of the data previously described it appears that indeed the TD may be the most important cause of the bow shock motion and speed.

When comparing solar wind data with actual shock crossing, however, we should expect a statistical agreement rather than a one-to-one correlation, unless we are sure we may state if the shock was moving or not.

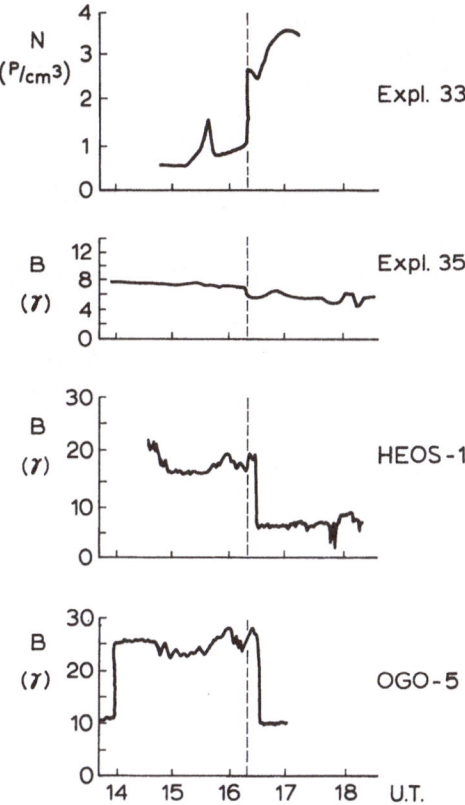

Fig. 4. Magnetic field from HEOS 1, OGO 5 and Explorer 35 are shown together with time shifted (23 min) plasma density from Explorer 33 for the bow shock crossing of 1628 UT day 43, 1969.

In one particular case when the shock speed is measured with two satellites or deduced by mass conservation, a detailed comparison with Völk and Auer theory shows that a transmitted discontinuity (probably a shock) is indeed formed, but has a rather short life time. The bow shock speed predicted may be a little lower than measured but is certainly of the same order. The difference may be due to the fact that when a TD hits the shock we have not only the motion described by Völk and Auer (1973), but also, along the flanks, an adjustment of the shock surface to the new Mach cone (see Formisano, 1973). This motion has a speed linearly increasing along the flanks, so the speed predicted by Völk and Auer is a minimum (valid in the nose region) while at 90° from the subsolar point the shock speed may be larger.

In conclusion we may state that the interactions of TD with the bow shock is probably the main cause of the shock motion, and that there are in the solar wind enough large TD to explain, according to the Völk and Auer theory, the large observed average shock speeds.

Acknowledgements

We are indebted to the other people of the Laboratorio per il Plasma nello Spazio, who have participated in the development of the solar wind experiment.

We acknowledge useful discussions with E. W. Greenstadt and G. Moreno. This research was supported by the Consiglio Nazionale delle Ricerche of Italy.

References

Bonetti, A., Moreno, G., Cantarano, S., Egidi, A., Marconero, R., Palutan, P., and Pizzella, G.: 1969, *Nuovo Cimento* **X**, 64B, 307.

Formisano, V.: 1973, *J. Geophys. Res.* **78**, 6787.

Formisano, V., Hedgecock, P. C., Moreno, G., Palmiotto, F., and Chao, J. K.: 1973, *J. Geophys. Res.* **78**, 3731.

Formisano, V., Hedgecock, P. C., Moreno, G., Sear, J., and Bollea, D.: 1971, *Planetary Space Sci.* **19**, 1519.

Greenstadt, E. W., Hedgecock, P. C., and Russell, C. T.: 1972, *J. Geophys. Res.* **77**, 1116.

Guha, J. K., Judge, D. L., and Marburger, J. H.: 1972, *J. Geophys. Res.* **77**, 604.

Shen, W. W. and Dryer, M.: 1972, *J. Geophys. Res.* **77**, 4627.

Smit, G. R.: 1968, *J. Geophys. Res.* **73**, 4990.

Volk, H. J. and Auer, R. D.: 1973, Max Planck Inst. preprint 1973, *J. Geophys. Res.*, in press 1974.

THE CALM AND TURBULENT STATE OF THE EARTH'S MAGNETOSHEATH FROM DECEMBER 1968 THROUGH MARCH 1970

E. AMATA

Space Science Dept. (ESLAB), European Space Research and Technology Centre, Noordwijk, The Netherlands

1. Introduction

In the past years several authors have studied the magnetic field oscillations in the magnetosheath and developed different but basically complementary ideas about their origin. This is quite understandable if one bears in mind the most outstanding feature of the magnetosheath structure, i.e. its large variability: although at times the magnetic field undergoes rather ordered oscillations, in other occasions these are very chaotic; moreover, it is by no means unusual that both 'calm' and 'turbulent' periods alternate many times during one magnetosheath traversal by a satellite.

Generation of magnetic field oscillations at the magnetopause has been suggested for example by Southwood (1967) and indeed some evidence of enhanced turbulence close to the magnetosphere boundary has been found among others by Mariani *et al.* (1970) and Fairfield and Ness (1970). The latter also show evidence of a similar process in the proximity of the bow shock. Another possibility is that at least some of the observed turbulence be generated inside the magnetosheath.

In this paper, however, the emphasis is put on the possible transmission and amplification of waves from interplanetary space through the bow shock (McKenzie and Westphal, 1970), checking the hypothesis, put forward, e.g. by Fairfield (1969), Barnes (1970), and Formisano *et al.* (1973) that upstream waves may play an important role in this process. Section 2 of the paper is devoted to illustrate the experimental data which have been used in this work. Section 3 will deal with the 'turbulent' and 'calm' states of the magnetosheath, and a discussion of the results is presented in Section 4.

2. The Experimental Data

The bulk of this work is based on the power spectra obtained from the data supplied by the three-axis fluxgate magnetometer mounted on the ESRO satellite HEOS 1.

Some hourly averages of solar wind parameters from the plasma experiment (Bonetti *et al.*, 1969) on board the same satellite have also been used, together with magnetic field data from Explorer 35 and 41 and hourly averages of plasma data from Explorer 33 (NSSDC information packet on Explorer 33, 35 and 41, published by NASA, Goddard Space Flight Center, Greenbelt, MD., U.S.A.).

The HEOS-1 magnetometer with the two modes of operation has been described by Hedgecock (1970). Data from both modes of operation have been used to calculate

V. Formisano (ed.), The Magnetospheres of the Earth and Jupiter, 25–38. All Rights Reserved
Copyright © 1975 by D. Reidel Publishing Company, Dordrecht-Holland

power spectra in the two ranges 2.6×10^{-4}–1.04×10^{-2} and 7.0×10^{-3}–3.5×10^{-1} Hz for real time (RTD) and core memory (CMD) data respectively, the real time ones referring, usually, to a period of several hours around the CMD data. Digitization noise, approximately $2.0\ \gamma^2\ \mathrm{Hz}^{-1}$ for CMD and approx. $0.063\ \gamma^2\ \mathrm{Hz}^{-1}$ for RTD, does not contribute appreciably as in both cases it is well below the power densities of the spectra hereafter displayed. However, because of the effect of zero level errors, a narrow peak is sometimes noticeable around 0.17 Hz in the core memory data spectra, while the real time data are less affected. A fixed number of spectral estimates has been used, namely 50 for the core memory data and 40 for the real time data. The number of degrees of freedom for spectral estimate is generally around 20.

The frame of reference in which the spectra have been computed is not the usual GSE but a system in which the Z axis is aligned to the local average field vector and the X axis is perpendicular to the plane containing the Sun-Earth line and the average field vector. B is the field vector intensity. This choice allows to distinguish transverse from longitudinal oscillations.

Figure 1 shows the satellite orbit segments for the 32 magnetosheath traversals used in the present study, 19 on the dawn side and 13 on the dusk side, with a total of 35 core memory periods (see figure caption). The orbit segments shown do not begin and end with the bow shock and magnetopause crossing, but closely match the periods used for RTD power spectra. When the latter could not meaningfully be calculated, the segment lengths bear no special meaning.

3. The 'Calm' and 'Turbulent' Magnetosheath

3.1. Two states of the magnetosheath

The variable structure of the magnetosheath can be well exemplified by the two transitions shown in Figure 2, day 363, 1968, corresponding to the segment A on the dusk side in Figure 1, and in Figure 3, day 3, 1969, corresponding to segment C on the dawn side. On the inbound pass of day 363, the magnetometer detects what appears to be a series of multiple bow shock crossings around 16:00 UT and then experiences very large field oscillations through the magnetosheath before entering the magnetopause after 19:00 UT. Noticeably, the magnetic field in front of the bow shock is by no means steady as oscillations of several gammas are present in the un-shocked solar wind for several hours. These are most probably bow-shock associated upstream waves (Fairfield, 1969; Greenstadt et al., 1970a) and the index I displayed in the upper panel ($I = 1$ when upstream waves are observed) strongly supports this interpretation. On the outbound pass of day 3, on the contrary, the spacecraft, after crossing the magnetopause boundary, with the magnetic field changing smoothly through it, goes through a rather calm transition region, where the oscillations seem to have an amplitude less than a half that detected on day 363. Finally a very clean bow shock crossing is recorded soon after 05:00 UT with a sudden jump of the field intensity from about $13\ \gamma$ to about $3.5\ \gamma$, while the direction of the field (unlike on day 363) changes only by about 30°.

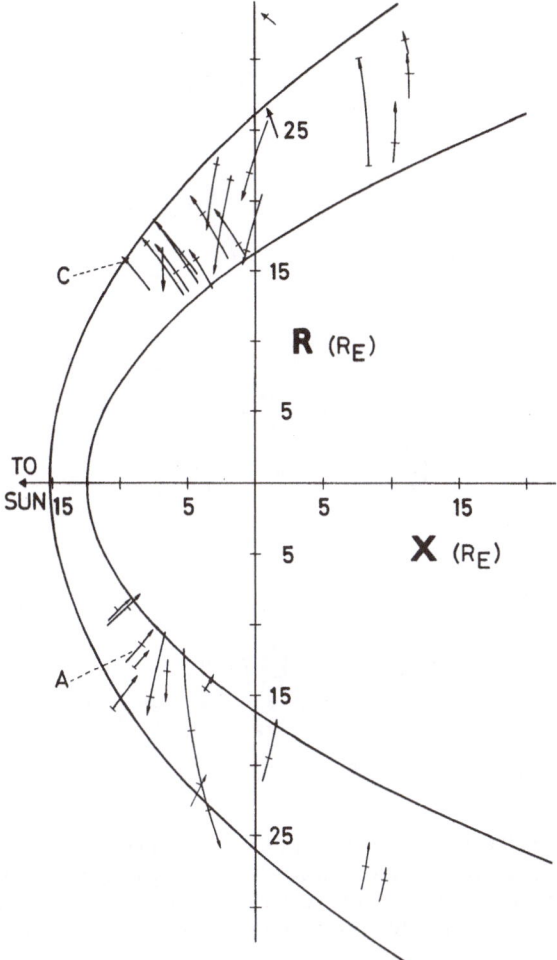

Fig. 1.   Segment orbits of HEOS 1 during the 32 magnetosheath transitions used in the work are drawn in the ecliptic plane after having been rotated around the X axis. The quantity R displayed on the ordinates is $\sqrt{Y^2+Z^2}$; when Y is positive the segment is drawn on the left of the Earth–Sun line, when Y is negative on the right. The bars which cross the segments correspond to the core memory replay periods. The distances are in Earth radii (R_E). The two curves are the intersections with the ecliptic plane of the best fit paraboloids for the bow shock and the magnetopause by Egidi *et al.* (1970).

A further interesting difference is that no fluctuations comparable to those of day 303 are observed in front of the shock (and the index I is now never equal to 1). The large difference between the two cases is confirmed by the core memory periods that were recorded in the magnetosheath, Figures 4 (day 363, hour 17.94) and 5 (day 3, hour 4.08), and in interplanetary space, Figures 6 (day 363, hour 11.47) and 7 (day 3, hour 10.45). The time of core memory observation is represented by a horizontal bar in Figures 2 and 3. A visual estimate suggests that in the sheath the amplitude of the oscillations on day 363 is higher than that of day 3 by more than a factor of 2, while the frequencies appear also to be different. In interplanetary space (Formisano *et al.*, 1973) on day 363 (Figure 6) we observe oscillations of at least 4–5 gammas

E. AMATA

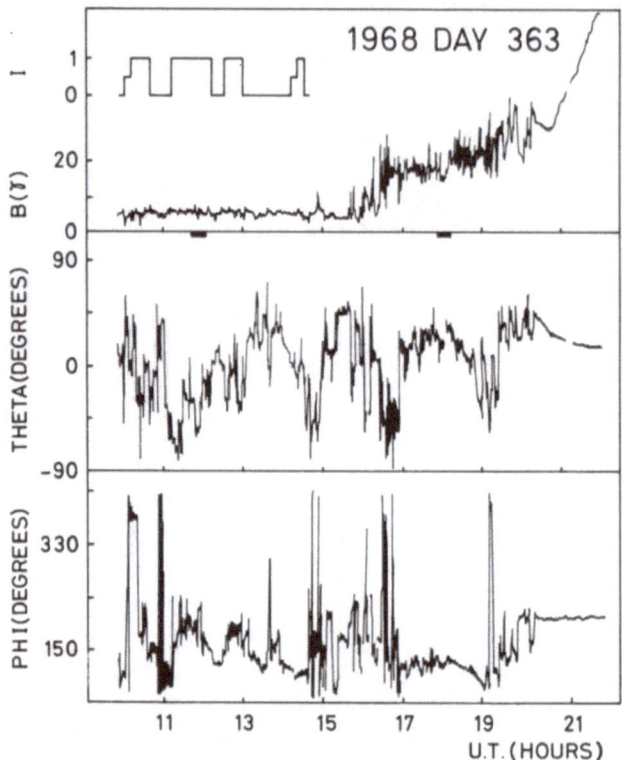

Fig. 2.   Plot (in spherical GSE and on a compressed time scale) of magnetic field real time data in interplanetary space through the magnetosheath and magnetopause for part of day 363, 1968. The two horizontal bars beneath the upper panel correspond to core memory data periods. The index I displayed in the
upper panel is equal to 1 when upstream waves should be detected.

and Doppler shifted periods of tens of seconds while on day 3 (Figure 7) the small
fluctuations we see are very likely due to experimental effects. These qualitative considerations have been checked by calculating the power spectra for these periods.

Table I shows the power densities in the various components, on days 363 and 3
outside and inside the magnetosheath, in the frequency ranges 2.6×10^{-4}–7.8×10^{-4}
Hz (RTD), and 7.0×10^{-3}–2.1×10^{-2} Hz (CMD). It is clear that in interplanetary
space substantial differences exist between day 363 and day 3 only in the CMD power
spectra; in fact while in the real time data the order of magnitude is roughly the same
and only the partition in the different components is different, in the core memory
data the power on day 363 is higher than that on day 3 by a factor of 10 for the X
and Y and by a factor of 40 for the Z and B components. In both cases the oscillations
are mainly transverse, except at low frequency (RTD) on day 363.

In the magnetosheath the same pattern shows up again: at low frequencies the
order of magnitude of the spectra on days 363 and 3 is roughly the same (although
some differences are still to be noted, as on day 3 the power appears to be equally
distributed among X, Y, Z and B, while on day 363 X's power is one order of mag-

TABLE I

Total power densities (γ^2 Hz^{-1}) in interplanetary space and the magnetosheath in the frequency ranges 2.6×10^{-4}–7.8×10^{-4} Hz (RTD) and 7×10^{-3}–2.1×10^{-2} Hz (CMD)

Day	Com-ponent	Interplanetary space		Magnetosheath	
		RTD	CMD	RTD	CMD
363	X	5.76×10^3	1.2×10^2	1.42×10^5	1.2×10^3
	Y	9.58×10^3	2.19×10^2	2.66×10^4	2.4×10^2
	Z	1.31×10^4	7.8×10	7.85×10^4	9.6×10^2
	B	1.6×10^3	6.7×10	2.74×10^4	5.2×10^2
3	X	9.5×10^3	1.82×10	1.24×10^4	1.05×10^2
	Y	2.67×10^3	1.21×10	1.01×10^4	6.3×10
	Z	1.25×10^3	1.8	1.01×10^4	6×10
	B	4.3×10^3	1.5	1.52×10^4	2.3×10

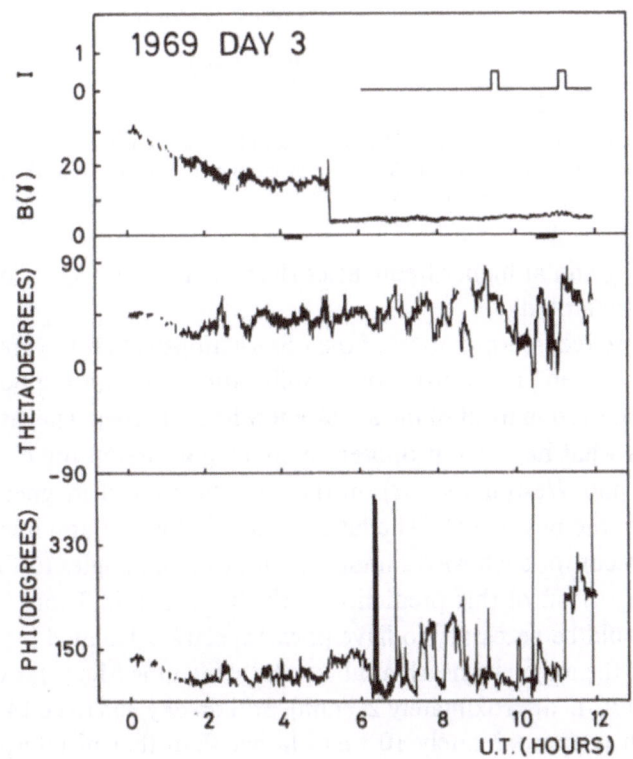

Fig. 3.   The same as in Figure 2, for day 3, 1969.

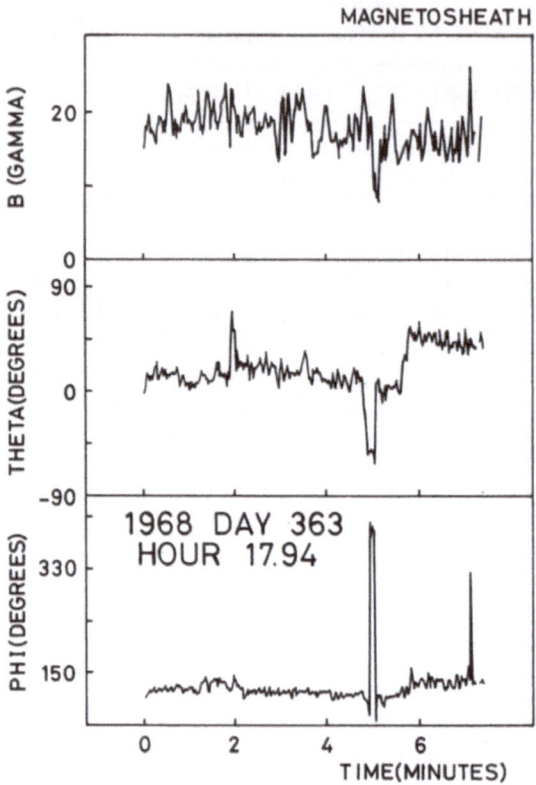

Fig. 4.   Magnetic field vector plot (in spherical GSE) for the core memory replay on day 363, 1968, recorded when HEOS 1 was in the magnetosheath. The time in minutes is counted from the start of the
replay period (17.94 h UT).

nitude higher than *B*'s), and at higher frequencies the power on day 363 is one order
of magnitude higher than on day 3.

A possible and attractive interpretation of the observations is that the greater turbulence observed on day 363 arises from the amplification through the bow shock
of waves like those observed in front of the shock a few hours earlier. This hypothesis
is in accordance with what has been proposed by many authors on the basis of the
results of McKenzie and Westphal (1968) on the amplification of magneto-hydro-
dynamic waves across the bow shock. The latter conclude that an amplification by
a factor of 3–4 may occur for such waves under small incidence angles to the shock.
Unfortunately a comparison of this prediction with the results in Table I can not
be conclusive as it would be necessary to have power spectra calculated just outside
the bow shock and in the magnetosheath at about the same time (the solar wind can
travel a distance of 10 R_E in approximately 2.5 min). Still on day 363 the CMD power
in the magnetosheath is approximately 10 times larger than that in interplanetary
space, which corresponds to a factor of 3 in amplitude (the same result is obtained
by visual scrutiny of Figures 4 and 6 (Formisano *et al.*, 1973a).

3.2. CLASSIFICATION OF THE OBSERVED MAGNETOSHEATH TRAVERSALS

In order to establish the correlation between magnetosheath turbulence and up-stream waves more firmly all the HEOS-1 magnetosheath traversals from December 1968 through January 1970 have been examined. After neglecting many of them either because no CMD were available or because of data gaps, only 32 cases have been retained and classified according to the following criteria:

(1) Visual inspection of real time data magnetic field plots (such as those shown in Figures 2 and 3) to determine whether or not waves are present in front of the bow shock.

(2) Calculation, when data from the HEOS-1 plasma experiment are available in interplanetary space, of an index I, similar to the one introduced by Greenstadt (1972): $I=0$ means no upstream waves, $I=1$ means that upstream waves should be detected. The plasma data have been used to determine the bow shock position and a minimum value (p_m) has been calculated for the ratio $(p = V_p/V_{sw})$ of the reflected protons velocity to the solar wind velocity for upstream waves to be generated at the HEOS-1 position in interplanetary space. When this value is less than 3, I is set equal to 1, when it is between 3 and 5, I is set equal to $\frac{1}{2}$ and, when $p_m > 5$, $I=0$. This is less stringent than just using the $p = 1.6$ empirical value found by Greenstadt (1972),

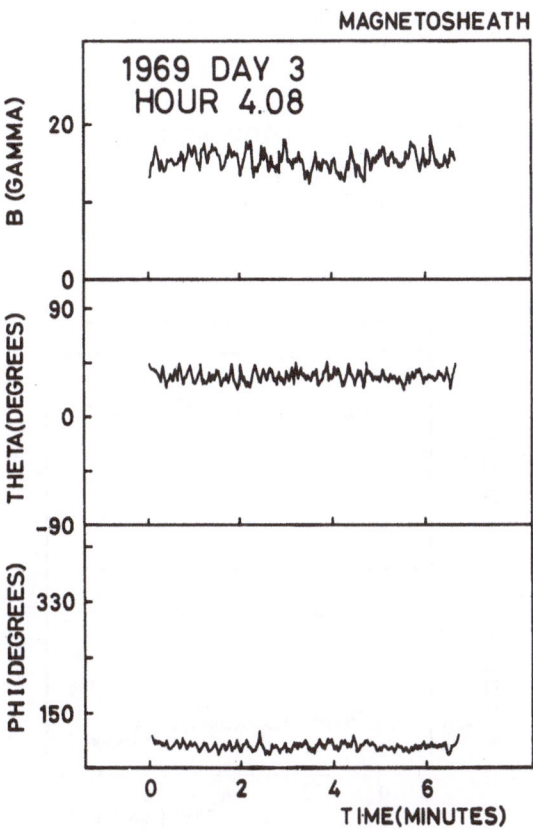

Fig. 5.   The same as in Figure 4 for the core memory replay at 4.08 h UT, day 3, 1969 (in the magnetosheath).

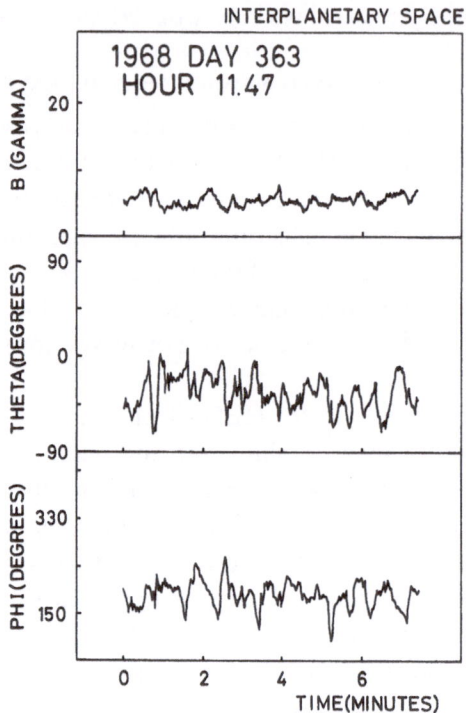

Fig. 6.   Same format as in Figure 4 for the core memory replay at 11.47 h UT, day 363, 1968
(in interplanetary space).

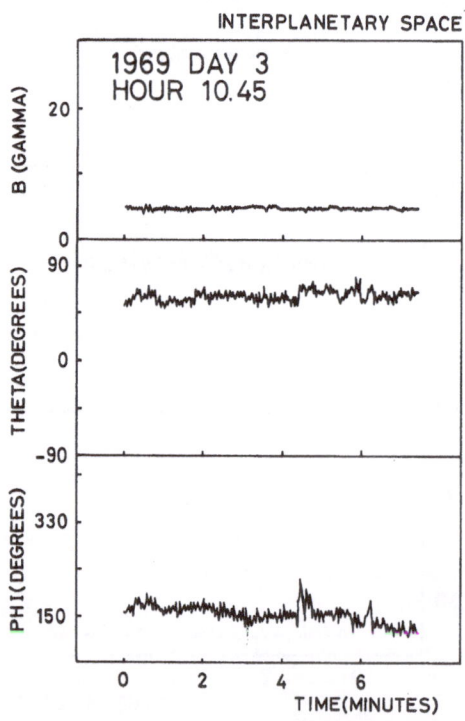

Fig. 7.   Same format as in Figure 4 for the core memory replay at 10.45 h UT, day 3, 1969
(in interplanetary space).

TABLE II

Bow shock geometry reconstruction (year 1969)

Day	Explorer 33 position in GSE			P_{m1}	P_{m2}	With upstream waves
	$X(R_E)$	$Y(R_E)$	$Z(R_E)$			
21	35	30	12	2.1		YES
104	−40	−51	−30	0.8		YES
105	−23	−46	−18	1.5		YES
108	15	0	10	0.8		YES
108	14	7	10	1.0		YES
111	−12	44	−10	0.7		YES
256				0.8	1.1	YES
261				12.4	1.6	YES
96	−91	−46	−65	16.1		NO
99	−77	−50	−55	5.3		NO
110	− 6	40	− 5	5.1		NO
253	–		–	3.5	3.1	NO
254	–		–	6.3	4.5	NO

but it is in accordance with a value $p = 2$–3 quoted by Barnes (1970) and a value $p = 2.7$ found earlier by Fairfield (1969). The index has been displayed for days 363, 1968 and 3, 1969 in the upper panel of Figures 2 and 3.

(3) Calculation of the same index using magnetic field data from Explorer 35 and 41 and plasma data from Explorer 33 at the same time when HEOS 1 was measuring inside the magnetosheath. Table II shows from left to right the day number, Explorer-33 position, p_{m1} (minimum value of p determined using Explorer-35 data), p_{m2} (determined using Explorer-41 data) and the final choice made (YES = case 'with upstream waves', NO = 'without upstream waves'). Explorer-33 plasma data have not only been used to determine the position of the bow shock but also to calculate the time elapsed between the observations at HEOS 1 and at the three other satellites. The largest corrections were rather to be made for Explorer-33 itself, apogee up to approx. 80 R_E, and Explorer 35 in orbit around the Moon, than for Explorer 41, apogee approx. 40 R_E. For the four cases in which Explorer-33 orbit was not available, the results based on Explorer-41 data have been taken as decisive.

TABLE III

Dawn-dusk asymmetry seen by HEOS-1 and Explorer 41

Year	Day	Hour	HEOS-1	Magnetic field	Explorer 41	Magnetic field	Region
1969	254	6.13	E	calm	W	disturbed	IS
	256	22.13	E	disturbed	W	calm	IS
	261	17.63	E	disturbed	W	calm	IS
1970	71	10.62	W	disturbed	E	calm	MS
	78	12.23	W	disturbed	E	calm	MS

The cases 'with upstream waves' summed up to 15 and those 'without upstream waves' to 9, while in 11 cases no clear choice could be made.

3.3. DAWN-DUSK ASYMMETRY

In some of the cases Explorer 41 was in interplanetary space very close to the bow shock just when HEOS 1 was in the magnetosheath. This coincidence has given a further opportunity of checking the validity of the chosen classification.

Table III gives a list of these events with the day, the starting hour, in UT of the core memory replay from HEOS 1, the indication whether the satellites were on the east (E) or the west (W) of the Sun-Earth line, the state of the magnetic field (calm or disturbed), the region of space where Explorer 41 was at that time (IS = interplanetary space, MS = magnetosheath). In the five cases available, when one of the satellites was on the dawn side the other one was on the dusk side and vice versa. The magnetic field was always 'disturbed' at either satellite and 'calm' at the other one. Now this is just what one might expect from the idea of amplification of waves through the bow shock: for a given magnetic field direction the angle between the field vector and the bow shock surface is very different on the dawn and dusk side; if this angle is near 90° on one side it will be near 0° on the other side; if on one side upstream waves may be generated in the so-called 'knowing region' and then amplified through the bow shock, that will not happen on the other side (Barnes, 1970).

3.4. THE POWER SPECTRA FOR THE CASES 'WITH UPSTREAM WAVES' AND 'WITHOUT UP-STREAM WAVES'

An examination of the power spectra for the cases with or without upstream waves leads to the conclusion that the power is generally higher in the cases when upstream waves are present just outside the bow shock. It is however practical not to study in detail the single cases, but to try to extract the common features of the two groups of spectra. The reason for this is that, as it has already been pointed out, many processes are probably at work at the same time in the sheath, while we want to isolate only one of them.

Figure 8 shows the average power spectra in the two frequency ranges 2.6×10^{-4}–

TABLE IV

Total power density (γ^2 Hz^{-1}) for cases with and without upstream waves in the frequency ranges 2.6×10^{-4}–7.8×10^{-4} Hz (RTD) and 7.0×10^{-3}–2.1×10^{-2} Hz (CMD)

Component	With upstream waves		Without upstream waves	
	RTD	CMD	RTD	CMD
X	1.29×10^5	1.32×10^3	5.03×10^4	1.9×10^2
Y	7.10×10^4	1.32×10^3	3.78×10^4	1.2×10^2
Z	5.08×10^4	1.32×10^3	2.55×10^4	2.1×10^2
B	2.43×10^4	5.8×10^2	9.20×10^3	1.8×10^2

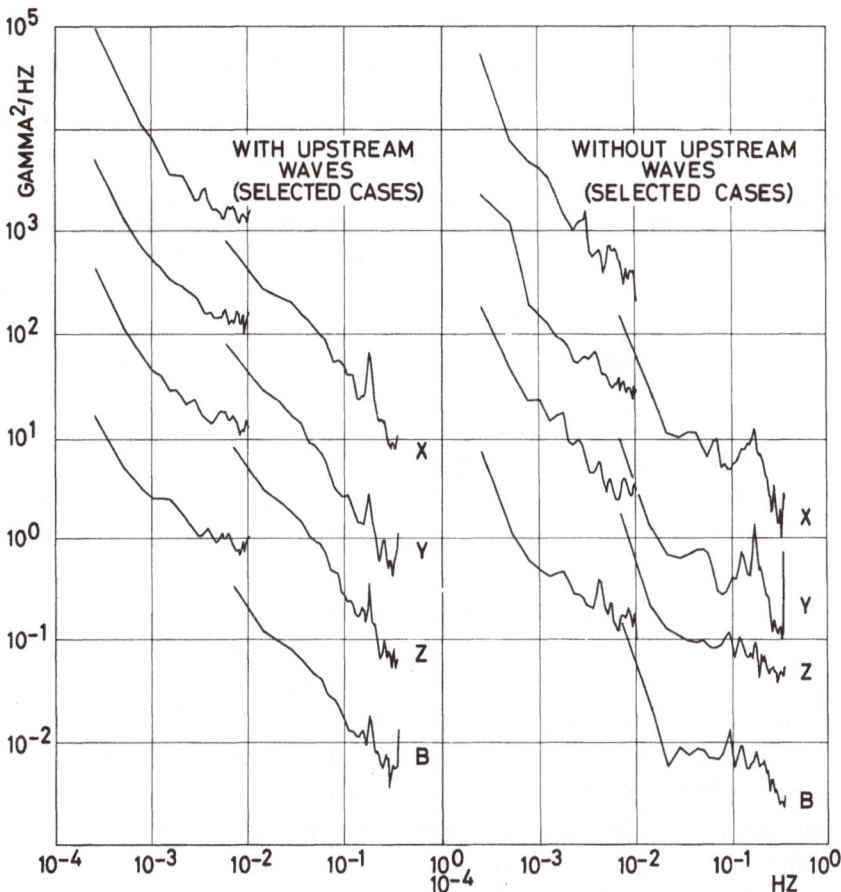

Fig. 8. Average power spectra in X, Y, Z, and B for the two groups of cases 'with upstream waves' (left), and 'without upstream waves' (right). Only the cases selected by means of bow shock geometry reconstruction with data from Explorer 33, 35, 41 have been used. The Y, Z, and B power curves have been offset by a factor of 10^{-1}, 10^{-2}, 10^{-3} respectively.

1.04×10^{-2} Hz (RTD), 7.0×10^{-3}–3.5×10^{-1} Hz (CMD) for the cases in which the third criterion was used. On the left hand side are the spectra 'with upstream waves' on the right hand side are those 'without'. At first sight it is clear that when upstream waves are present the power is substantially higher. This can neither be due to uncertainties in the computation of the spectra nor to aliasing. In fact although aliasing does occur for the HEOS-1 magnetometer, it can be shown that in this case it is important only near the Nyquist frequency (in fact some evidence of the spectra folding up can be seen in Figure 8).

Table IV displays the total power density for the different components in the frequency intervals 2.6×10^{-4}–7.8×10^{-4} Hz (RTD) and 7.0×10^{-3}–2.1×10^{-2} Hz (CMD) for the two cases. While the low frequency values are the same (within a factor of 2), the high frequency ones are higher in the case with upstream waves by a factor of 10 for the X and Y components and a factor 5–3 for Z and B. These values are, however, not the same at other frequencies. As one may notice, the spectra on

TABLE V

Average power densities (γ^2 Hz^{-1}) around six frequencies for all the cases

	X	Y	Z	B	Frequency (in Hz)
Real time data	1.02×10^5	5.17×10^4	2.30×10^4	2.46×10^4	2.6×10^{-4}
(22 cases)	3.25×10^3	1.98×10^3	2.19×10^3	1.31×10^3	2.3×10^{-3}
	2.66×10^3	2.11×10^3	2.29×10^3	1.36×10^3	3.6×10^{-3}
Core memory data	8.34×10^2	4.93×10^2	6.56×10^2	2.86×10^2	7.0×10^{-3}
(35 cases)	6.60×10	4.43×10	5.93×10	3.60	6.4×10^{-2}
	4.14×10	2.26×10	2.87×10	1.83×10	9.9×10^{-2}

the left side of Figure 8, fall rather smoothly from approx. $10^5 \, \gamma^2$ Hz^{-1} to approx. $10 \, \gamma^2$ Hz^{-1} over about three decades in frequency. The spectra on the right hand side on the other hand fall off from 3–$5 \times 10^4 \, \gamma^2$ Hz^{-1} to approx. $1 \, \gamma^2$ Hz^{-1}. Moreover there is a kink in the spectra at frequencies between 10^{-2} and 10^{-1} Hz^{-1}.

In conclusion the ratio of the left hand side spectra to the others is low at low frequencies (order of 2), it increases up to more than one order of magnitude around 0.02 Hz and then falls to a factor of 7–10 at the upper frequency limit where the effects of aliasing make the result less reliable. The slope ranges from approx. 1.31 (X, Y, Z curves) to approx. 1.13 (B curve) for the case 'with upstream waves'; and from approx. 1.59 (X, Y curves) to approx. 1.08 (Z, B curves) for the case 'without upstream waves' (the calculation of the slope for the latter is rather ambiguous because of the already mentioned 'kink').

The statistic has been improved by taking also into account the cases classified using only the first two criteria. It has been found that all the considerations made in the previous case still hold. Moreover, the process of averaging over a larger number of cases makes the spectra smoother, so that it is easier to calculate the slope. This ranges from approx. 1.4 (X, Y, Z curves) to approx. 1.16 (B curve) for the case 'with upstream waves', and from 1.62 (X, Y, Z curves) to approx. 1.22 (B curve) for the case 'without upstream waves'. This suggests that the B curve is always less steep than the others and that the case 'without upstream waves' has generally steeper spectra than the other one. It is well known that in the magnetosheath, unlike in interplanetary space, longitudinal oscillations are largely present (e.g. Siscoe *et al.*, 1967). In fact, Table IV and Figure 8 show that the power in the Z component, aligned to the average magnetic field, is comparable to the transverse power. It is also interesting to note that the power in the Z component is higher in the case 'with upstream waves' than in the case 'without' just as it happens for the other components. This suggests that the process of amplification of upstream waves through the bow shock can account not only for part of the transverse oscillations in the magnetosheath, but also for some of the longitudinal waves, contrary to what has been suggested in the past (Barnes, 1970).

A further interesting observation based on Figure 8 is that the Y component appears to contain less power than X and Z. This has been checked by averaging over all the cases (including those listed as 'undecided'), which amount to 35 for the CMD and to 22 for the RTD. Table V displays the resulting power densities around six frequencies. Except at the lowest frequency, the Y component does carry less power than X and Z, although the effect is not dramatic. To attempt to interpret this phenomenon let us recall that in the field aligned system of co-ordinates used for the computation of the spectra X would actually be parallel to the GSE Z axis if the magnetic field lay in the ecliptic, while the Y axis would lie in the ecliptic at a right angle to the magnetic field. On the average one would expect the magnetic field vector to be parallel to the ecliptic plane perpendicular to the Sun–Earth line. Thus, the interpretation of the results, on a statistical basis, is that generally more power is found in the magnetosheath transverse to the ecliptic plane than along the Sun-Earth line.

4. Discussion

The model for generation of magnetosheath turbulence which has been studied is essentially based on a three step process. First, some of the solar wind protons that should generate the shock going through a thermalization process are reflected at the shock front and travel upstream against the solar wind along the magnetic field lines. Upstream waves are then generated in the 'knowing region', e.g. through the process suggested e.g. by Barnes (1970) and are eventually carried through the bow shock by the solar wind, undergoing the already mentioned amplification (McKenzie and Westphal, 1968).

The classification of the magnetic field power spectra in the magnetosheath in two cases ('with upstream waves' and 'without upstream waves') and the comparison between them shows that in fact two states of the magnetosheath can actually be isolated, in connection with different conditions in interplanetary space next to the bow shock, or, to put it in another way, in connection with a different structure of the bow shock.

The main observed difference between the two cases is that the power density is substantially higher when upstream waves are observed, by a factor of 2, 10–15 and 7–10 as the frequency increases from 2.6×10^{-4} to 5×10^{-2} to 3.5×10^{-1} Hz respectively. It has also been found that generally in the magnetosheath transverse and longitudinal waves are equally present and that the enhanced power in the case 'with upstream waves' shows up for all the components of the spectrum. The latter result suggests that longitudinal oscillations are amplified effectively through the bow shock. A side result is that the magnetosheath oscillation seems to be more enhanced perpendicularly to the ecliptic plane than along the Sun-Earth line. Finally, the discussion of the magnetosheath transitions used in this study has also provided evidence of the so-called dawn-dusk asymmetry.

To conclude the discussion let us try to assess the relevance of the present results to the study of the magnetosheaths of other planets. Greenstadt (1970b) has found, using data from Mariner 4 and 5, evidence for a variable structure of Venus and

Mars bow shocks, just as at the Earth. The same author has also made some speculation about the situation at Jupiter (Greenstadt, 1973). In all of the three cases it should
be possible, if a large enough amount of data were available, to duplicate the results
presented in this paper, apart perhaps from the dawn-dusk asymmetry which is
strictly connected to the particular orientation of the average interplanetary magnetic field at 1 AU. In fact, a 'garden hose angle' very close to 45° corresponds in this
case to an oblique bow shock on the dawn side and a perpendicular shock on the
dusk side. On the other hand at a distance of 5.2 AU from the Sun, i.e. close to Jupiter's
orbit, the 'garden hose angle' is approximately 80°. This practically should exclude
the presence of upstream waves near the nose of the bow shock and favour their
presence on the flanks of it; the only remnant of the dawn-dusk asymmetry observed
at the Earth should consist in a bigger extension of the 'knowing region' toward the
subsolar point on the dawn side.

Acknowledgements

The magnetic field data from Explorer 35 and 41 and the Explorer-33 plasma data
were provided by the Goddard Space Flight Centre, and the HEOS-1 plasma data
by the Laboratorio Plasma nello Spazio (LPS), Frascati, Italy.

I would like to thank Dr V. Domingo and Dr P. C. Hedgecock for valuable discussions and the latter also for the use of two already existing computer programmes
and for providing the HEOS-1 magnetic field data.

Finally, I would like to thank for their hospitality Prof. H. Elliot, Head of the
Cosmic Ray group at Imperial College, London, where the bulk of this work was
done under a Fellowship from the Italian Council of Researches C.N.R. (Bando
203.2.4 del 23-12-1971), and Dr E. A. Trendelenburg, Head of the Space Science
Department at ESTEC, Noordwijk, where the work was completed under an ESRO
Fellowship.

References

Barnes, A.: 1970, *Cosmic Electrodyn.* 1, 90.
Bonetti, A., Moreno, G., Cantarano, S., Egidi, A., Marconero, R., Palutan, F., and Pizzella, G.: 1969,
    Nuovo Cimento **64B**, 307.
Egidi, A., Formisano, V., Palmiotto, F., and Saraceno, P.: 1970, *J. Geophys. Res.* **75**, 6999.
Fairfield, D. H.: 1969, *J. Geophys. Res.* **74**, 3541.
Fairfield, D. H. and Ness, N. F.: 1970, *J. Geophys. Res.* **75**, 6050.
Formisano, V., Moreno, G., Palmiotto, F., and Hedgecock, P. C.: 1973, *J. Geophys. Res.* **78**, 3714.
Greenstadt, E. W., Green, I. M., Inouye, G. T., Colburn, D. S., Binsack, J. H., and Lyon, E. F.: 1970a,
    Cosmic Electrodyn. 1, 279.
Greenstadt, E. W.: 1970b, *Cosmic Electrodyn.* 1, 380.
Greenstadt, E. W.: 1972, *J. Geophys. Res.* **77**, 5467.
Greenstadt, E. W.: 1973, Internal Note No. 21333-RU-OO, TRW Systems Group.
Hedgecock, P. C.: 1970, in V. Manno and D. E. Page (eds.), *Intercorrelated Satellite Observations Related
    to Solar Events*, D. Reidel Publ. Co., Dordrecht-Holland, p. 419.
Mariani, F., Bavassano, B., and Ness, N. F.: 1970, *J. Geophys. Res.* **75**, 6037.
McKenzie, J. F. and Westphal, K. O.: 1968, *Phys. Fluids* 1, 2350.
Siscoe, G. L., Davis, L., Smith, E. J., Coleman, P. J., and Jones, D. E.: 1967, *J. Geophys. Res.* **72**, 1.
Southwood, D. J.: 1967, *Planetary Space Sci.* **6**, 587.

THE RESPONSE OF THE MAGNETOSPHERE
TO THE SOLAR WIND

CHRISTOPHER T. RUSSELL

Institute of Geophysics and Planetary Physics, University of California, Los Angeles, Calif. 90024, U.S.A.

1. Introduction

Almost seventeen years have elapsed since the first probes were sent into the Earth's magnetosphere. Since that time, our understanding of the magnetosphere has gradually evolved to the point that now we have a clear, although at times qualitative, understanding of the dominant physical mechanisms at work in the terrestrial magnetosphere. Many of these processes undoubtedly occur in the magnetosphere of Jupiter also. While their relative importance will be altered by the different scales of the parameters in the Jovian magnetosphere, our understanding of Jupiter will be hastened by our experience in studying the Earth.

To set the stage for the Pioneer 10 results, we will first review the terrestrial magnetosphere, its topology, morphology dynamics and how the solar wind transfers energy to the magnetosphere. Next we review attempts to predict magnetospheric response to interplanetary conditions, and finally discuss the possibility of inferring interplanetary parameters from geomagnetic records.

2. Magnetospheric Topology and Morphology

The principal controversy of magnetospheric physics in the 1960's concerned the topology of the terrestrial magnetosphere, whether it was open or closed. In the open magnetosphere field lines from the polar cap, which are blown back behind the earth and form the geomagnetic tail, do not return to Earth. Rather, they enter the solar wind. In the closed magnetosphere, all field lines cross the surface of the Earth twice.

It is now commonly accepted that the magnetosphere is indeed open. The evidence for the open magnetosphere comes from many sources, the strongest of which is the control of the magnetospheric dynamics by the direction of the interplanetary magnetic field, and the nature of energetic particle access to the polar caps. The former evidence has been reviewed by Russell and McPherron (1973a) and the latter by Morfill and Scholer (1973) and will not be repeated here. However, because we have every reason to believe the Jovian magnetosphere is also open, we will discuss the topology of the magnetosphere at some length.

The principal architect of the open magnetosphere was J. W. Dungey (1961, 1963). Figure 1 shows Dungey's open magnetospheric model for a southward directed interplanetary magnetic field in the top panel and for northward directed fields in the bottom panel. In the top panel, the southward directed interplanetary magnetic field is carried by the highly electrically conducting solar wind towards the magneto-

V. Formisano (ed.), The Magnetospheres of the Earth and Jupiter, 39–53. All Rights Reserved
Copyright © 1975 by D. Reidel Publishing Company, Dordrecht-Holland

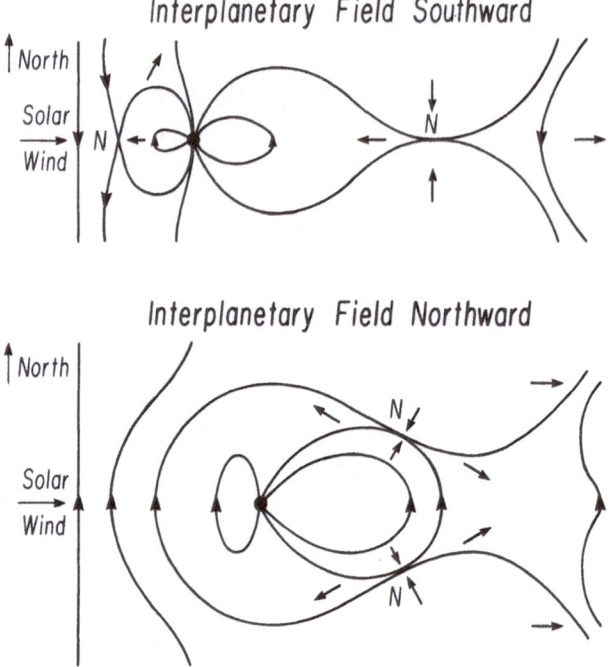

Fig. 1.   Schematic illustration of the open magnetosphere, for southward, top panel and northward, bottom panel, interplanetary fields (Dungey, 1963). These sketches do not attempt to show the correct length of the tail nor the magnitude of the magnetic field normal to the magnetopause. Arrows illustrate plasma flow directions.

pause, the boundary of the terrestrial magnetic field. If the magnetopause were a perfect superconductor, currents would be set up to cancel this field inside the magnetopause and the interplanetary field would be shielded from the interior. However, the magnetopause is not perfectly conducting and the two fields merge, or, reconnect. At one point on the nose there is a neutral point where the field strength goes to zero. This should not be mistaken for the merging region which is much bigger and which occurs along a line across the nose of the magnetosphere.

The interplanetary magnetic field lines are carried along by the solar wind exerting a stress on the magnetospheric field lines which were connected in the merging process, and these lines become draped behind the Earth forming a tail. The flux in the tail cannot build up forever, so that eventually merging or reconnection must occur here too. Again, there is a neutral point and a merging line which is a continuation of the dayside line.

We note that this topological discussion should also hold for Jupiter. Since Jupiter's dipole moment is opposite the Earth's, it interacts with northward fields. The relative importance of merging in the Jovian magnetosphere may be quite different. For example, the rapid rotation of Jupiter distorts the Jovian magnetosphere into a tail-like configuration at all local times. Thus, the magnetosphere of Jupiter might be responsive to changes in the stress of the solar wind in ways not found on Earth.

Returning to Figure 1, the lower panel shows merging for a northward field. This mechanism adds flux to the dayside of the magnetosphere and removes it from the tail. There is only weak evidence for the existence of this mechanism.

Figure 2 shows the morphology of the magnetosphere. First, there is the bow shock which deflects the solar wind, slows it and heats it so it can flow around the Earth. The bow shock is a complex and poorly understood region requiring of much more detailed study. The region of diverted and heated solar wind is called the magnetosheath. Magnetosheath plasma is also found in the polar cusps which at times appear to be extensions of the magnetosheath into the ionosphere. This region falls between the region of open and closed field lines. The tail consists of two bundles of flux, or lobes, with field towards the Earth in the north lobe and away from the Earth in the south lobe. These two lobes are separated by the plasma sheet. The plane that separates field away from the Earth from that towards the Earth has been termed the neutral sheet even though it is far from neutral.

In the interior of the magnetosphere is a field aligned 'doughnut' of cold plasma called the plasmasphere. This is the region of corotating cold plasma. The cold plasma in the magnetosphere moves along equipotential lines of the electric fields due to corotational forces and convective forces. Corotational forces are due to the drag of the neutral particles on the ionospheric particles causing them to rotate with the Earth. Convective forces are ultimately due to the tangential stresses of the solar wind on the boundary and in the open magnetosphere cause a flow that intersects the boundary. When these field lines encounter the boundary and merge with the interplanetary field, they open up and lose their cold plasma. Thus, the field lines in the convection dominated region are depleted in cold plasma by this periodic emptying, while the corotation zone can fill up to saturation. On Jupiter the corotation zone is thought to dominate the magnetosphere.

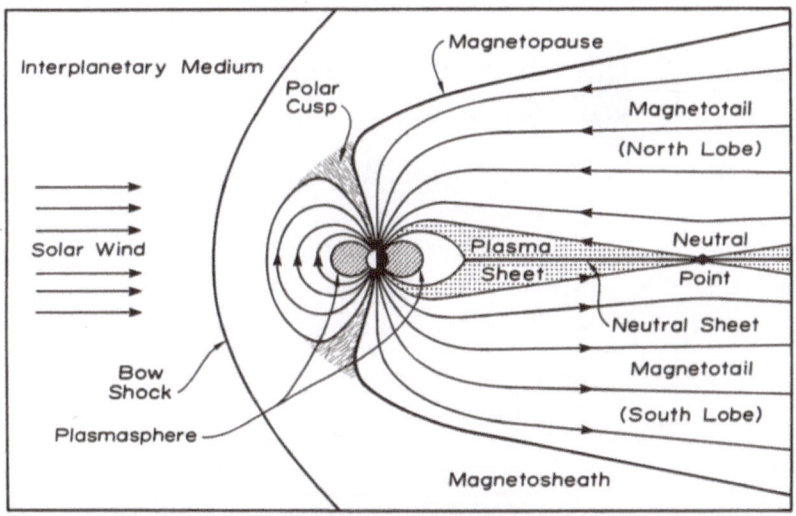

Fig. 2. A schematic noon-midnight meridian cross-section of the terrestrial magnetosphere (Russell, 1972).

The same physical processes should act in the Jovian magnetosphere as in the terrestrial magnetosphere. There should be radial diffusion caused by electric and magnetic fluctuations at the particle drift period. There should also be pitch angle diffusion caused by a variety of wave-particle interactions and plasma instabilities. One important process that is presently receiving a lot of attention in magnetospheric physics is field-aligned currents. These currents transmit stresses from the outer magnetosphere to the ionosphere. If the current density becomes too large, these currents develop instabilities and become resistive. Figure 3 shows an example of field-aligned currents in the polar cusp accompanied by a VLF emission (Fredricks *et al.*, 1973). If such wave amplitudes accompanied this current all the way down to the ionosphere a potential drop of 2 kV would have been present on this occasion.

3. Magnetospheric Dynamics

While the solar wind dynamic pressure determines the overall size and shape of the

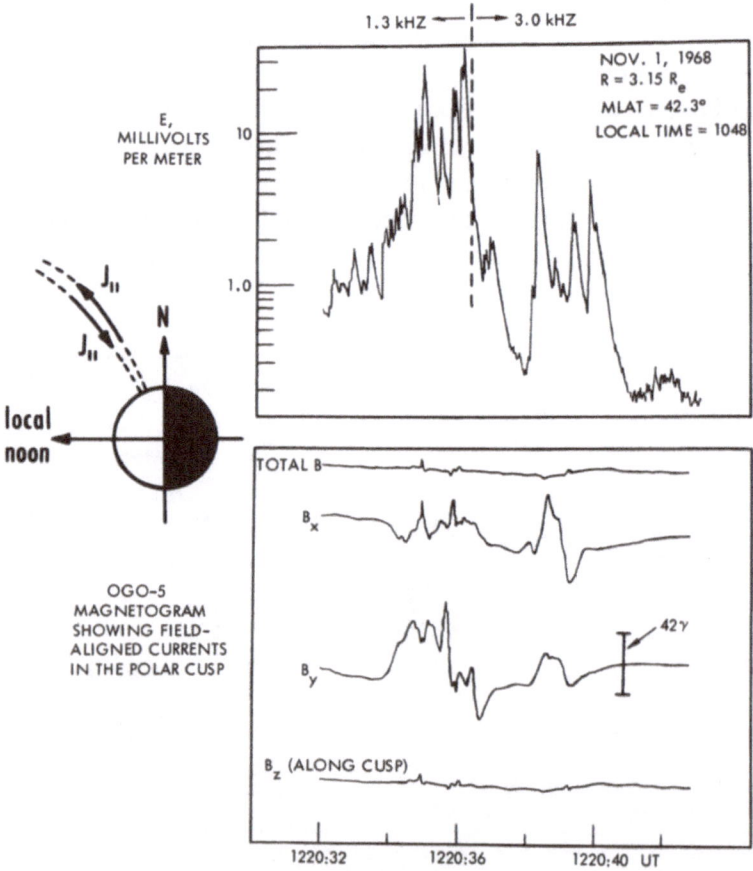

Fig. 3.   VLF electric field amplitude (top panel) seen in conjunction with field-aligned currents (bottom panel) during a traversal of OGO 5 through the polar cusp (Fredricks *et al.*, 1973).

magnetosphere, tangential stresses play an essential role in the dynamics of the magnetosphere, particularly of the magnetospheric tail. Since we have recently reviewed the evidence for the role of tangential stress on the magnetosphere and the sequence of events which ensue upon a sudden increase in this stress (Russell and McPherron, 1973a; Russell, 1974a), we will not repeat the detailed evidence here. However, we will briefly review the sequence of events associated with what has been called an isolated magnetospheric substorm.

Figure 4 summarizes the phenomena observed when the interplanetary magnetic field suddenly changes from north-pointing to south-pointing. This has been called the growth phase. At the nose of the magnetosphere, the magnetopause moves inward. In the tail the boundary increases its flare angle, and the resulting increase in pressure on the boundary results in an increased field strength in the lobes. The plasma sheet thins, tail currents strengthen and the midnight magnetosphere becomes tail-like deep into the magnetosphere.

Figure 5 illustrates what is commonly thought to occur at the onset of the expansion phase, when auroral breakup occurs. First, the plasma sheet begins to thin fastest close to the Earth so that the neutral points forms or moves to a position close to the Earth. Once the neutral point is formed, reconnection takes place at a rapid rate and the midnight magnetosphere returns to a more dipolar configuration. Eventually, when the 'demand' for merged flux by the magnetosphere is satiated, the neutral point moves tailward. This behavior explains the appearance of southward magnetic fields in the plasma sheet, the thickening of the plasma sheet with variable

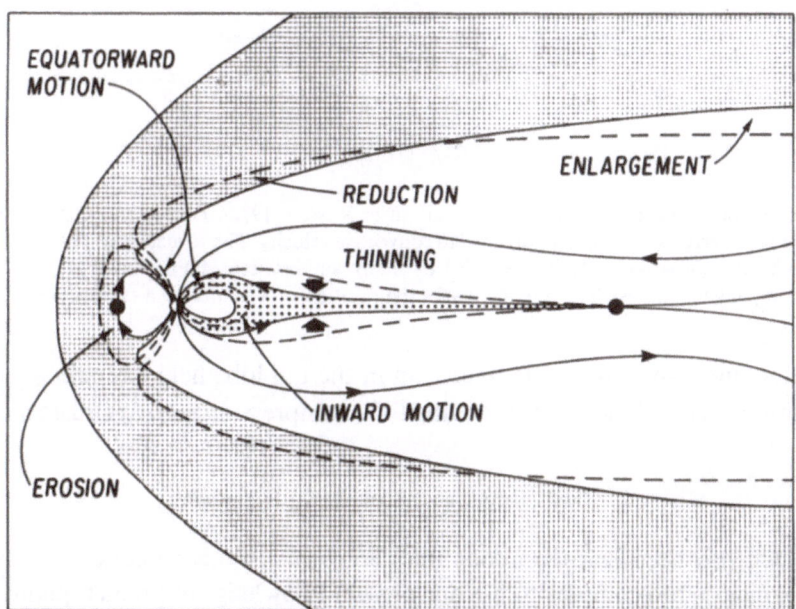

Fig. 4.   Summary of the magnetospheric changes observed during the growth phase of
substorms (Russell, 1974a).

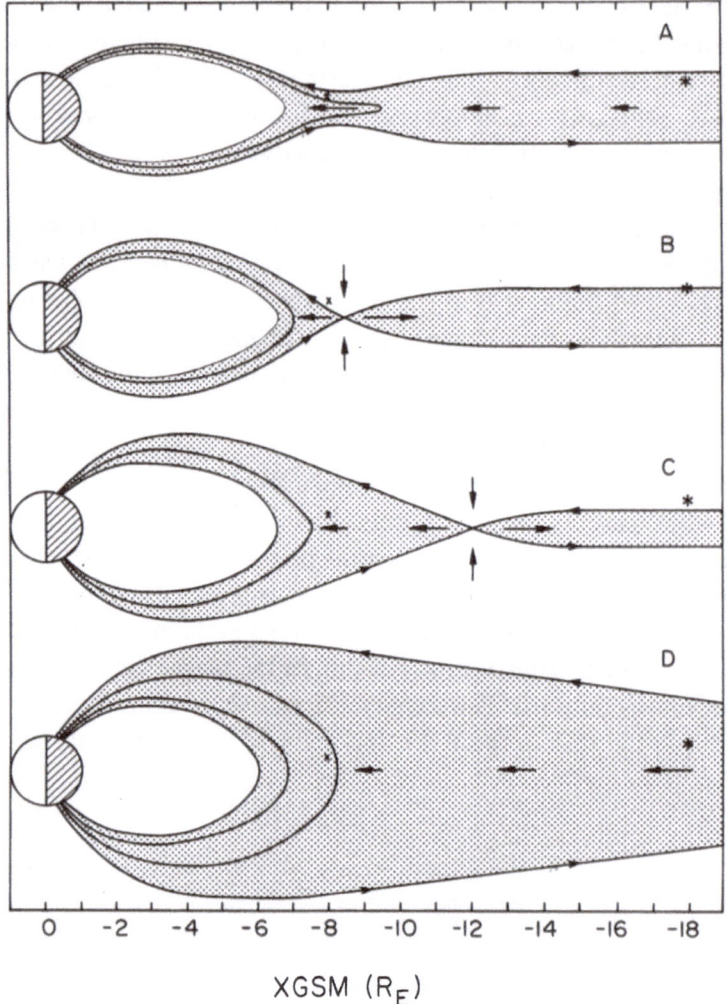

XGSM (R_E)

Fig. 5.   Model of the expansion phase in the tail (after Russell, 1972). Dotted area represents plasma sheet and length of arrows roughly proportional to convective velocity. The × marks the position of OGO 5 during a substorm on August 15, 1968 studied by many authors (see McPherron *et al.*, 1973 and accompanying papers). The asterisk marks the position of Vela during a typical event.

delays after substorm onsets, the reduction in the tail lobe field strength at the time of substorm onset and the inward moving field compression seen at substorm onsets.

4. Predicting Magnetospheric Response

While much has been learned about the details of magnetospheric processes, the magnetosphere is too complex to use our present knowledge to predict quantitatively the entire sequence of cause and effect within the magnetosphere, given a change in the magnetospheric boundary conditions. Yet it is important for many purposes to be able to predict the magnetospheric response. Thus, several workers have treated the magnetosphere as a 'black-box' whose transfer function is to be determined. The inputs to this black-box are the solar wind parameters: number density and velocity, field strength and direction. The output is usually taken to be one of the geomagnetic indices.

One of the first successful predictions of magnetospheric response was that of Arnoldy (1971) who used hourly integrals of the north-south component of the interplanetary field to predict the AE index (auroral electrojet index). In his integrations he assumed that northward fields were non-interacting and gave them zero weight. In this way, he acheived an 80% correlation between the integrated southward component and the hourly AE index one hour later. Several important points noted by Arnoldy in this study are:

(1) that use of solar magnetospheric coordinates returns higher correlation coefficients than solar ecliptic (see also Hirshberg and Colburn, 1969)

(2) that the coefficients relating the AE index and the integrated field varied with time

(3) that the interplanetary electric field rectified in an analogous manner gave a like correlation.

A similar approach has been followed by Garrett et al. (1974), who have attempted to model the K_p index and the A_p index. When they use the same integrated southward component as Arnoldy, they obtain an 80% correlation coefficient. When they use the rectified interplanetary electric field, the coefficient increases to 84%, and when a second term is added such as the variance of the interplanetary magnetic field the correlation coefficient increases to 90%. However, they also show that other functional forms involving the southward field, e.g., $\int V^2 B_s \, dt$ can correlate just as well or even better and that the addition of dynamic pressure terms also improve the correlation.

The prediction of the D_{st} index has been undertaken by Burton et al. (1974). First, they divide the index into two parts: that due to the magnetopause surface currents and that due to the ring current. The former currents are taken to be proportional simply to the square root of the solar wind dynamic pressure. The latter currents are considered to be a balance between a source whose strength is a function of the interplanetary electric field and a sink whose strength is proportional to the strength of the ring current. In this model, if the source is turned off, the ring current decays exponentially with a time constant of 8 hours, and if the Y-solar magnetospheric component of the interplanetary electric field suddenly increases to positive values the ring current will strengthen until a saturation level is reached at which the loss rate balances the new source rate.

Figure 6 shows an example of the actual and the predicted D_{st} index using this technique.

The prescription used here was:

$$\frac{d}{dt} D_{st_0} = F(E) - a \, D_{st_0},$$

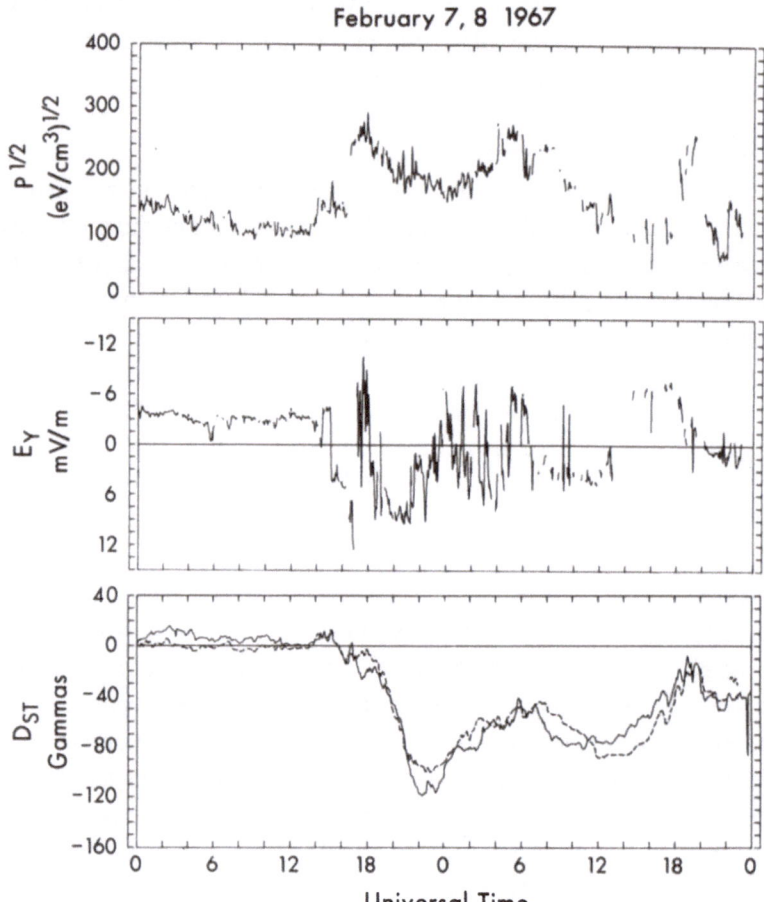

Fig. 6. Comparison of predicted (dashed line) and measured (solid line) D_{st} using equations given in text. Upper two panels give the square root of the dynamic pressure and the Y-solar magnetospheric component of the interplanetary electric field.

where

$$D_{st_0} = D_{st} - b\sqrt{p} + c$$
$$F(E) = 0 \qquad\qquad E < 0.5\ \text{mV-m}^{-1}$$
$$ = d(E - 0.5) \qquad E > 0.5\ \text{mV-m}^{-1}$$

with

$$a = 3.5 \times 10^{-5}\ \text{s}^{-1}$$
$$b = 0.2\gamma - (\text{eV-cm}^{-3})^{-1/2}$$
$$c = 20\gamma$$
$$d = 1.5 \times 10^{-3}\ \gamma\,(\text{mV-m}^{-1})^{-1}\ \text{s}^{-1}$$

P is the solar wind dynamic pressure and E is the Y-solar magnetospheric component of the interplanetary electric field. We note that the usual rectification found in such models is centered on 0.5 mV-m^{-1} here and not on 0.

The fact that the magnetosphere acts as a rectifier and seemingly does not interact with northward magnetic field (or equivalently negative Y-solar magnetospheric electric fields) has been used to explain the observed semiannual variation of geomagnetic activity (Russell and McPherron, 1973b) and to predict many of the properties of this variation, some of which were known but unexplained at the time of the conception of the model and some of which were discovered later. Most importantly it predicted the separation of the semiannual variation of geomagnetic activity into two annual waves one for each polarity of the interplanetary field. This neatly explains the occasional observations of a 12 month wave in geomagnetic activity (Meyer, 1972). It also explains the 22-yr variation discovered by Chernosky (1966) and predicted the separation of the diurnal variation of activity according to interplanetary polarity (Mishin *et al.*, 1973). However, the magnetic field alone cannot account for all the properties of the semiannual variation and Murayama (1974) has demonstrated that the velocity must play a role. This will be discussed in more detail in the following section.

Finally, we note that the fact that it is the interplanetary electric field which is the important driving function for geomagnetic activity, rather than simply the southward component, explains quite well the correlation observed between the solar wind velocity and the K_p index by Snyder *et al.* (1963). The principal feature of this correlation was an average increase in K_p with increasing solar wind velocity, but with such scatter in the correlation that almost every possible K_p value could be found at any solar wind velocity.

5. Inferring Interplanetary Conditions from Geomagnetic Records

The inverse to the problem discussed in the previous section is also of some interest. Geomagnetic records have been kept for over 100 years. If we can infer interplanetary conditions from these records, we can determine, for example, whether the properties of the solar wind measured during the present solar cycle are indeed typical. We may even be able to infer some properties of the solar magnetic field in the extended period in which good optical measurements were taken, but before the solar magnetograph was invented.

The first step in this direction was taken by Friis-Christensen *et al.* (1971, 1972) who showed that terrestrial polar cap magnetic variations are sensitive indicators of the interplanetary magnetic field. This effect has, in turn, been used to infer the interplanetary magnetic polarity from 1926–1969 (Svalgaard, 1972). However, this technique has its limitations. First, the measurement responds to the Y-solar magnetospheric component of the interplanetary magnetic field not the projection of the field along the expected Archimedean spiral direction which defines the interplanetary polarity and which is ordered in solar equatorial coordinates. (For a discussion of coordinate systems and the transformations from one to another see Russell, 1972). In a test using actual interplanetary data, the sign of the Y-solar magnetospheric component agreed with the spiral polarity of the field only about 85% of the time (Russell and Rosenberg, 1974).

In applications of the technique to ground-based data, Campbell and Matshusita (1973) found that this upper limit to the accuracy was approached only in the sunlit polar cap, northern summer in the data they used, and only when the signature was large. Overall they found an accuracy of only 60% in predicting interplanetary polarity. Svalgaard (1972) nevertheless, has published an index of interplanetary polarity from 1926–1969. While the above discussion indicates that this index should not be used for the identification of the polarity of specific days or times of sector boundary passages, the index might be useful for statistical purposes if the errors were random. Unfortunately this is not true.

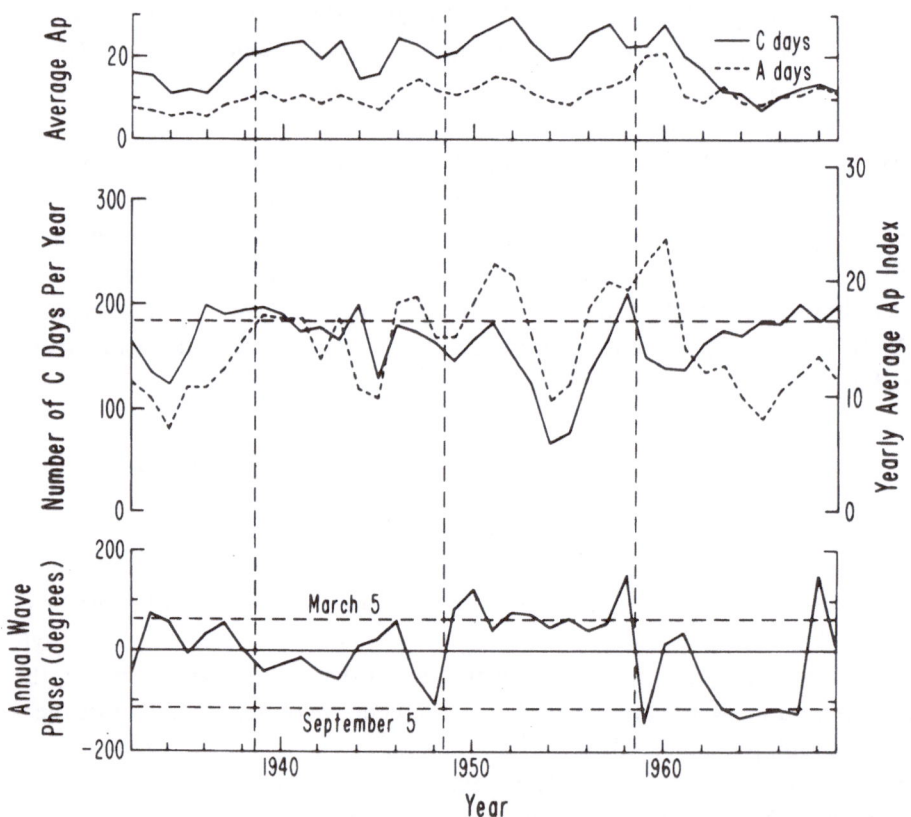

Fig. 7.   Average A_p index for C and A days (top panel); yearly average A_p index and number of C days per year (middle panel) and phase of the annual wave in C days (bottom panel). (Russell and Rosenberg, 1974).

The top panel of Figure 7 shows the A_p index for C days, which are days of inferred towards polarity, and A days which are days of inferred away polarity, from 1932 to 1969. Up to about 1962 when *in situ* measurements of the interplanetary field began, A and C days are quite different geomagnetically. However, after 1962 the traces join and the C/A designation no longer orders geomagnetic activity. This correlation is also shown in the middle panel in a different way. This shows the average

A_p index and the yearly number of C days. We note the strong correlation until 1962. This correlation with geomagnetic activity has been found independently both by Fougere (1974) and Russell *et al.* (1974).

The bottom panel shows the phase of the annual wave in the C days. A heliographic latitude dependence in the interplanetary polarity has been found by Rosenberg and Coleman (1969). If it were present in the index the phase of the C days should be either $-115°$ or $65°$ and should change every solar cycle at the time of the dashed vertical line. It does, indeed, show up for the last two solar cycles but it is not present from 1932 to 1949. For reasons discussed later we interpret this as a decrease in the amplitude of the heliographic dependence of the sector structure rather than a further decay of the index prior to 1949. In light of the correlation of the index with geomagnetic activity, and its poor predictive ability even in the satellite era, we urge extreme caution in the use of this index and suggest that it be reworked, perhaps under the auspices of IAGA.

Another method of deducing the average solar wind properties in the past is to use the average properties of geomagnetic activity. For example, the discussion of the driving function of geomagnetic activity in the previous section indicates that geomagnetic activity is proportional to the product of the solar wind velocity and the rectified southward interplanetary magnetic field, or some similar function. Thus, we could interpret the decrease in the solar cycle average aa index (Mayaud, 1973) from 1868 to 1907, its subsequent rise to a peak during the IGY solar cycle and then its recent decline as a long term, perhaps periodic, variation in the solar wind velocity, or the strength or variance of the interplanetary field. Examination of high latitude polar cap magnetograms could provide the answer as to whether the interplanetary strength had changed since the signature of Y-solar magnetospheric effect is proportional to the strength of the component (Friis-Christensen *et al.*, 1972).

The amplitude and phase of the semiannual variation of geomagnetic activity can also be used in this regard. Figure 8 shows predictions of the semiannual variation according to various idealized models. The top model shows the variation expected assuming a fixed 5γ field along the Archimedean spiral direction, (Russell and Mc-Pherron, 1973b). This curve maximizes about April 5 and October 5. If there is a modulation of the dominant polarity of the interplanetary magnetic field as a function of heliographic latitude then this will either enhance or diminish the interaction as sketched in the second panel. The phase will also change. Such a modulation is seen in the interplanetary magnetic field and it switches about 2 years after solar maximum (Rosenberg and Coleman, 1969). This can also be seen in Figure 7, bottom panel. We can remove this effect in the data by averaging data from two successive sunspot cycles.

The next panel shows the semiannual variation when the variability of the field is added (Murayama, 1974). Here an amplitude of 2.3γ in B_z has been assumed. The phase of the maxima are not altered by the introduction of the variability of the field. To alter the phase we must modulate the solar wind velocity, the magnetic field strength or the variability of the field with heliographic latitude. In the bottom

panel we show the semiannual variation of the rectified interplanetary electric field
from a 5 γ magnetic field with 2.3 γ variability and a 25% velocity gradient from 0°
to ±7.25° heliographic latitude. Both the phase and amplitude are altered by the
addition of the latitudinal variation.

Murayama (1974) has shown that the phase and amplitude of the semiannual variation varies with sunspot number. A harmonic dial of these results is shown in the

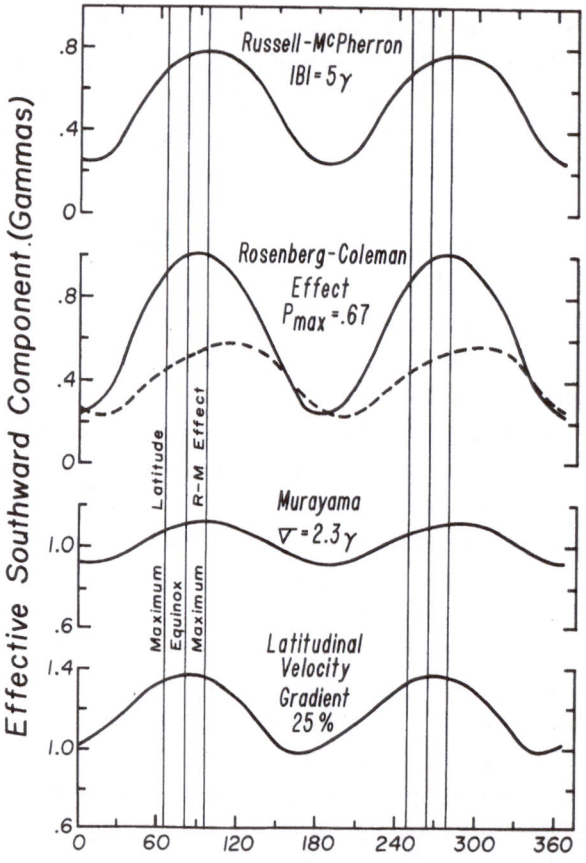

Fig. 8.   The semiannual variation of the effective southward component of the interplanetary magnetic
field, i.e., the strength of the solar wind-magnetospheric interaction, for several models. The numbers on
the abscissa are days of the year.

lower right-hand corner of Figure 9. Roughly as sunspot number *decreases* the amplitude of the semiannual variation *increases* and the variation maximizes earlier. By
modulating the solar wind velocity and introducing differing amounts of variability
we can attempt to model this behavior. The upper left dial shows the results when
the variability is set equal to zero. This obviously disagrees with observations. The
other two dials do mimic the observations quite well. However, a quantitative comparison should await analyses of linear measures of activity such as A_p.

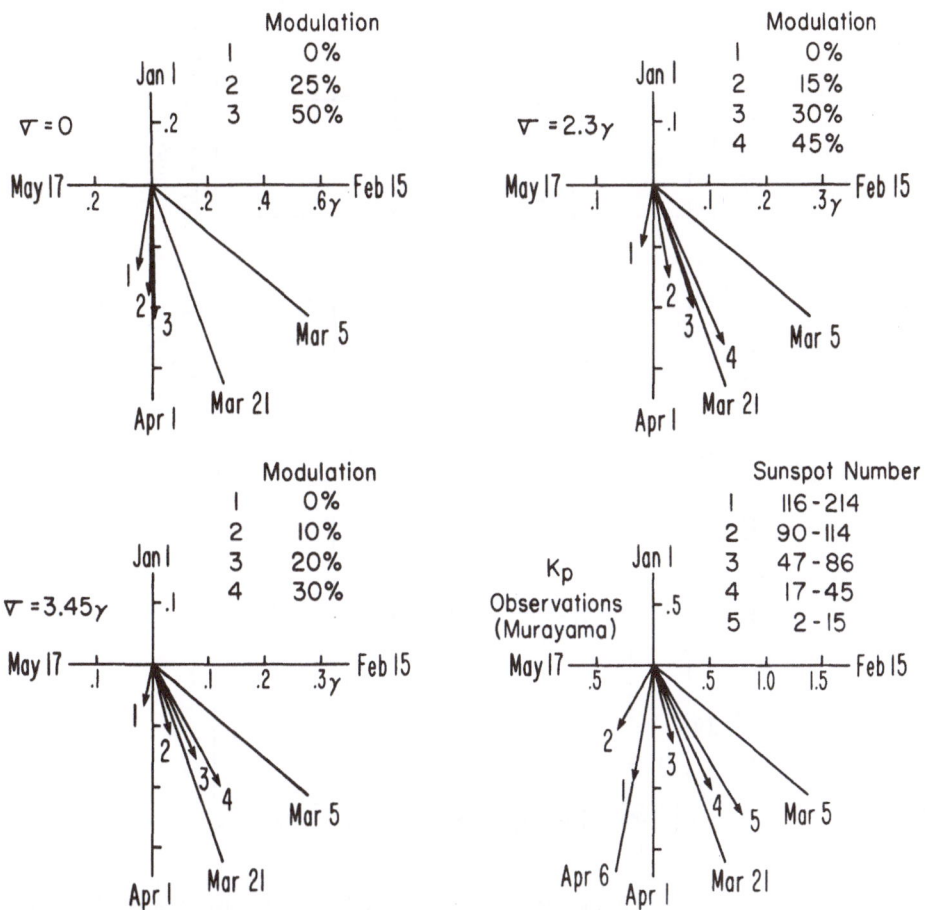

Fig. 9. Harmonic dials of the semiannual variation of geomagnetic activity. The lower right-hand dial shows the observations of Murayama (1974). The other three dials show several model variations.

Finally, the 22-yr variation of geomagnetic activity (Chernosky, 1966) can be used to measure the amplitude of the heliographic latitude dependence of the dominant polarity of the magnetic field (Russell, 1974b). As discussed above this effect either diminishes or enhances the interaction. Thus, alternate solar cycles, starting two years after solar maximum, should be alternately more active and less active. The bottom panel in Figure 10 shows this effect in the C_i index. The middle panel shows the amplitude of this cycle to cycle oscillation formed by taking the ratio of each value to the average amplitude of its neighbours and inverting every second ratio. This then should reflect the amount of heliographic latitude dependence of the polarity of the interplanetary field. It has been suggested that this variation is in fact due to the overall dipole moment of the Sun (Rosenberg and Coleman, 1969) and thus it appears that the solar polar field strength has changed with time reaching a minimum early this century. In the top panel, we show a smoothed C_i index for this period. It also shows a long term trend. As discussed above this could be due to changes in the solar wind velocity, or the magnetic field strength or its variability. We should add

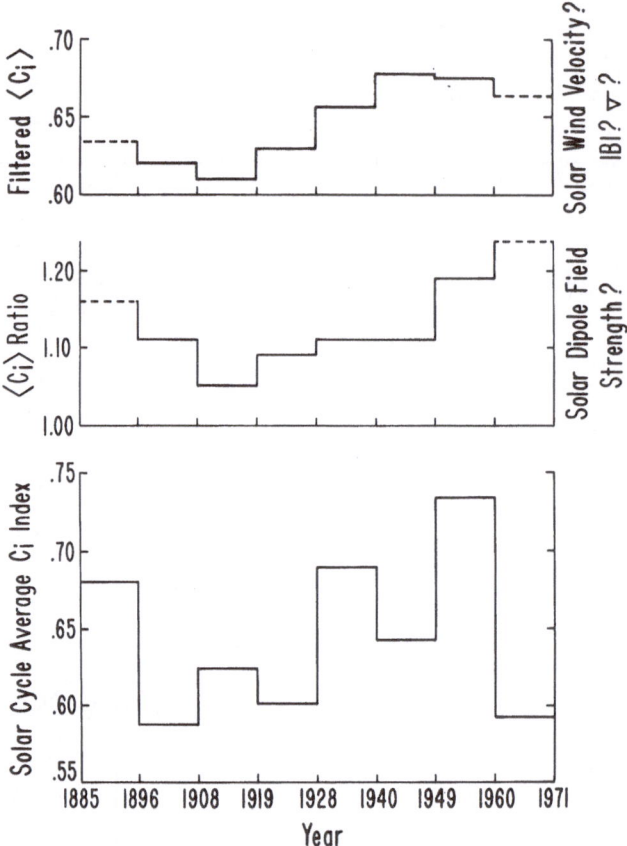

Fig. 10. The 22-yr cycle as seen in the geomagnetic C_i index. The bottom panel shows the solar cycle average C_i index starting 2 years after sunspot maximum. The middle panel shows the ratio of the C_i index for each cycle to that of its neighbours inverting every second ratio. The top panel shows the average C_i index with the 22-yr ripple removed.

a warning here that C_i is a subjective index which may itself have varied somewhat with time.

6. Summary and Conclusions

Nearly two decades of magnetospheric research has revealed the terrestrial magnetosphere to be a complex and dynamic region. The processes occurring therein are in many cases quantitatively and most cases at least qualitatively understood. Merging of the terrestrial magnetic field with the interplanetary magnetic field is a key element in the dynamics of the magnetosphere and ultimately provides the driving force for most magnetospheric processes. Presently, the interaction is understood well enough to predict geomagnetic indices rather accurately, and this understanding can be turned around to infer interplanetary conditions.

The Jovian magnetosphere should topologically resemble that of the Earth, and the same physical processes should act there as in the terrestial magnetosphere. However, the relative importance of the various processes will be different on Jupiter primarily because its corotation electric field is stronger relative to the merging electric field than in the terrestrial case. Another important difference between Jupiter studies and terrestrial studies, is that we expect a much more rapid advance in our understanding of the Jovian magnetosphere because of our terrestrial experience. Of course this will transpire only if we use this experience. The most important area to which this experience should be used is the proper selection of magnetospheric instrumentation for future Jupiter missions.

Acknowledgement

This work was supported by the National Science Foundation under NSF grant GA 34148-X.

References

Arnoldy, R. L.: 1971, *J. Geophys. Res.* **76**, 5189.
Burton, R. K., Russell, C. T., and McPherron, R. L.: 1974, unpublished manuscript.
Campbell, W. H. and Matsushita, S.: 1973, *J. Geophys. Res.* **78**, 2079.
Chernosky, E. J.: 1966, *J. Geophys. Res.* **71**, 965.
Dungey, J. W.: 1961, *Phys. Rev. Letters* **6**, 47.
Dungey, J. W.: 1963, in C. DeWitt, J. Hieblot, and A. Lebeau (eds.), *Geophys., The Earth's Environment*, Gordon Breach, New York.
Fougere, P. F.: 1974, *Planetary Space Sci.* **22**, 1173.
Fredericks, R. W., Scarf, F. L., and Russell, C. T.: 1973, *J. Geophys. Res.* **78**, 2133.
Friis-Christensen, E., Lassen, K., Wilhelm, J., Wilcox, J. M., Gonzalez, W., and Colburn, D. S.: 1972, *J. Geophys. Res.* **77**, 3371.
Friis-Christensen, E., Lassen, K., Wilcox, J. M., Gonzalez, W., and Colburn, D. S.: 1971, *Nature Phys. Sci.* **233**, 48, September 20.
Garrett, H. B., Dessler, A. J., and Hill, T. W.: 1974, *J. Geophys. Res.*, in press.
Hirshberg, J. and Colburn, D. S.: 1969, *Planetary Space Sci.* **17**, 1183.
Mayaud, P. N.: 1973, 'A Hundred Year Series of Geomagnetic Data 1868–1967', *IAGA Bull.*, No. 33, 255 pp.
McPherron, R. L., Russell, C. T., and Aubry, M. P.: 1973, *J. Geophys. Res.* **78**, 3131.
Meyer, J.: 1973, *J. Geophys. Res.* **77**, 3566.
Morfill, G. and Scholer, M.: 1973, *Space Sci. Rev.* **15**, 267.
Mishin, V. M., Bazarzhapov, A. D., Nemtsova, E. I., Popov, G. V., and Shelomentsev, V. V.: 1973, unpublished manuscript.
Murayama, J.: 1974, *J. Geophys. Res.* **79**, 297.
Rosenberg, R. L. and Coleman, P. J., Jr.: 1969, *J. Geophys. Res.* **74** (24), 5611–5622.
Russell, C. T.: 1971, *Cosmic Electrodyn.* **2**, 184.
Russell, C. T.: 1972, in E. R. Dyer (ed.), *Critical Problems of Magnetospheric Physics*, IUCSTP Secretariat, Washington, D.C., p. 1.
Russell, C. T.: 1974a, in D. E. Page (ed.), *Correlated Interplanetary and Magnetospheric Observations*, D. Reidel, Dordrecht-Holland, p. 3.
Russell, C. T.: 1974b, *Geophys. Res. Letters* **1**, 11.
Russell, C. T. and McPherron, R. L.: 1973a, *Space Sci. Rev.* **15**, 205.
Russell, C. T. and McPherron, R. L.: 1973b, *J. Geophys. Res.* **78** (1), 92.
Russell, C. T. and Rosenberg, R. L.: 1974, *Solar Phys.* **37**, 251.
Russell, C. T., Burton, R. K., and McPherron, R. L.: 1974, *J. Geophys. Res.*, in press.
Snyder, C. W., Neugebauer, M., and Rao, U. R.: 1963, *J. Geophys. Res.* **68**, 6361.
Svalgaard, L.: 1972, *Danish Meteorological Institute, Geophysical Papers*, R-29.

SUBSTORM EFFECTS ON THE NEUTRAL SHEET
INSIDE 10 EARTH RADII

B. T. THOMAS and P. C. HEDGECOCK

Imperial College, London, SW7, England

1. Introduction

The major features of the Earth's geomagnetic tail have been known for some time (Heppner *et al.*, 1963; Cahill, 1965; Ness, 1965). On the night side of the Earth the field lines are stretched out away from the Sun. The solar and anti-solar directed field lines are separated by a neutral sheet in which the field magnitude is small and across which the field direction reverses. The neutral sheet region is inflated by a plasma sheet, several R_E thick, which diamagnetically reduces the field magnitude near to the neutral sheet and provides the pressure to maintain the solar and anti-solar field separation. There is usually a small component of the field normal to the sheet indicating that the field lines pass through it (Speiser and Ness, 1967). However most sightings of a well defined neutral sheet have been at fairly large distances down the tail and there is therefore some controversy as to how close to the Earth the neutral sheet may be found. It is generally regarded that the Earth's field is tail-like beyond 10 R_E behind the Earth and that the neutral sheet is parallel to the Earth–Sun line and stretches away behind the Earth from a point hinged at 10 or 11 R_E from the Earth on the geomagnetic equator. Away from the midnight meridian it is assumed to curve towards the geomagnetic equator with an approximately circular cross section (Russell and Brody, 1967). To our knowledge, there are no reports of neutral sheet encounters inside 10 R_E and it is generally regarded that the neutral sheet does not approach nearer to the Earth than this. In this paper we provide evidence that during the growth phase of a substorm or during very disturbed periods the neutral sheet may be observed as close to the Earth as $\sim 7\,R_E$. ATS 1, in geostationary orbit at 6.6 R_E, observes more tail-like fields prior to a substorm onset but no neutral sheet observations have been reported. During the growth phase of a substorm the plasma sheet at $\sim 8\,R_E$ from the Earth has been observed to become very thin (McPherron, 1973; McPherron *et al.*, 1973a; Kivelson *et al.*, 1973). It has been proposed that coincident with the substorm onset the plasma sheet may neck to form a localized neutral point which then propagates down the tail (McPherron *et al.*, 1973b). This hypothesis is deduced from OGO-5 observations at 8 R_E from the Earth, just above the geomagnetic equator and our own observations in this region strongly support their findings.

2. Data Presentation

The experimental data presented here are provided by a 3-axis fluxgate magneto-

V. Formisano (ed.), The Magnetospheres of the Earth and Jupiter, 55–70. All Rights Reserved
Copyright © 1975 by D. Reidel Publishing Company, Dordrecht-Holland

meter of Imperial College, London (Hedgecock, 1974) providing vector measurements every 48 seconds.

The magnetic data are represented in geocentric magnetospheric polar co-ordinates.

The satellite position data are given in a slightly different Cartesian system, the Z
co-ordinate being taken as the distance above the predicted neutral sheet position
measured in the solar magnetospheric Z direction. The distance of the predicted
neutral sheet above the solar magnetospheric equator, Z_n, is given by the equation:

$$Z_n = (11^2 - (11^2/15^2) \, Y_{gsm}^2)^{1/2} \sin \lambda,$$

where λ = geomagnetic latitude of Sun (Fairfield and Ness, 1972; Russell and Brody,
1967). Thus the distance of the satellite above the predicted neutral sheet, Z, used in
this paper is given by $Z = Z_{gsm} - Z_n$.

In the midnight meridian this equation predicts that the neutral sheet is a line parallel to the X_{gsm} axis anchored at $11 \, R_E$ from the Earth on the geomagnetic equator,
and at other local times the neutral sheet meets the geomagnetic equator at a slightly
greater distance than $11 \, R_E$. This equation will evidently be inaccurate for observation within $11 \, R_E$ behind the Earth and therefore Z acts only as a guide, rather than
a prediction of the neutral sheet position. The neutral sheet would be expected to lie
nearer to the geomagnetic equator inside $11 \, R_E$ than is predicted by this equation
and this is in fact indicated by our observations.

The trajectories of the satellite for the days presented here are shown in Figures
1 and 2. In Figure 1 the trajectories are projected on the XY (gsm) plane and in
Figure 2 are projected on the local meridian plane. The circles denote the hourly
positions of the satellite which was moving towards the Earth on all occasions. It
can be seen from Figure 2 that the satellite is only close to the predicted neutral sheet
when it is near to the Earth. The field data for six days, which are shown on the left
hand side of Figures 3 and 4, are plotted against the satellite's radial distance from
the Earth projected in the XY (gsm) plane (i.e. $(X^2 + Y^2)^{1/2}$). Since the satellite is fairly
close to the midnight meridian at the time of a sheet encounter, this distance is approximately the distance of the satellite behind the Earth at this time. The time in
UT is shown above each plot.

The data are also illustrated as 15-min vector averages on the right hand side of
Figures 3 and 4. The vector averages are projected into the local meridian plane and
are plotted along the satellite trajectory. These plots illustrate the satellite position
at the time of any significant field changes and give a more graphic picture of the
field direction.

3. Observations

The neutral sheet is characterised in this data by a large depression of the field associated with a 180° reversal of the azimuthal angle. The field turns northwards
($\theta \rightarrow 90°$) as the satellite passes through the sheet indicating the small northward
component normal to the plane of the sheet. The width of the neutral sheet is taken

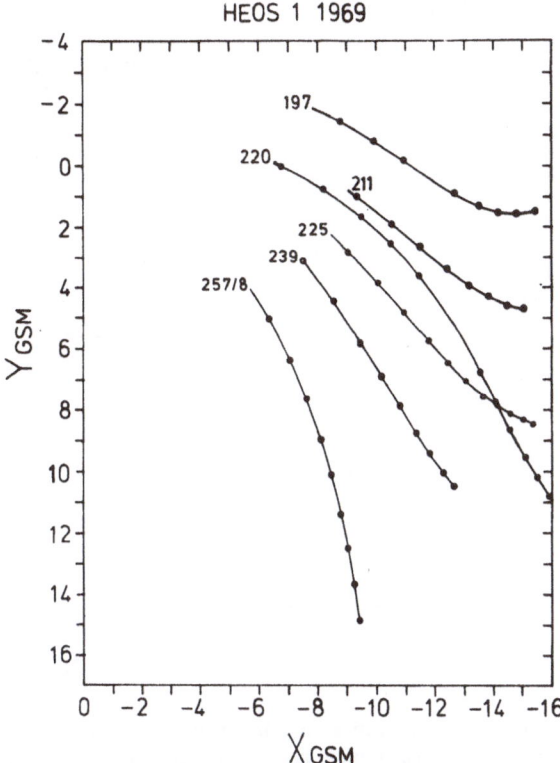

Fig. 1.   Satellite trajectories projected on geocentric solar magnetospheric XY plane. Dots give hourly positions while data are being received.

in this paper as the Z_{gsm} distance between the two points at which the field is inclined at $45°$ to the magnetospheric XY plane with the field magnitude falling to a relatively small value between these points.

Day 220 (Figure 3) and Day 239 (Figure 3) are fine examples of neutral sheet crossings. On Day 220 the field falls from $\sim 40\,\gamma$ to $\sim 7\,\gamma$ in a period of 1 h consistent with an increasing plasma pressure around the satellite as it nears the neutral sheet.

We will loosely describe the plasma sheet as the region over which the diamagnetic field depression is noticeable in the field signature. The field on this day becomes slightly more inclined as it approaches the neutral sheet and then quickly switches over to an anti-solar field in a period of ~ 15 min. The field does not return to a tail like configuration but becomes steadily more dipolar as the satellite approaches the Earth. In the hour during which the field magnitude was falling the satellite had moved through a Z_{gsm} distance of $\sim 0.7\,R_E$. From our definition of the width of the neutral sheet we can say that the satellite has moved a vertical distance of $\sim 0.2\,R_E$ during the neutral sheet crossing. These distances clearly represent the sheet thicknesses only if the sheet planes are not themselves moving. Assuming that they are not in motion we estimate the neutral sheet and plasma sheet half-thicknesses as $\sim 0.1\,R_E$ and $\sim 0.7\,R_E$ respectively. There is evidence that the plasma sheet was fluctuating in thickness over this time but we shall return to this in the next section.

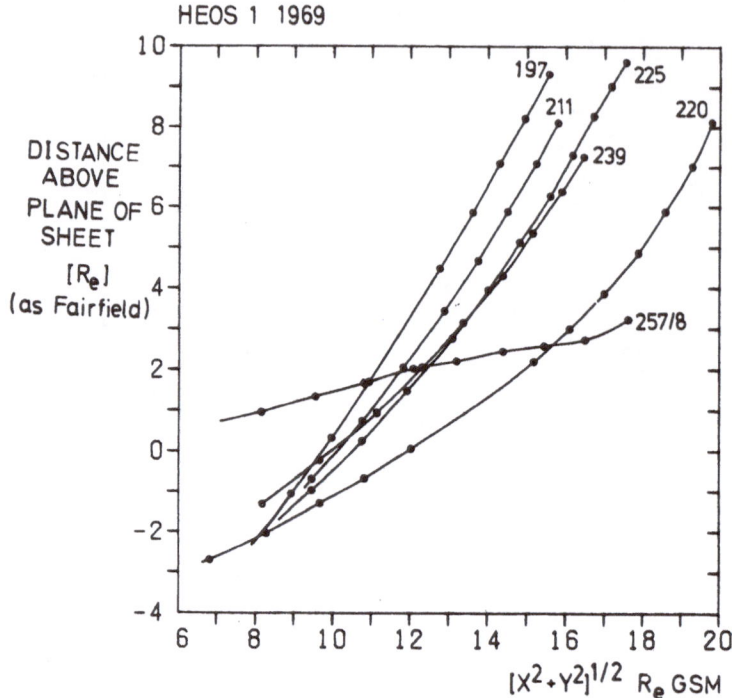

Fig. 2.   Satellite trajectories projected on local meridian plane. Vertical axis is $Z_{gsm} - (11 - (11^2/15^2) \times$
$\times Y_{gsm}^2) \sin \lambda$, where λ = geographic latitude of Sun. Dots as for Figure 1.

Day 239 shows a similar sort of signature but it is perhaps a slightly more interesting day. The field magnitude was falling during the period 0610 UT to 0745 UT which
would suggest that the satellite had entered the plasma sheet but it then increased
for about 1 h indicating that the plasma was retreating ahead of the satellite (this is
discussed in the next section in relation to substorms). At ~0850 UT the field started
to fall rapidly from 70 γ to 14 γ in a period of 20 min. The change in θ was slightly
broader in this case but here θ returned to a small value indicating tail-like fields
both before and after crossing the sheet. The half-thicknesses of the neutral sheet and
plasma sheet are estimated at ~0.2 R_E and ~0.5 R_E respectively. For both Day 220
and Day 239 the satellite was at a little over 9 R_E behind the Earth when the neutral
sheet is encountered.

Days 257/8 show a very disturbed field but, nevertheless, with a similar field signature to the previous two cases. A sudden field depression at 2055 UT associated
with direction changes towards a dipolar field suggests a plasma sheet encounter as
do all the earlier fluctuations on this day adding to the evidence that the plasma sheet
may be expanding and contracting. However, the field appears to settle down to a
moderately tail-like configuration by 2300 UT, followed by typical dense plasma
sheet and neutral sheet encounters. The field magnitude falls rapidly from 2350 UT
for ~50 min and the neutral sheet is clearly sighted at 0038 UT just before a data
break prevents us from being absolutely certain. However, an estimate for the half-
thickness of the dense plasma region and of the neutral sheet are ~0.3 R_E and $\leqslant$0.2 R_E

8/8/69 (DAY 220)

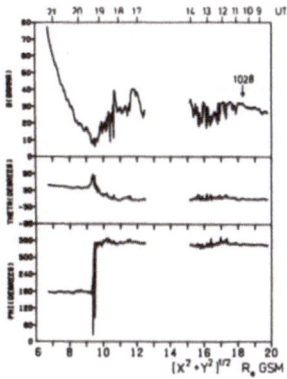 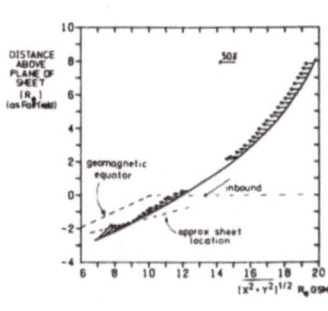

27/8/69 (DAY 239)

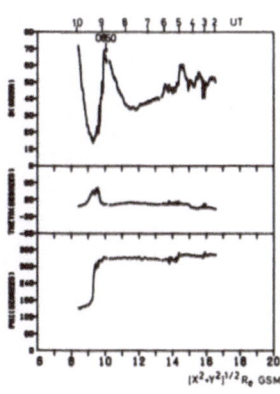 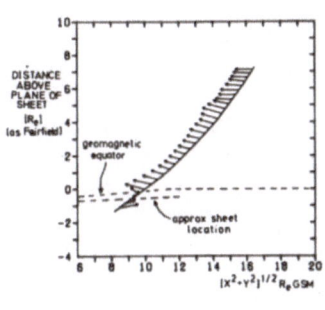

14/9/69 - 15/9/69 (DAYS 257/8)

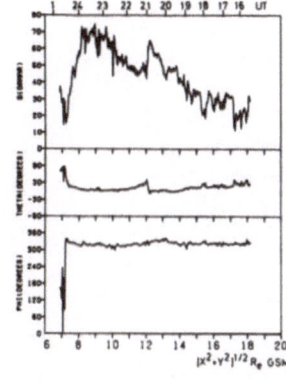 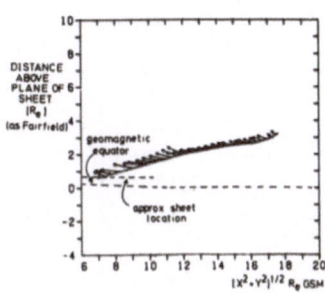

Fig. 3.

16/7/69 (DAY 197)

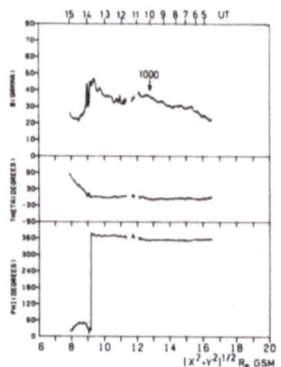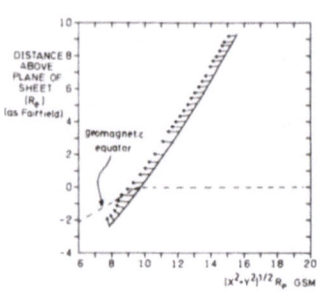

30/7/69 (DAY 211)

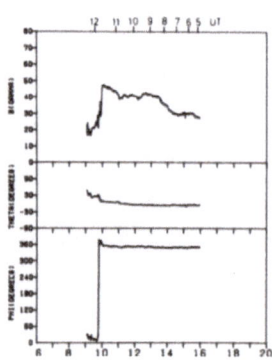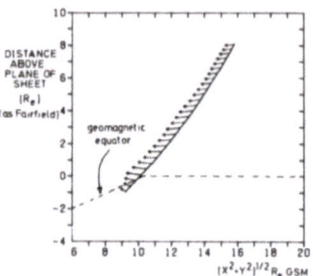

13/8/69 (DAY 225)

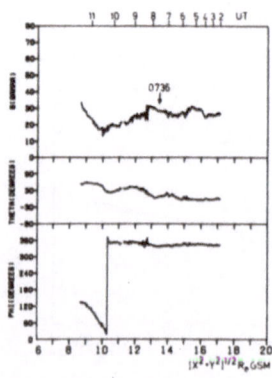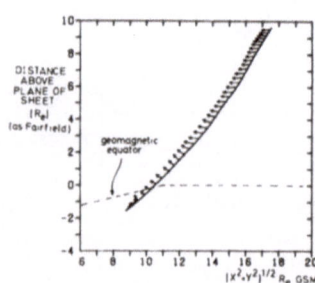

Fig. 4.
Figs. 3 and 4.   Field data plotted in solar magnetospheric polars; B, θ, ϕ (θ is measured from XY plane).
Also 15-min average vectors projected on local meridian plane and plotted along satellite trajectories
(as Figure 2.)

respectively. This is our closest neutral sheet sighting as the satellite is just over 7 R_E from the Earth and is less than 6 R_E behind the Earth (i.e. in the $-X$-direction).

We can compare the position of these neutral sheet observations with the predicted position for these days and also with the position of the geomagnetic equator, taking into account the geomagnetic latitude λ of the Sun. On Day 220 $\Delta Z = -1.3\ R_E$ and $\lambda = 25°$, on Day 239 $\Delta Z = -0.6\ R_E$ and $\lambda = 6°$ and on Days 257/8 $\Delta Z = +0.7\ R_E$ and $\lambda = -3°$. The indication is that the sheet is below the predicted position when λ is positive and above the predicted position when λ is negative. Comparing the sheet position with the geomagnetic equator in the vector plots (Figure 3) we see that the neutral sheet is also below the geomagnetic equator when λ is positive and above when λ is negative. This pattern suggests that the neutral sheet lies on a surface which curves from the plane of the geomagnetic equator near the Earth into the elliptical surface of Fairfield and Ness (1972) at, or at some distance beyond 11 R_E.

Day 197 (Figure 4), Day 211 (Figure 4) and Day 225 (Figure 4) do not reveal a neutral sheet. The plasma sheet is seen in all three cases but the angular changes are slow and no clearly defined reversal region is seen.

We summarize this section by saying that it is not unusual to find a narrow and clearly defined neutral sheet as close to the Earth as 7 R_E whilst on other passes through this region no neutral sheet is observed. It seems likely that whether the neutral sheet extends this close to the Earth depends on some macroscopic state of the magnetosphere. In the next section we look at these six days again in more detail with relation to substorm activity.

4. Substorm Effects

We shall look first at Day 239 as on this day there was a chance near-alignment of three satellites; OGO 5, HEOS 1 and ATS 1.

The OGO-5 data are shown in Figure 7 and uncorrected ATS-1 data in Figure 8. The auroral electrojet magnetic indices are shown in Figure 5. Figure 6 gives several mid-latitude, low-latitude and high-latitude magnetograms for this day. The AE index is included with the mid-latitude stations for comparison.

From Figure 6 we see that the magnetosphere was reasonably quiet from 0600 UT to ~0745 UT. At ~0745 UT the growth phase of a small substorm begins, with the AE index showing an increase in activity. The substorm onset is timed as ~0850 UT from the abrupt change in slope of the AE index and the small disturbance seen at Sitka, which is near to local midnight. The other mid-latitude stations see the onset a short time later. The high-latitude stations College and Gt. Whale see a clear bay at this time whilst the other high-latitude stations, some distance from local midnight, see very little.

The HEOS field magnitude in the tail lobes is considerably higher than usual at these distances. This is also observed at OGO 5 where the field is also higher than usual, apart from the period 0410 UT to 0730 UT when OGO is itself near the neutral sheet region and the field is depressed by plasma. This high field is almost certainly

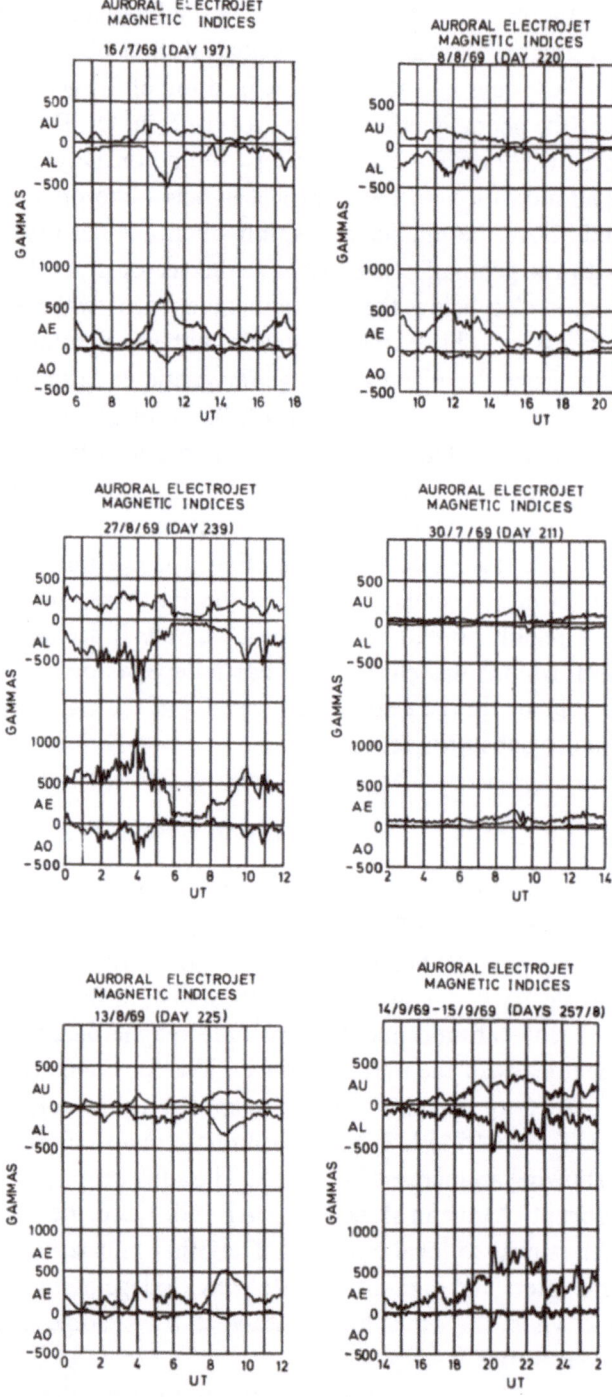

Fig. 5.   Auroral Electrojet Magnetic Indices.

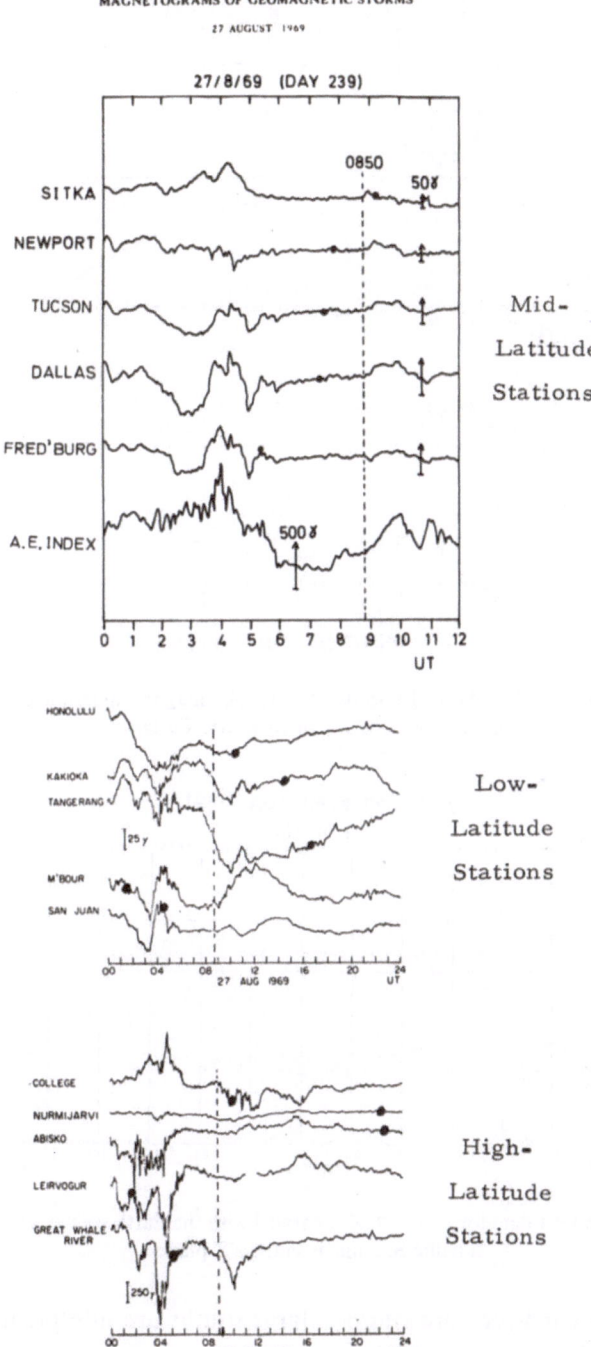

Fig. 6.   Ground magnetograms from various low-, mid-, and high-latitude stations. The AE index is included with the mid-latitude stations for comparison.

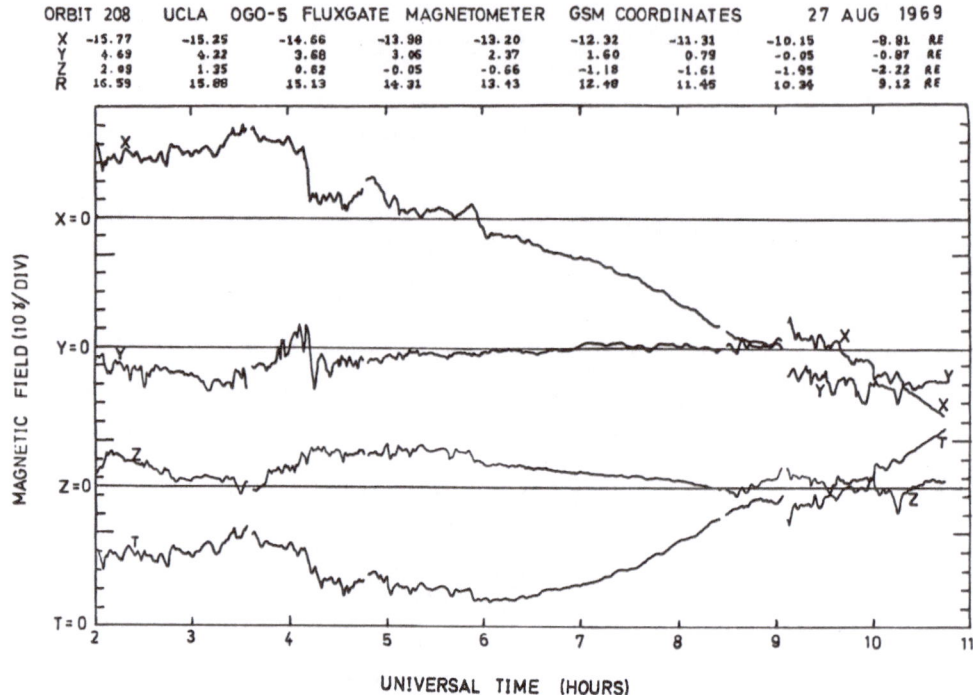

Fig. 7.   OGO-5 data for Day 239. The field co-ordinates are solar magnetospheric and the satellite position
is in geocentric solar magnetospheric Cartesians.

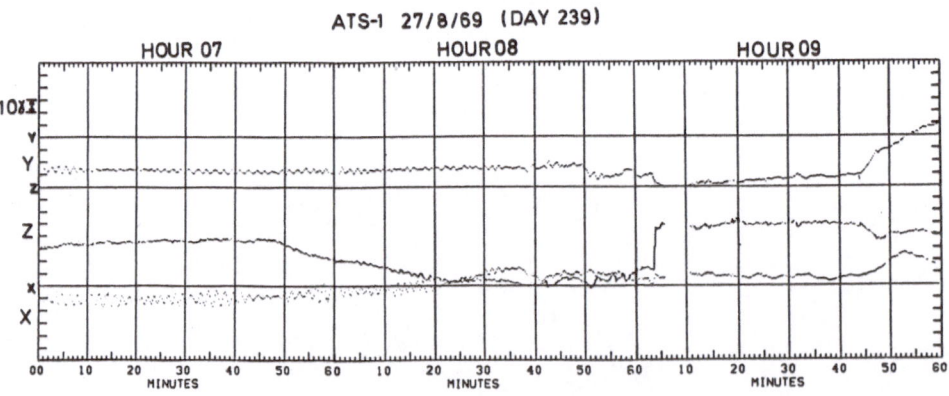

Fig. 8.   Uncorrected ATS-1 data for Day 239. Z is parallel with the Earth spin axis (northwards) and the
satellite-Sun line is in the XZ plane.

due to a high solar wind pressure and/or a large southward interplanetary field component for much of this day, resulting in more flux in the tail together with a smaller
flaring angle of the magnetopause. This will also cause the field to be tail-like closer
to the Earth than is usual.

From 0600 UT to 0743 UT the field magnitude at HEOS is falling despite the fact that the satellite is approaching the Earth. This suggests that the satellite has entered the plasma sheet at $\sim$0600 UT and the field is being increasingly diamagnetically depressed as HEOS approaches the neutral sheet region. At 0600 UT HEOS is $\sim$3.2 R_E above the predicted neutral sheet, suggesting a plasma sheet half-thickness of about this value. At 0743 UT the field starts to increase very rapidly suggesting that the plasma is retreating in front of the satellite, so that the plasma sheet appears to be thinning down. This time corresponds very closely with our estimate of the start of the growth phase of the substorm. At 0851 UT, when the field has reached a value of 69 γ, we can conclude that most of the plasma is below the satellite. The field magnitude now falls extremely quickly suggesting that the plasma may have expanded at this time, coincident with the substorm onset, engulfing the satellite very quickly in a region of high plasma pressure. This hypothesis is supported by OGO 5 (Figure 7) where a sharp fall in field magnitude at $\sim$0907 UT preceding an increase in field turbulence implies that the expanding plasma sheet has reached OGO at this time. OGO was $\sim$3 R_E below the predicted neutral sheet position at this time and assuming an initial plasma sheet half-thickness of 0.5 R_E the plasma has expanded by 2.5 R_E in 17-min, giving an estimated vertical expansion velocity of $\sim$15 km s^{-1}. Kivelson *et al.* (1973) report that the plasma sheet expansion velocity is typically 12–16 km s^{-1} so our figure is in excellent agreement.

The plasma pressure required to support a field of $\sim$70 γ outside the plasma sheet and 14 γ inside is given by:

$$\text{Field pressure} = (B_{\text{out}}^2 - B_{\text{in}}^2)/8\pi = 1.88 \times 10^{-8} \text{ dyn cm}^{-3},$$

where B_{out} and B_{in} are in gauss.

$$\text{Plasma pressure} = 0.667 \times (\text{Kinetic Energy Density})$$
$$= 0.667 \times (E \times n \times 1.6 \times 10^{-12}) \text{ dyn cm}^{-3},$$

where n (number density) is in particles cm^{-3} and E (energy) in eV. Thus we require a kinetic energy density of 18 keV cm^{-3} to support this field depression. The proton energy density in the plasma sheet at 18 R_E varies between 0.1 and 7 keV cm^{-3} with the electron energy density reaching 2 keV cm^{-3} (Akasofu and Chapman, 1972) so a value of 18 keV cm^{-3} at 9 R_E seems to be reasonable.

At 0920 UT the satellite passes through the neutral sheet and as we have said, this implies a half-thickness for the plasma sheet prior to the substorm expansion of $\sim$0.5 R_E. After passing through the neutral sheet the field returns to a quite tail-like orientation and this is a little surprising as after a substorm onset one would expect the field at this distance to become more dipolar. This may be due to the generally high fields in the tail maintaining a more tail-like configuration in this region than might be expected after a substorm.

Now however we look at the ATS-1 data for this period (Figure 8). The X, Y and Z components are a right handed cartesian system with the z-axis parallel with the Earth's spin axis (northwards) and the satellite-Sun line contained in the XZ plane (Coleman and McPherron, 1969). At $\sim$0745 the Z component at ATS begins to decrease and the X component to increase. This is what one would expect to see during the growth phase of a substorm with the field tilting towards a more tail-like orientation. At 0850 OGO 5 is very near local midnight and $\sim 10\ R_E$ behind the Earth, HEOS is at a local time of ~ 2215 and $\sim 10\ R_E$ behind the Earth and ATS is at a local time of ~ 2300 and 6.6 R_E from the Earth. Yet, as we see from the data, there is no significant onset signature at ATS 1.

This fact may explain why a substorm onset was observed in the high-latitude and mid-latitude magnetograms but was not seen in the low latitude stations at this time. It is possible that we have observed a small quasi-substorm confined to high L values seen by HEOS 1 and OGO 5 beyond 9 R_E but not at ATS at 6.6 R_E. This may also explain why HEOS sees continuing tail-like fields after the onset as HEOS is moving into the near Earth tail inside of 9 R_E.

Next we consider Day 197 (Figure 4), for which the auroral electrojet magnetic indices are shown in Figure 5. There appear to be two encounters with the plasma sheet on this day, one at ~ 1025 UT and another which appears to be a double encounter at ~ 1345 UT and at 1400 UT. The first encounter at 1025 UT occurs shortly after a substorm onset at ~ 1000 UT seen in the AE index. At this time we are 12.5 R_E behind the Earth and $\sim 4\ R_E$ above the predicted neutral sheet. If we assume that the initial plasma sheet half-thickness is $\sim 1\ R_E$, then the plasma sheet has expanded $\sim 3\ R_E$ in 25-min giving an expansion velocity of ~ 12 km s^{-1}, again in good agreement with Kivelson (1973). The field magnitude is increasing between 1230 UT and 1335 UT which may suggest that the plasma sheet is thinning as the substorm subsides. The entry into the dense plasma sheet occurs first at 1345 UT and again at 1400 UT. This double encounter does not suggest a plasma sheet expansion but a fairly static plasma sheet that is fluctuating slightly. An estimate of the vertical velocity of the fluctuation from the vertical distance travelled by the satellite gives ~ 1.5 km s^{-1}. The data break occurred just as θ was approaching 90° and so we assume that the satellite is near the centre of the plasma sheet. The dense plasma region is a little over 1.5 R_E in half-thickness if this assumption is correct.

Day 220 (Figure 3) is a fine example of a neutral sheet crossing as the angles change rapidly and the field falls to the very low value of 7 γ. The AE indices show a substorm onset at ~ 1028 UT shortly after which the plasma sheet is encountered at ~ 1055 UT when HEOS is 18 R_E down the tail and $\sim 5\ R_E$ above the predicted neutral sheet position. The substorm appears to reach its peak intensity at 1130 UT as measured by the AE indices, so the plasma sheet is observed well before this time. This contrasts with the observations of Hones *et al.* (1967) and Nishida and Hones (1974) in which the plasma sheet expansion at 18 R_E is found to correspond most closely with the peak of intensification of the substorm rather than with the onset. It is interesting to note however that the neutral sheet encounter on this day occurs when a small substorm is reaching its peak intensity at 1900 UT.

Day 225 (Figure 4) shows a similar pattern to the other days with the plasma sheet encountered at 0545 UT and again at ~ 0823 UT. This second encounter occurs a short time after a substorm onset timed at ~ 0736 UT from the AE indices. At 0740

UT, very shortly after the onset, the field at HEOS starts to become more dipolar and is inclined to the magnetospheric equator by 20° at the time when the plasma sheet is encountered. The field remains quite dipolar from 0823 UT until 1030 UT when the field magnitude reaches its lowest value. The angular changes are very slow and are quite consistent with entry into a dipole field configuration.

Days 257/8 (Figure 3) are very disturbed over the whole period with the field magnitude much greater than average. The AE indices (Figure 5) show that there are several substorms over this time and it is difficult to define any clear onsets. Due to the shallow angle of approach to the sheet region on this day, the satellite is never further than 3 R_E above the predicted neutral sheet position. The field is fluctuating due to many encounters with the plasma sheet indicating many expansions and contractions or a large amplitude flapping motion of the plasma sheet as indicated by the observations of Hones *et al.* (1971). A large depression of the field at ~2058 UT, associated with a return to a dipole like field configuration is almost a classical signature of what is expected due to a substorm expansion but no individual onset can be seen in the AE index. This is followed by two hours of steady return to a tail-like configuration with a dense plasma sheet and neutral sheet encounter. It is difficult to make any clear deduction from such a disturbed day as the magnetosphere is unlikely to be behaving in a well ordered way.

Day 211 (Figure 4) shows another clear plasma sheet encounter. The AE indices (Figure 5) show that there are no substorms over this period and so we should be seeing a quiet-time field configuration. Before 1142 UT, when the satellite is at 10 R_E from the Earth, the field is quiet tail-like with θ increasing slightly in the last hour or so as the satellite approaches the Earth. The dense plasma sheet is encountered at 1142 UT and the field inside is quite dipolar. On such a quiet day it is unlikely that the neutral sheet is to be found this close to the Earth but a fairly thin plasma sheet is clearly observed.

We summarize these observations as follows. On Day 239 we observe the contraction of the plasma sheet during the growth phase of a small substorm, followed by a rapid expansion coincident with the substorm onset. The neutral sheet near midnight is found to exist close to the Earth during disturbed periods on Day 239, Day 220 and Days 257/8. On Day 220, Day 197 and Day 225 the plasma sheet is observed shortly after substorm onsets suggesting plasma sheet expansions and giving an estimate of the expansion velocities. In the late recovery stages of two substorms on Day 197 and Day 225, the plasma sheet is encountered by the satellite in the region near to 10 R_E from the Earth but the field was quite dipolar. This is similar to the situation observed during a very quiet period on Day 211. This suggests a picture of the processes occurring in this region during substorms as follows. During the growth phase of a substorm the plasma sheet contracts and becomes very thin, the field becomes tail-like and the neutral sheet moves closer to the Earth. Coincident with the onset of the substorm the plasma sheet expands in the region near to 10 R_E at velocities in the range 10–20 km s^{-1}. The field becomes more dipolar during the expansion phase and the neutral sheet retreats back down the tail.

5. Conclusions

We have presented evidence in this paper that a clear and well defined neutral sheet can be observed near midnight at geocentric distances as small as 7–8 R_E. Comparing our observations with the AE indices we find that the neutral sheet was observed inside 10 R_E from the Earth following the onset of a substorm and during generally disturbed periods. It was observed to be absent during the late recovery phase of substorms or during quiet periods with a more dipolar field configuration existing at these times. The position of the neutral sheet on these occasions is consistent with it lying on a surface that curves from the near-Earth geomagnetic equator into the surface of Fairfield and Ness (1972) at some distance beyond 11 R_E. Our observations also indicate that during the growth phase of a substorm the plasma sheet thins at geocentric distances near 10 R_E and prior to the onset can be less than 0.5 R_E in half

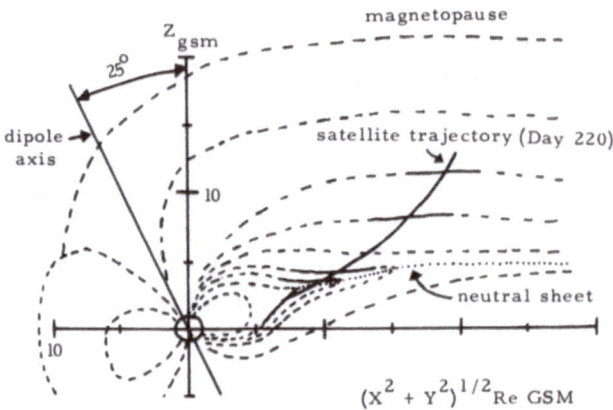

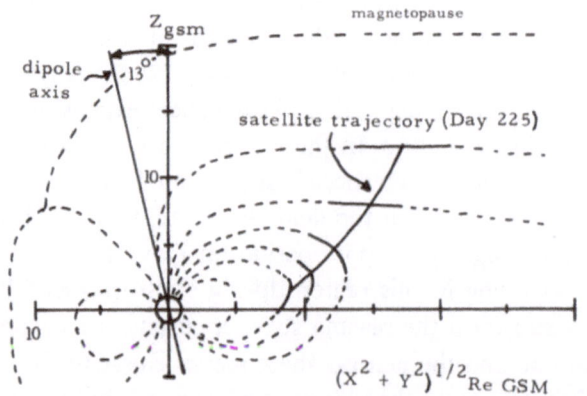

Fig. 9. Local meridian sketches of extrapolated field line configurations consistent with field directions observed by HEOS 1 on Day 220 and Day 225.

thickness. At the substorm onset the plasma sheet is observed to expand with velocities normal to the sheet in the range 12–15 km s^{-1} (these values depend on the assumed initial conditions), in good agreement with Kivelson *et al.* (1973).

In Figure 9 we have presented a freehand extrapolation of a field configuration consistent with the directions observed on two days. The plots are local meridian projections which in these cases are approximately equivalent to midnight meridian projections. For Day 220 which is quite disturbed over this period we observe a very pinched or stretched tail-like configuration in the region 8–10 R_E with the neutral sheet dotted in as close as 8 R_E. On Day 225 the satellite is near this region in the late recovery phase of a substorm and the field is considerably more dipolar.

McPherron *et al.* (1973b) postulate a model for magnetospheric substorms in which the plasma sheet thins in a localized time sector 8 to 10 R_E behind the Earth during the growth phase of the substorm. The plasma sheet then necks to form a neutral point at the onset of the substorm after which the plasma sheet expands rapidly. One of our days of data, Day 239, fits very well with the event studied by McPherron and is not inconsistent with his interpretation. The other two days on which the neutral sheet was observed, Day 220 and Days 257/8 are very disturbed days with multiple onsets and rapid plasma sheet fluctuations. A simple interpretation of these days is not easy and this emphasises the point, made by McPherron, that it is important to discriminate in substorm studies between isolated substorms and substorms with multiple onsets.

We note also that our neutral sheet sightings occur at local times as much as 2.5 hours before midnight. This is consistent with the observations of Clauer *et al.* (1973) who report substorms centred at local times many hours from midnight.

We conclude that the region between 7 and 10 R_E behind the Earth and near to the geomagnetic equator plays an important role in the development of magnetospheric substorms. It seems probable that the substorm is triggered in this region and the disturbance then moves down the tail.

Acknowledgements

We would like to thank Margaret Kivelson for many helpful discussions during the course of this work and also Chris Russell and Bob McPherron for providing the OGO-5 and ATS-1 data for comparison. Prof. H. Elliot was Principal Investigator for the HEOS magnetometer experiment which was supported by the British Science Research Council.

References

Akasofu, S. and Chapman, S.: 1972, *Solar Terrestrial Physics*, published by Oxford Press.

Cahill, L. J., Jr.: 1965, in C. C. Chang and S. S. Huang (eds.), *Proc. of the Plasma Space Science Symposium*, D. Reidel Publishing Co., Dordrecht, Holland, p. 227.

Clauer, C. R. and McPherron, R. L.: 1973, paper presented at the *Chapman Memorial Symposium on Magnetospheric Motions*, Boulder, Colo., June 1973.

Coleman, P. J., Jr. and McPherron, R. L.: 1969, in B. M. McCormac (ed.), *Particles and Fields in the Magnetosphere*, D. Reidel Publ. Co., Dordrecht-Holland, p. 171.
Fairfield, D. H. and Ness, N. F.: 1972, *J. Geophys. Res.* **75**, 7032.
Hedgecock, P. C.: 1974, Imperial College preprint.
Hones, E. W., Jr., Asbridge, J. R., Bame, S. J., and Strong, I. B.: 1967, *J. Geophys. Res.* **72**, 5879.
Heppner, J. P., Ness, N. F., Scearce, C. S., and Skillman, T. L.: 1963, *J. Geophys. Res.* **68**.
Hones, E. W., Jr., Asbridge, J. R., and Bame, S. J.: 1971, *J. Geophys. Res.* **76**, 4402.
Kivelson, M. G., Farley, T. A., and Aubry, M. P.: 1973, *J. Geophys. Res.* **78**, 3079.
McPherron, R. L.: 1973, *J. Geophys. Res.* **78**, 3044.
McPherron, R. L., Aubry, M. P., Russell, C. T., and Coleman, P. J., Jr.: 1973a, *J. Geophys. Res.* **78**, 3068.
McPherron, R. L., Russell, C. T., and Aubry, M. P.: 1973b, *J. Geophys. Res.* **78**, 3131.
Ness, N. F.: 1965, *J. Geophys. Res.* **70**, 2989.
Nishida, A. and Hones, E. W., Jr.: 1974, *J. Geophys. Res.* **79**, 535.
Russell, C. T. and Brody, K. I.: 1967, *J. Geophys. Res.* **72**, 6104.
Speiser, T. W. and Ness, N. F.: 1967, *J. Geophys. Res.* **72**, 131.

THE PLASMA SHEET BOUNDARY AND K_p

J. W. FREEMAN

Dept. of Space Physics and Astronomy, Rice University, Houston, Tex. 77001, U.S.A.

1. Introduction

The concepts discussed in this paper were originally inspired by Brice from his early discussions on plasma flow in the magnetosphere (Brice, 1967).

Freeman and Maguire (1967) first drew attention to the intrusion of energetic plasma from the tail to the geostationary orbit during geomagnetically disturbed times. Vasyliunas (1968b) reported a correlation between the inward extension of the inner boundary of the plasma sheet and the K_p index and pointed out that the plasma sheet could occasionally reach the geostationary orbit distance. More recently Mc-Ilwain (1972) using the more refined detectors aboard the ATS-5 geostationary spacecraft, has emphasized the correlation between the location of the plasma sheet boundary along the ATS orbit and geomagnetic activity. Using some older but unpublished data from the Suprathermal Ion Detector aboard the ATS-1 geostationary satellite, we report here the relation between the local time occurrence of the plasma sheet at the geostationary orbit (6.6 R_E) and the K_p index and then derive a relationship for the shift in the plasma sheet radial position.

Freeman and Maguire (1967) described data from the ATS-1 suprathermal ion detector (SID). They reported the frequent observation of an enhanced response of the instrument near or just past local midnight. Figure 1 illustrates such an observation. They attributed this enhancement to 'an energetic diamagnetic plasma'. Frank (1971) has shown that this SID response was due principally to energetic electrons associated with the plasma sheet. He also drew attention to the earthward motion of plasma sheet inner boundary during geomagnetic storms.

2. Analysis

Data such as that shown in Figure 1 are typical of rising K_p intervals when the K_p index is at low or moderate values; usually the initial phase of magnetic storms. As the K_p index increases the point of the initial enhancement or onset moves to an earlier local time along the geostationary orbit. Figure 2 shows the result of a study of the K_p index vs local time of the onset for 37 orbits. Also shown is the longitude of the onset ϕ measured in degrees eastward from the dusk meridian. The line drawn through the data points has the equation

$$K_p = 6 - 0.042\phi \quad \text{or} \quad LT = 27.6 - 1.6 K_p. \tag{1}$$

Equation (1) agrees well with a similar result from ATS-5 data (Mauk and McIlwain,

V. Formisano (ed.), The Magnetospheres of the Earth and Jupiter, 71–75. All Rights Reserved
Copyright © 1975 by D. Reidel Publishing Company, Dordrecht-Holland

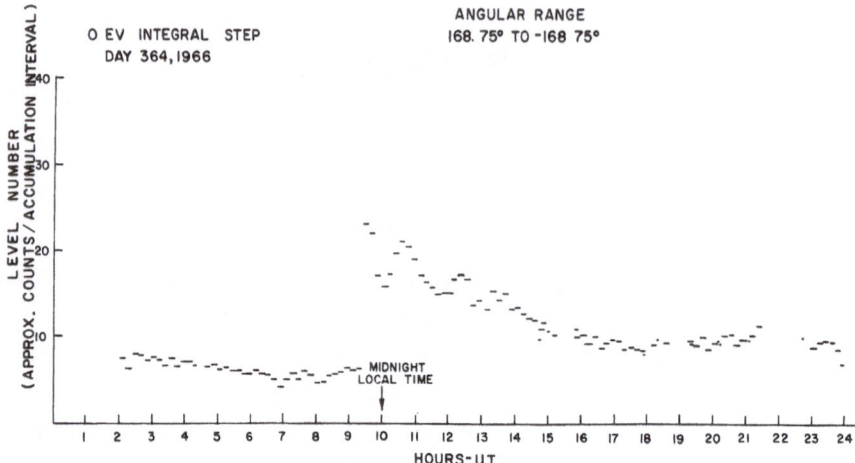

Fig. 1.   Typical count-rate response of the ATS-1 Suprathermal Ion Detector as a function of universal time. The detector is an integral device responding in this mode to positive ions above 0 eV or electrons above approximately 3 keV.

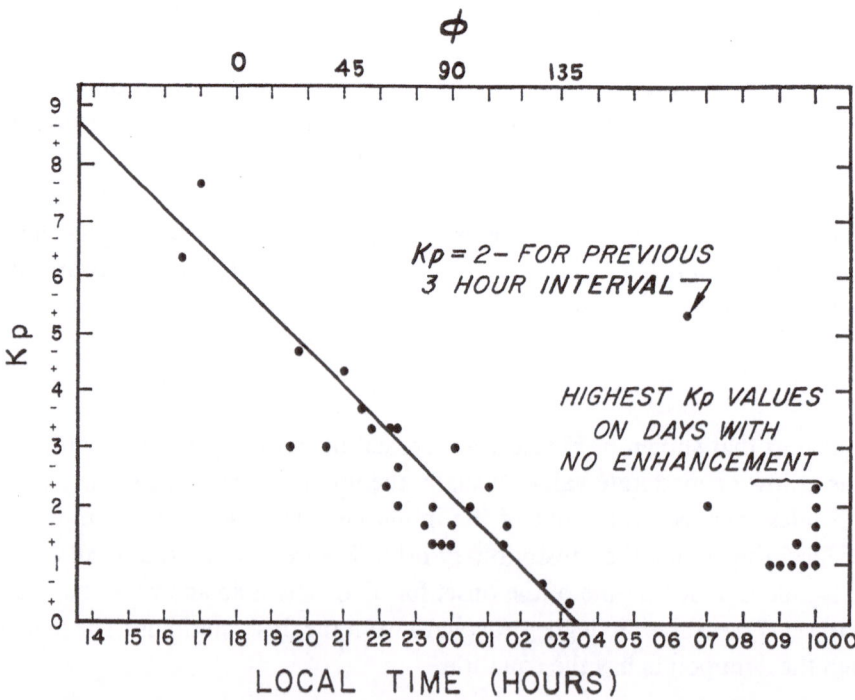

Fig. 2.   The K_p dependence of the local time onset of the enhanced fluxes for 37 orbits similar to those shown in Figure 1 or in which no enhancement was observed. Orbits for which the onset of enhancement is ill-defined are deleted from the study. Invariably the deleted orbits are high or falling K_p days and the enhancement begins early in local time.

1974), when differences in the satellite location and detector response are taken into account.

Following the spirit of Brice (1967), Kavanagh et al. (1968) calculated particle drift paths in the magnetospheric equatorial plane using idealized electric and magnetic field models. Allowing plasma to enter the magnetosphere from the tail they showed several principal features of the magnetosphere to be realistically reproduced including the abrupt inner boundary of the plasma sheet (the Alfvén layer) or the existance of a 'forbidden' region for inward moving plasma. Alfvén and Fälthammer (1963) had provided an analytic solution for the same problem only with the omission of the co-rotation electric field and a reversed sign for the convection electric field. Their solution is sufficiently similar to the results of Kavanagh et al. (1968) that it will be employed here (with the correct sign) to extend the usefulness of Equation (1).

Figure 3 shows drift paths of an electron moving under the combined influence

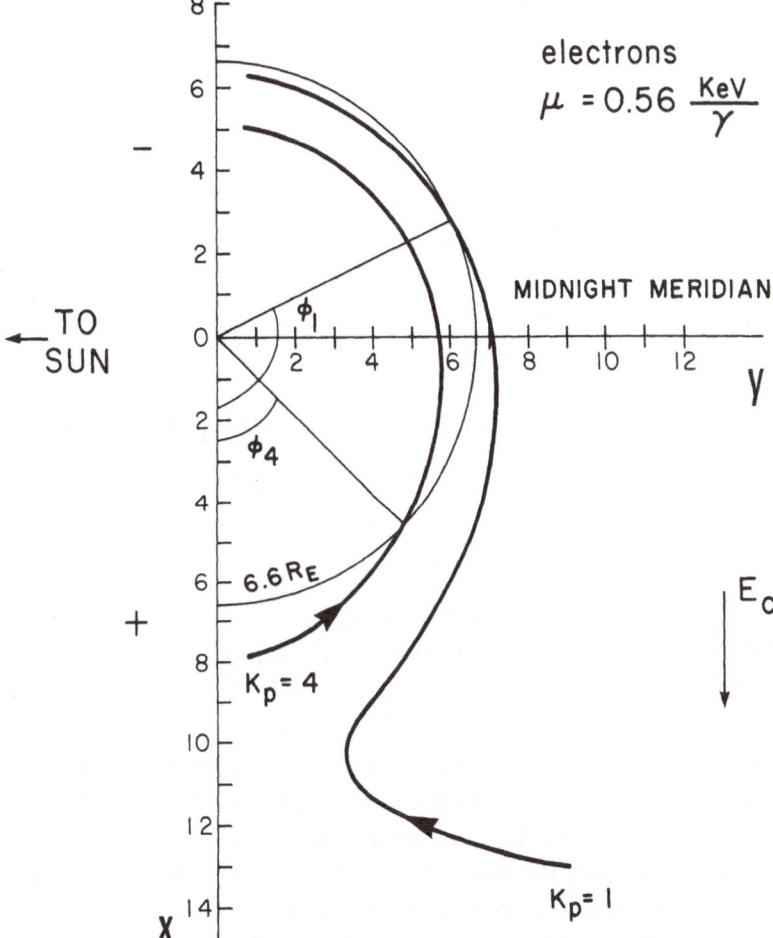

Fig. 3. Equatorial plane guiding center trajectories of electrons of $\mu = 0.56$ keV γ^{-1} ($1\,\gamma = 10^{-5}$ G or 1 nT (nanotesla)) and 90° pitch angle at $K_p = 1$ and $K_p = 4$. Note that the coordinate system used is not the conventional solar ecliptic or magnetospheric.

of a homogeneous electric field E_c and a dipole magnetic field centered at the origin. These drift paths represent the inner boundary of the plasma sheet. The point of intersection between the geostationary orbit and the boundary is given by the angle ϕ. According to Alfvén and Fälthammar (1963) the equation of such trajectories is

$$x_0 - x = P^4 (x^2 + y^2)^{-3/2}, \tag{2}$$

where

$$P = \left(\frac{\mu a}{|e| E} \right)^{1/4} \tag{3}$$

with μ the particle invariant magnetic moment, a the dipole magnetic moment, e the electron charge and E the homogeneous electric field. x_0 is the particle x position at $y = \infty$.

Equation (2), together with the expression for $\cos \phi$ and the equation of the geostationary orbit, can be solved for x_0 for the particle intersecting $6.6\,R_E$ at ϕ. This in turn allows the calculation of y at $x = 0$, the midnight meridian point, by

$$y_{x=0} = \left\{ 3.48 \times 10^{-3} + \frac{6.6 \cos \phi}{\dfrac{\mu a}{|e| E}} \right\}^{-1/3}. \tag{4}$$

We take a to be the geomagnetic dipole moment and E the convection electric field, E_c. Vasyliunas (1968a) has derived an approximate expression for E_c, which reduces to

$$E_c \approx 4.5 \times 10^{-4} \left(1 - \frac{K_p}{10} \right)^{-2}, \tag{5}$$

where E_c is in volts/meter. Using this and Equation (1) Equation (4) becomes

$$y_{x=0} = \left\{ 3.48 \times 10^{-3} + \frac{9.7 \times 10^{-11} \cos (143 - 24\,K_p)}{\mu \left(1 - \dfrac{K_p}{10} \right)^2} \right\}^{-1/3}, \tag{6}$$

where y is in Earth radii and μ is in mks units.

Table I lists values of $y_{x=0}$ for several μ and K_p values. From Table I we see that for the conditions applicable to Figure 2 i.e. K_p equal 2 with the boundary at geostationary orbit near local midnight, a wide range of electron magnetic moments is allowable. From the independent experimental evidence (e.g., Frank, 1971; McIlwain, 1972) the lower end of this range is probably applicable to the SID response. For this detector these particles define the inner boundary of the plasma sheet. Table I and Equation (6) then give the manner in which this boundary can be expected to vary radially with K_p along the midnight meridian.

It should be emphasized that these results are valid only when K_p is small or moderate ($\lesssim 4$) and increasing.

Finally, Figure 3 gives two representative trajectories showing the inward motion

TABLE I

Values of the midnight meridian approach distance of trajectories
of electrons of magnetic moment μ for several K_p

$\mu\left(\dfrac{keV}{\gamma}\right)$ \ K_p	0	1	2	3	4	5
0.16	13.0 R_E	9.4	7.0	5.5	4.5	3.9
0.31	8.1	7.6	6.8	5.9	5.2	4.5
0.63	7.2	7.0	6.7	6.2	5.7	5.2
1.25	6.9	6.8	6.6	6.4	6.1	5.8

of the boundary as K_p increases from 1 to 4. This figure illustrates how the geo-stationary orbit is such a sensitive indicator of increasing geomagnetic activity and in particular the difficulty of resolving temporal changes at a single spacecraft from changes brought about by the shift in the plasma sheet position. Better models of the magnetospheric electric field are available (e.g., McIlwain, 1972; Jaggi and Wolf, 1973; and Volland, 1973), however, this general result is probably correct regardless of details of the model.

Acknowledgements

Helpful discussions with C. E. McIlwain, A. J. Dessler, F. C. Michel and R. A. Wolf are acknowledged. This research was supported by NASA grant NGR 44-006-12.

References

Alfvén, H. and Fälthammar, C.-G.: 1963, *Cosmical Electrodynamics*, Oxford University Press, p. 55.
Brice, N. M.: 1967, *J. Geophys. Res.* **72**, 5193.
Frank, L. A.: 1971, *J. Geophys. Res.* **76**, 2265.
Freeman, J. W., Jr. and Maguire, J. L.: 1967, *J. Geophys. Res.* **72**, 5227.
Jaggi, R. K. and Wolf, R. A.: 1973, *J. Geophys. Res.* **78**, 2852.
Kavanagh, L. D., Jr., Freeman, J. W., Jr., and Chen, A. J.: 1968, *J. Geophys. Res.* **73**, 5511.
McIlwain, C. E.: 1972, Paper presented at the *Conf. of Magnetospheric Substorms*, Rice University, Houston, Texas, October, 1972.
McIlwain, C. E.: 1972, in B. M. McCormac (ed.), *Earth's Magnetospheric Processes*, D. Reidel Publishing Co., Dordrecht, Holland, p. 268.
Mauk, Barry H. and McIlwain, C. E.: 1974, to be published in *J. Geophys. Res.*
Vasyliunas, V. M.: 1968a, *J. Geophys. Res.* **73**, 2529.
Vasyliunas, V. M.: 1968b, *J. Geophys. Res.* **73**, 2839.
Volland, H.: 1973, *J. Geophys. Res.* **78**, 171.

THE AURORA

BENGT HULTQVIST

Kiruna Geophysical Institute, S-981 01 Kiruna 1, Sweden

1. Introduction

The visualization of the interaction between the hot magnetospheric plasma and the Earth's atmosphere in the form of aurora has played an important role in the development of the understanding of the Earth's environment outside the lower atmosphere. For a very long time only the observations of the aurora by the naked eye and of the variations of the geomagnetic field by means of a compass needle were at the disposal of those pioneers who tried to understand that most spectacular phenomenon in the skies. The close connection between the aurora and the geomagnetic variations were discovered as long ago as in 1741 by Celsius and Hiorter in Sweden.

Around the turn of the century the radiowaves and the photographic plate became available as new tools for the study of the aurora and a lot more was learned about the processes in the upper atmosphere. Still the experimental knowledge was limited to a thin region near the Earth. The ideas of the phenomena that take place well outside the ionosphere proper were very vague and largely wrong. The revolution of our views of the phenomena in the magnetosphere and in the interplanetary space outside it that has been brought about in the last 15 years by instruments carried on board spacecraft of variations kinds, from sounding rockets to probes into outer space sent out from the Earth for no return, is a very thorough one indeed. We have explored the magnetosphere fairly intensely but are still only in the early phase of the detailed study of the physics involved.

For each big step that has been taken in the investigation of the aurora the complexity of the phenomenon under study has appeared to have increased rather than decreased. Still, in recent years a certain amount of coherence has developed in our knowledge of various aspects of the aurora and associated phenomena in the magnetosphere. We are beginning to understand some of the dominating physical processes. I will try to report here on some of the more recent developments. Far from all physical processes will be discussed. I will concentrate on those where new data has increased our understanding significantly and my own personal interest has also played a part in the selection.

Before going into the various aspects of the aurora let us consider for a moment what we mean by aurora. It was originally defined by the characteristics of the naked eye. This is sensitive only in a limited wavelength range, it is much more sensitive to contrasts in the field of view than to the absolute level of luminosity and it is particularly sensitive to motions of the luminous object. The all-sky camera has similar characteristics as the eye in the two first-mentioned respects but not in the last one. The aurora that has been studied by means of direct observations and all-sky cameras

V. Formisano (ed.), The Magnetospheres of the Earth and Jupiter, 77–111. All Rights Reserved
Copyright © 1975 by D. Reidel Publishing Company, Dordrecht-Holland

is therefore mainly the discrete aurora. The auroral oval, in particular, is defined for discrete auroral forms. Only recently have satellite-borne scanning photometers produced records where a diffuse unstructured aurora of large extension appears as, in certain respects, the dominating form of the auroral phenomenon (see e.g. Anger and Lui, 1973; Lui and Anger, 1973; Lui et al., 1973). This diffuse aurora is the one that is most easy to relate to the satellite observations of the hot plasma in the upper ionosphere. Most of this report will deal with the diffuse aurora, but towards the end we will also discuss the discrete auroral forms.

2. Processes Determining the Geographic Distribution of the Diffuse Aurora

2.1. DISTRIBUTION OF AURORAL ELECTRON AND PROTON PRECIPITATION IN LATITUDE AND LOCAL TIME

The distribution in corrected geomagnetic latitude and eccentric dipole time of the average rate of precipitation of 6 keV electrons and protons into the atmosphere during quiet to moderately disturbed conditions ($K_p = 1-3$) are illustrated by Figure 1a and b (after Riedler and Borg, 1972). Electrons of keV energies play the dominating role in the production of the most common kind of aurora, the one with a greenish colour. Six keV protons are in the lower range of energies responsible for the generation of proton aurora but the 6 keV flux gives a rough measure of the proton aurora intensity (Deehr et al., 1973). We thus expect the distributions shown in Figure 1 to be closely related to the average distribution of the aurora. They certainly deviate a lot from the well-known auroral oval, particularly on the dayside. The reason for this is, as indicated in the introduction, that the auroral oval has been defined on the basis of discrete auroral forms. These occur primarily in the polar cusp on the dayside (Akasofu and Kimball, 1973; Vorobjev et al., 1973) and it has been possible to identify these forms with the entry of soft magnetosheath plasma into the atmosphere (see e.g. Heikkila and Winningham, 1971; Eather and Mende, 1971, 1972; Hoffman, 1972; Hultqvist et al., 1974). It seems clear, therefore, what the relations are in the noon sector. In the evening and midnight sector the entire region of keV electron precipitation coincides with the auroral oval fairly well and in the other time sectors the oval tends to follow the poleward boundary of the precipitation zone for electrons of energies above 1 or 2 keV.

The satellite measurements of keV electrons and protons show that some keV particle fluxes are always present at all local times in the auroral oval (see e.g. Hultqvist et al., 1974). In this respect the auroral oval is a continuous formation. Whether it is or not in terms of visible aurora seems mainly to be a question of sensitivity of the detectors, which have mostly been all-sky cameras. Because the sources of the auroral particles in various time sectors probably are different, the meaningfulness of the auroral oval concept has been disputed (see Eather, 1973). No stand in this controversy will be taken here, but we will only emphasize that the auroral oval constitutes only a part of the circumpolar zone in which aurora producing keV electrons and protons are found. It seems fairly clear that the distributions in Figure 1

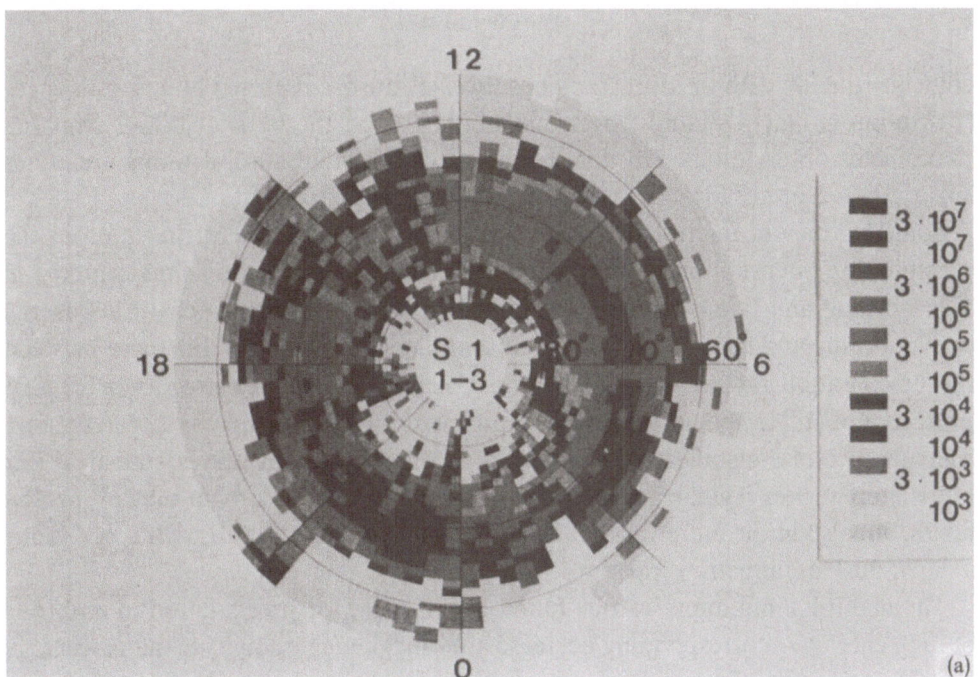

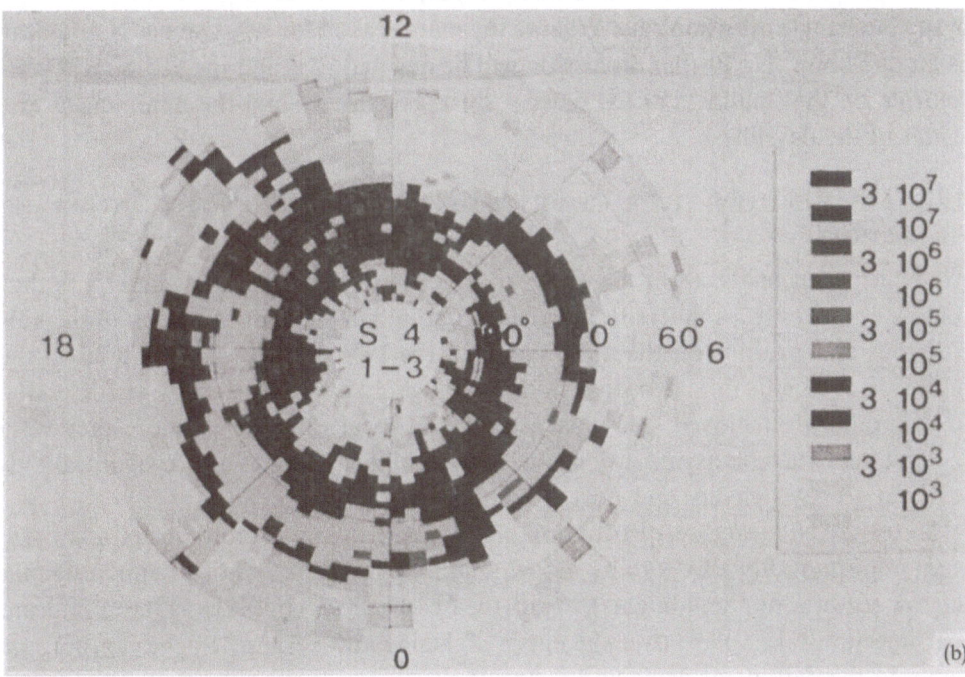

Fig. 1. Average fluxes of precipated electrons (a) and protons (b) of 6 keV energy for fairly low activity level ($K_p = 1$–3) as measured during a 20-month period from October 1968 to June 1970 with the ESRO-1A satellite. The coordinates are corrected geomagnetic latitude (practically identical with invariant latitude) and eccentric dipole time. Original is coloured. The highest intensities are found for the electrons (a) in the dark region surrounded at higher and lower latitudes by lighter areas, whereas for the protons (b) the light regions indicate the highest precipitation fluxes. After Riedler and Borg (1972).

coincide roughly with the diffuse, often subvisual, aurora reported by Lui and Anger (1973), Lui *et al.* (1973) and Anger and Lui (1973) on the basis of ISIS-2 scanning photometer observations. On this diffuse aurora discrete auroral forms are either superimposed or are located just poleward or just equatorward of it.

Some features of interest in Figure 1 are that the local time distribution of the precipitation intensity for keV electrons and protons has broad maxima centered in late morning and late evening with minima in between at 0100–0200 EDT in the early morning and near noon (EDT) and that there is hardly any difference between the precipitation rate at dawn and at dusk neither for keV electrons nor for keV protons. For still lower electron energies no dawn-dusk asymmetry at all can be seen. This absence of a significant dawn – dusk asymmetry is not only a statistical fact but is more or less regularly seen in individual passes particularly in the keV proton fluxes, but also in the more variable keV electrons. For energies above the keV range a dawn-dusk asymmetry is present.

The nighttime minimum can be found also in data for precipitation of electrons of high energy. A corresponding decrease after magnetic midnight of the percentage hourly occurrence of aurora as function of the local time has been reported (Davis, 1962). The postmidnight minimum in keV electron and proton precipitation may be related to special features of the convection pattern or to the large-scale distribution of the pitchangle diffusion rate. What is the main reason for its existence is not clear as far as I know. No further discussion will be devoted to it, but instead we will concentrate on the smallness of the dawn – dusk asymmetry and the minimum in the center of the dayside.

2.2. ABSENCE OF DAWN – DUSK ASYMMETRY FOR PRECIPITATION OF keV ELECTRONS AND PROTONS

An evident conclusion from the observed absence of dawn-dusk asymmetry is that there is no, or only a slight, effect of the azimuthal magnetic drift motion of the keV electrons and protons on the distribution of the keV particle precipitation on the nightside. A significant dawn-dusk asymmetry would appear if the keV particles due to azimuthal drift moved over a large sector of local time angle before they were precipitated. Such an asymmetry can be seen at higher energies and is significant already at 13 keV (Riedler and Borg, 1972; Riedler, 1972).

The observed local time distribution of keV electron and proton precipitation rate is an important difficulty for any model assuming injection of hot plasma only in a narrow sector around midnight. Instead, the observations shown in Figure 1 indicate that injection takes place over the entire nightside and perhaps reaches an hour or two into the dayside at dawn and dusk.

In the evening the L dependence of the proton energy spectrum in the equatorial plane has been found to show in some cases a 'nose feature', with protons of energies just above 10 keV reaching closer to the Earth than those of lower and higher energies as shown in Figure 2 (Smith and Hoffman, 1974). This can be understood as an effect of the motion of the charged particles in the magnetic and electric fields around

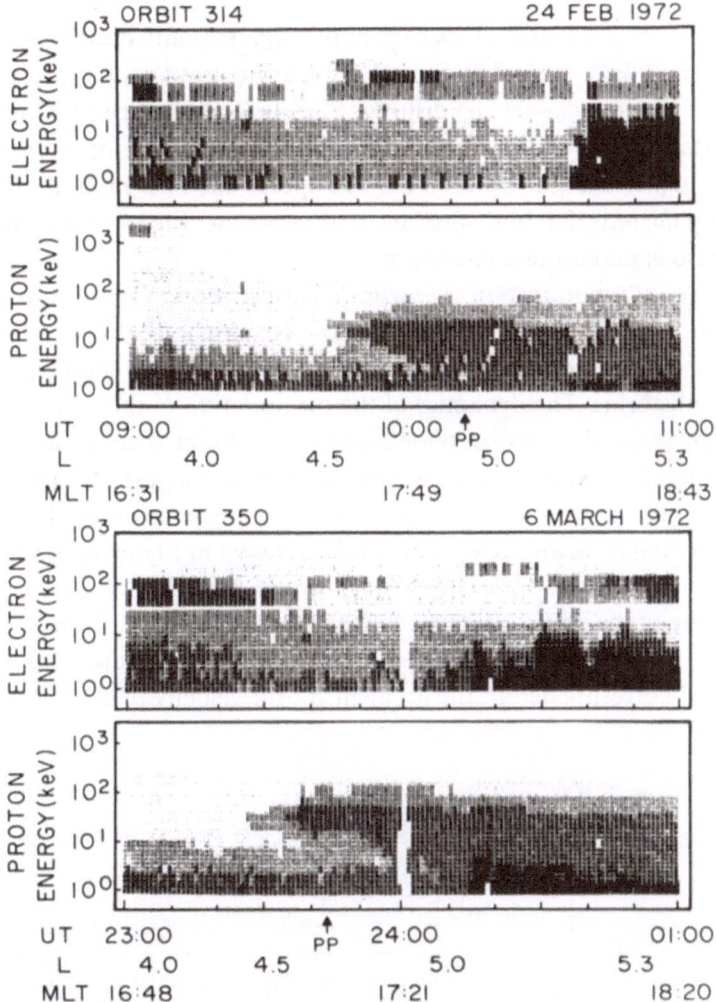

Fig. 2. Proton and electron energy spectrograms for two S³-A orbits during the development of the main phase of two magnetic storms. The grey shading is a measure of the differential flux of equatorially mirroring particles. Black represents the most intense flux. The plasmapause, as determined by S³ instrumentation, is indicated by PP. After Smith and Hoffman (1974).

the Earth. For one specific energy the corotation electric field gives rise to an eastward drift of equal magnitude but opposite direction to that which the magnetic field gradient and field line curvature causes. This occurs at the 'nose energy'. For lower energies the corotation drift brings the protons toward the morning side and for higher energies the deviation towards west occurs at greater L values than at the nose energy.

As the corotation drift and the gradient B drift are fairly stable processes inside 5–6 R_E one would expect that there should be a strong dawn-dusk asymmetry in the precipitation of 6 keV protons into the atmosphere. This is not the case in Figure 1b.

Again, the simplest interpretation seems to be that the entire energy range of protons in general is brought in to those regions of the magnetosphere which are magnetically connected with the auroral oval over the entire nightside and dawn and dusk sectors, where they then drift around the Earth as indicated above and are continuously scattered into the loss cone while drifting. For the lower energies the distribution in magnetic local time of precipitation is mainly defined by the injection distribution, whereas for higher energies with higher driftspeeds the drift motion affects the distribution significantly for both protons and electrons, Figure 3 illustrates this for 13 keV electrons under quiet conditions.

The relative unimportance of the azimuthal drift motion of auroral particles for the distribution of the aurora on the nightside is supported also by observations of the temporal development of precipitation of electrons with energy above 15 keV, say during substorms. The dynamics of the precipitation of the more energetic electrons mentioned which give rise to ionization in the lower E layer and in the D layer and thereby to absorption of radio waves has been investigated for 60 substorms by Berkey *et al.* (1974). They found that the eastward expansion of the precipitation from the initiation area, an example of which is shown in Figure 4, in a large fraction of the cases occurred at appreciably higher speeds than that of the gradient B drift of the majority of the electrons involved which cannot reasonably have energies well above 100 keV in normal substorms. In the extreme case studied, the eastward expansion speed corresponded to the gradient B drift velocity of electrons with energy

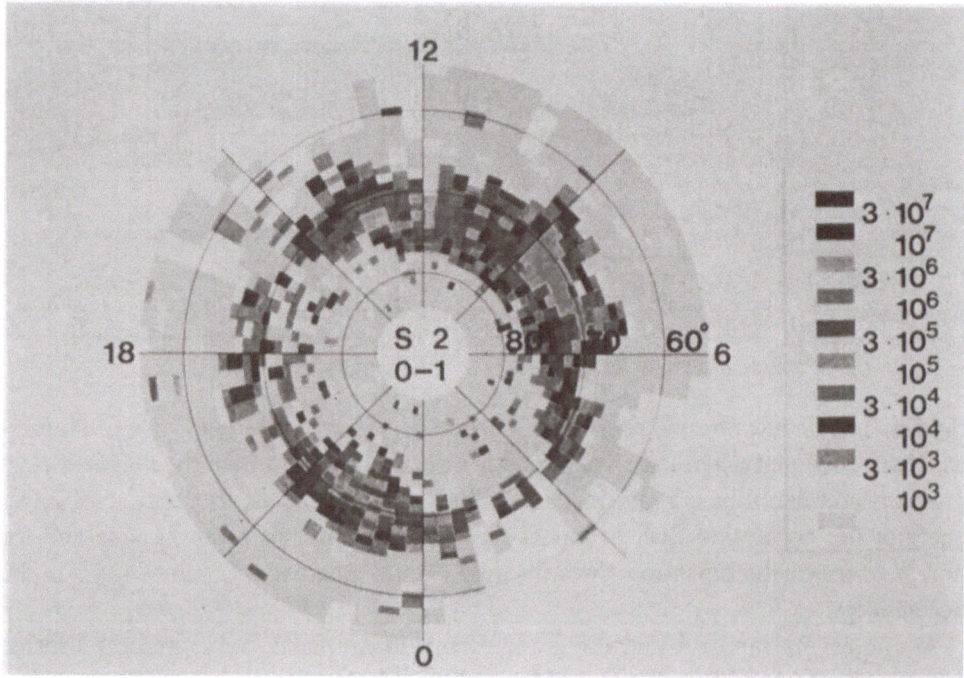

Fig. 3.   The same as in Figure 1 but for electrons of 13 keV energy. After Riedler and Borg (1972).

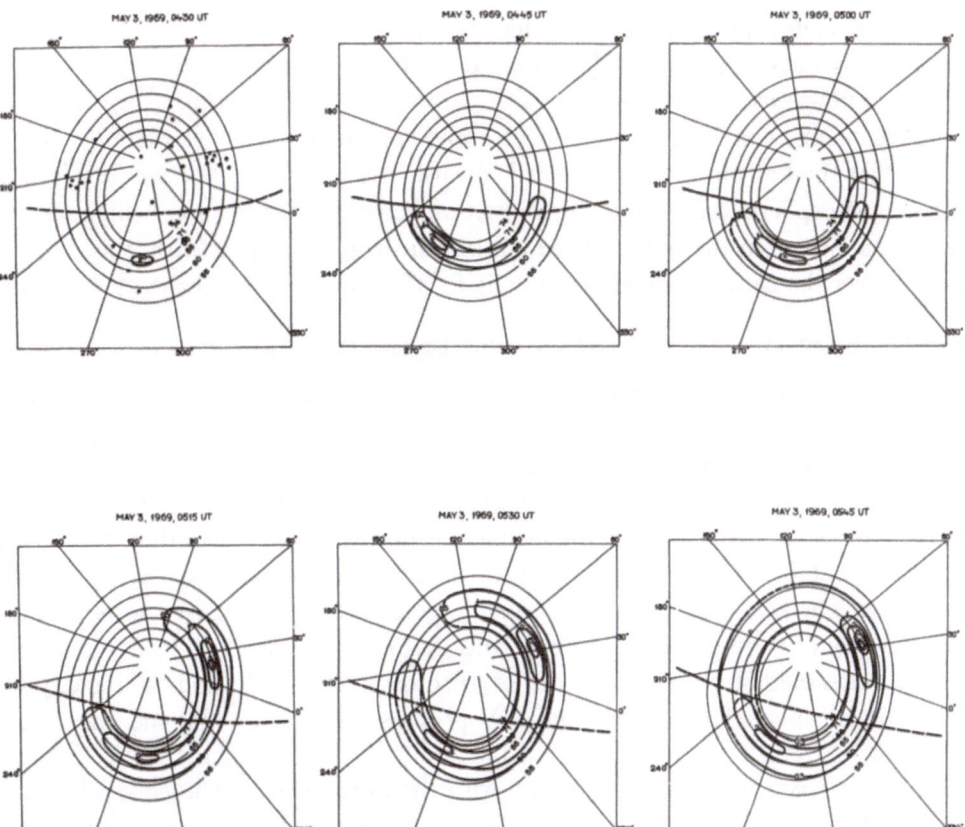

Fig. 4. Synoptic maps of auroral absorption for a substorm starting around 0430 UT on May 3, 1969 (after Berkey *et al.*, 1971), for the first $1\frac{1}{4}$ h after the start. The points in the first diagram shows the riometer stations on which the synoptic maps are based.

of order of magnitude 1 MeV. The only reasonable conclusion, therefore, seems to be that the expansion of the precipitation region in the eastward direction during substorms is due mainly to the temporal development of the extension of the injection region.

The difficulties in interpreting the temporal development of the distribution of energetic electron precipitation during substorms in terms of drifting auroral particles is illustrated particularly well by the westward expansion of the precipitation region which was seen in about half of the substorms studied by Berkey *et al.* (1974) (see Figure 4 for an example). The westward expansion can certainly not be due to drifting electrons, since unreasonably high electric fields would be needed to drive these fairly energetic electrons westward against the forces due to grad B and the curvature of field lines. As energetic protons can be ruled out as source of the radio-wave absorption (see Berkey *et al.*, 1974), it seems clear that the westward expansion also is caused by the extension of the injection region.

That injection takes place over the entire nightside does not mean that the injection rate is independent of local time. For higher energies the injected fluxes are generally appreciably higher near midnight than at other local times; for lower energies (keV) the local time variation is small.

A reasonable conclusion from all the above mentioned evidence therefore seems to be that the hot plasma is convected towards the Earth over the whole width of the magnetospheric tail, with the precipitation generally starting and being most intense somewhere in the center of the tail in transient events like substorms, but occuring over the entire nightside eventually for most of the auroral energy spectrum.

2.3. ORIGIN OF AURORAL PARTICLES AT SUBCUSP LATITUDES ON THE DAYSIDE

After injection the particles start drifting in the magnetic and electric fields around the Earth. Most of the dayside seems to be populated in that way. The keV electron distribution, with diminishing fluxes from late morning over noon to the afternoon, where frequently no keV electron fluxes at all are seen, is easily interpretable in terms of feeding the electrons into the nightside auroral oval from where they drift to the dayside while being continuously reduced in number density by scattering into the loss cone. The keV electrons in the evening sector are then supposed to be directly injected from the outer magnetosphere into that region.

It appears possible to interpret the keV proton EDT distributions on the dayside of the Earth in similar simple terms. keV protons are injected into the inner magnetosphere over the whole width of the tail and precipitate along the entire nightside when they reach the region connected with the auroral oval. The generally lower proton fluxes seen in the central parts of the dayside may then be understood as due to deterioration of the proton populations in moving toward noon along both sides of the Earth.

The effect of substorms on the auroral particle precipitation on the dayside equatorward of the polar cusp seems to be composed of two opposing processes. One is the increase of the flow of particles from the nightside at a drift speed given by the $E \times B$ drift in the lowest part of the energy range, by grad B and field line curvature drift at the higher energies, and by both kinds of forces in the intermediary range. The other is an increased loss rate for the auroral particles which are on the dayside equatorward of the polar cusp when the substorm starts, presumably due to an increased rate of pitch angle scattering into the loss cone.

The first mentioned effect is illustrated for the high energy tail of auroral particles in Figure 4. For the keV electrons the increased flux of particles on the dayside after substorms has been reported by Hoffman (1972) and De Forest and McIlwain (1971). The disappearance of the keV electrons over large fractions of the dayside during the expansive phase of a substorm has been reported by Hultqvist et al. (1974). The characteristic times of the two effects are so different that for a disturbance following a long period of quiet conditions the keV particles can be seen first to disappear from the noon sector and after a few hours return, with the higher energy electrons arriving before the lower energy ones (Hultqvist et al., 1974). An example of the disappearance of the keV electrons at the onset of a substorm is shown in Figure 5, where the data in (a) were taken before the substorm, those in (b) in the expansive phase and those in (c) 100 minutes later. In the last satellite pass shown in Figure 5 electrons of energies above 14 keV already produced measurable amounts of radiowave absorption where the satellite was (see Figure 6) but not a single electron count was obtained below 14 keV.

2.4. THE POLAR CUSPS(CLEFTS) AS SOURCES OF AURORAL PARTICLES

The importance of the polar cusps for the latitudinal distribution of the precipitation of auroral particles in the keV energy range on the dayside is illustrated by Figure 5 and 7. The polar cusps are marked by the high fluxes of soft keV protons, whereas electrons of energies above 1–2 keV occur in measurable amounts only equatorward of the cusp, if they occur at all in the noon sector (the <1 keV electrons characteristic of the polar cusps were not measured by the ESRO-1 satellites).

This relation between keV proton and electron latitude profiles are regularly found in the central part of the dayside (from about 0900 to 1500 EDT). An interesting observation in the ESRO-1 data is that the relation shown in Figures 5 and 7 is found, somewhat modified, also at greater distances from noon out to the dawn and dusk sectors (Hultqvist *et al.*, 1974). The modification mentioned is that the >2 keV electron precipitation can be observed not only at latitudes equatorward of the keV proton cusp but also somewhat poleward of the low-latitude cusp border, up to about

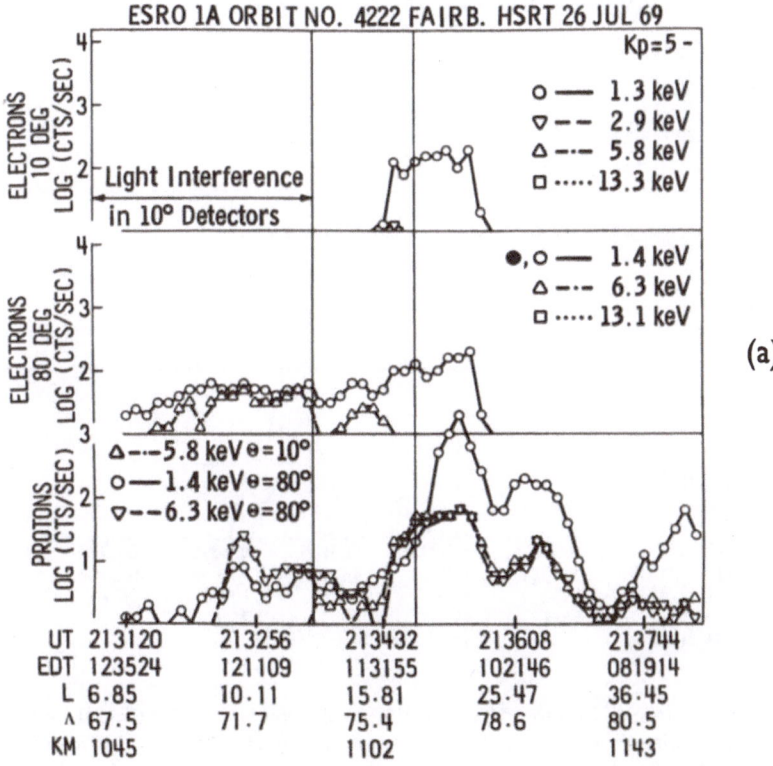

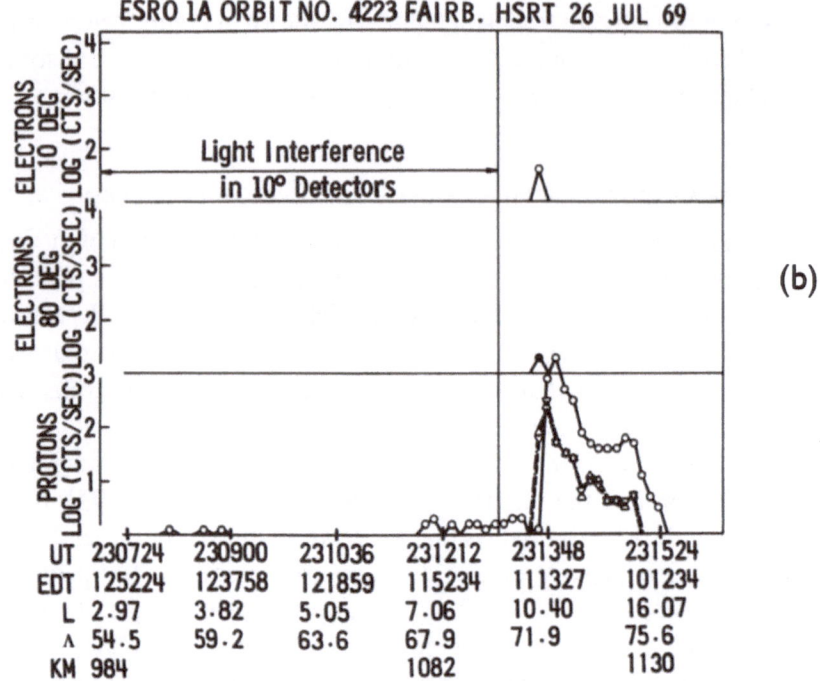

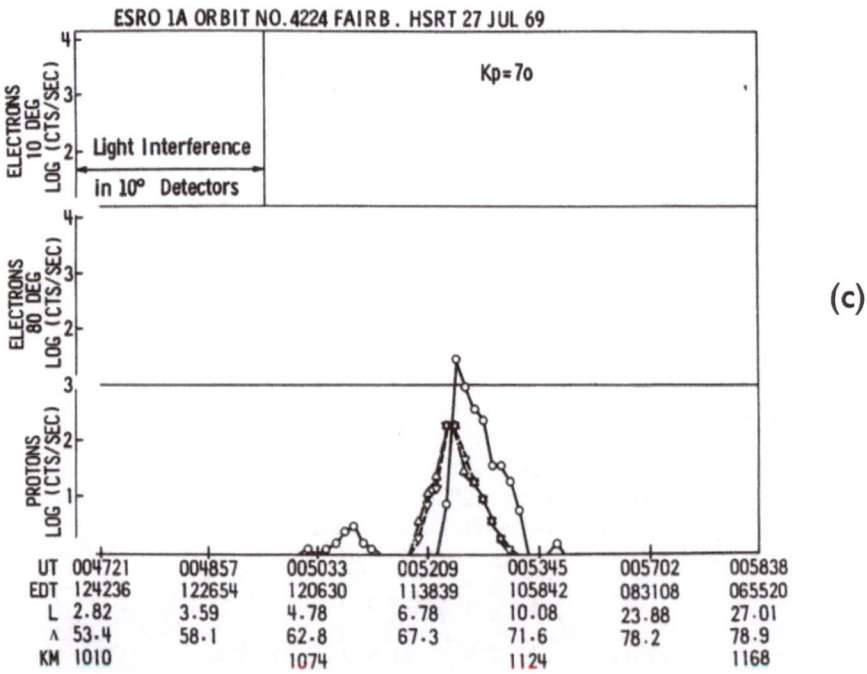

Fig. 5. keV electron and proton data from ESRO-1A orbit No. 4222, before the onset of a substorm, (a); No. 4223, during the expansion phase of a substorm, (b); and No. 4224, 100 minutes after the data in b, (c). The calibration factors can be found in the caption of Figure 10. After Hultqvist *et al.* (1974).

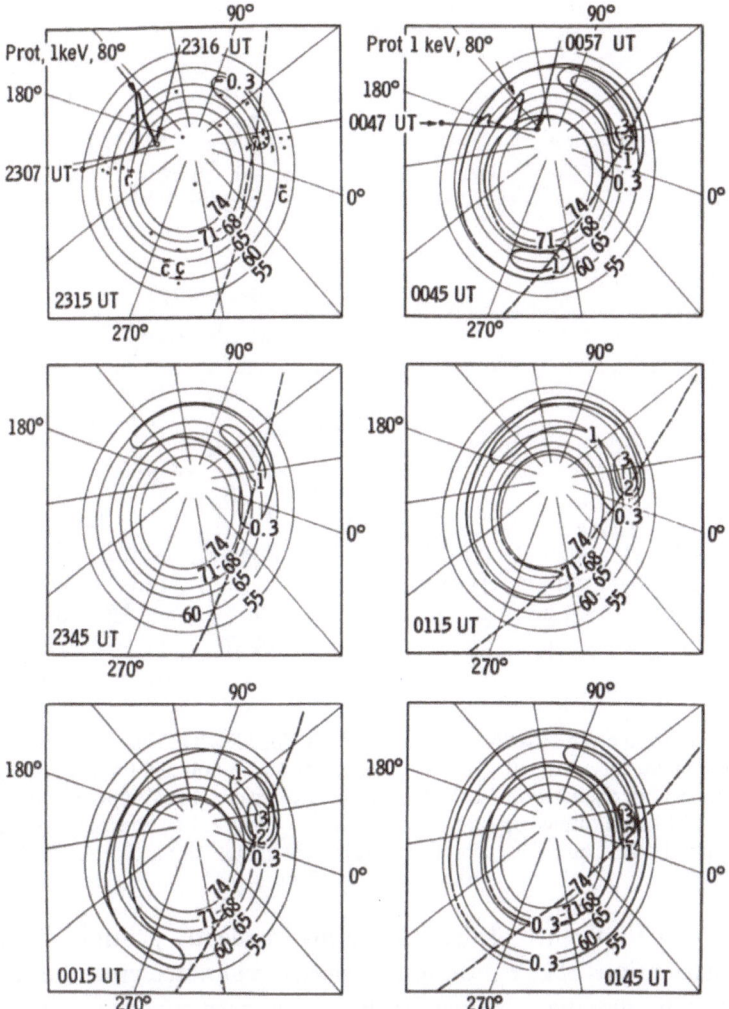

Fig. 6.   Synoptic maps of auroral absorption for a substorm starting around 23 h on July 26, 1969 (after Berkey *et al.*, 1971) with the ESRO 1A passes and associated keV particle measurements indicated. The points in the first diagram shows the riometer stations on which the synoptic maps are based. At 0045 UT on July 27 no electrons of energies below 14 keV were detectable, but higher energy electrons were precipitated and produced weak cosmic noise absorption.

the latitude of maximum proton precipitation at dawn and dusk. Sometimes the 3, 6, and 13 keV electrons measured by ESRO 1 also on the nightside reached only as far poleward as to the peak of the keV proton latitude profile. The kind of profile mentioned is illustrated by Figure 8, for the dawn, and by Figure 9 for the dusk sector. Also in this respect there is no very marked dawn-dusk asymmetry. There is, however, some asymmetry between the morning and evening sides. Significant poleward displacement of the keV protons relative to the electrons is seen more frequently on the morning side than in the evening.

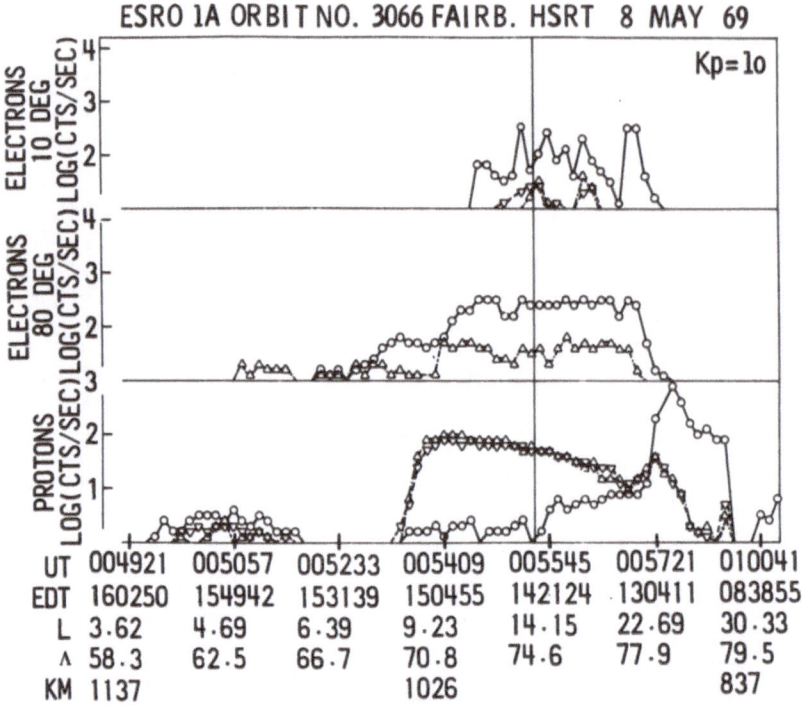

Fig. 7.   keV electron and proton data from ESRO-1A orbit No. 3066. The symbols in the diagrams have the following meanings. Electrons, 10°: circles 1.3 keV; downward pointing triangles 2.9 keV; upward pointing triangles 5.8 keV; squares 13.3 keV. Electrons, 80°: circles 1.4 keV; upward pointing triangles 6.3 keV; squares 13.1 keV. Protons: upward pointing triangles 5.8 keV, 10°; downward pointing triangles 6.3 keV, 80°; circles 1.4 keV, 80°. The calibration factors can be found in the caption of Figure 10. After Hultqvist *et al.* (1974).

In summary, the influence of the cusp seems frequently possible to trace in the latitudinal distributions of keV protons and electrons at much greater distances from the noon magnetic meridian than where the typical cusp conditions are seen. There appears to be two injection processes contributing, one, at the highest latitudes, connected with the polar cusps, and the other, at somewhat lower latitudes, associated with the plasma sheet. The contribution of the cusp-connected injection sometimes reaches well into the nightside. These results are in agreement with Frank's (1971) proposal of convection of magnetosheath plasma from the dayside toward the night. Such a process appears however, to be responsible for only part of the spatial distribution of hot plasma in the inner magnetosphere. Equatorward of the cusp-connected plasma convection from day towards nightside, plasma from the plasma sheet convects from night towards day. These observations are also consistent with the measurements of the electrostatic field as a function of latitude in the auroral regions (see e.g. the review by Gurnett, 1972), which have shown a sharp reversal of the electric field direction within the particle precipitation region.

In the late evening sector the latitudinal distributions are frequently quite different from those in other time sectors. This will be discussed somewhat further in Section 4.4.

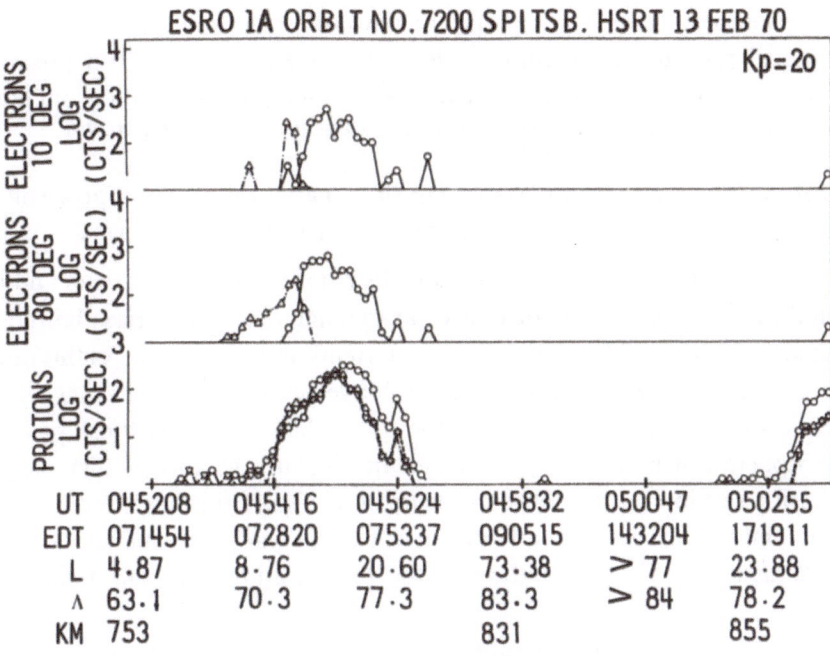

Fig. 8. keV electron and proton data from ESRO-1A orbit No. 7200. For the meanings of the symbols see caption of Figure 7. The calibration factors can be found in the caption of Figure 10. After Hultqvist *et al.* (1974).

2.5. GENERAL COINCIDENCE OF REGIONS OF PRECIPITATION FOR AURORAL ELECTRONS AND PROTONS

The great similarity of the magnetic latitude-local time distribution for keV electrons and protons shown in Figure 1 and the general coincidence of latitude profiles and large scale correlation in flux variations in the keV energy range observed by the ESRO-1 satellites outside the cusp (Hultqvist *et al.*, 1974) suggests that the processes which result in the precipitation of both keV protons and electrons are active simultaneously along the same field lines over (almost) identical regions of the magnetosphere. This result obviously is in conflict with models with preferential precipitation of electrons in the evening and protons in the morning, recently proposed by Heikkila (1972) and originally presented at least as early as 1939 as a consequence of Alfvén's (1939) auroral theory.

At energies above the keV range there seems to be larger differences between the electron and proton precipitation patterns, although no thorough studies of larger amounts of particle data from satellites seem to have been published. From optical observations the conclusion has been drawn that the hydrogen aurora normally is located equatorward of the electron aurora in the evening and poleward of it in the morning (see Wiens, 1968 and the reviews by Eather, 1967; Eather and Carovillano, 1971; Omholt, 1971; and Eather and Mende, 1972). It seems likely that the mentioned separation in the evening is largely due to protons of energies above 10 keV being precipitated from their drift orbits on the evening side where they constitute the ring current. There are, however, some minor dawn-dusk asymmetries also in the keV electron and proton precipitation profiles. The average keV proton precipitation maximum is slightly south of the electron maximum in the late evening and poleward relative to the electron precipitation peak in the morning (Riedler, 1972).

2.6. SUMMARY OF CONCLUSIONS ABOUT PHYSICAL PROCESSES DRAWN FROM THE GEOGRAPHICAL DISTRIBUTION OF AURORAL PARTICLE PRECIPITATION

We have already concluded above that the distributions of keV electron and proton precipitation in magnetic latitude (corrected geomagnetic or invariant latitudes) and local time (excentric dipole time), as well as the temporal distribution of this precipitation in substorms show that hot plasma is convected into the inner magnetosphere over more or less the entire width of the tail. When it reaches those parts of the magnetosphere which are connected magnetically with the auroral regions, particle precipitation sets in over the entire nightside of the Earth, the dawn and dusk sectors included. Only at energies above 10 keV, say, has the gradient B and curvature drift a strong influence on the distribution of the precipitation over the nightside. This

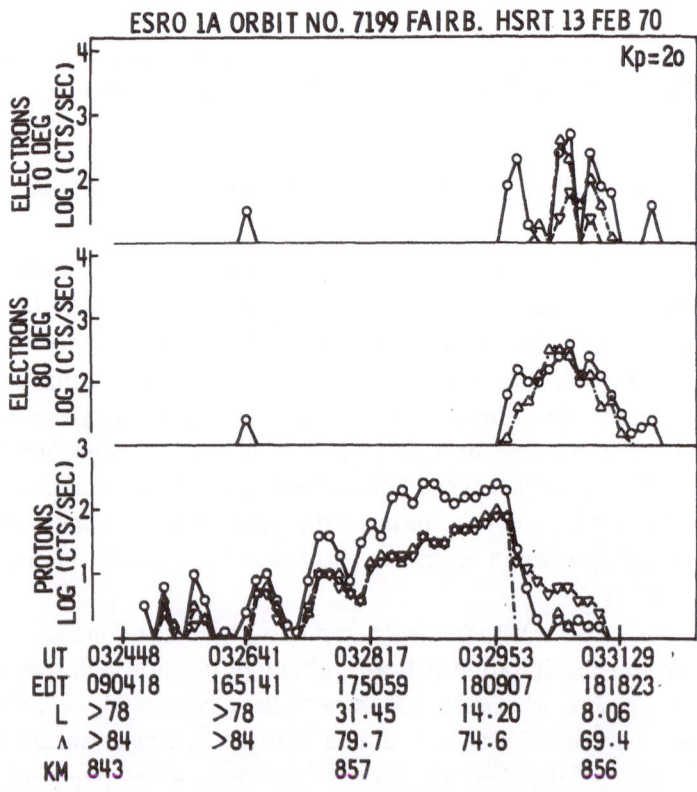

Fig. 9.   keV electron and proton data from ESRO-1A orbit No. 7199. For the meanings of the symbols see caption of Figure 7. The calibration factors can be found in the caption of Figure 10. After Hultqvist *et al.* (1974).

indicates that particles of lower energies drift only small distances along the nightside before being precipitated or being removed by an electric field.

At subcusp latitudes on the dayside, gradient B and curvature drift from the nightside appears to be the only source mechanism needed for consistency between observations of substorm onset and observations of particle appearance on the dayside.

Magnetosheath plasma is injected through the polar cusps (clefts). The latitude profiles of keV proton and electron precipitation indicate that it is also convected along the flanks of the magnetosphere toward nightside, as do also measurements of the horizontal electric field in the poleward part of the auroral regions (see e.g. Gurnett, 1972). This convection takes place poleward of the convection in the opposite direction of hot plasma originating in the magnetospheric tail.

The above conclusions can be drawn from the magnetic latitude – local time distributions of either auroral electron or auroral proton precipitation. From the close coincidence of these precipitation patterns for keV electrons and protons outside the polar cusps a further important conclusion can be drawn, and has already been drawn above; processes precipitating both keV electrons and protons are active simultaneously along the same field lines in almost identical regions of the magnetosphere. The fact that both electrons and protons *in the same energy range* are affected indicates that no single resonance effect (instability) is responsible for the precipitation. Neither seems an electric acceleration with a component in the direction of the magnetic field lines able to explain the observations nor the adiabatic acceleration associated with the convection of the plasma from the tail to the inner magnetosphere. The observations may be interpreted in terms of a turbulence containing a complex set of elementary processes, which scatters both electrons and protons in the entire energy band of auroral particles into the atmosphere. We shall devote the next section of this report to a closer review of what can be derived from the observations about the properties of such a turbulence in the magnetosphere.

3. Some Characteristics of the Turbulence that Scatters Auroral Particles into the Atmosphere

3.1. AURORAL PARTICLE PITCH ANGLE DISTRIBUTIONS

In the keV energy range the auroral protons measured near the atmosphere have a roughly isotropic pitchangle distribution (within a factor of 3, say) in most of the precipitation region (Hultqvist *et al.*, 1971; Hultqvist *et al.*, 1974; Bernstein *et al.*, 1974). This roughly isotropic condition, indicating a fairly strong to quite strong pitchangle diffusion of keV protons in the central part of the auroral particle precipitation zone, is observed always, at all local times and for all levels of magnetospheric disturbance.

At the low L boundary of keV proton precipitation a region with a loss cone distribution (much larger fluxes at 90° pitch angle than at 0°) is regularly found irrespective of disturbance level. It generally extends several degrees equatorward of the transition from isotropy to anisotropy.

The width of the isotropic region shows a marked K_p dependence with an equatorward broadening with increasing disturbance level.

The transition from isotropic to anisotropic distribution is usually observed when the keV proton fluxes have already decreased somewhat from their maximum levels in the isotropic zone. There is thus generally no enhancement in the keV proton flux at the transition from isotropy to anisotropy.

For the keV electrons the pitchangle distribution is somewhat more, but not very much more, variable than for the protons in the same energy interval, when the spatial resolution is limited to 50 km (see Figure 10). Mostly the loss cone is fairly well filled in most of the precipitation zone. A narrow zone with loss cone distribution is frequently seen at the low L edge of the keV electron precipitation zone too. An example of keV electron and proton measurements at $\sim 80°$ and $\sim 10°$ pitch angles illustrating the above points is shown in Figure 10.

When higher spatial resolution is available, the pitchangle distribution for the keV electrons shows still higher degree of structure in comparison with the keV protons. Some of this structure is caused by deviation of the pitch angle distribution from iso-

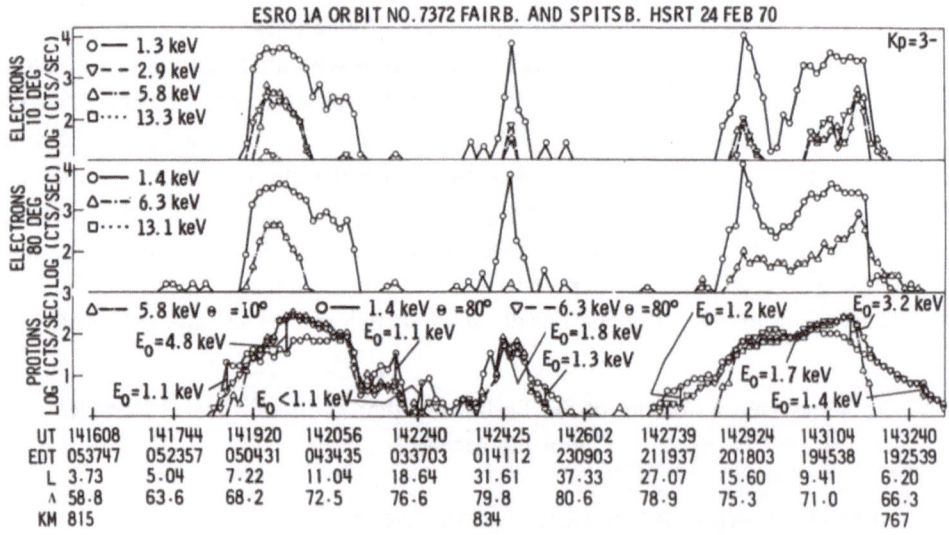

Fig. 10.   keV electron and proton data from ESRO-1A orbit No. 7372. After Hultqvist *et al.* (1974). The calibration factors for transforming count rates into particles $cm^{-2} s^{-1} sr^{-1} keV^{-1}$ are the following for the various detectors:

$$
\begin{aligned}
&10°, \text{protons,}\quad 5.8\ \text{keV: } 0.37 \cdot 10^4\\
&\quad\quad \text{electrons,}\quad 1.3\ \text{keV: } 12.7\ \cdot 10^4\\
&\quad\quad \text{electrons,}\quad 2.9\ \text{keV: } 50\quad\cdot 10^4\\
&\quad\quad \text{electrons,}\quad 5.8\ \text{keV: } 0.48 \cdot 10^4\\
&\quad\quad \text{electrons,}\ 13.3\ \text{keV: } 0.50 \cdot 10^4\\
&80°, \text{protons,}\quad 1.4\ \text{keV: } 13.5\ \cdot 10^4\\
&\quad\quad \text{protons,}\quad 6.3\ \text{keV: } 0.74 \cdot 10^4\\
&\quad\quad \text{electrons,}\quad 1.4\ \text{keV: } 10.6\ \cdot 10^4\\
&\quad\quad \text{electrons,}\quad 6.3\ \text{keV: } 0.94 \cdot 10^4\\
&\quad\quad \text{electrons,}\ 13.1\ \text{keV: } 4.5\ \ \cdot 10^4
\end{aligned}
$$

tropy occuring in discrete forms. We will discuss such deviations in a later section.

For auroral protons of energies above the keV range the situation seems to be only slightly different from that in the keV range. Data for >20 keV protons (Mizera, 1974) and for >100 keV protons (Søraas, 1972) thus appear to be consistent with the keV proton data in that the proton precipitation is characterized by a zone of isotropic pitchangle distribution bounded on the equatorward side by a region with depleted loss cone. The energy dependence of the location of transition regions has not been fully established. The existing data indicates that an energy dependence exists in such a way that the transition from isotropy to loss cone distribution shifts poleward with increasing energy.

For the high-energy tail of the auroral electrons (>15 keV, say) the pitchangle distribution shows much more temporal and spatial structure than in the keV range. Only in regions with very intense precipitation does the pitchangle distribution approach isotropy, as found already by O'Brien with the Injun-3 satellite (O'Brien, 1964). Most common for these electrons is a loss cone distribution. In velocity space, the region of strong pitch angle diffusion of the electrons appears to reach well above the speed of protons of a few hundred keV energy.

3.2. BOUNDARIES OF PRECIPITATION OF keV PROTONS AND ELECTRONS

In Figure 11 the dependence of the location of the transition region on K_p (Bernstein *et al.*, 1974) is compared with the dependence of the location of the plasmapause as

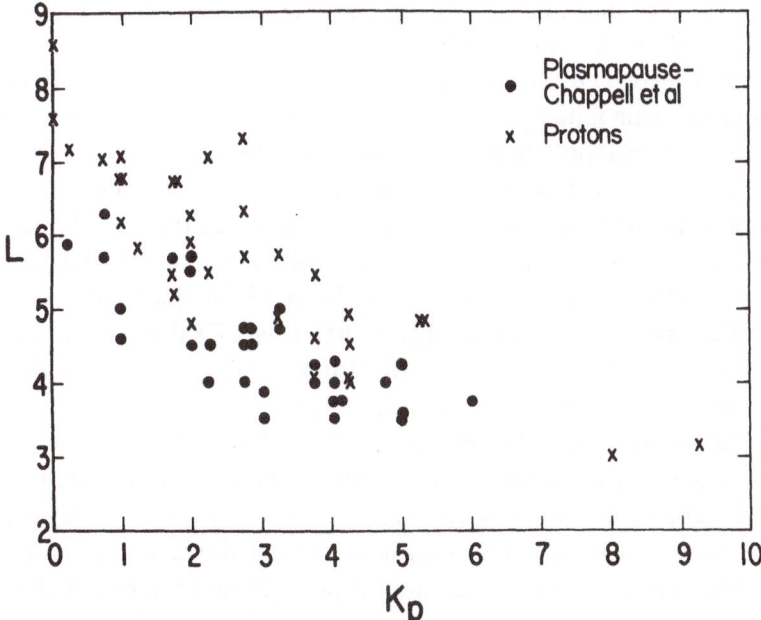

Fig. 11.   Dependence of the latitudinal location (L) of the point where the 6 keV precipitated (10°) proton flux falls below the detection threshold on geomagnetic activity (K_p). Also shown are plasmapause locations determined by Chappell *et al.* (1970) for equivalent geomagnetic conditions. After Bernstein *et al.* (1974).

reported by Chappel *et al.* (1970). The proton data shown were obtained in the evening local time sector whereas the measurements of the plasmapause location were made in the midnight to morning sector. The apparent poleward displacement of the proton pitch angle transition relative to the plasmapause shown in Figure 11 is consistent with the existence of the plasmasphere bulge in the evening sector. Figure 11, therefore, is consistent with the existence of a close spatial relationship between the plasmapause and the transition region. Such a relationship has been reported also for <15 keV protons during an isolated geomagnetic storm by Winningham (1972). Burch (1973) has associated the point of disappearance of 7 keV protons observed at 0° pitchangle at low latitudes during several geomagnetic storms with an independent measurement of the plasmapause location.

Also for protons of energies above the keV range the plasmapause has been found to constitute the inner boundary of precipitation in the midnight sector (Mizera, 1974).

The observations of proton latitude profiles referred to above together with the general absence of an enhancement of the precipitation rate at the low L edge of the isotropic precipitation region, expected in the case of ion cyclotron instability being the cause of the pitchangle diffusion, suggest that the inner boundary of the ring current is produced not by an enhanced precipitation loss process but rather by the transport process in the static electric and magnetic fields of the magnetosphere. The theoretical models of the trajectories of particles which originate in the tail and enter the inner magnetosphere (Chen, 1970; Jaggi and Wolf, 1973) predict that the entering particles can penetrate into the magnetosphere until they reach a forbidden region, which they cannot enter and are subsequently transported back to the magnetopause. In general, the boundary of the forbidden region coincides with the plasmapause for particles which initially have very low energies and it shifts to higher L values with increasing initial particle energy, all of which appears to be in agreement with the observations reported above. The particles are subject to heavy precipitation during the inward convection but the convection speed is evidently sufficiently large that a major fraction of the entering particles can arrive at the outer boundary of the forbidden region. Estimates show that only a small fraction of the proton population is precipitated into the atmosphere (Mizera, 1974). The majority of it is transported away from the inner magnetosphere. The ESRO-1 data (Hultqvist *et al.*, 1974; Bernstein *et al.*, 1974) indicate that some inward diffusion of protons through the boundary of the forbidden region can occur (see Section 3.3).

The conclusion that precipitation loss of keV auroral particles is not important in defining the boundary of precipitation is valid for the low L boundary on the nightside but it is not valid for the local time boundary on the dayside. The daytime minimum in precipitation rate shown in Figure 1 frequently manifests itself so strongly in the keV electron fluxes that no measurable precipitation can be found in the afternoon equatorward of the cusp. During substorms the rate of pitchangle diffusion increases and the (approximately meridional) boundary of the subcusp region of keV electron precipitation may withdraw into the forenoon sector as mentioned earlier. For the protons the lifetime even in the presence of strong pitchangle diffusion is so long that measurable fluxes can be found at all local times also on the dayside, although their values decrease towards noon (where the dayside minimum is seen in Figure 1). The few reported data of the substorm effect on the precipitation of auroral particles on the dayside (Hultqvist *et al.*, 1974) thus indicate that the precipitation into the atmosphere is the important process in determining the distribution and sometimes the extension in local time of the precipitation region. Changes in the convection would be expected to affect keV protons and keV electrons more equally.

3.3. CONCLUSIONS ABOUT THE PITCH ANGLE SCATTERING MECHANISM FROM THE OBSERVATIONS OF PITCH ANGLE DISTRIBUTION FOR keV PROTONS

The proton data described above indicate that the ring current is subject to fairly strong or strong pitchangle diffusion outside the plasmasphere.* The latitudinal extension of the isotropic region is generally larger the lower the proton energy. If the transition to loss cone distribution occurs near the plasmapause for keV protons, as the data indicate, pitch angle diffusion appears to be inhibited rather than enhanced by the presence of cold plasma.

The role of the self-generated electromagnetic ion cyclotron instability driven by an anisotropic particle distribution in the equatorial plane, in determination of the spatial distribution of ring current protons has been considered by Cornwall *et al.* (1971) and in a more generalized sence by Thorne (1972). Their theoretical model suggest that the protons should be unstable and subject to pitch angle and energy diffusion just within the high cold plasma density region at the plasmapause and at high L values where the magnetic field is weak. In between these two regions the ring current should be stable because the particle energy required for resonance with the cyclotron waves exceeds that present in the ring current distribution (unless the anisotropy is very large). This stable region predicted by the ion cyclotron instability thus coincides with the observed region of main precipitation of keV protons. It, therefore, seems fairly clear that the 'component resonance' that scatters the keV protons, in the complex turbulence which affects both electrons and protons over wide energy ranges, is not primarily the ion cyclotron instability.

Coroniti *et al.* (1972) have shown that an electrostatic ion loss cone instability, with frequencies near the proton plasma frequency, can be generated in the ring current throughout the low density region outside the plasmapause; the associated pitch angle and energy diffusion rates would be comparable to those predicted for the ion cyclotron instability. Furthermore, they suggest that the growth rates for this instability decrease with increasing cold plasma density and that it would be inoperative in the high density within the plasmapause. The ESRO-1 data described above therefore seem to suggest that this instability mode occurs in the turbulent region of the magnetosphere. Some recent observations by means of a particle experiment on board the low orbit polar satellite ESRO 4 (Holmgren, personal communication) of the detailed structure of the region of transition from isotropy to anisotropy of keV

---

* Strong pitch-angle diffusion is defined as one which causes completely isotropic pitch-angle distribution.

protons at the low L edge of the precipitation zone have shown that the isotropic region at least in the early part of a storm period extends to somewhat lower L values for 8.5 keV protons than for 2.2 keV protons; in other words, the precipitated keV protons show a hard low – latitude edge, which is opposite to what has been reported for keV electrons (Liszka *et al.*, 1970; Sharp and Johnson, 1971). It seems not to be clear from the paper of Coroniti *et al.* (1972) if the mentioned energy dependence for the pitch angle distribution for keV protons is consistent with the occurrence of ion loss cone instability at the edge of the plasmasphere. The anisotropic protons observed equatorward of the isotropic region, which are in weak pitch angle scattering, may be unstable to the ion cyclotron instability as proposed by Cornwall *et al.* (1971) or to the ion loss cone instability of Coroniti *et al.* (1972). Because of the very great decrease in the rate of inward transport associated with diffusion across the boundary region when compared to that produced by convection outside the boundary, the observed weak pitch angle diffusion would be sufficient to prevent the buildup of the proton flux within the plasmasphere.

For the auroral electrons outside the plasmasphere the degree of anisotropy is generally larger the higher the energy (see e.g. Hultqvist *et al.*, 1974). This is a characteristic which is consistent with an electrostatic cyclotron instability but not with an electromagnetic (whistler) instability (Lyons, 1974). The whistler instability is, however, possibly of importance there at higher energies. Inside the plasmapause the observations of electrons in the equatorial plane seem to be well interpretable in terms of this instability (Williams and Lyons, 1974). Still it has to be remembered that in most part of the auroral particle precipitation zone the turbulence simultaneously affects both electrons and protons of high as well as of low energies. The processes discussed above appear, therefore, to be only components of a complex set of processes. The more or less general coincidence in both space and time of these various 'elementary' processes indicates that they are generated together as parts of a complex turbulence. We will discuss in the following sections a few more characteristics of this turbulence, which can be derived from available experimental information.

3.4. Relations between spatial/temporal variations in the auroral electron and proton latitude profiles

Detailed correlation between spatial and/or temporal variations in keV electron and proton fluxes has been found by means of sounding rockets (see e.g. Bernstein *et al.*, 1969; and Chase, 1970). Large scale correlations have been demonstrated by the ESRO-1 satellites (Hultqvist *et al.*, 1974). Even though there is such a large scale general correlation, the details and scale factors mostly vary very much for keV proton and electron latitude profiles as seen by low orbit satellites. An example is shown in Figure 12a. Correlated and anticorrelated proton and electron flux variations may be seen close to each other in Figure 12b. Anticorrelation in the variation of keV proton and electron fluxes on a still smaller scale have be reported from sounding rocket measurements by e.g. Bernstein and Wax (1970) and Rème and Bosquet

(1971). The spacecraft measurements alone unfortunately do not suffice for separating time and space variations.

On the whole the variation in latitude profiles of the keV proton fluxes is much less than in those for the keV electrons (Hultqvist *et al.*, 1974). Also the temporal variations in the proton latitude distribution are smaller than those in the electron profiles. Unfortunately we cannot say very much about the rate of change of the flux on the basis of satellite measurements which are repeated in about the same region only every 100 minutes. On the whole, however, these satellite results agree with optical observations of the variability of the hydrogen aurora.

Hultqvist *et al.* (1974) have not seen any cases of structure in the keV proton latitude profiles without any corresponding structure in the keV electrons in the main part of the energetic particle zone. At the edges of it, however, the opposite may very well be true. Also well inside the zones one finds occasionally strong variations in the keV electron populations without any corresponding structure in the keV protons. Practically without exception the variations in the latitude profiles are larger in the electrons than in the protons as mentioned. Also this is in agreement with ground observations of hydrogen aurora.

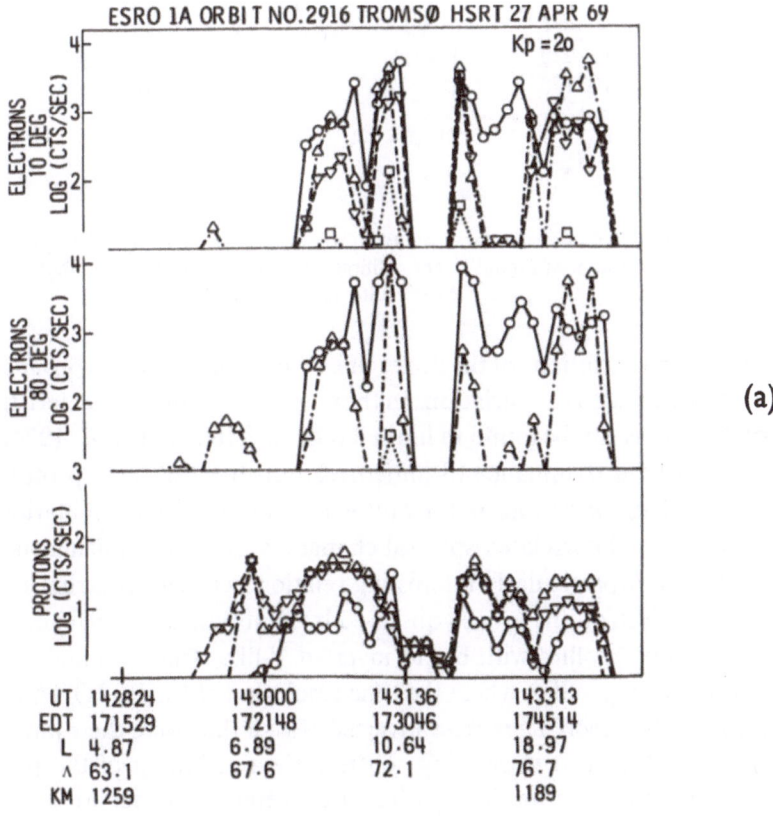

(a)

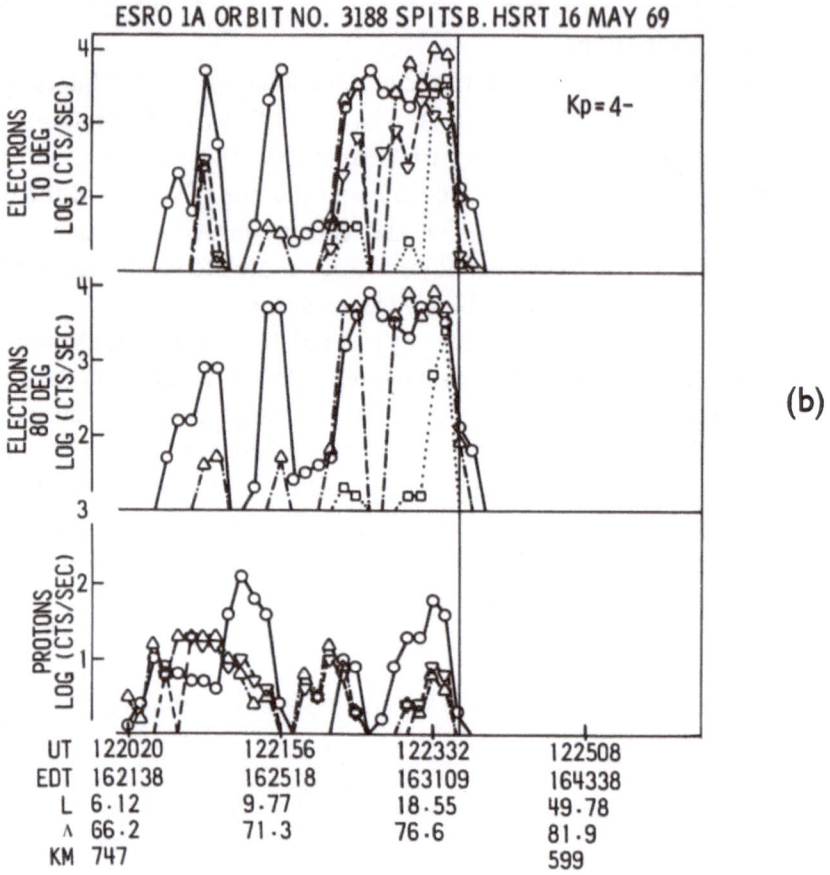

Fig. 12. keV electron and proton data from ESRO-1A orbits Nos. 2916 (a) and 3188 (b). For the meanings of the symbols see caption of Figure 7. The calibration factors can be found in the caption of Figure 10. After Hultqvist *et al.* (1974).

Not only the variations of the fluxes of keV protons and electrons may show different relations but also the variations in the energy spectrum characteristics. Practically all combinations can be found in limited regions (Hultqvist *et al.*, 1974). There seems, however, to be a dominance of anticorrelation between keV proton and electron spectral hardness variations in the keV energy range. The combination of correlated flux increases and correlated spectral changes is the most difficult one to find in the ESRO-1 data. A particularly interesting relation between spectrum hardness of keV protons and electrons have been observed by Hultqvist *et al.* (1974) in the late evening sector. Latitude profiles with broad inverted *V*-like structures and such with highly variable electron profiles, where the time resolution of the ESRO-1A experiment was not sufficient for resolving narrow inverted *V*'s – if they were there (and they probably were) – were found to show very consistently a softening of the proton spectrum where the electron one was hardest (i.e. at the center of the inverted *V*'s). An example is shown in Figure 13. These regions sometimes also show field aligned proton fluxes

(Hultqvist *et al.*, 1974) as well as field aligned electron pitch angle distributions (Frank and Ackerson, 1971; Lundin, 1973). Evidently these observations are difficult to fit into a simple physical model such as acceleration in a magnetic-field aligned electric field, unless oppositely directed electric fields exist very close to each other or hemispherical asymmetry with respect to the total potential drops along the magnetic field lines exist. We will discuss this further in a later section.

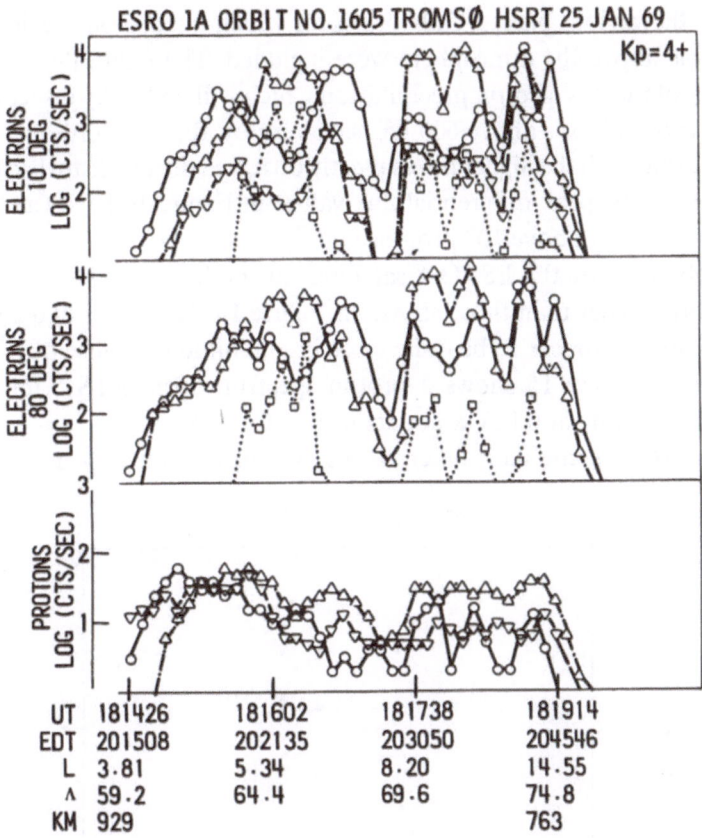

Fig. 13. keV electron and proton data from ESRO-1A orbit No. 1605 showing inverted-V like regions. For the meanings of the symbols see caption of Figure 7. The calibration factors can be found in the caption of Figure 10. After Hultqvist *et al.* (1974).

One possible way of interpreting the anticorrelation of the 'temperature variations' of keV electrons and protons in the latitudinal distributions determined by means of ESRO 1A and B is in terms of a turbulence which transfers energy from the protons to the electrons. Similar ideas have been put forward by Johnstone (1971) in order to explain anticorrelations found in sounding rocket data.

4. Interaction of the Hot Plasma with the Ionosphere

We defer to Sections 4.3 and 4.4, the discussion of the anisotropic fluxes, peaking in the direction of the magnetic field lines, which for electrons seem to occur mainly in discrete auroral forms. Figure 14 shows the peak flux of 6 keV protons at about 80° pitchangle in a number of ESRO 1A and B passes through the auroral regions, one point for each pass, as function of the satellite height. The K_p had widely different values during the various passes. Only orbits for which the proton precipitation was nearly isotropic above the atmosphere were included. The figure illustrates several characteristics of the keV proton precipitation which will be briefly discussed here.

The 6 keV proton fluxes above 600 km, say, vary by about an order of magnitude from orbit to orbit, with the higher fluxes mostly observed during disturbed conditions (high K_p values). The peak differential flux values in Figure 14 have values between $(2\text{--}3) \times 10^5$ and slightly above 10^6 protons $\mathrm{cm}^{-2}\,\mathrm{s}^{-1}\,\mathrm{sr}^{-1}\,\mathrm{keV}^{-1}$. The highest 6 keV proton flux observed by the ESRO-1 satellites during heavily disturbed conditions was only slightly higher than those shown in Figure 14. The fluxes of 6 keV protons near the atmosphere appear to be quite close in magnitude to those observed in the equatorial plane. Figure 15 shows a proton spectrum from ATS 5 for which the differential number flux at 6 keV amounts to 8×10^5 protons $\mathrm{cm}^{-2}\,\mathrm{s}^{-1}\,\mathrm{sr}^{-1}\,\mathrm{keV}^{-1}$. Figures 16a, b show some data taken in the equatorial plane by means of the S^3

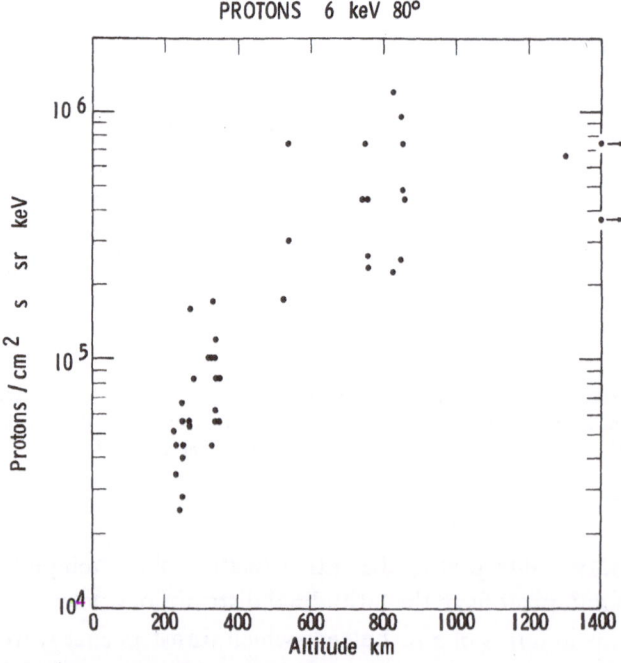

Fig. 14.   Maximum observed flux values of 6 keV, 80° protons observed in a number of ESRO 1A and 1B passes through the particle precipitation zone at different altitude. After Hultqvist *et al.* (1974).

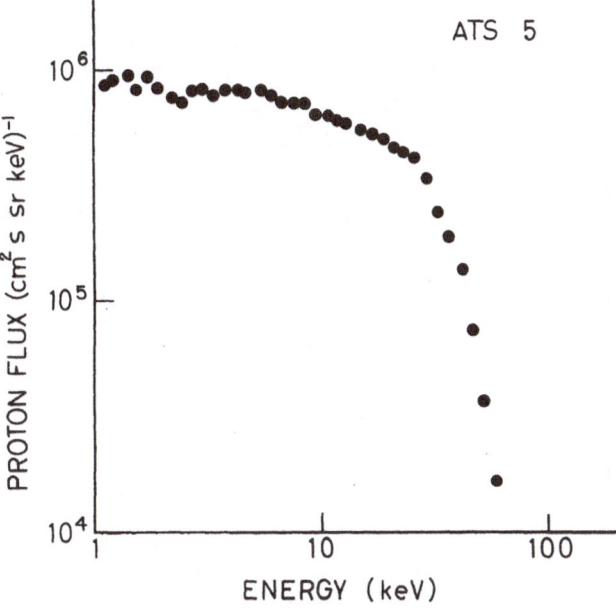

Fig. 15.   Proton energy spectrum measured by DeForest and McIlwain on ATS5. After Whalen and
McDiarmid (1973).

satellite during various levels of magnetic disturbance (Smith and Hoffman, 1974;
Williams *et al.*, 1974). They show 6 keV fluxes in the range of Figure 14. Considering
the limited accuracy in the absolute fluxes it seems reasonable to conclude that the
keV proton flux values observed in the equatorial plane and near the atmosphere do
not differ by more than a small fraction of an order of magnitude at keV energies.

For proton energies above 10 keV the precipitated flux – in spite of the isotropy
found in the region of maximum precipitation and an energy spectrum similar in
shape to the average equatorial distribution given by Pizzella and Frank (1971) –
appears to be about one order of magnitude lower than the flux of protons mirroring
at the equator (Mizera, 1974). No measurement carried out simultaneously close to
the same magnetic field line in the equatorial plane and near the atmosphere with
well intercalibrated instruments have been reported, however.

The proton observations described above are consistent with the absence of significant energization of the protons between the equatorial plane and the atmosphere,
the particles only being scattered into the loss cone with an efficiency that decreases
with increasing energy.

For the auroral electrons an attempt has been made by Sharp *et al.* (1971) to
compare temporally coincident fluxes in the loss cone near the atmosphere with those
at pitch angles well outside the loss cone in the equatorial plane. The number flux in
the ionosphere several hundred km from the estimated conjugate point was found to
be roughly the same as in the equatorial plane for keV energies (within a factor of two,
in general). (A significant difference in the spectral shape, with a monotonous decrease
with energy at the equator but a peak at a few keV in the ionosphere, was, however,

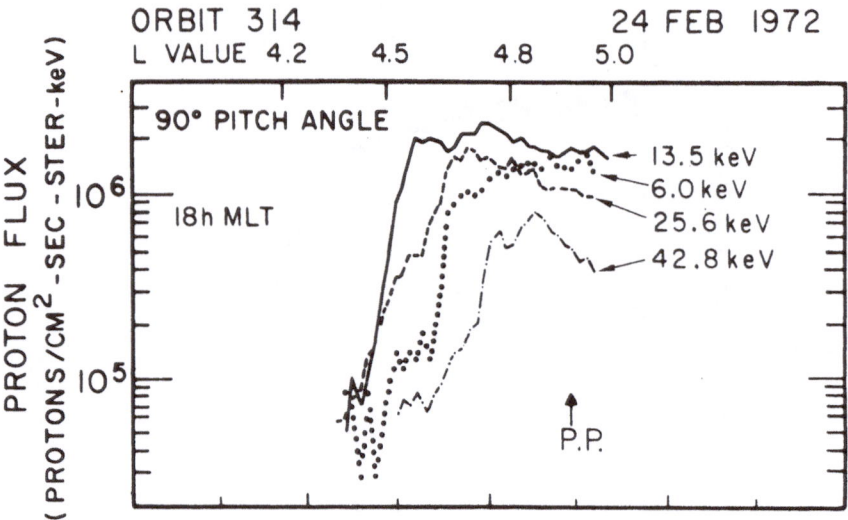

Fig. 16a. L profiles of four energy bands of protons, whose widths are about $\pm 15\%$ around the center energies listed, obtained with the S^3-A satellite during the storm of February 24, 1972. After Smith and Hoffman (1974).

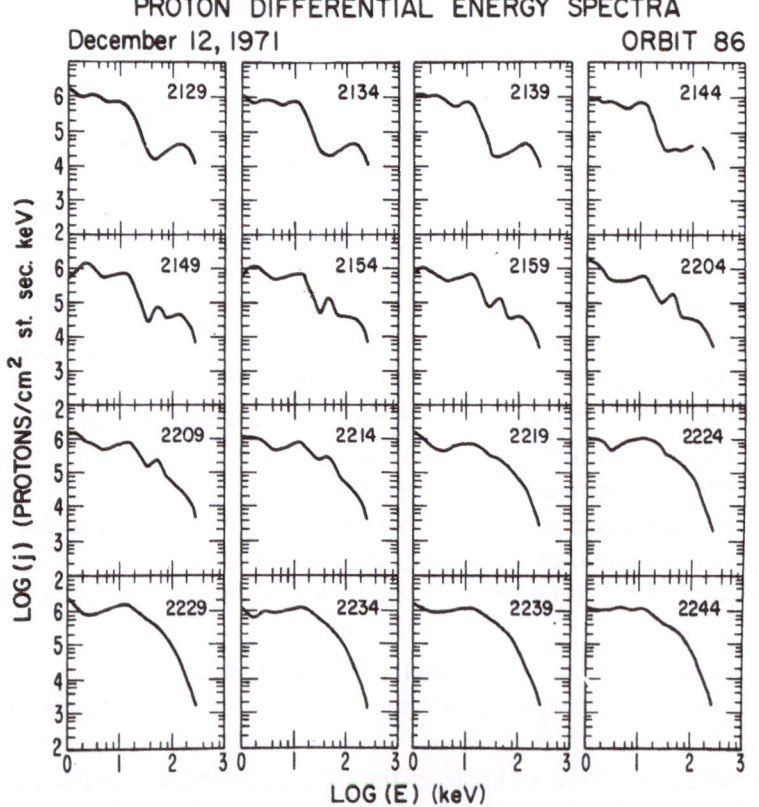

Fig. 16b. Five-minutes averages of proton spectra taken by S^3-A (Explorer 45) before, during, and after the substorm of 2200 ± 10 UT on Decembre 12, 1971. Times given are in universal time. S^3 exited the plasmasphere at ~ 2221 UT. After Williams et al. (1974).

seen in most of the cases studied.) Also the average fluxes of keV electrons observed near the atmosphere (Riedler and Borg, 1972; Riedler, 1972) are similar to the typical values given by DeForest and McIlwain (1971) for the geostationary orbit. The conclusion that keV electron fluxes are the same in the equatorial plane and near the atmosphere in regions with no discrete aurora has recently been drawn also by Johnstone *et al.* (1974).

For energies well above the keV range it is no longer true that the electron fluxes observed in the ionosphere are approximately the same as what is observed in the magnetically conjugate region of the equatorial plane. During both quiet and disturbed conditions the equatorial electron flux of energies above 40 keV at large pitch-angles is more than a factor of ten larger than the loss cone flux. For energies above 230 keV the ratio of high pitch angle to loss cone flux are generally of the order of a hundred and only in very disturbed conditions the loss cone flux appears to approach the flux values of higher pitch angles, as pointed out before.

An important conclusion can be drawn from the observation that the roughly isotropic auroral particle fluxes at keV energies are closely similar in the equatorial plane and near the atmosphere: there is no accelaration of the auroral particles between the equatorial plane and the atmosphere. This is, however, not true in many of the discrete auroral forms, nor in regions of field aligned proton pitch angle distribution.

4.2. VARIATIONS IN ISOTROPIC keV PROTON FLUXES BETWEEN THE EXOSPHERE AND THE LOWER F-LAYER

A further interesting conclusion that can be drawn from Figure 14 is that there is no energization, but only charge exchange affecting the protons between 1500 and 250 km in the majority situation with roughly isotropic pitch angle distribution at keV energies.

The fluxes below 300 km are a factor of 5 to 10 less than those measured above 600 km. This is approximately the ratio expected from the assumption that proton charge exchange and charge stripping reactions are responsible for the flux changes with altitude.

The data in Figure 14 are in Figure 17 compared with 90° pitch angle fluxes at different altitudes, deduced from detailed measurements of the pitch angle distribution at about 1000 km on board ESRO 4 (Holmgren, personal communication). As can be seen, the two sets of data are in good agreement with each other. The altitude values for the ESRO-4 data have been computed from the pitchangle on the basis of a simple r^{-3} dependence of the magnetic field intensity.

The agreement between the measurements shown in Figure 17 and the values from charge exchange theory indicates that in general the dominant modification of the incident proton flux at low altitude is simply the change in charge state produced by charge exchange and charge stripping reactions. Although special cases could occur where modifications of these *isotropic* incident proton fluxes could be imposed at low altitudes such effects have not been clearly identified yet.

In a minority of observations of keV auroral particles *anisotropic* field-aligned

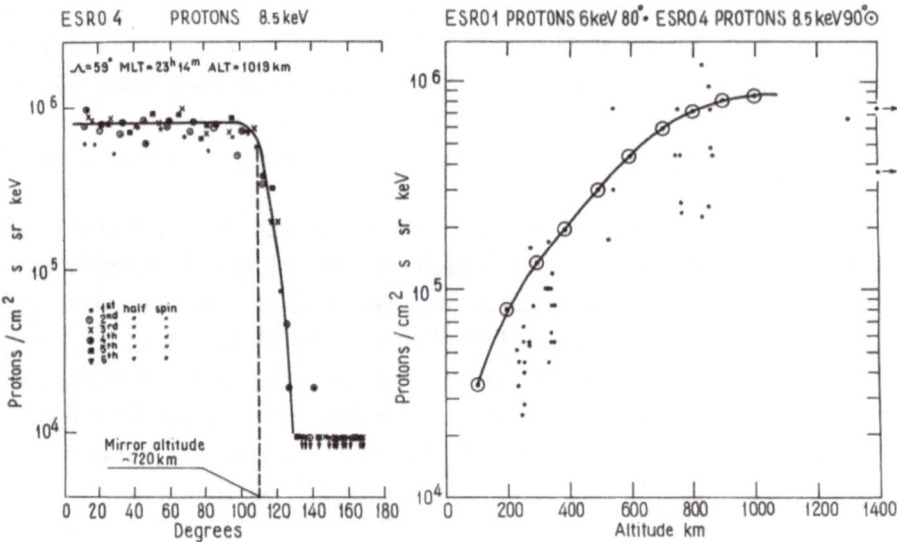

Fig. 17.   Comparison of the altitude dependence of keV proton flux obtained by ESRO 1 with the one obtained from a detailed pitchangle distribution measured by ESRO 4, assuming an r^{-3} dependence of the geomagnetic field (Holmgren, personal communication).

pitch angle distribution have been observed for which the above conclusion is not valid. These effects are discussed in the following sections.

4.3. Observations of keV proton fluxes affected by energizing processes close to or in the ionosphere

Observations of keV proton fluxes with higher fluxes in the loss cone than at high pitch angles, have been reported by Rème (1969) and Rème and Bosqued (1971) from a sounding rocket experiment and by Hultqvist *et al.* (1971) from measurements with the ESRO-1 satellites.

The observations may be summarized in the following way:

Field-aligned ion fluxes at keV energies

– occur fairly frequently (in about 10% of the satellite passes in the evening sector for 6 keV protons);

– occur at all local times and in a wide band of latitudes centered at about 70° invariant latitude;

– occur in both quiet and disturbed conditions;

– may cover large areas;

– are not a transient type of phenomenon;

– occur only in regions of fairly strong electron precipitation.

On the basis of the observations reported by Hultqvist *et al.* (1971) they concluded the following about the accelerating force which gives the ions their field aligned pitch angle distribution:

– The force is directed along the geomagnetic field lines.

 - The acceleration takes place in or near the upper ionosphere.
 - The acceleration is associated with fluxes of energetic electrons.
 - The acceleration is not necessarily associated with field aligned currents.

One possible interpretation of these field-aligned ion fluxes is as a manifestation of a potential difference between the cold ionospheric plasma and the hot plasma in the outer magnetosphere, which interact with each other freely along the geomagnetic field lines. The mechanism is briefly discussed in Section 4.6.

4.4. DISCRETE AURORAL ARCS

Diffuse aurora occurs in the entire zone of precipitation of the keV electrons and protons. Discrete auroral forms are either superimposed on the diffuse aurora or are located just poleward or just equatorward of it. Are then the discrete auroral forms only the peak of the iceberg? In the last few years a lot of evidence against such a simple state of affairs has been presented. A large number of sounding rocket measurements have shown that the electron energy spectrum mostly has a pronounced peak in the keV range in the auroral arcs, sometimes superimposed on a broader spectrum (Albert, 1967; Westerlund, 1968; Choy *et al.*, 1971; Frank and Ackerson, 1971; Evans *et al.*, 1972; Whalen *et al.*, 1972; Lundin, 1973; Reasoner and Chappel, 1973). Some rocket experiments have shown more than one peak in the spectrum (see e.g. Chase, 1970; Choy *et al.*, 1971; Reasoner and Chappel, 1973). Another group of measurements over auroral arcs have resulted in spectra with a much wider peak than those referred to above, spectra closely resembling a Maxwellian distribution (Rearwin, 1971; Bryant *et al.*, 1973).

Ackerson and Frank (1972) were on one occasion able to show that the inverted *V* type of events found in the Injun 5 data coincided with an auroral arc. The inverted *V* events (Frank and Ackerson, 1971) are characterized by a peaked energy spectrum with the peak energy highest at the very center of the region occupied by the inverted *V* and declining in both poleward and equatorward direction from there. The latitudinal width is typically 200 km. They are located just poleward of the trapping boundary and occur primarily in the dusk to midnight sector according to Frank and Ackerson (1971). The latitudinal width mentioned refers to spectrograms covering energies down to about one hundred eV. When the low-energy threshold of the measuring instrument is higher the width is smaller. Inverted-*V* events of only a few tens of kms latitudinal span measured above ~1 keV have been reported by Lundin (1973).

The large width of an inverted *V* region may invite one to believe that multiple arc situations cannot be of this type, but they can. Reasoner and Chappel (1973) from sounding rocket measurements found an inverted *V* event of order 100 km wide containing multiple arcs with the peak energy at the center of the arcs 9–10 keV and between the arcs 2–3 keV. The event may be considered as the superposition of a number of inverted *V* events, slightly displaced in latitude. It is usually not possible to resolve this kind of internal structure in inverted *V*'s in data taken with low orbiting satellites.

The pitch angle distribution of the energetic electrons in the inverted *V* events has been found to be sometimes peaked in the direction along the magnetic field lines (e.g. Frank and Gurnett, 1971; and Lundin, 1973).

In the electron precipitation with Maxwellian type of spectra reported by Rearwin (1971) and Bryant et al. (1973) the pitch angle distribution has been demonstrated to be approximately isotropic.

A pitch angle distribution strongly aligned with the magnetic field is perhaps the most unambiguous signature that can be imposed upon the auroral particle population by an electric field parallel to the line of force. Hoffman and Evans (1968), Rème (1969), Frank and Gurnett (1971), Hultqvist et al. (1971), Rème and Bosqued (1971), Whalen and McDiarmid (1972), Arnoldy and Choy (1973), Holmgren and Aparicio (1973), Maehlum and Moestrue (1973), Paschmann et al. (1974), and others have reported such magnetic field aligned particle fluxes and proposed explanations in terms of an electrostatic field component in the direction of the magnetic field. Bryant et al. (1973) found that the arcs, where they measured approximately Maxwellian spectra, occurred in the boundary layer between two magnetospheric plasma of different phase-space density.

We thus see that arcs may show quite different characteristics. Whereas the inverted V related arcs mostly occur just poleward of the trapping boundary, it is not known if the observed 'Maxwellian arcs' were located poleward or equatorward of the trapping boundary. A possibility may be that the arcs on the high latitude side of the trapping boundary are generally of the inverted V type with sharply peaked spectrum and field-aligned pitch angle distribution, and arcs on the low latitude side show isotropic fluxes with Maxwellian energy spectrum. This would then also fit the finding of Rearwin (1971) and Bryant et al. (1973) that the energetic electron population measured in the ionosphere could have been produced by adiabatic compression of the plasma sheet population, which is believed to reach the ionosphere below the latitude of the trapping boundary by e.g. Frank (1971) (the relation of the 40 keV electron trapping boundary to the plasma sheet is still somewhat controversial).

From the above one is inclined to conclude that the discrete auroral forms are not just one phenomenon but rather are caused by several different processes. It is therefore very interesting that recently a suprisingly simple model has been developed which appears to be able to include practically all the various observations of auroral arcs. That model will be discussed in the next section.

4.5. Effects of a Magnetic Field Aligned Potential Difference in Auroral Arcs

Evans (1974) has taken into account that a potential difference along the magnetic field lines which accelerates electrons downward also acts as a barrier for upgoing backscattered electrons of an energy insufficient to surmount the potential. Such upgoing electrons are, therefore, reflected downwards and appear as 'precipitating' electrons.

Evans computed the complete precipitation electron population that would be observed in the presence of a region with an electric field above the atmosphere, located at about 2000 km latitude. B was taken to be a factor of two smaller in the electric field region than at auroral heights. He assumed that the shape and intensity of the secondary electron spectrum is independent of the incident electron spectrum. The spectrum was based on the total backscatter spectrum given by Banks *et al.* (1974) and was truncated at the incident electron energy when applied to low energy primary electrons. He further assumed that the backscatter spectrum is independent of the pitch angle of the incident electron and that the backscattered electron fluxes are isotropic. Taking as an example the incident electrons to have a Maxwellian distribution, with a characteristic energy of 800 eV and a number density of 1.5 cm^{-3}, and a total potential difference of 2000 V, the primary electron beam would have a peak in the spectrum at 2 keV, a total number flux of 1.4×10^9 electrons cm^{-2} s^{-1} at the low altitude boundary of the electric field and an energy flux of 7.6 erg cm^{-2} s^{-1}, all of which is characteristic of the electrons in auroral arcs. The differential energy spectrum of the beam itself and of the backscatter electron population produced by this beam (tertiary backscatter taken into account) which would be observed at the top of the atmosphere at a pitch angle of 0° is shown in Figure 18.

The separation between that part of the electron population which is primary and that part which originates from the atmosphere is clearly exposed by the discontinuity in the spectrum. A physical particle detector would smooth these discontinuities so that experimental data would simply show a prominent peak in the

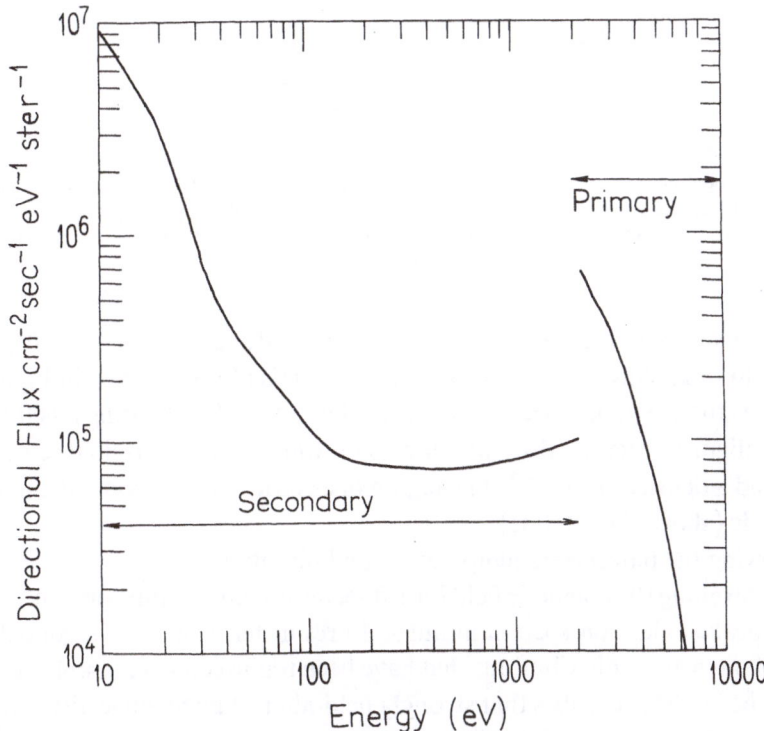

Fig. 18.   A model energy spectrum appropriate to 0° pitch angle precipitating electrons observed just above the atmosphere. The discontinuity separates electrons of atmospheric origin from magnetospheric. After Evans (1974).

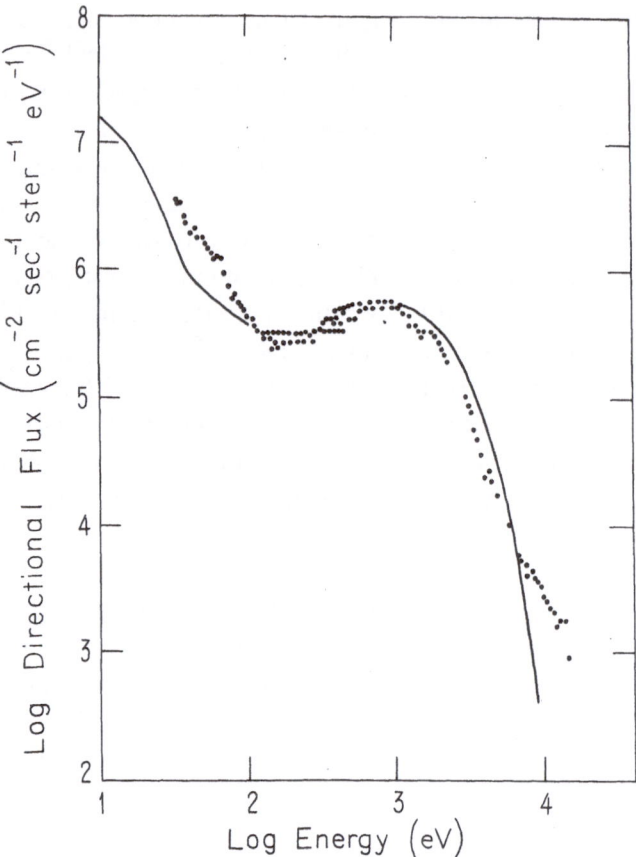

Fig. 19. A model electron energy spectrum computed assuming a 400 V potential difference along a magnetic field line and an unenergized Maxwellian electron distribution of temperature 800 eV and density 5 cm^{-3}. The data represent an electron spectrum observed by Frank and Ackerson (1971). After Evans (1974).

spectrum as shown in Figure 19. A set of experimental data obtained at $\sim 0°$ pitch angle by the Injun-5 satellite (Frank and Ackerson, 1971) is plotted as dots in Figure 19. The solid line is the $0°$ energy spectrum computed by Evans (1974) using a 400 V value for the field aligned potential drop and a plasma which had an original temperature of 800 eV and a density of 5 cm^{-3}. The region of electric field was assumed to be at 2500 km altitude (above the satellite).

By varying the parameters temperature and density of the hot plasma and the potential drop along the magnetic field lines it seems possible to interpret most observations of peaked electron spectra in auroral arcs in terms of Evans' model. Also the variety of pitch angle distributions that have been reported may possibly be consistent with it. The model prescribes that strongly field-aligned pitch angle distributions shall exist for electron energies just above the electric potential drop. At lower energies approximate isotropy is expected and at higher energies the field alignment decreases until isotropy is reached at about twice the electric potential energy.

4.6. On the interaction processes

Signatures of a field-aligned electric field can, according to earlier sections, be found in fairly stable situations or in variable auroral forms, they are found in regions of large or of small latitudinal extension, they correspond to a potential drop along the magnetic field lines varying by at least one order of magnitude, and to upward as well as downward directed electric field. It seems unlikely that one single process can cause the potential drop in all conditions in which its signatures have been found. Probably a number of processes are involved. Several have been proposed in the literature. We will here only summarize them briefly.

A potential difference accelerating positive ions downward may, under certain conditions, exist between the hot magnetospheric plasma and the ionosphere (Hultqvist, 1971). The conditions to be met can be described roughly as follows: the density of the upper ionosphere must be so low that the influx of hot electrons into the ionosphere is larger than the outflux of cold electrons from the ionosphere in some height interval in the absence of an electric field. The potential difference between the plasma on the two sides of the mentioned altitude range is then determined by the temperature of the hot plasma, i.e. of the order of kiloelectron volts. This process is a possible cause of the field-aligned proton fluxes observed by means of the ESRO-1 satellites. It can only accelerate electrons upward and can, therefore, be responsible for field-aligned electron fluxes only if the potential difference between the hot and the cold plasma is significantly different at the two ends of the field line.

Other sources of field-aligned electric fields, which may produce upward pointing as well as downward pointing fields, are difference in pitch angle distributions between trapped electron and ion populations (Persson, 1963; Alfvén and Fälthammar, 1963), potential double layers (sheats) (see Block, 1972) and voltage drops through current-carrying regions with anomalous resistivity (see e.g. Kindel and Kennel, 1971).

The electric field at the edge of a plasma with different pitch angle distributions for electrons and ions is of a character similar to that of a sheath near a wall. The voltage drop over the sheat is expected to be related to the characteristic energies of the two populations.

The potential double layers are pourly understood. They may be produced by currents along the magnetic field lines of values above a threshold, and possibly also by strong gradients in plasma parameters such as density and temperature.

Anomalous resistivity may be produced by various kinds of wave-particle interactions. The current-carrying particles are scattered by the waves and experience, therefore, a hindrance in their motion similar to that caused by ordinary Coulomb collisions.

There are problems with all the different mechanisms mentioned. In order that the potential difference between the hot and the cold plasma shall be given by the temperature of the hot one, the pressure and temperature distributions in the transition region have to meet strict requirements. A problem with the different pitchangle distributions is that it is far from clear how they can be upheld and how long they may persist if they are produced. The potential double layer caused by currents require a high current value. The anomalous resistivity dissipates so large amounts of power that there are difficulties in finding a reasonable energy sink outside the dense ionosphere. We, therefore, conclude this report with the observation that we are still quite far from an understanding of the roles of the various possible mechanisms mentioned in the interaction between the hot plasma and the ionosphere.

Acknowledgements

The research programme at Kiruna Geophysical Institute on which a large part of this report is based has been supported economically by the Swedish Natural Science Research Council, The Swedish Board for Space Activities and the European Space Research Organization.

References

Ackerson, K. L. and Frank, L. A.: 1972, *J. Geophys. Res.* **77**, 1128.
Akasofu, S.-I. and Kimball, D. S.: 1973, *Planetary Space Sci.* **21**, 696.
Albert, R. D.: 1967, *J. Geophys. Res.* **72**, 5811.
Alfvén, H.: 1939, *Kgl. Sv. Vetenskapsakad. Handl. III* **18**, No. 3, Stockholm.
Alfvén, H. and Fälthammar, C.-G.: 1963, *Cosmical Electrodynamics*, Oxford, At the Clarendon Press.
Anger, C. D. and Lui, A. T.: 1973, *Planetary Space Sci.* **21**, 873.
Arnoldy, R. L. and Choy, L. W.: 1973, *J. Geophys. Res.* **78**, 2187.
Banks, P. M., Chappell, C. R., and Nagy, A. F.: 1974, *J. Geophys. Res.* **79**, 1459.
Berkey, F. T., Driatskiy, V. M., Henriksen, K., Jelly, D. H., Shchuka, T. L., Theander, A., and Yliniemi, J.: 1971, World Data Center A, Upper Atmosphere Geophysics, Report UAG-16.
Berkey, F. T., Driatskiy, V. M., Henriksen, K., Hultqvist, B., Jelly, D. H., Shchuka, T. I., Theander, A., Yliniemi, J.: 1974, *Planetary Space Sci.* **22**, 255–307.
Bernstein, W., Ionoye, C. T., Saunders, N. L., and Wax, R. L.: 1969, *J. Geophys. Res.* **74**, 3601.
Bernstein, W. and Wax, R. L.: 1970, *J. Geophys. Res.* **75**, 3915, 1970.
Bernstein, W., Hultqvist, B., and Borg, H.: 1974, *Planetary Space Sci.* **22**, 767.
Block, L.: 1972, *Cosmic Electrodynamics* **3**, 349.
Bryant, D. A., Courtier, G. M., and Benett, G.: 1973, *Planetary Space Sci.* **21**, 165.
Burch, J. L.: 1973, *J. Geophys. Res.* **78**, 6579.
Chappell, C. R., Harris, K. K., and Sharp, G. W.: 1970, *J. Geophys. Res.* **75**, 50.
Chase, L. M.: 1970, *J. Geophys. Res.* **75**, 7128.
Chen, A. J.: 1970, *J. Geophys. Res.* **75**, 2458.
Choy, L. W., Arnoldy, R. L., Potter, W., Kintner, P., and Cahill, L. J. Jr.: 1971, *J. Geophys. Res.* **76**, 8279.
Cornwall, J. M., Coroniti, F. V., and Thorne, R. M.: 1971, *J. Geophys. Res.* **76**, 4428.
Coroniti, F. V., Fredericks, R. W., and White, R.: 1972, *J. Geophys. Res.* **77**, 6243.
Davis, T. N.: 1962, *J. Geophys. Res.* **67**, 59.
Deehr, C. S., Egeland, A., Aarsnes, K., Amundsen, R., Lindalen, H. R., Søraas, F., Dalziel, R., Smith, P. A., Thomas, G. R., Stauning, P., Borg, H., Gustafsson, G., Holmgren, L. A., Riedler, W., Raitt, J., Skovli, G., Wedde, T., and Jaeschke, R.: 1973, *J. Atmospheric Terrest. Phys.* **35**, 1979.
DeForest, S. E. and McIlwain, C. E.: 1971, *J. Geophys. Res.* **76**, 3587, 1971.
Eather, R. H.: 1967, *Rev. Geophys.* **5**, 2–8.
Eather, R. H. and Carovillano, R. L.: 1971, *Cosmic Electrodyn.* **2**, 105.
Eather, R. H. and Mende, S. B.: 1971, *J. Geophys. Res.* **76**, 1746.
Eather, R. H. and Mende, S. B.: 1972, in K. Folkestad (ed.), *Magnetosphere-Ionosphere Interactions*, Universitetsforlaget, Oslo.
Eather, R. H.: 1973, *Rev. Geophys. Space Phys.* **11**, 155.
Evans, D. S., Jocobsen, T., Maehlum, B. N., Skovli, G., and Wedde, T.: 1972, *Planetary Space Sci.* **20**, 233.
Evans, D. S.: 1974, *J. Geophys. Res.* **79**, 2853.
Frank, L. A.: 1971, *J. Geophys. Res.* **76**, 5202.
Frank, L. A. and Ackerson, K. L.: 1971, *J. Geophys. Res.* **76**, 3612.

Gurnett, D. A.: 1971, in E. R. Dyer (ed.), *Critical Problems of Magnetospheric Physics*, p. 123, National Acad. Sciences, Washington, D.C.

Heikkila, W. J. and Winningham, J. D.: 1971, *J. Geophys. Res.* **76**, 883.

Heikkila, W. J.: 1972, in E. R. Dyer (ed.), *Critical Problems of Magnetospheric Physics*, p. 67, National Acad. Sciences, Washington, D.C.

Hoffman, R. A.: 1972, in K. Folkestad (ed.), *Magnetosphere-Ionosphere Interactions*, Universitetsforlaget, Oslo.

Holmgren, L.-A. and Aparicio, B.: 1973, *Space Research XIII* (COSPAR), p. 555, Akademie-Verlag, Berlin.

Hultqvist, B.: 1971, *Planetary Space Sci.* **19**, 749.

Hultqvist, B., Borg, H., Christophersen, P., and Riedler, W.: 1971, *Planetary Space Sci.* **19**, 279.

Hultqvist, B., Borg, H., Christophersen, P., Riedler, W., and Bernstein, W.: 1974, NOAA, ERL, Space Environment Laboratory, Boulder, Colorado, U.S.A., Technical Report ERL 305 SEL 29.

Jaggi, R. K. and Wolf, R. A.: 1973, *J. Geophys. Res.* **78**, 1187.

Johnstone, A. D.: 1971, *J. Geophys. Res.* **76**, 5259.

Johnstone, A. D., Boyd, J. S., and Davis, T. N.: 1974, *J. Geophys. Res.* **79**, 1403.

Kindel, J. M. and Kennel, C. F.: 1971, *J. Geophys. Res.* **76**, 3055–78.

Liszka, L., Borg, H., and Riedler, W.: 1970, *Phys. Norv.* **4**, 121.

Lui, A. T. Y. and Anger, C. D.: 1973, *Planetary Space Sci.* **21**, 799.

Lui, A. T. Y., Perreault, P., Akasofu, S.-L., and Anger, C. D.: 1973, *Planetary Space Sci.* **11**, 857.

Lundin, R.: 1973, paper presented at the *Symposium on European Sounding Rocket and Related Research at High Latitudes at Spatind*, Norway, April, ESRO SP-97.

Lyons, L. R.: 1974, *J. Geophys. Res.* **79**, 575.

Maehlum, B. N. and Moestue, H.: 1973, *Planetary Space Sci.* **21**, 1957.

Mizera, P. F.: 1974, *J. Geophys. Res.* **79**, 581.

O'Brien, B. J.: 1964, *J. Geophys. Res.* **69**, 13.

Omholt, A.: 1971, *The optical aurora*, Springer-Verlag, Berlin.

Paschmann, G., Johnson, R. G., Sharp, R. D., and Shelley, E. G.: 1974, *J. Geophys. Res.*, in press.

Persson, H.: 1963, *Phys. Fluids* **6**, 1756.

Pizzella, G. and Frank, L. A.: 1971, *J. Geophys. Res.* **76**, 88.

Rearwin, S.: 1971, *J. Geophys. Res.* **76**, 4505.

Reasoner, D. L. and Chappell, C. R.: 1973, *J. Geophys. Res.* **78**, 2176.

Rème, H., Thesis, CESR Toulouse, 1969.

Rème, M. and Bosqued, J. M.: 1971, *J. Geophys. Res.* **76**, 7683.

Riedler, W.: 1972, in B. M. McCormac (ed.), *Earth's Magnetospheric Processes*, D. Reidel Publ. Co., Dordrecht-Holland, p. 133.

Riedler, W. and Borg, H.: 1972, *Space Research XII*, 1397, Akademie-Verlag, Berlin.

Sharp, R. D., Carr, D. L., Johnson, R. G., and Shelley, E. G.: 1971, *J. Geophys. Res.* **76**, 7669.

Sharp, R. D. and Johnson, R. G.: 1971, in B. M. McCormac (ed.), *The Radiating Atmosphere*, D. Reidel Publ. Co., Dordrecht-Holland, p. 239.

Smith, P. H. and Hoffman, R. A.: 1974, *J. Geophys. Res.* **79**, 966.

Søraas, F.: 1973, in B. M. McCormac (ed.), *Earth's Magnetospheric Processes*, D. Reidel Publ. Co., Dordrecht-Holland, p. 120.

Thorne, R. M.: 1972, in K. Folkestad (ed.), *Magnetosphere-Ionosphere Interactions*, Universitetsforlaget, Oslo, p. 185.

Vorobjev, V. G., Gustafsson, G., Starkov, G. V., Feldstein, Y. L., and Shevnina, V. F.: 1973, paper presented at the *Substorm Symposium of the Second General Scientific Assembly of IAGA*, Kyoto, Japan, September 1973.

Westerlund, L. H.: 1968, Ph.D. Thesis, Rice University, Houston, Texas.

Whalen, B. S. and McDiarmid, I. B.: 1972, *J. Geophys. Res.* **77**, 191.

Whalen, B. A., Miller, J. R., and McDiarmid, J. B.: 1972, *Science Research Council*, Ottawa, Preprint.

Wiens, R. H.: 1968, Ph.D. Thesis, University of Saskatchewan.

Williams, D. J., Barfield, J. N., and Fritz, T. A.: 1974, *J. Geophys. Res.* **79**, 554.

Williams, D. J. and Lyons, L. R.: 1974, *J. Geophys. Res.*, in press.

Winningham, J. D.: 1972, *EOS* **53**, 489.

WAVE-PARTICLE INTERACTIONS IN THE OUTER
MAGNETOSPHERE: A REVIEW

R. W. FREDRICKS

TRW Systems Group, Systems Groups Research Staff, One Space Park, Redondo Beach, Calif. 90278, U.S.A.

1. Introduction

The application of wave-particle interactions to magnetospheric phenomenology is so extensive and wide-ranging that to attempt to review it in its entirety would result in the production of a sizeable monographic book. For this reason, I have chosen to restrict the present review to only wave-particle interactions and their implication in outer magnetospheric phenomena. Thus, large bodies of low-altitude applications are not covered at all.

Even within the arbitrarily chosen boundaries of the outer magnetosphere, I have left out some areas in which wave-particle interactions play crucial roles. Since these omitted topics may be dear to the hearts of some readers, I apologize for not having included them. However, time and space limitations produced the necessity of selectivity; I therefore apologize to everyone offended by my omissions, and wish to assure that I would have enjoyed reviewing everything.

Wave-particle interactions are important to the detailed understanding of many geophysical phenomena. In Section 2, their role as currently understood in the Earth's bow shock is discussed, with the conclusion that their role in proton thermalization remains a mystery. In Section 3, I touch on the magnetopause region, and point out that practically nothing is understood there on sound theoretical grounds.

However, in Sections 4 and 5 I have reviewed the role of wave-particle interactions due to VLF and ELF electromagnetic waves and electrostatic waves present in the plasmaspheric and exo-plasmaspheric regions of the outer magnetosphere. Here, one sees a relative triumph of intellectual understanding of many observed properties of the inner trapped electron belt based on careful application of wave-particle interactions to pitch-angle diffusion theory. One also sees a glimmering of incipient understanding of the ring current morphology, again utilizing concepts involving wave-particle interactions. Many problems remain to be solved, but much of the basic information to formulate these problems appears to have revealed itself.

In Section 6 I have reviewed the state of knowledge of current-driven turbulence, and anomalous resistivity produced by it (a wave-particle interaction) in the outer magnetosphere. Here, high-altitude field-aligned currents appear to produce local anomalous resistivity and extrapolations of isolated spacecraft measurements indicate that this resistivity may support electric potential drops along auroral zone L-shells. If these parallel electric fields can accelerate charged particles over a significant segment of the field line, kilovolt electrons may be delivered to the auroral ionosphere to produce arcs, electrojets and auroral forms. However, the latter state-

V. Formisano (ed.), The Magnetospheres of the Earth and Jupiter, 113–152. *All Rights Reserved*
Copyright © 1975 by D. Reidel Publishing Company, Dordrecht-Holland

ment is speculative at the present time. Section 7 contains mention of some neglected areas and the rationale for their omission.

It is hoped that the reader of this review may derive benefit from its very general nature, and that the rather extensive bibliography and reference list will aid him in searching out important research papers of more or less recent vintage. In this regard, if I have offended any author by omitting a publication, let me assure that individual that the omission was not deliberate and I apologize for it.

2. Wave-Particle Interactions in Earth's Bow Shock

According to classical MHD flow concepts, the first encounter of the solar wind plasma with the Earth occurs at the collisionless MHD bow shock. Because of the 'collisionless' nature of this shock (that is, particle-particle collisional mean free path very large compared to the characteristic structural lengths, such as density, velocity and magnetic field jump gradients), the formation and maintenance of the shock structure must depend upon thermodynamically dissipative processes in which wave turbulence interacts with particle populations in such a way that an effectively irreversible particle heating occurs in the shock structure over lengths very much smaller than particle-particle mean free paths.

In actual practice, the leading sentence of the previous paragraph is incorrect. Under a wide variety of solar wind parametric values, both protons and electrons may be 'reflected' from the bow shock by an as yet ill-understood mechanism. These reflected protons and electrons then backstream into the oncoming solar wind plasma, and produce sub-low-frequency MHD waves (Greenstadt *et al.*, 1967; Fairfield, 1969; Russell *et al.*, 1971; Scarf *et al.*, 1970a) via proton beam-plasma interactions, and also higher frequency electron plasma oscillations via an electron beam-plasma interaction (Scarf *et al.*, 1971; Fredricks *et al.*, 1971).

There has been considerable experimental data analysis and interpretation carried out in an effort to discover the primary wave-particle interactions in bow shock structures. In order to attain this goal, it is really necessary to have available a host of diagnostic data from magnetometers, particle probes, and plasma wave (electric field) detectors, at high data rates, and preferably simultaneous data bases from multiple satellites, with one of these upstream to provide solar wind input parameters. Unfortunately, this favorable arrangement of satellites only occurs with rarity, thus rendering an adequate data base difficult to acquire. Further compounding this difficulty is the fact that the complete instrumental combination and the high data rates simply do not coexist on all the available spacecraft used to date. This situation should be alleviated by the extensive instrumentation and coordinated positioning of the Mother, Daughter and Heliocentric spacecraft comprising the joint NASA/ESRO International Magnetospheric Explorer project.

To the present date, the only high data rate plasma wave (electric and magnetic fluctuations) measurements in the bow shock have been provided by detectors aboard the OGO-5 and IMP-6 spacecraft. A great deal of interpretation of these measurements in terms of wave-particle interactions in the bow shock has been published, and we can expect much more in the future as the data are extensively gathered and analyzed by Greenstadt and by Formisano and his colleagues. This topic is more extensively covered by Formisano in this symposium, so we shall make no attempt to review the waveparticle interactions in bow shocks here. We simply wish to point out that there remain some mysteries even after more than six years of data analysis since the launch of OGO 5 in March, 1968.

For a reasonably complete but brief review of what is currently understood about wave-particle interactions and plasma wave instabilities in the bow shock, the reader is referred to the article by Greenstadt and Fredricks (1974) in the *Proceedings of the Summer Advanced Study Institute* in Sheffield in 1973. A rather complete bibliography of publications so far available describing experimental or experimentally-guided theoretical conclusions concerning wave-particle interactions and their roles in bow shock structures is given at the end of that reference.

It is clear that certain unanswered questions remain regarding the wave-particle interactions as dissipative processes in collisionless shocks. The most important of these is the unknown wave-particle interaction leading to rapid proton heating in a wide variety of bow shock structures. Available evidence points to proton heating over scale lengths comparable to a fraction of the main magnetic field jump gradient in many bow shocks, while in some rather ill-defined oblique shock structures, proton heating appears to occur only gradually and over a very large scale length.

A second mysterious phenomenon is that of reflected protons (and electrons). What, if any, roles do wave-particle interactions play in the reflection process? How do the upstream waves stimulated by the reflected particle beams influence or pre-condition the solar wind plasma in the upstream region in the near vicinity of the bow shock itself?

The role of plasma instabilities and the wave-particle effects, such as anomalous resistivity, which result from them, is better understood with regard to electron heating in bow shocks. As further studies are completed, there is good reason to believe that a nearly complete understanding of the wave-electron interactions will come about. With respect to a complete understanding of the wave-proton interactions, it would be premature to predict that a similar understanding will be achieved without significant improvements in the measurement techniques, both from the standpoint of instrumentation as well as of satellite deployment.

3. Wave-Particle Interactions and the Magnetopause

It is clear from the viewpoint of plasma physics that a boundary between a vacuum magnetic field and a generally obliquely flowing plasma carrying along a quasi frozen-in magnetic field must contain significant wave-particle interactions. This sort of boundary approximately describes the Earth's magnetopause.

The classical analyses of such a boundary began with the famous Chapman-Ferraro solution, which modeled the interaction by a one-dimensional vacuum magnetic field region in one half-space, in contact with a half space filled with an unmagnetized flowing plasma whose bulk velocity vector upstream was normal to the plane containing the vacuum field lines. Even in this simplest of models, the diamagnetic drift currents required to produce a self-consistent magnetic field jump across the boundary were found to be large enough to produce plasma oscillations (Piddington, 1960; Bernstein *et al.*, 1964) whose immediate effect would be to broaden the Chapman-Ferraro 'sheath' (or magnetic field gradient) beyond the electron inertial scale length c/ω_{pe} predicted by Chapman-Ferraro theory. In addition, MHD instability could be expected for oblique flow conditions. For example, Kelvin-Helmholtz instabilities may occur (Southwood, 1968). In fact, no reliable extension of a Chapman-Ferraro solution to cases of oblique incidence (flow velocity at an arbitrary angle to the boundary normal) has ever been carried out, even neglecting the frozen-in magnetic field in the flowing plasma. A number of relevant theoretical considerations of this problem are listed in the bibliography for this section at the end of the paper. The reader may draw his own conclusions after reading the cited body of work. However, it is the author's opinion that the problem remains unsolved, and that no available theoretical calculation accounts adequately for the complete physics of such an MHD interaction.

This sad state of affairs in theoretical development is only partly alleviated by satellite observations of the magnetopause structure. Such a condition of very incomplete understanding of the magnetospheric boundary is extremely unfortunate, since the entire energy transfer from solar wind plasma to the magnetosphere (and consequently many major magnetospheric phenomena such as magnetic substorms and storms, and indeed the aurora) must be determined by boundary processes involving wave-particle dynamics.

As a concrete example, we note that Axford and Hines (1961) and Axford (1964) have introduced the concept of a 'viscous' interaction between solar wind flow and the magnetopause boundary in order to sustain internal electric fields necessary to explain their convection model of the magnetosphere. From the point of view of MHD theory as used by these authors, the macroscopic transport coefficient labeled 'viscosity' is simply inserted at the appropriate point in the equations. The microscopic origin of such 'viscosity' was never investigated by Axford and Hines.

The only real attempts to explain 'viscosity' in the collisionless boundary layer of the magnetopause appear to be those of Tsuda (1967) and Eviatar and Wolf (1968). These latter authors invoked quasi-linear theory to obtain a saturated spectrum of ULF and ELF fluctuations by assuming that conditions for the generation of two-stream ion cyclotron instabilities were satisfied. The wave spectrum so deduced is then used to estimate the diffusion of protons across the boundary layer, and also the statistical momentum transfer which leads to an effective 'collisionless' coefficient of viscosity for the boundary layer. Use of suitable typical values of plasma and field parameters, an effective coefficient of viscosity on the order of 8×10^{12} cm^2-s^{-1} was computed by Eviatar and Wolf, a number close to that which Axford (1964) had estimated as necessary macroscopically to explain powers dissipated during substorms.

The other major MHD phenomenon expected to contribute or even control the rate at which energy and momentum may be transferred to the magnetosphere by solar wind flow is magnetic field merging, or 'reconnection'. There exist several largely phenomenological models for merging (or reconnection) based on macroscopic MHD concepts. Unfortunately, the 'merging rates' in these models, and consequently the energy and momentum transfer rates, are not calculated from first principles; that is, they are not calculated from a kinetic theory description of the basic plasma physics processes in the region where magnetic field lines are 'destroyed' in the sense that magnetic field energy is converted to plasma particle energy. Thus, in all the reference papers found in our bibliography treating the reconnection problem, one will find widely varying estimates of merging rates (that is, dissipation rates). It is my opinion that if any of the quoted rates are nearly correct, it is simply a fortuitous circumstance. The realization of a realistic model for magnetic merging at the magnetopause is without doubt the most important single unsolved problem in magnetospheric physics. Its solution will remain to be obtained until such time as the proper microscopic plasma physics of the merging (or diffusion) region have been formulated, and the wave-particle interactions leading to anomalous transport coefficients in this collisionless MHD interaction are understood.

4. VLF and ELF Electromagnetic Waves in the Magnetosphere

4.1. CYCLOTRON RESONANCE INTERACTIONS

Among the best known of those phenomena called 'wave-particle interactions', and certainly the most extensively studied, is the case of cyclotron resonance between a moving charged particle (electron or positive ion) and the electric component of an electromagnetic wave. We shall make no attempt to review in any depth the work in this field prior to about 1970, since Gendrin (1972) has provided an excellent review of the subject up to about that time. We shall emphasize only some of the work covered by Gendrin as it relates to the developments after 1970, which developments we report in the present review.

In order to make the present review sufficiently self-contained, we shall remind the reader that cyclotron resonance (or gyroresonance) between a charged particle and an electromagnetic wave occurs when the particle has a component of velocity $v_{\parallel}$ parallel to a steady background magnetic field line, and interacts with an electromagnetic wave field of propagation vector $\mathbf{k}(k_{\parallel}, k_{\perp})$ and frequency ω in such a way that

$$\omega - k_{\parallel} v_{\parallel} + n\Omega_j = 0, \qquad n = \pm 1, \pm 2, \pm 3, \dots,$$

where $\Omega_j = eB_0/M_j c$ is the gyrofrequency in $\mathbf{B}_0$ of the jth species of charged particle. It is clear that in a *distribution* of particles described by some $f(\mathbf{v})$, exact resonance occurs mathematically over a set of zero measure in velocity space for monochromatic waves $(\omega, \mathbf{k})$. In a real plasma, $f(\mathbf{v})$ has a width, and any wave phenomenon of geophysical interest has an effective bandwidth, leading to a nonpathological wave spectrum $F(\omega, \mathbf{k})$.

Thus, in assessing the net results of wave-particle interactions in the magnetosphere, one attempts to either measure $f(\mathbf{v})$ and $F(\omega, \mathbf{k})$, or to construct these functionals from whatever measurements or theoretical models are available. The most familiar tool for doing this is the quasi-linear theory (Kennel and Engelmann, 1966) or weak turbulence theory (Kadomtsev, 1965). One constructs from plasma instability and quasi-linear theory a set of 'collisionless' transport quantities, the most important being 'diffusion coefficients', and 'amplification factors' or 'gain factors'.

In the case of 'diffusion' coefficients, one can discriminate among several types of processes. First, diffusion in configuration space can be caused by waveparticle interactions which lead to violation of one or more of the classical 'adiabatic invariants' of particles trapped between turning points (mirror points) in the geomagnetic field.

An excellent example of diffusion in configuration space is that of radial diffusion of energetic trapped particles in the Van Allen belt. In the theory, first suggested by Kellogg (1959) and further pursued by Fälthammer (1965), one assumes that the first two adiabatic invariants, namely the magnetic moment $\mu = M_j V_\perp^2 / 2B_0$ and longitudinal action $J_\parallel = \int v_\parallel \, \mathrm{d}s$ (which is related to the bounce period $T_B = \int \mathrm{d}s / v_\parallel$) are conserved, while the third invariant associated with azimuthal drift of particles around the earth is violated. These azimuthal drift and bounce periods in a dipole field are approximately (Dungey, 1965)

$$T_{\mathrm{d}} \approx \frac{2\pi \Omega_0 R_{\mathrm{E}}^2}{3v^2 L (0.35 + 0.15 \sin \alpha_0)} \tag{1}$$

$$T_{\mathrm{B}} \frac{4L R_{\mathrm{E}}}{v} (1.30 - 0.56 \sin \alpha_0), \tag{2}$$

where v is the total velocity of the particle of energy W, R_{E} is one Earth radius, L is McIlwain's shell parameter, Ω_0 is the rest mass gyrofrequency of the particle, and α_0 is the equatorial pitch angle of the particle.

Diffusion in configuration space occurs readily whenever electromagnetic fluctuations are present in the magnetosphere with spectral distributions $F(\omega)$ having significant power at frequencies such that $\omega \sim T_{\mathrm{d}}^{-1}$ (breaking the third adiabatic invariant) or $\omega \sim T_{\mathrm{B}}^{-1}$ (breaking the second adiabatic invariant). In the latter case, there may also be a *diffusion in velocity space* because breaking the invariant associated with bounce motion usually leads to eventual precipitation of trapped particles by lowering mirror points, and thus depleting certain regions of velocity space, in turn leading to plasma instabilities which tend to replenish these depleted regions.

Diffusion in velocity space always accompanies the breaking of the first and second adiabatic invariant, which occurs for electromagnetic perturbations spanning frequencies comparable to the gyrofrequency or the bounce frequency. It readily occurs for the doppler-shifted resonance involving particles in the tail of a distribution, and leads to *pitch-angle scattering*. In the magnetosphere, it is very likely that trapped radiation belt particles more often than not experience wave field perturbations over a sufficiently broad frequency spectrum that radial diffusion in configuration space occurs simultaneously with pitch-angle and energy diffusion in velocity space. This eventuality has been discussed by Haerendel (1970), who points out that the entire source-sink properties of radial (or L-shell) diffusion as observed in the magnetosphere cannot be understood without assuming both configuration and velocity space diffusion simultaneously.

A second important effect due to cyclotron resonant wave-particle interactions is that of natural amplification of ULF, ELF and VLF electromagnetic waves by particle distributions having *velocity space anisotropy*. Application of this theory to ULF is found in the review of Gendrin (1972); much original work was done by Liemohn (1967) to explain multiple-hop whistlers and by Cornwall (1965, 1966). In the following sub-sections, we shall concentrate on surveying newer material which has appeared since Gendrin's review.

4.2. WHISTLER MODE WAVES AND TURBULENCE

The first adiabatic invariant (magnetic moment) for trapped electrons may be destroyed by the presence of significant levels of VLF or ELF whistler mode turbulence. This was apparently first pointed out by Dungey (1963) and Cornwall (1964). In a now classical paper, Kennel and Petschek (1966) carried out an extensive investigation in which they invoked the property of pitch angle anisotropy in the trapped electron population as an amplification mechanism for whistler mode waves. The model they constructed contained basically a feedback mechanism which balanced pitch-angle diffusion due to wave turbulence against precipitation and wave growth and convection out of the amplifying volume, assumed to reside near the geomagnetic equator. Thus, they were led to expressions for stably trapped limits on trapped electron flux populations. These stable limits resulted from an exact balance between loss of particles due to precipitation and wave gain due to residual velocity space anisotropy.

Using an almost identical analysis, Liemohn (1967) has shown that VLF and ULF whistlers may be amplified on field lines in near-equatorial regions at high altitudes, where pitch-angle anisotropies in particle distributions may be present. He presents this as a mechanism to largely offset ionospheric reflection, and to provide an effective 'guidance' mechanism along field lines, since the amplification factor sharply peaks for waves with $k_\perp \to 0$. This, Liemohn points out, is an alternative to ionization ducting as a guiding mechanism for lightning whistlers, and also allows one to understand the many- (up to 40) hop whistlers observed at times at ground stations.

A much-improved theory of the basic wave-particle interaction between magnetospheric ELF whistler mode wave turbulence and the trapped electron distributions has been published very recently by Etcheto *et al.* (1973). As first suggested by Kennel and Petschek (1966), the equilibrium trapped flux profiles and precipitation patterns (if magnetospheric dynamics ever admit of any true, $\partial/\partial t = 0$, equilibrium) must result as a balancing of the wave turbulence spectrum, the detailed shape of the trapped particle velocity distribution function, and the source and loss mechanisms for the particles and the waves.

Gendrin (1972) later pointed out that some of the concepts in the Kennel and Petschek (1966) theory of turbulent diffusion and trapped flux equilibrium were not sharply defined, for example 'critical anisotropy' and 'critical energy' of resonantly interacting particles. He suggested, in that paper, that one should strictly solve a set of quasi-linear equations for the self-consistent wave spectral density and trapped particle distribution function (TPDF) using a pitch-angle diffusion equation to define the TPDF, coupled to a wave kinetic equation which defines the wave spectral density function. Source terms in the pitch-angle diffusion equation are empirically defined, while loss terms are either neglected or are due to (p, L) scatterings in the ionosphere.

Etcheto et al. (1973) extended the qualitative ideas in Gendrin's (1972) paper, and performed self-consistent analytical and numerical solutions to the coupled pitch-angle diffusion and wave kinetic equations for an assumed functional form of the source term (injected particle spectrum), and a set of boundary conditions on wave reflection at the ionosphere and at the borders of the wave-particle interaction region. They only treated parallel-propagating ELF whistler waves $(k_\perp = 0)$, an admittedly questionable assumption whose validity they discussed in their paper, to which the reader is referred for details.

One of the basic assumptions of Etcheto et al. (1973) is that in equilibrium, the injection rate of fresh electrons (dn_2/dt) in their notation) must be precisely offset by the observed precipitation rate on any given L-shell. Thus, dn_2/dt, the unknown injection rate, can be related to precipitated flux, a measurable quantity. Therefore, only a model of the shape factor of the TPDF as a function of pitch angle and velocity had to be assumed by Etcheto et al. They demonstrated that their equilibrium results are relatively intensitive to this shape factor, insofar as pitch-angle distribution is concerned.

The main results of this important paper are as follows. A self-consistent TPDF and wave spectrum were calculated. The peak frequency of the wave spectrum inside the plasmasphere was found to be approximately one-half the frequency at which the electrons with $v_\| = (2E_0/m_e)^{1/2}$ would resonate with whistler waves, where E_0 is the characteristic energy of the injected electrons. The peak intensity of this selfconsistent wave spectrum was found to be directly proportional to source intensity $(dn_2/dt$, and thus to precipitation flux) and to cold plasma density.

Etcheto et al. (1973) also found that the 'limiting flux' expression introduced by Kennel and Petschek (1966) is a zero-order approximation, which was recognized previously by Gendrin (1972b). They found that increased injection of fresh anisotropic electrons would indeed lead to enhanced pitch-angle diffusion and a regulation of the flux of trapped particles. However, they pointed out that there is no unique 'trapped flux limit', but rather they obtained an expression which is energy-dependent (exponentially) and thus have quantified the ideas of Kennel and Petschek (1966), who also had stated qualitatively that their zero-order flux limitation would be invalidated for intense injection sources.

With respect to whether wave intensity or injection source intensity should yield the criterion for definition of strong and weak pitch-angle diffusion regimes, Etcheto *et al.* (1973) find that one can define a source intensity above which diffusion becomes so strong that the limiting flux concept no longer applies. Thus, in this sense, they answer the question by stating that source, not wave, intensity defines the 'boundary' between strong and weak pitch-angle diffusion. The reason for this behavior appears to be related to the fact that the mean or effective anisotropy factor of the equilibrium TPDF decreases markedly in the strong diffusion limit. Thus, even though wave intensity increases in proportion to dn_2/dt, the effective anisotropy factor defined by the self-consistent TPDF decreases, and thus offsets the increase in $B_W^2(\omega)$.

The final important result of their theory is the finding that equilibrium directional and omnidirectional fluxes computed from the self-consistent TPDF's do not depend on the cold plasma density. This means that these flux profiles are not discontinuous across the plasmapause. This is a result of having calculated self-consistent quasi-linear TPDF's and effective self-consistent growth rates and anisotropies.

All of these latter analytical results of Etcheto *et al.* (1973) are consistent with the results of numerical simulations reported by Cuperman *et al.* (1973). Cuperman *et al.*, performed a particle-in-cell computer simulation of the whistler instability under various ratios of cold-to-hot plasma. They found that, consistent with the ideas of linear stability theory, initial growth rates of the whistler instability for initial pitch-angle (or temperature $T_\perp/T_\parallel >$ critical) anisotropies greater than critical (Kennel and Petschek, 1966), were indeed enhanced by addition of cold plasma. However, the nonlinear effects of diffusion in velocity space caused the anisotropy to decrease continuously and thus reduced growth rates. They found that, on time scales available to the simulation, all systems tested under different initial conditions on cold-to-hot plasma ratio n_C/n_H relaxed to the same final anisotropy ($T_\perp/T_\parallel \sim 1.35$, which is *linearly stable*), and the only apparent influence of n_C/n_H was to produce a final wave energy spectrum whose peak intensity was proportional to n_C/n_H. This trend of the numerical results is indeed consistent with the analytical results of Etcheto *et al.* (1973).

One of the drawbacks of the self-consistent theory of Etcheto *et al.* (1973) is their assumption that the wave spectrum contains only parallel-propagating whistlers. As they clearly point out, Landau effects and higher order resonances with harmonics of the gyrofrequency have been neglected, and thus higher energy components of the equilibrium TPDF are not correctly treated ($\geqslant 200$ keV). However, since their theory contains an 'interaction length' or equatorial arc segment which is an adjustable parameter, they argue that exclusion of oblique wave normals is justified.

A theory of trapped flux profiles which does include the Landau resonance and gyrofrequency harmonic resonances for electrons interacting with oblique whistler waves has been presented by Lyons *et al.* (1972). The theory is not self-consistent in the sense of Etcheto *et al.* (1973); that is, Lyons *et al.* did not use coupled pitch-angle diffusion and wave kinetic equations in a full quasi-linear approach to find a wave spectrum $B^2(\omega)$ consistent with their TPDF, nor did they explicitly introduce a source function for injected particles into the diffusion equation.

Lyons *et al.* (1972) assumed a constant level and spectral shape of ELF hiss through-out (independent of L) the plasmasphere. They also assumed a distribution of oblique waves with normals distributed according to a law $\exp(-\tan^2\theta/\tan^2\theta_\omega)$, where $\theta_\omega \sim 80$ deg (measured away from $\mathbf{B}_0$). This distribution function of wave normals is quite flat out to about 70 deg, then decays rather rapidly. They demonstrate that exact distributions of wave normals are probably not important as long as wave energy is spread over a wide range, up to ~ 80 deg, of normal directions.

Under such assumptions, Lyons *et al.* introduced a bounce-averaged pitch-angle diffusion equation governing the equatorial pitch-angle distribution function $f_0(\alpha, t)$. This equation involved the total pitch-angle diffusion coefficient $\bar{D}_\alpha$, which was summed over all Landau and gyrofrequency harmonic resonances as well as averaged over the bounce period. The technique for including all resonances is described in Lyons *et al.* (1971).

Lyons *et al.* (1972) then convert the diffusion equation to an integral equation which is subsequently solved numerically for the pitch-angle distribution under the assumption of separability $f_0(\alpha, t) = F(t) g(\alpha)$, i.e., steady-state (Roberts, 1969). The solution $g(\alpha)$ is then used to compute precipitation lifetimes, using $\bar{D}_\alpha$ coefficients for pitch-angles near the edge of the loss cone.

The lifetimes so computed are shown to depend both on electron energy and on L-value. Lyons *et al.* compute equatorial pitch-angle distributions for several energies $3.3 \leqslant L \leqslant 4$, and compare these with experimental distributions [furnished by private communication] found by H. West from OGO-5 data. The computed shapes show some discrepancies in detail, but they generally possess the trends characteristic in the measurements. The major triumph of the theory of Lyons *et al.* is its ability to predict the existence of the electron slot region, and to approximately reproduce the general energy dependence of the minimum lifetime (slot) locations in L-value, although this agreement becomes progressively worse beyond $L = 3.5$.

Perhaps the most recent and convincing application of the Kennel-Petschek concept, combined with Haerendel's (1970) suggestions about coupling radial and pitch-angle diffusion, has been made by Lyons and Thorne (1973). These authors combined radial diffusion (source for pitch-angle diffusion) with pitch-angle diffusion (sink for radial diffusion) and Coulomb scattering (a common sink), but treated the wave spectral density function self-inconsistently, i.e., as a known function. Following the same procedures as Lyons *et al.* (1972), they included a broad angular distribution of oblique ELF whistler turbulence, computed diffusion coefficients for combined Landau and cyclotron harmonic resonances, averaged over bounce periods, and thus obtained electron flux distributions at several typical energies.

A main result of their treatment, even though it does not deal with the wave spectrum self-consistently, is that the wave intensity must be significantly controlled by the convection process (which drives radial diffusion). Enhanced convection apparently would be associated with enhanced spectral density of wave turbulence. This qualitatively is almost self-evident, since enhanced convection may reasonably be assumed to accompany a sharpening of the pitch-angle distribution of the injection spectrum, as well as an enhanced density rate dn/dt of the injected particles. This would lead to increased wave levels and perhaps transient strong diffusion, with quick precipitation loss, and rapid restoration of more reasonable levels of trapped fluxes.

According to Lyons and Thorne (1973), their equilibrium flux curves are quite sensitive to fluctuations in the ratio of $\langle E \rangle / \langle B_W \rangle$, i.e., the average fluctuation field $\langle E \rangle$ of convection, during the radial diffusion, and $\langle B_W \rangle$ the average whistler mode turbulence amplitude. They point out that changes in this ratio of a factor as small as 3 would be detectable over the solar cycle, as nonpersistence of the electron slot. Based on the measurements, this appears not to be the case, leading Lyons and Thorne to postulate that radial diffusion due to $\langle E \rangle$ exerts effective control over $\langle B_W \rangle$. During major storms, they do not expect inner zone flux profiles to satisfy their model, but rather to become transiently enhanced, with subsequent decay to prestorm levels.

In conclusion to this discussion of whistler turbulence, it seems fair to say that extensions of the pioneering ideas of Kennel and Petschek (1966) have now produced a large degree of qualitative understanding of the equilibrium structure of the electron belts as a result of the interaction of whistler turbulence with trapped particles. We are clearly on the verge of having a reasonably acceptable, quantitative and self-consistent theory of trapped flux distributions as the ideas of Etcheto *et al.* (1973) are combined with those of Lyons *et al.* (1972). We are clearly further away from the final quantitative theories of dynamical behavior, or if one prefers, transient behavior, of the belts after strong injection events have occurred. We are also still in the qualitative stages of understanding wave-particle interactions and their effects on particles beyond the plasmapause.

4.3. ION CYCLOTRON TURBULENCE

The turbulence theories such as that of Kennel and Petschek also have been invoked for waves in the ion cyclotron band. Cornwall (1965, 1966) has used such concepts to explain Pc 1 pearl micropulsations and other ULF emissions. However, the application of ion cyclotron turbulence theory drawing the most attention recently has been to the precipitation of ring-current protons (Cornwall *et al.*, 1970) and to the hypothesis of an energy source for sub-auroral red arcs (SAR arcs) (Cornwall *et al.*, 1971).

The precise nature of the source of ring current protons remains undetermined. It is known that clouds of 10 to 50 keV protons can appear quite suddenly in the midnight-to-dusk sector near synchronous orbit (DeForest and McIlwain, 1971), and their presence is felt in ground magnetometer variations, from which the D_{st} index is derived. These proton injection events are associated with the development phases of both geomagnetic storms and substorms. The 10 to 50 keV protons drift duskward, while electrons of comparable energies in the injection clouds drift dawnward (DeForest and McIlwain, 1971), under combined gradient and curvature drift forces. Just how the injection clouds are created at onset is still unknown.

However, it seems plausible that such clouds, at least in the case of substorm events, may well be the result of forces acting on plasma sheet particles at greater distance than 6.6 R_E. If they are produced by drift towards the Earth under the influence of a cross-tail electric field ($\mathbf{E} \times \mathbf{B}_0$ drift), and conserve their first two adiabatic invariants as they proceed into stronger geomagnetic field regions, one may expect that they will naturally develop a distorted pitch-angle distribution, with an increase of kinetic energy in their motion perpendicular to the magnetic field which is greater than the increase in parallel energy; that is, the velocity component $v_\perp(R)$ will increase faster than the component $v_\parallel(R)$ as R decreases. The same effect may be produced by time variations in the 'convection' electric field $\mathbf{E}$ on scales comparable to the drift time, or by inhomogeneous $\mathbf{E}$.

Thus, the fresh injected plasma can be expected to arrive at the plasmapause boundary region with a pitch-angle distribution peaked perpendicular to $\mathbf{B}_0$. Such an anisotropic pitch-angle distribution contains free energy. If the distortion exceeds certain critical values, one expects unstable wave generation to occur. In the case of ion cyclotron wave generation in a multicomponent plasma, the linear mode instability criteria have been discussed by Cornwall and Schulz (1971), especially with regard to the influence of non-negligible artificially injected densities of light ion tracers (such as lithium).

The ion cyclotron wave turbulence hypothesized by Cornwall *et al.* (1970), is assumed by them to occur in a thin layer just inside the plasmapause boundary. Russell and Thorne (1970) had already deduced from the ring current proton fluxes measured by Frank (1967, 1970), and simultaneous low-energy (plasmaspheric) hydrogen measurements by Taylor *et al.* (1968) on the OGO-3 satellite, that the inner edge of the ring current proton fluxes coincided with the steep plasmaspheric hydrogen gradient, i.e., the plasmapause. Cornwall *et al.* (1970) argued that convective forces alone could not account for this simultaneous ring current-plasmapause boundary during the main and recovery phases of geomagnetic storms. Thus, they suggested that another loss mechanism, based on ion cyclotron wave turbulence, could be effective at this mutual boundary.

Cornwall *et al.* (1970) argued that the high energy (Davis-Williamson) component of the ring current proton flux could possibly produce ion cyclotron wave instability outside of the plasmasphere, but that the lower energy component (< 50 keV) comprising the bulk of the ring current fluxes, would be stable there. On the other hand, this < 50 keV component could be unstable to ion cyclotron wave generation upon mixing with the cold plasmasphere distribution just inside the boundary, since the cold dense plasma there produces a much lower Alfvén speed, and lowers the proton energy required to resonate with the ion cyclotron waves. Thus, growth rates for ion cyclotron waves could be significant in this mixing region, provided the anisotropy factor for < 50 keV protons exceeds the critical value (Kennel and Petschek, 1966). In their paper, Cornwall *et al.* (1970) discuss the various transport mechanisms involved in their model, and estimate the pitch-angle diffusion coefficient and precipitation rates expected due to generation of ion cyclotron waves. They show that under the assumptions made concerning proton anisotropy factors, cold plasma densities, and wave convection losses out of the growth region, that one could expect strong diffusion in the turbulent layer. The proton lifetimes (< 50 keV) are estimated at

~ 1 h, which predicts complete proton loss prior to drift to the 1200 meridian, and ring current decay times $\lesssim 1$ day.

An alternative theory of plasma wave turbulence leading to ring current proton precipitation and ring current decay has been published by Coroniti et al. (1972). This same sort of instability has also been suggested by Nambu (1973). Since this theory is based on electrostatic wave generation, it will be discussed in the next section, along with the experimental measurements which appear to support it. It should be pointed out here that the two theories (ion cyclotron wave and electrostatic wave turbulence) are not necessarily incompatible nor mutually exclusive. As always, the choice between the two, if indeed either is correct, can only be made by comparison with adequate data.

A second important application of ion cyclotron wave turbulence has been made to the model for generating stable auroral red (SAR) arcs, in a paper by Cornwall et al. (1971). Cole (1965) had already established that ring current protons could provide sufficient energy to supply SAR arcs. His hypothesis was that ring current protons transfer energy to cool plasmaspheric electrons via Coulomb collisions in the region of mutual boundary between plasmasphere and ring current. This heat, according to the model, is then conducted down to low altitudes by electron heat conduction along magnetic field lines, thus providing the necessary energy input to excite the SAR arc radiance. This process requires excitation by electrons of energy less than a few eV, rather than by precipitation fluxes which commonly are much more energetic.

Cornwall et al. (1971) pointed out that recent experimental results, along with some theoretical considerations, indicates that such Coulomb collisions between ring current protons and plasmaspheric electrons is perhaps insufficient to produce heating rates required to drive SAR arcs. Thus, they suggested that the same process of ion cyclotron wave turbulence described in their 1970 paper could do sufficient heating of plasmaspheric electrons to allow subsequent heat conduction to produce the SAR arc. They thus invoke the same concept used by Cole (1965), but use a wave turbulent heating mechanism in place of Coulomb collisions.

Their basic idea is that ion cyclotron waves generated by the ring current proton pitch-angle anisotropy, while assumed to have $\mathbf{k}$ parallel to $\mathbf{B}_0$ in the generation region, will upon propagation away from this region into the inhomogeneous geomagnetic field and changing plasmaspheric density, suffer a turning of $\mathbf{k}$ away from colinearity with $\mathbf{B}_0$ (Kitamura and Jacobs, 1968). This propagation effect produces a component $k_\perp$ perpendicular to $\mathbf{B}_0$. Thus, the ion cyclotron wave becomes obliquely propagating, and this leads to development of an appreciable Landau damping (absorption) decrement which allows energy transfer to plasmaspheric electrons. Cornwall et al. (1971) evaluated the path-integrated Landau absorption under reasonable assumptions on the magnetospheric parameters, but pointed out that since they did not use a self-consistent quasi-linear saturation spectrum of ion cyclotron waves, their Landau absorption estimate is a somewhat uncertain approximation. They combined the Landau absorption and Coulomb collision heating rates to obtain a total electron heating rate. They then used a strong diffusion heat conduction law ($T^{3/2}$ law) and a classical collisional heat conduction law ($T^{7/2}$ law) to compute the electron heat flux incident on the ionosphere, and showed that sufficient energy is delivered by strong turbulent diffusion heat conduction to supply even rather intense SAR arcs. Thus, the theory of Cornwall *et al.* (1971) predicts the location (just inside the plasmapause boundary at ionospheric altitudes) of SAR arcs; it also indicates a latitudinal width of a few hundred km; no preferred local time dependence of SAR arc location; persistence times of 0.5 to 1 day during the recovery phase of storms; SAR arc intensities proportional to ring current proton anisotropy factors.

The theory just discussed appears to be the only one based on wave turbulence so far invoked to explain energetic input to SAR arcs. It requires generation of mean wave field amplitudes of 1 or 2 γ just inside the plasmapause and within some ± 20 deg of the equator. To date, no direct measurements have been reported of such ion cyclotron waves in the region $3 \lesssim L \lesssim 5$ during the recovery phase of a storm. Furthermore, low altitude (~ 1500 km) precipitation patterns of protons with energies greater than about 6 keV have recently been examined by Bernstein *et al.* (1974) and this evidence seems to lend an overwhelming statistical argument that ring current protons precipitate along L shells mapping out to near-equatorial regions *beyond* the plasmapause boundary. Thus, although one cannot unequivocally rule out the ion cyclotron turbulence mechanism of Cornwall *et al.* (1970; 1971), it is fair to say that this theory still awaits experimental confirmation strong enough to be convincing.

In fact, recent data from Explorer 45 instruments has shed additional light on this ion cyclotron wave theory. Williams and Lyons (1974) (see also the paper by Williams in this volume) have now examined ring current proton flux measurements from storm times just inside and just outside the local plasmapause boundary. The distribution outside was found to be sharply peaked near 90 deg pitch angle, and to have an empty loss cone, while the distribution inside was more rounded and had precipitating components in the loss cone. However, this distribution was one corresponding to *weak* pitch-angle diffusion, rather than the strong diffusion predicted by Cornwall *et al.* (1971). It is hoped that many more such observations in this interesting region will be added, so that we may gain better statistics to use in verifying the ion cyclotron wave theory of ring current decay and SAR arc energization.

To conclude this topic, it is noteworthy that Cladis (1973) has explored the effects of a magnetic field gradient on the resonance of ions with ion cyclotron waves. He showed that due to gradient effects, an ion resonant with a parallel-propagating ion cyclotron wave may gain large energy in even a single resonant interaction with a monochromatic wave. This energization would be reduced by a spectrum of waves. Cladis also showed how an ion would diffuse in velocity space due to gradient interaction coupled with cyclotron resonance. He used these results to provide a possible explanation for the observation of Shelley *et al.* (1972) of kilovolt O^+ ions at 800 km during a magnetic storm period, provided some other, initial acceleration mechanism exterior to his theory, can operate to push O^+ ions into the equatorial regions at high altitudes where ion cyclotron waves are assumed to operate.

4.4. CHORUS

VLF and ELF discrete emissions from the outer magnetosphere have been observed at ground level at latitudes corresponding to magnetospheric regions just beyond the plasmapause for more than a decade (Carpenter, 1973). The most recent study of such emissions, commonly called chorus, from an Earth-orbiting satellite (OGO 5) is that of Burton and Holzer (1974). For many years, it has been postulated that such emissions are whistler-mode waves of finite duration which are triggered either by lightning-generated whistlers, by powerful VLF transmissions from the ground, or are spontaneously generated by some wave-particle interaction involving cyclotron resonance. There is no disagreement among the various workers in this field that chorus emissions are generated predominantly in the vicinity of the geomagnetic equator.

Chorus emissions are important since the wave-particle interactions associated with them cause precipitation of electrons whose energy is appropriate to resonate with the VLF or ELF waves. They are furthermore interesting because they are generated by wave-particle interactions. In the following, we shall discuss some of the recent theoretical and experimental work in this still fertile area of research.

Discrete chorus emissions were indirectly associated with bursts of electron precipitation measured by VLF and electron detectors aboard the INJUN-3 satellite [Oliven and Gurnett, 1968] at low altitude ($\sim$1000 km). Enhanced fluxes of $\geqslant$40 keV electrons were observed along the geomagnetic field during VLF chorus bursts.

Another indirect measurement by Rosenberg et al. (1971) correlated short bursts of >30 keV X-rays seen by detectors aboard high-altitude balloons above Siple Station, Antarctica (L=4.1) and bursts of VLF emissions with $f \sim 2.5$ kHz recorded by the ground antenna at Siple Station. Presumably, the X-rays are bremsstrahlung from precipitating electrons with primary energy $\sim$60 keV and distributed over the range 30 to 100 keV. The VLF bursts were observed to *precede* the X-ray bursts in time by some tenths of a second ($\sim$0.3 to 0.4 s). By assuming that VLF emissions are stimulated in the vicinity of the equator, and a model of plasma distribution along the relevant flux tubes passing through that region and Siple Station, the delay times between the arrival of chorus bursts and the electrons causing bremsstrahlung were explained by Rosenberg et al. (1971).

Another low-altitude correlation of precipitating electrons and ELF chorus using search coil magnetometer (10–1000 Hz) data and simultaneous E>45 keV electron spectrometer data from the OGO-6 satellite (400 to 1100 km altitude range) has been presented by Holzer et al. (1974). In this study, no attempt was made to reconcile the individual delay times between ELF chorus occurrence and precipitation. The argument used by Holzer et al. (1974) is that precipitating electrons must necessarily follow field lines from the near equatorial region in which wave-particle interactions caused their pitch angles to diffuse into the loss cone, while the waves from that interaction region need not be guided down the same flux tube. In fact, unless ionization ducting occurs, the waves upon propagation away from the region of their unstable generation will develop a $k_\perp$, and hence an $E_\parallel$ component along $\mathbf{B}_0$. Thus,

Landau damping of these oblique whistlers can occur much as in the process of oblique ion cyclotron wave damping invoked by Cornwall *et al.* (1971) to explain SAR arcs.

Even if ducting is assumed, Holzer *et al.* (1974) point out that the duct will effectively terminate in the high ionosphere, above their satellite, so that waves may exit the duct at angles up to their final internal reflection angle, producing effectively an endfire waveguide antenna irradiating the region below the duct exit. Thus, electron precipitation and low-altitude wave patterns need not match in general, but rather have a systematic tendency which locates wave patterns equatorward of the flux tube on which they were generated, i.e., equatorward of the precipitation pattern.

The study of Burton and Holzer (1974) presents clear evidence that chorus originates in the outer magnetosphere (beyond the plasmapause) and near the geomagnetic equator. They show that, by measuring local wave normal directions of the chorus, that propagation may be either ducted or unducted. They further show that dayside and nightside chorus have distinctly different characteristics. The most important differences were: (1) dayside chorus was always comprised of rising tones (risers); (2) nightside chorus could be either rising or falling tones (never mixed); (3) dayside chorus was found at all geomagnetic latitudes; (4) nightside chorus was detected only within about 10 degrees of the geomagnetic equator; (5) dayside chorus was detected over a range of K_p from quiet to moderately disturbed (up to 4); (6) nightside chorus was only observed under magnetically active conditions and thus is most probably associated with substorm activity.

The difference in characteristics (1), (2), (3) and (4) indicates that the generation mechanism for dayside and nightside chorus elements may be different, and that propagation effects also may differ. The differences (5) and (6) clearly demonstrate that sources of dayside and nightside chorus can be different. For example, the region beyond the plasmapause on the nightside consists of the plasma sheet in the 'near-equatorial' region, and an extremely low density region in the lobes of the magnetotail containing polar cap field lines. The plasma sheet intensifies, thins, and moves earthward during the 'growth phase' of substorms. Thus, one can expect the region within some 10 deg of the geomagnetic equator to contain an earthward-convecting plasma sheet at these times. The appearance of the energetic plasma clouds at synchronous orbit (6.6 R_E) as seen by ATS 5 (DeForest and McIlwain, 1971) in association with substorm events indicates that copious fluxes of 10 to 20 keV electrons and protons on the order of twice this mean energy occur in this region. They have pitch-angle distributions favorable to the development of the whistler mode instability, and the energetic electrons (~ 10 to 20 keV) are expected to preferentially drift into the region 0000 to 0300 LT where the chorus measurements of Burton and Holzer (1974) were made. Under quiet conditions, this region of the outer magnetosphere contains little if any 10 to 20 keV electrons, so that no chorus would be expected, especially beyond the trapping boundary for >40 keV electrons.

In any event, Burton and Holzer (1974) have shown that at least in one observed chorus generation event, the region of maximum electron pitch-angle anisotropy coincided with the origin of wave generation as determined by a wave normal analysis, and present this as experimental evidence confirming the theory of Kennel and Petschek (1966). Another such bit of evidence is offered in Section 5.

Structural details of chorus emissions in the outer magnetosphere have been studied extensively by Burtis and Helliwell (1969). Another, perhaps somewhat controversial, study of very fine structure of banded chorus and other emissions using electric field data, has been presented by Coroniti et al. (1971). It is interesting to note that these latter authors discuss an observation from OGO 5 near 0600 LT, at a magnetic latitude near 8 deg, at $L \sim 6$, of a short duration sequence of falling-tone chorus preceded and followed by rising-tone chorus. No believable explanation of this phenomenon was given, but the observation does not necessarily refute the statements of Burton and Holzer (1974) that nightside chorus always consists, on the same pass of the satellite, of either rising or falling tones, but never both, since 0600 LT is on the boundary between night- and dayside.

4.5. Discussion

In concluding this section on wave turbulence and wave-particle interactions with trapped particle populations, it seems appropriate to point out the general state of development as of the time of this writing.

With respect to the understanding of the equilibrium trapped electron fluxes, at least in the plasmaspheric region, great advances have been made by the calculations of Lyons et al. (1972) and Lyons and Thorne (1973) in explaining both the electron slot formation, as well as the coupling between radial and pitch-angle diffusion. Even though their calculations employ an ad hoc, rather than a self-consistent, wave spectrum, the general features of the trapped electron and precipitation patterns in 'equilibrium' seem well-explained. The remaining detail work yet to be accomplished is to extend the $k_\perp = 0$ self-consistent wave spectrum and trapped electron distribution work of Etcheto et al. (1973) to oblique waves, and then to perform the self-consistent radial and pitch-angle diffusion calculation coupled to the self-consistent wave spectrum. This would probably complete the understanding of the inner zone (plasmaspheric region) electron profiles.

The wave-particle interactions and consequent effects on outer zone (exo-plasmaspheric) distributions, especially on the Earths dayside, are yet to be fully understood. It appears that nightside chorus (and also electrostatic wave) generation may well be confined to substorm-associated plasma sheet dynamics in the exo-plasmaspheric region (Tsurutani and Smith, 1974; see also, Scarf et al., 1973).

The theoretical understanding of proton populations in the radiation belts is still very incomplete. The ion cyclotron turbulence theories of Cornwall et al. (1970, 1971) are intellectually appealing, but experimental verification as yet remains elusive. On the other hand, Williams and Lyons (1974) present some limited indirect evidence for weak ion cyclotron turbulence just inside the plasmapause. The electrostatic wave turbulence theory of Coroniti et al. (1972), which would apply outside the plasmapause, is at least consistent with observations presented by Bernstein et al. (1974) of precipitation patterns falling *outside* the plasmapause. This is discussed more fully in the next section. However, only very weak experimental evidence for the presence of electrostatic waves outside the plasmapause is available (Anderson and Gurnett, 1973). Similarly, only weak evidence is available for possible ion cyclotron waves just within the plasmapause from the OGO-5 storm-time measurements of Scarf *et al.* (1972, Figure 5). The amplitudes of $\langle B_W \rangle$ from this latter observation would correspond to something less than 0.1 mγ rms.

5. Electrostatic Wave Turbulence in the Magnetosphere

In the previous section, electromagnetic wave turbulence in the whistler and ion cyclotron modes was discussed, especially in relationship to wave-particle interactions which can control particle diffusion and precipitation, thus influencing fluxes in the trapped radiation belts.

In addition to such electromagnetic modes, a plasma can also support a variety of electrostatic modes, such as electron plasma oscillations (Langmuir waves), ion acoustic oscillations, certain loss-cone or velocity-space driven modes, and electrostatic ion cyclotron waves. The electrostatic modes generally are not seen within the plasmasphere above the F_2 ionization peak because of heavy Landau damping (an electrostatic effect even for oblique electromagnetic waves), and because appropriate conditions are usually absent. The possible existence and consequences of an electrostatic ion cyclotron instability in the topside ionosphere has been discussed by Kindel and Kennel (1971). Thus, the entire outer volume of the plasmasphere is probably a region in which the electromagnetic mode turbulence plays the dominant role in wave-particle interactions. However, in the upper ionosphere, in the auroral L-shell regions, and polar cusp, this is not necessarily the case.

There are at least two important cases of observations of electrostatic waves in the outer nightside magnetosphere which have important bearing on particle dynamics through wave particle interactions. The first such observation is that reported by Kennel *et al.* (1970), and extended measurements were later presented by Fredricks and Scarf (1973). These waves are narrow-band electrostatic emissions primarily at approximately 1.5 times the local electron gyrofrequency often also at 3.5, and rarely at 5.5 f_{ce}. They are found to be substorm-associated and to occur on L-shells beyond the plasmapause boundary primarily on the nightside. Thus, these waves appear to be associated with plasma sheet dynamics and particle injection events during disturbed times.

The second important observation, although limited, is that of Anderson and Gurnett (1973) cited in the previous section. The third important observations, as yet unpublished (Scarf *et al.*, private communication), involves measurements of plasma waves in the distant plasma sheet ($\sim 35\ R_E$) in what appear to be an ELF whistler mode and an $(n + \frac{1}{2})\ f_{ce}$ electrostatic mode. These observations, and some theories which they have stimulated, will be discussed in the subsequent sub-sections.

5.1. OBSERVATIONS OF $(n+\frac{1}{2})f_{ce}$ ELECTROSTATIC MODES

Kennel *et al.* (1970) originally reported discovery of strong $(1-10 \text{ mV m}^{-1})$ electric field emissions at large L-values $(4 < L < 10)$ in the morning (0000–1200 LT) sector of the magnetosphere, from data acquired by electric dipole antennas on OGO 5. The most common emission was found to occur with a very narrow frequency bandwidth $(\Delta f/f_0 \lesssim 0.07$, Coroniti *et al.*, 1971) and to have little or no structure within this narrow spectrum. No magnetic component, to within the sensitivity of the magnetic loop sensors $(\lesssim 1 \text{ m}\gamma)$ was detected accompanying these electric field oscillations. The electrostatic nature of these $(n+\frac{1}{2})f_{ce}$ emissions has been confirmed by a more sensitive plasma wave detector experiment aboard IMP 6 (D. A. Gurnett, private communication), and the emissions have been ascribed to wave-particle interactions in the plasma sheet (Gurnett and Frank, 1973).

The spatial distributions of the $(n+\frac{1}{2})f_{ce}$ emissions as mapped by OGO 5 unfortunately depend largely on the volume of the magnetosphere mapped by an eccentrically orbiting spacecraft. In addition, as pointed out in the paper by Fredricks and Scarf (1973), the broad-band analog channel measurements employed in the original study of Kennel *et al.* (1970) suffered from a simultaneous power-sharing scheme with two other sensors on the payload. This power-sharing had the effect of setting minimum detectable thresholds for the $(n+\frac{1}{2})f_{ce}$ signals at $\gtrsim 1 \text{ mV m}^{-1}$ levels. Thus, weaker signals were possibly obscured. Secondly, the orbital coverage of OGO 5 involved in the data set used by Kennel *et al.*, simply excluded large areas of L vs local time space. These two effects influenced the L vs L.T. and the occurrence vs magnetic latitude distributions presented by Kennel *et al.* As an example, Kennel *et al.* reported occurrence of $(n+\frac{1}{2})f_{ce}$ electrostatic waves to be confined to approximately $-10° < \lambda_M < 10°$ for values in the range $4 \lesssim L \lesssim 10$.

However, a different data set, taken later in the life of OGO 5, was used by Fredricks and Scarf (1973) to re-examine the spatial distribution of $(n+\frac{1}{2})f_{ce}$ emissions. Kennel *et al.* had found such emissions in the near 0000 to near 1200 L.T. sector. The study of Fredricks and Scarf encompassed the near 1900 to near 0700 L.T. sector. There has been no orbital coverage of the remaining ~ 1200 to ~ 1900 afternoon sector, so one cannot make any statement about occurrence or lack thereof of the $(n+\frac{1}{2})f_{ce}$ waves in that region.

Furthermore, the study of Fredricks and Scarf benefited from use of data taken on the broad-band analog channel when the other two sensors sharing it had been turned off. Thus, weaker signals became detectable. A third advantage in data analysis enjoyed by Fredricks and Scarf was due to the realization, subsequent to the study of Kennel *et al.* (1970), that the $(n+\frac{1}{2})f_{ce}$ emissions were clearly related to magnetic activity, i.e., they appeared to be associated with substorm and storm activity. Thus, the data set used in Fredricks and Scarf was pre-selected to pay special attention to orbits during which the K_p index indicated activity. The relationship between K_p, ground magnetograms or other substorm indicators, and occurrence of the $(n+\frac{1}{2})f_{ce}$ emissions has never been determined in a systematic way. However, a clear qualitative correlation exists, as the results of the study of Fredricks and Scarf show.

The important feature evident from the results in Fredricks and Scarf (1973) is that for the range $4.5 < L \lesssim 14$, the $(n+\frac{1}{2}) f_{ce}$ emissions may occur over the entire range $-50° \lesssim \lambda_M \lesssim 40°$, although the strongest and most densely distributed emissions appear to occur on $5 \lesssim L \lesssim 7$, between about $-35° \lesssim \lambda_M \lesssim 5°$. This skew in the magnetic latitude distribution probably reflects the geometry of the plasma sheet, since the data were taken during the winter months in the northern hemisphere (October 1970 to April 1971), when the plasma sheet at larger L-values would be expected to tilt into the southern lobe of the magnetotail. However, another possible explanation based on orbital geometry has been given in Fredricks and Scarf (1973). Furthermore, one should probably view with some caution the use of the L parameter on the nightside in the larger L part of the plots in Fredricks and Scarf, since the nature of the field there is significantly non-dipolar.

5.2. Theories of the Origin of $(n+\frac{1}{2}) f_{ce}$ Emissions

The electrostatic nature of the $(n+\frac{1}{2}) f_{ce}$ emissions appeared to be well established by the OGO-5 plasma wave detector data, and has been confirmed by IMP-6 observations. A very early and crude attempt to explain this type of electrostatic mode was advanced by Fredricks (1971), who recognized that such modes may only be generated by an instability driven by free energy contained in an electron distribution having a positive slope, $\partial f/\partial v_\perp > 0$ over some range $v_\perp(\min) < v_\perp < v_\perp(\text{peak})$. In the absence of operating plasma instruments on board OGO 5 capable of measuring the relatively low energy (~ 1–20 keV) electron population, Fredricks (1971) examined the stability of a very unrealistic model 'ring' distribution function with the form factor $\delta(v_\parallel)$ times a shifted maxwellian $\exp[-(v_\perp - b)^2/c^2]$ with appropriate normalization constant. This distribution was found to be non-resonantly unstable for a wide range of wave numbers $\mathbf{k}(k_\perp, k_\parallel)$ in a wedge centered around the direction normal to the magnetic field of some sizable ($\gtrsim 20°$) width, and to peak for perpendicular wave numbers satisfying $k_\perp b/\omega_{ce} \sim 4.6$, for local ratio of plasma-to-electron gyrofrequency $\omega_R/\omega_{ce} \sim 3$.

The main criticism of this admittedly crude theory is obvious: the $\delta(v_\parallel)$ factor is equivalent to an infinite pitch-angle anisotropy, so that growth rates calculated from it would be excessively overestimated. However, the main point of the paper by Fredricks (1971) remains valid, namely that one *must* assume that there exists a warm or hot electron component producing a region of $\partial f/\partial v_\perp > 0$ in a plasma having $\omega_{pe}/\omega_{ce} > 1$, in order to explain the $(n+\frac{1}{2}) f_{ce}$ observations. Therefore, one could assume that such a plasma with distorted pitch-angle distribution must necessarily occur on the L-shells where the emissions were seen.

A vastly improved and much more realistic model of the plasma distribution function was examined in great detail by Young (1971) in his doctoral thesis. This modeling has been extended also in Young et al. (1971) and the application to the observations in the magnetosphere specifically has been made by Young et al. (1973). In this theory, the plasma is modeled by a variable mixture of cold plasma and a warm plasma with a maxwellian distribution parallel to $\mathbf{B}_0$, but with a distorted distribution in the form of a mirror (or loss cone) function (Guest and Dory, 1965) which has free energy due to a region where $\partial f/\partial v_\perp > 0$. Young *et al.* (1973) perform a complete analysis of this model to obtain criteria for and linear growth rates of the linear instability as functions of hot-to-cold plasma density, wave number, frequency, and ω_{pe}/ω_{ce}.

Because of the parallel velocity dispersion in the model distribution function used by Young *et al.* (1973), Landau resonance plays a significant part in the instabilities they find. Thus, the non-resonant instability limitation of the results of Fredricks (1971) has been removed by Young *et al.*. Young *et al.* find the frequencies, linear growth rates, and threshold criteria for a sequence of $(n+\frac{1}{2})\,f_{ce}$ instabilities.

Young *et al.* (1973) conclude on the basis of their analysis that the most likely cause of the $(n+\frac{1}{2})\,f_{ce}$ generation is instability driven by a non-monotonic perpendicular velocity distribution having a region of $\partial f/\partial v_\perp > 0$ for $v_\perp > 0$, rather than a simple temperature anisotropy ($T_\perp/T_\parallel >$ critical). The growth rates were found to vary with both $k_\parallel$ and $k_\perp$, as well as with the ratio n_c/n_H of cold-to-hot electrons, and on the steepness and location (in $v_\perp$) of the velocity gradient $\partial f/\partial v_\perp (>0)$.

Based on the incomplete data set presented by Kennel *et al.* (1970), Young *et al.* (1973) have drawn an unwarranted conclusion that the $(n+\frac{1}{2})\,f_{ce}$ emissions predominantly occur in the local morning time region of the plasma sheet. They did question the confinement to $\pm 10°$ of the magnetic equator of the emissions reported by Kennel *et al.*, and the answer to their question has been given by the high-latitude observations of Fredricks and Scarf (1973).

Young *et al.* (1973) also roughly estimate a pitch-angle diffusion coefficient for the $(n+\frac{1}{2})\,f_{ce}$ waves, and note that energy diffusion would also be important. However, in view of the much more complete diffusion analysis of Lyons (1973) discussed in the next section, this topic will not be pursued further here.

Another possible theoretical explanation of the generation of emissions at $(n+\frac{1}{2})f_{ce}$ has been given by Oya (1972). He applied a wave-wave interaction theory previously developed (Oya, 1971) to explain a series of 'diffuse resonances' (Oya, 1970) stimulated in the ionosphere by the sounding antenna pulses from Alouette 2.

According to the mechanism advocated by Oya (1972), there must exist an initially large temperature anisotropy $T_\perp/T_\parallel \gtrsim 5$. This initiates an electron cyclotron harmonic wave instability at a frequency close to nf_{ce} with $n \geqslant 2$. As this wave grows to nonlinear amplitude, a three-wave decay process produces two additional emissions at frequencies satisfying the three-wave decay selection rules, and the allowed cyclotron harmonic waves depend on the size of the parameter ω_{pe}/ω_{ce}. As this latter parameter increases, the number of propagating electron cyclotron harmonic waves increases, since the bands above that band $n_{max} < n < n_{max}+1$ containing the upper hybrid frequency $\omega_{UH} = (\omega_{pe}^2 + \omega_{ce}^2)^{1/2}$ are cut off, while those below propagate. Wave decay relations may be satisfied for cyclotron harmonic waves up to $\omega = \omega_{UH}$ down to branches below. The decays allowed are thus dependent on ω_{pe}/ω_{ce} as well as $T_\perp/T_\parallel$.

Oya (1973) has shown that details of some of the frequency-time ionograms of sequences of $(n+\frac{1}{2})\,f_{ce}$ emissions measured by the OGO-5 plasma wave detector do indeed fit his theoretical predictions, in that the observed frequencies are correlated quite well with the measured values of ω_{pe}/ω_{ce} deduced from cold plasma density measurements and magnetometer measurements made simultaneously.

There is so far no way to choose between the two different theories of generation of $(n+\frac{1}{2})f_{ce}$ emissions suggested by Young *et al.* (1973) and by Oya (1972). The satellite data are insufficient to discriminate between these two theories, and indeed the available data are consistent with either. Of the two, the nonlinear decay theory at least is capable of predicting simultaneous generation of several $(n+\frac{1}{2})f_{ce}$ bands, whereas linear theory has some difficulty in explaining such simultaneous emissions. It is probable that under suitable conditions, both generation mechanisms may occur.

The two theories of generation of $(n+\frac{1}{2})f_{ce}$ electrostatic turbulence begin with the assumption that the electron distribution function is distorted, i.e., has a pitch-angle anisotropy. The remaining question is, how does the plasma sheet electron distribution develop such a pitch-angle distribution? The occurrence of these emissions has been found to coincide with substorm activity, which in turn implies that it is probably associated with earthward convection of the plasma sheet.

The first attempt to quantitatively approach this problem was published by Men'shutina and Pudovkin (1971), who noted that plasma sheet particles under inward convection into stronger magnetic field regions due to imposition of a cross-tail electric field must develop a gradual sharpening of their pitch-angle distribution. They correctly pointed out that, depending upon the initial pitch-angle distribution at some injection point at large L-value, this pitch-angle anisotropy would develop with decreasing L-value until it exceeded some critical value for wave generation, and at this L-value, wave-particle pitch-angle scattering into the loss cone would begin, thus providing auroral particle precipitation. The unstable mode they assumed to be an electromagnetic one driven by a 'temperature' anisotropy $T_\perp/T_\parallel >$ critical, although the exact mode was never specified.

The important feature of the paper by Men'shutina and Pudovkin is not contained in their detailed computations of critical thresholds, pitch-angle expressions, nor in the precise unstable wave mode chosen; it rather is the recognition that the convection of plasma across the field lines earthward during substorm periods naturally leads to pitch-angle distributions with considerable free energy, and that this free energy increases with decreasing L-value, thus leading to the conclusion that plasma instabilities must begin at some L-value and lead to auroral zone precipitation.

Another attempt at modeling the convective processes in the plasma in the magnetotail has been presented by Tamao (1972). In Tamao's model, the cross-tail near dc electric field of long-time scale is imposed as a result of the polarization of the magnetopause during passage of a solar wind containing a significant southward interplanetary magnetic field component. He then considers the distribution functions of plasma sheet particles to satisfy a zero-order guiding center drift-kinetic transport equation in an inhomogeneous tail magnetic field and the potential due to the cross-tail electric field. In addition to the 'steady-state' or long-time scale solutions for penetration depths of protons, he also analyzes some of the problems of short-term instabilities of a local nature. However, explicit expressions for pitch-angle distributions on the various L-shells are not given, nor has Tamao considered the observations of the $(n+\frac{1}{2}) f_{ce}$ waves and their implications in his model. There are, however, a number of valuable qualitative ideas discussed by Tamao, and his paper should be read and understood by all workers in magnetospheric physics.

A very detailed attempt to calculate the pitch-angle distributions of plasma sheet particles under enhanced convection across dipole field lines in the region $1 < L \leqslant 10$ has been presented by Ashour-Abdalla and Cowley (1974). In their model, they conserve the first two adiabatic invariants $\mu = \text{const.} = m v_{\perp}^2 / 2B$ and $I = \frac{1}{2} \oint v_{\parallel} \, ds$, where $\oint$ implies integration between symmetric mirror points. They take as an initial condition an isotropic equatorial pitch-angle distribution injected at $L = L_0 = 10$. The treatment is nonrelativistic.

The pitch-angle distributions as a function of L computed by Ashour-Abdalla and Cowley exhibit the expected behavior, i.e., a continuous formation of a deepening cusp centered on $v_{\perp} = 0$ as L decreases. This produces a gradually increasing positive slope region $\partial f / \partial v_{\perp} > 0$ in a region $0 \leqslant v_{\perp} \leqslant v_{\max}$, potentially providing free energy for certain classes of velocity-space instabilities by which wave generation may occur. Thus, as is intuitively obvious, the adiabatic convection alone will result in pitch-angle distortions increasing with decreasing L-value.

However, the results of the calculations of Ashour-Abdalla and Cowley (1974), just as those of Men'shutina and Pudovkin (1971) mentioned previously, yield a dominantly $T_{\perp}/T_{\parallel}$ anisotropy, rather than a generalized pitch-angle anisotropy. This leads to their result that the threshold for the $(n+\frac{1}{2}) f_{ce}$ instability is higher than their computed anisotropy in $T_{\perp}/T_{\parallel}$ can overcome in the region $L > 3$, thus leading to the theoretical conclusion that the plasma would be stable, contrary to the copious experimental evidence for the waves presented by Kennel *et al.* (1970) and Fredricks and Scarf (1973). In addition, Ashour-Abdalla and Cowley examine their solutions for conditions suitable to generate electron whistlers and ion cyclotron waves, with generally negative results except under assumptions of rather large cross-tail electric fields.

It would be difficult to improve upon the wording found in the paper by Ashour-Abdalla and Cowley to summarize the state of affairs in the outer zones $(5 \lesssim L \lesssim 10)$ of the magnetosphere. As they so deftly stated, one must first understand the gross macroscopic flow dynamics of this region in order to explain the production of the large pitch-angle anisotropy in (at least) the electrons. The wave measurements appear to require such an anisotropy and the simple adiabatic flow model does not provide it. Secondly, a self-consistent calculation perhaps along the lines of Etcheto *et al.* (1973) but coupled to appropriate diffusion equations would be in order.

The next obvious step is to replace the assumption of a steady, constant cross-tail electric field by some better time-dependent model, and also to account for the high-beta character of the plasma sheet distribution being injected. In addition to the time-dependence, one also probably has to account for some extra L-dependence of this cross-tail field. The time-dependence can allow for a betatron process to occur, with consequent production of additional energy-dependent pitch-angle distribution.

The problem remains to be solved by the theoreticians.

Even without an understanding of the processes by which large enough pitch-angle anisotropies to drive the observed waves, considerable progress has been made in assessing the phenomenological consequences of the waves on particle diffusion in the outer zone. We turn to a discussion of this progress in the next section.

5.3. Consequences of the $(n + \frac{1}{2}) f_{ce}$ Electrostatic Turbulence

One of the important observations made by Young *et al.* (1973) was that of noting that the measured electron energy distributions in the plasma injection clouds reported by DeForest and McIlwain (1971) contained free energy in the peak at $v_\perp > 0$. This free energy was sufficient to satisfy the instability criteria computed by Young *et al.*

This lead Lyons (1974) to carry out an extensive examination of the pitch-angle diffusion of the electrons subjected to the wave fields as measured by the OGO-5 experimenters (Kennel *et al.*, 1970; Scarf *et al.*, 1973; Fredricks and Scarf, 1973). Following the same procedures of bounce-averaging the diffusion equation as those used for whistler interactions within the plasmasphere (Lyons, 1971; Lyons *et al.*, 1971, 1972; Lyons and Thorne, 1973; Lyons, 1973), and using an electrostatic wave spectrum in the definition of the diffusion coefficients, Lyons was able to show that pitch-angle diffusion and energy diffusion are brought about by these electrostatic waves.

In fact, Lyons' calculations show that the observed amplitudes of 1 to 10 mV m^{-1} of the typical $3f_{ce}/2$ turbulence is sufficient to put electrons of energies up to several keV on strong diffusion, while to put ~ 100 keV electrons on strong diffusion would require wave turbulence in the 100 mV m^{-1} range. This is consistent with the results shown in the paper by Scarf *et al.* (1973) during a substorm event, where fluxes of ~ 80 keV electrons were observed to isotropize in the presence of $3f_{ce}/2$ waves of ~ 100 mV m^{-1} amplitude.

As an additional support for the importance of the electrostatic $(n + \frac{1}{2}) f_{ce}$ turbulence on large L-shells, Lyons (1974) shows that the observations from a sounding rocket, passing through a post-breakup aurora at $L \sim 8$, reported by Whalen and McDiarmid (1973) are consistent with his theoretical bounce-averaged diffusion coefficients using a wave amplitude of $3f_{ce}/2$ turbulence of 1.7 mV m^{-1} at the equator. His calculated fluxes, at $\alpha = 42°$ for 1 to 20 keV electrons also matches the measured flux quite well. It thus would appear that some of the low-altitude, high-latitude electron precipitation is clearly due to the high-altitude electrostatic wave turbulence putting electrons of moderate energies (1 to tens of keV) on strong diffusion.

In conclusion, it is fair to say that the electrostatic wave turbulence at $(n + \frac{1}{2}) f_{ce}$ found in the outer magnetosphere beyond the plasmapause is of extreme importance to the understanding of the high-latitude precipitation fluxes in the auroral and perhaps even in the polar cap region during disturbed times. The (unpublished) occurrence of this electrostatic turbulence primarily during storm or substorm activity periods has been loosely established, and should be as thoroughly documented as possible in the future.

5.4. Drift Waves in the Outer Magnetosphere

In addition to the instabilities driven by velocity-space anisotropies of naturally-occurring particle-distribution functions in the outer zone (beyond the plasmapause), there may be others driven by different sources of free energy. One such other source may be provided by density or temperature gradients in the plasma. Such spatial gradients can produce drift wave instabilities. For an elementary tutorial treatment of drift wave generation, and the underlying plasma physics of instabilities driven by small inhomogeneities in a thermal collisionless plasma, the reader is referred to Rosenbluth (1965).

Application of drift wave theory to actual space observations in the outer magnetosphere has been made by LaQuey (1973) and by Baxter and LaQuey (1973). They used the ATS-5 plasma probe data of McIlwain and examined the energy-time spectrograms taken during plasma cloud injection events similar to those studied by DeForest and McIlwain (1971).

LaQuey (1973) noted that these injection clouds near local midnight possess electric, gradient and curvature drifts of the same order of magnitude, but also possess gradients in number density and 'temperature' significantly larger than local magnetic field gradient. In such regions, high time resolution analysis of the energy-time spectrograms for electrons in the range 50 eV to 50 keV revealed clearly defined oscillations of the local electron density, about an equilibrium value.

These oscillations were fit in some detail to an electrostatic drift wave theory, using a particular form of the linearized drift-kinetic transport equation. Spatial gradient lengths, and fluctuation amplitudes of the density oscillations were taken directly from the plasma probe data. Constancy of the first two adiabatic invariants, and an equilibrium distribution function isotropic in pitch-angles (or magnetic moments) was assumed. The observations were fitted to an electrostatic drift wave of amplitude $\langle E \rangle \sim 1.75$ mV m^{-1}.

A different set of data were also fit to the drift wave theory by Baxter and LaQuey (1973), and the amplitude of the electric field at $\sim 2 \times 10^{-3}$ Hz in the electrostatic drift wave required to fit the data was found to be $\langle E \rangle \sim 1$ mV m^{-1}, i.e., the same order of magnitude as that found previously by LaQuey (1973).

Baxter and LaQuey (1973) have noted that the frequency of these drift wave, or density oscillation, phenomena (~ 2 mHz) lies in the Pc5 micropulsation band, and suggest that such oscillations may be related in some way to such micropulsations seen by ground-based magnetometers; however, they do not suggest any specific mechanism. It is tempting to postulate that there indeed is a mechanism which could generate such ground Pc5.

One notes that the plasma cloud in which the drift wave oscillations are found is high-beta, i.e., $\beta \sim 1$. Thus, the density oscillations should carry along a frozen-in magnetic field line oscillation. Since the plasma loading on the field lines must extend to significantly high latitudes, the effect could be that of plucking a string on a musical instrument with a broad pick (that is, over a large segment of the string). Since 2 mHz is below the resonance of a field line at $L = 6.6$, the foot of the field line would try to vibrate at the same frequency. The only way the oscillations of the field line at the ground could be inhibited from following this high-altitude 'plucking' mechanism due to the drift wave would be through enhanced ionospheric conductivity of sufficient magnitude to anchor the field line in the topside ionosphere, thus providing an immovable boundary or end condition there. Since the substorm associated precipitation fluxes into the auroral zone at $L \sim 6.6$ indeed lead to enhanced height-integrated conductivity, it is not clear that field-line vibrations generated at ATS-5 altitude by drift-wave plucking of the field lines could be transmitted through the ionosphere.

As a final comment, we note that the frequency range (mHz) and amplitudes (few mV m^{-1}) of the drift waves inferred by LaQuey (1973) and Baxter and LaQuey (1973) correspond to the type of fluctuation often assumed *ad hoc* to drive radial diffusion (Fälthammar, 1965; Cornwall, 1972; Birmingham, 1969).

5.5. ELECTROSTATIC LOSS CONE MODES

Coroniti *et al.* (1972) have attempted to explain some of the decay properties of the enhanced ring current proton injection by postulating the existence of an electrostatic ion loss cone instability due to sharp pitch-angle anisotropy of protons outside the plasmapause. They noted that if the initial injection spectrum of ring current protons in the outer zone is sufficiently peaked near 90° and has an almost empty loss-cone, there is a quasi-electrostatic instability which may occur [for a rather complete discussion of this instability, see Rosenbluth (1965)].

An analysis based on a 'mirror' distribution (Dory *et al.*, 1965) with an anisotropy $A \sim 2$ was performed by Coroniti *et al.*, which included finite $k_\parallel$ and a mixture of hot ring current and cold plasmasphere particles. It was shown that increasing the cold-to-hot population ratio quenched the electrostatic loss cone instability, so that one would expect it to be important (if it occurs) primarily outside the plasmapause. Typical growth rates for this type of instability scale to the ion plasma frequency $\omega_{\mathrm{pi}} \gg \omega_{\mathrm{ci}}$, and thus grow faster than electromagnetic ion cyclotron waves (which may also be unstable beyond the plasmapause according to Coroniti, *et al*). A phenomenological diffusion coefficient of $v_\perp$-scattering was presented by Coroniti *et al.*, and evaluated for typical instability parameters and local plasma parameters in the exoplasmaspheric ring current. In some sense, this diffusion coefficient describes pitch-angle scattering due to the electrostatic mode. Scattering times ~ 100 s were estimated, which are an order of magnitude shorter than minimum lifetimes (~ 1500 s) for ring current protons at $L = 4$.

For this reason, Coroniti *et al.*, suggest that such loss cone turbulence outside the plasmapause would act rapidly to reduce the anisotropy of the ring current protons, and compute the necessary convection speed which would be required to create anisotropy at a high enough rate to offset its destruction by turbulence and resulting wave-particle interactions. As Coroniti *et al.* point out, the reduction of proton pitch-angle anisotropy to values which correspond to marginal stability of the electrostatic loss cone waves still leaves sufficient anisotropy to produce some electromagnetic ion cyclotron turbulence. They show, in fact, that the combination of the loss cone and ion cyclotron turbulence could be expected to produce velocity-space diffusion over almost the entire $v_{\parallel} - v_{\perp}$ plane. A similar independent analysis by Nambu (1973) has led to the same general conclusions as those of Coroniti *et al.*

The experimental results of low altitude precipitation patterns of ring current protons for energies $\gtrsim 6$ keV have been examined recently by Bernstein *et al.* (1974). These results are quite indicative of a powerful precipitative turbulence source at high altitudes beyond the plasmapause. Bernstein *et al.* found that the precipitation fluxes equatorward of the trapping boundary were isotropic and bounded on the equatorward border by a region of anisotropic (loss cone) precipitation. This transition from anisotropic to isotropic precipitation appears to be a consistent feature of the data, and to nearly coincide with the plasmapause boundary. They conclude that these results are most likely consistent with the existence of electrostatic turbulence such as that postulated by Coroniti *et al.* (1972) in the exo-plasmaspheric region. The low energy patterns of Bernstein *et al.*, appear to be very close to those of > 100 keV protons reported by Söraas (1971). In addition, Søraas (private communication, oral presentation at the Summer Advanced Study Institute, *Earth's Particles and Fields*, Sheffield, England, 1973) has computed the resonant energy profile $E_R = B^2/4\pi N$ as a function of L from magnetometer and cold ion density measurements in the region outside the plasmapause for times during which he had observations of low altitude precipitation data for protons > 100 keV. He demonstrated that the precipitation patterns for > 100 keV protons were entirely inconsistent with the resonant energy profiles, and concluded that such precipitation was inconsistent with ion cyclotron wave turbulence in the exo-plasmaspheric region. Recent results published by Mizera (1974) are consistent with the previous observations cited above.

Insofar as wave measurements are concerned, few data are available in the exo-plasmaspheric region in the frequency range of interest. The frequencies predicted by the model calculation of Coroniti *et al.* (1972) are $0.1 \leqslant \omega/\omega_{pi} \lesssim 0.5$. For typical ring current number densities, one has $f_{pi} \sim 200—600$ Hz. Thus, extremes on wave frequency are $20 \lesssim f \lesssim 300$ Hz. This band of electrostatic waves has been poorly covered by the OGO-5 plasma wave detector, and so far little has been reported from the IMP-6 plasma wave detector in this region. However, there is at least one reported measurement of wave activity in the frequency range 20 to 500 Hz from the Explorer 45 ($S^3 - A$) satellite just outside the plasmapause by Anderson and Gurnett (1973), who pointed out a possible relationship to the theoretical calculation of Coroniti *et al.* (1972). However, such an isolated and relatively low altitude ($L \sim 3.44$) measurement as that of Explorer 45 cannot be considered a verification of the existence of ion loss cone mode turbulence beyond the plasmapause. Such verification awaits future wave data from that region during injection events.

5.6. Discussion

In summary, it seems fair to state that the wave-particle interactions which dominate the particle-precipitation patterns during quiet times in the exo-plasmaspheric regions are still a matter of speculation. On the other hand, there is a growing body of experimental data pointing to the probable dominance of *electrostatic* wave turbulence in that region during injection events associated with geomagnetic storms and substorms.

The precipitation patterns of ring current protons ($\gtrsim 6$ keV) examined by Bernstein *et al.* (1974), by Søraas (1971) and the electron precipitation at low altitudes (see, for example, Whalen and McDiarmid, 1973; Lyons, 1974; and Hultqvist *et al.*, 1973) led the general conclusion (Bernstein *et al.*, 1974) that there must be an electrostatic turbulence in the outer zone during enhanced convection that acts on a broad energy band of both electrons and protons simultaneously. The combination of the relatively high frequency $(n+\frac{1}{2})\,f_{ce}$ turbulence (measured) and the ion loss cone ($\gtrsim f_{pi}$) turbulence (postulated) fits this requirement.

6. Current-Driven Instabilities and Anomalous Resistivity

The great bulk of the magnetospheric volume, above the E-layer of the ionosphere, for all practical purposes can be considered to be filled with a collisionless plasma. The term *collisionless* refers to the relative infrequency of the classical two-or-more body impacts. However, the plasma's charged particles are in constant Coulomb interaction, and thus will have at the very least a Coulomb-collision conductivity as calculated by Spitzer and Härm (1953) and Spitzer (1956). The mean-free path for this type of conductivity (or its reciprocal, resistivity) is commonly so large as to be ineffective in supporting meaningful potential differences between even points separated by many Earth radii.

An example of a phenomenon which requires certainly resistive and perhaps viscous dissipation mechanisms to operate over scale lengths much shorter than even a Spitzer-Härm mean-free path is the Earth's bow shock. The classical magneto-hydrodynamicist, in treating a problem such as the bow shock, simply *assumes* a thin layer in which the dissipation he requires is present, irrespective of the mean-free path requirement. However, the plasma physicist asks for descriptions of the *microscopic* processes producing such dissipative layers with scale lengths short compared to classically defined free paths.

Therefore, the plasma physicist is led to the conclusion that such layers must contain *plasma turbulence*, and that the true or effective mean-free paths or collision frequencies are produced by wave-particle interactions in the turbulent layer. The source of the plasma turbulence is usually presumed to be a plasma wave spectrum which results from the quasi-linear or nonlinear saturation limit of a plasma instability driven by free energy in the initial distribution function.

Unfortunately, even as of the date of this writing, no adequate theory of collisionless anomalous resistivity is available. A number of phenomenological models of anomalous resistivity are available, but none of these can be considered a self-consistent calculation of that important macroscopic transport coefficient called 'resistivity'. The reason for the theoretical difficulty lies in the inability to calculate slef-consistent wave electric field spectra in the saturated (nonlinear) limit.

I shall not bore the reader with any further extensive discussion of the now rather large literature on theoretical calculations aimed at explaining anomalous resistivity in collisionless plasmas. Let it suffice to say that all the relevant theoretical models for anomalous resistance due to plasma wave turbulence are reviewed in the book by Davidson (1972), and therefore the interested reader is referred to that source for the excruciating details from the viewpoint of the plasma physicist. I shall henceforth only cite papers having direct relevance to magnetospheric phenomena in which anomalous resistance is thought to play a crucial role in the remainder of this review.

In addition to the bow shock, other magnetospheric phenomena in which anomalous resistivity is now believed to play an important part (at least by *some* part of the scientific community) are: (1) reconnection; (2) field-aligned currents in the auroral and polar cusp (cleft) regions; (3) auroral electrojet processes; and (4) in regions of the auroral zone containing significant fluxes of low energy (0.01 to several keV) precipitating electrons. I shall discuss each of these phenomenological regions in arbitrary order.

6.1. NIGHTSIDE FIELD-ALIGNED CURRENTS AND RESISTIVITY

A number of observations, all from the OGO-5 satellite, have become available recently, indicating the occurrence of field-aligned currents and enhanced electrostatic wave turbulence at relatively high altitudes on high but closed L-shells on the nightside of the Earth. The first such observation to be reported in detail is that of Scarf *et al.* (1973). As stated in that paper, approximately a dozen other events of the same type had been identified in the OGO-5 data, with similar characteristics, but detailed analysis was incomplete.

The appearance of magnetometer signatures interpretable as field-aligned currents on the nightside had been published previous to the paper by Scarf *et al.* (1973). For example, Haerendel *et al.* (1971) analyzed HEOS-1 data and concluded that field-aligned currents were flowing in the 'horns' of the plasma sheet during a substorm-associated event. Their estimates of total current in the sheets are crude, but indicate that such field-aligned currents as deduced from the magnetometer excursions at HEOS-1 were of the same order of magnitude as those required by the ionospheric currents producing the observed bays at Great Whale. They concluded that their results were consistent with the postulate that plasma sheet sources in the outer magnetosphere indeed deliver the required currents to drive the auroral electrojets during substorm events.

Aubry *et al.* (1972) and Fairfield (1973) have examined magnetometer data taken on the nightside at high *invariant* latitudes and have concluded that field-aligned current systems are commonly present on the borders of the plasma sheet subsequent to substorm onsets, and during the 'thinning' phenomenon before onset. Their evidence points to single, double and multiple current sheets. Estimated currents at the altitudes where measurements were made would map down to low altitudes to yield currents on the order of those deduced by Zmuda et al. (1970), Armstrong and Zmuda (1970) and others. Apparently, all the measurements reported by Aubry et al. (except one) and by Fairfield corresponded to the outer border of the plasma sheet. Thus, it is perhaps speculative and controversial to assume that these currents were on closed field lines, although such an assumption is *probably* true.

Coleman and McPherron (1970) have reported ATS-1 measurements at $L=6.6$ which indicate inner-border currents along local field lines, and one example from Aubry et al. appears to perhaps be at a low enough L-value to be associated with the inner boundary. Therefore, the field-aligned currents examined in some detail by Scarf et al. (1973) are the first clear-cut observations at the inner boundary on the nightside.

The data presented by Scarf et al. (1973) were taken during a period of significant ground magnetometer activity, and fortuitously the trajectory of OGO 5 during the interesting times was crossing auroral L-shells at relatively low altitude ($\sim 2.5\ R_E$ at $L \sim 5.5$ and ~ 0330 LT). For some time prior to the boundary event discussed by Scarf et al., the cold proton density had been $\sim 1\ \mathrm{cm}^{-3}$ and gradually increasing, indicating that the spacecraft was beyond the plasmapause boundary but still in a region containing some cold plasma. The outer zone fluxes of >50 keV electrons was fairly steady at $\sim 10^6\ (\mathrm{cm}^2\ \mathrm{s\ sr})^{-1}$, and the low energy plasma probe indicated background. At about 1958 UT (on September 7, 1968), almost coincident with the peaking at $\sim 400\ \gamma$ of the AE index, a sudden flux of warm (0.4 to 12 keV) electrons of density $\lesssim 10\ \mathrm{cm}^{-3}$ appeared in the plasma probe, the trapped >50 keV electron flux dropped sharply by about two orders of magnitude, the cold proton density went below instrumental threshold, and the plasma wave detector measured broadband electrostatic fluctuations from at least 560 Hz up to 10 kHz with amplitudes over 100 mV m^{-1}. This event was over at roughly 2004 UT (i.e., its duration was some 5 min), at which time the AE index was decaying and the trapped >50 keV flux returned to its expected profile, warm electrons disappeared, cold protons reappeared, and electrostatic waves ceased.

Within this event, the magnetometer recorded the presence of a field-aligned current, but no magnitude of this current was deduced. The electrostatic wave turbulence was found to be loosely correlated with the field-aligned currents, but considerable wave turbulence occurred in this 5-min event without obvious correlation with the current system.

In any event, the electrostatic turbulence levels during this event of 10 to 100 mV m^{-1}, and covering a broad band of frequencies, would be expected to produce certain subsidiary effects, such as anomalous resistivity along the field lines with feet in the auroral ionosphere, and also considerable pitch-angle and energy diffusion of trapped particles. The 'drop-out' of two orders of magnitude in the >50 keV trapped fluxes along this L-shell could be due to strong diffusion simply emptying the shell of the lower energy components of > 50 keV flux, thus creating a transient slot.

In addition, the anomalous resistance along the L-shell involved could support significant potential difference between the equatorial plasma sheet and the ionosophere, providing means to support electron acceleration into the auroral zone at low L-values, or to feed auroral arc systems (Coroniti and Kennel, 1972). Since the measurement was made at $L = 5.5$ and at 2.5 R_E ,it is speculative to assume wave turbulence due to instabilities driven by the current would extend down the pertinent field lines to low altitudes. However, Koons et al. (1972), using data gathered by the OV3-3 satellite between 2000 and 4500 km, found occasional precipitation boundaries near $L \sim 5$ to 6 on the nightside, and these correlated with intense (up to 60 mV m)$^{-1}$ VLF electrostatic turbulence. Thus, the OGO-5 and OV3-3 measurements may indicate that such field-aligned current regions and electrostatic turbulence associated with them can produce anomalous resistivity over several Earth radii on high-latitude L-shells.

The origin of the event observed by OGO 5 is a matter of speculation, as are many peculiar events observed by a single spacecraft. However, the explanation advanced in Scarf et al. (1973) is that the sudden appearance of $\sim$ keV electrons indicates an injection event similar to those described by DeForest and McIlwain (1971) had occurred on the $L \sim 5.5$ shell. In the absence of a better hypothesis, one tends to accept that explanation. The ramifications of such events, as well as the gathering of data from similar occurrences, remain as topics for future studies.

6.2. FIELD-ALIGNED CURRENTS AND RESISTIVITY IN THE POLAR CUSP

The experimental discovery of the region now called the dayside polar cusp (or cleft) apparently was made simultaneously by plasma probe experimenters on ISIS 1 (Heikkila and Winningham, 1971) and INJUN 5 (Frank and Ackerson, 1971a), both in approximately polar orbits in the altitude range near 1000 km. High-altitude observations of this polar cusp were made by the IMP-5 satellite and reported by Frank (1971a, b). These pioneering investigations did much to clarify the morphology of the dayside magnetosphere and auroral region, both for quiet and disturbed times.

With the insight thus furnished, large bodies of dayside low altitude auroral zone data could be reinterpreted. For example, by use of high-latitude, low-energy electron precipitation data and correlative interplanetary field measurements, Burch (1972) was able to deduce the velocities at which the boundaries of the polar cusp moved as a function of the orientation of the north-south component of interplanetary field. The general results of Burch's study show that the equatorward boundary of the polar cusp moves towards the equator with equatorial speeds ~ 5 km s^{-1} during southward excursions of the interplanetary field. (This order of velocity is consistent with convection speeds deduced for plasma sheet motions in the tail. There is some connection between the two phenomena, of course, but its exact nature is not at all understood at this time.) This motion is believed to be the result of the sub-solar point reconnection phenomenon and the accompanying field-line erosion process described by Aubry *et al.* (1970), which 'eats away' the outermost closed field lines, thus reducing the radial distance of the magnetopause.

An extreme example of such equatorward excursions of the polar cusp occurred during the very large magnetic storm on November 1, 1968. During this storm, the trajectory of the OGO-5 spacecraft penetrated the cusp region and provided a very wide variety of data on particles and fields at the cusp boundary crossings and in the cusp plasma itself. The initial overview of some of the gross features of this unusual cusp encounter at magnetic latitudes as low as ∼44° has been given by Russell *et al.* (1971). The data from this single storm-time encounter are so rich in detail that they have led to a number of specialized papers in recent times.

For example, Kivelson *et al.* (1973) investigated in detail the dependence of the location and state of motion of the polar cusp in response to the north-south and south-north excursions of the interplanetary field as measured by Explorer 33 in the upstream solar wind. Detailed overviews of the plasma wave activity in the cusp and near its boundaries have appeared in two papers by Scarf *et al.* (1972, 1974). Specialized correlations of plasma wave data with details of the particle populations and magnetometer signatures indicating local field-aligned currents or cusp boundaries have been presented by Fredricks *et al.* (1973) and Fredricks and Russell (1973). It is basically the material in the latter two papers which will be discussed below. However, much useful and even essential background material is found in the other papers cited.

Fredricks *et al.* (1973) examined the magnetometer data on its fast time scale, and identified some 17 magnetic signatures interpretable as field-aligned currents. The most intense current sheet, as identified by the magnitudes of their associated magnetic field perturbations ($\Delta B_\perp \sim 50$ to $90\ \gamma$) all exhibited the simultaneous presence of electrostatic wave turbulence. (This phenomenon is similar to that described in the previous discussion of the nightside field-aligned current.) At other encounters when field-aligned currents whose magnetic field perturbations were smaller (~ 10–$20\ \gamma$) or at times when the plasma wave detector was in one of its higher-frequency channels (14.5, 30 or 70 kHz), little or no wave turbulence was noted.

Fredricks *et al.* concluded from these data that these currents and their associated electrostatic wave turbulence were on open cusp field lines, and that there was sufficient warm cusp plasma to support the current densities inferred from the data. They also concluded that the electrostatic turbulence spanned a frequency range from below 0.56 kHz to several kHz. They chose one particular current sheet crossing for detailed presentation and interpretation. In this sample, Fredricks *et al.* showed that the electrostatic turbulence was almost wholly confined to the current layer as defined by the $\Delta B_\perp$ profile. Using a double sheet current model, and interpreting thickness in terms of spacecraft velocity, they showed that an estimate of $J_\parallel$ of some 1.7×10^{-5} A m^{-2} was consistent with the data, and that this current would be capable of generating either ion-acoustic or Buneman instabilities in the cusp plasma.

Fredricks *et al.* then used the phenomenological formula for the effective wave-particle collision frequency (Sagdeev and Galeev, 1966), $v^* \sim \omega(\langle E^2 \rangle / 8\pi N\kappa T)$ in the usual definition of resistivity $\varrho^* = mv^* \, Ne^{-2}$ to compute a value of anomalous resistivity of ~ 7.9 ohm-m. They showed that if this order of magnitude of resistivity were to exist due to local wave turbulence from satellite altitude ($3.2 \, R_E$) down to $\sim 1 \, R_E$ in the polar ionosphere, then a drop of some 2 kV could be supported. The $E_\parallel$ implied could then accelerate electrons into the polar cusp auroral region.

Other implications of the anomalous resistance deduced from these measurements may be hypothesized. For example, the current layers appear in the data to be close to the equatorward boundary of the cusp, and are present at times when the magnetosheath field at the subpolar point of the magnetopause is expected to be southward. This implies that merging is occurring near the nose of the magnetosphere. It is tempting to associate the observed field-aligned current just outside the cusp boundary with a voltage source in the merging region. If this hypothesis were true, then the cross-polar-cap potential at ionospheric levels would not necessarily be the same as that between the subsolar merging zone and the last closed field line on the nightside, as is usually assumed. Using the extrapolated anomalous resistivity along the *entire* field line from the merging point to the ionosphere would imply a potential drop of some 10 kV. Furthermore, the inferred surface current density of 4.3×10^{-2} A m^{-1}, if it were driven by an equatorial merging source of extent $\sim 45°$ in longitude around the nose, would furnish $\sim 10^6$ A to the dayside ionosphere.

The paper by Fredricks and Russell (1973) contains a description of an entirely different system of waves found during the November 1, 1968 cusp encounter by OGO 5. High-time-resolution magnetometer data were examined, and found to contain, during certain short ($\sim$min) time periods, almost coherent trains of nearly transverse waves at measured frequencies in the range $0.67 \lesssim \omega/\omega_{ci} \lesssim 0.87$. The amplitudes of these trains varied from less than $1 \, \gamma$ up to as much as $10 \, \gamma$, in typical background fields of $340 \, \gamma$ to $550 \, \gamma$. By examining the > 50 keV trapped electron fluxes, it was inferred that these ion cyclotron band waves were on closed field lines, but that some warm magnetosheath plasma was also present (no cold plasmaspheric plasma appeared to be present). A detailed examination of external conditions revealed that at these times of observation, the cusp's equatorward boundary was moving poleward in response to a northward-pointing interplanetary field.

The warm plasma electrons of ~ 400 eV appeared to be mixed with a proton component ~ 2 keV, inferred from field inflations. Thus, the plasma beta was low to moderate, $0.07 \lesssim \beta \lesssim 0.15$ during the observations, with densities ~ 17 cm^{-3}. No field-aligned currents were obvious in the data during times the ion cyclotron waves were active. Fredricks and Russell concluded that the poleward excursion of the equatorward boundary involved a process by which some slippage, due to moderate β, between polar cusp plasma and closed field lines occurred, leading to a sort of diffusion region in a mixing layer containing mild diamagnetic effects.

The polarization of the observed waves indicated that they were obliquely propagating, rather than parallel propagating. It was not clear from the available data whether the waves were locally generated, or had propagated to their site of measurement from a source either farther up or farther down the field lines. It would seem unlikely that they were generated by a source far up the field line, since they were so close to the local gyrofrequency. That is, since the field strength decreases up the field line, the observed wave frequency would rapidly exceed the local ion gyrofrequency, a situation not favorable to propagation of cyclotron waves. With a single spacecraft observation using an ordinary triaxial magnetometer it is not possible to determine polarizations to within a 180° ambiguity; that is, the direction of **k** for the waves observed by Fredricks and Russell (1973) would have a projection *up* the field line for RH waves and *down* the field line for left-hand waves. The mode in question was thought to be LH and propagating *down* the field line by Fredricks and Russell. This geometrical interpretation, if correct, would imply that the waves were probably being observed near their source.

D'Angelo (1973), based on the early papers on OGO-5 fields data by Russell *et al.* (1971) and Scarf *et al.* (1972), postulated theoretically that the ULF magnetic field fluctuations reported by those authors could be due to a Kelvin-Helmholtz instability. He also considered the drift wave instability which one might initially invoke, since significant density gradients exist in the polar cusp plasma. However, D'Angelo points out that, using typical plasma and field parameters for the polar cusp, the drift wave theory does not explain observed frequencies nor the $B_\perp/B_\parallel$ ratio and frequency spectrum produced by this instability.

Recently, D'Angelo and Bahnsen (private communication, 1973) have examined search coil and plasma probe data from the HEOS-2 satellite during polar cusp encounters. In a paper not yet published at this writing (D'Angelo and Bahnsen, private communication) they show that fluctuations in the ULF range measured by the search coil magnetometer and velocity shears deduced from the plasma probe data indicate support for the Kelvin-Helmholtz instability theory of D'Angelo (1973). In the same paper, they postulate a clever Venturi tube model of plasma flow in the polar cusp.

In summary, it seems reasonable to say that the only experimental data available to this date in the open literature on the high-altitude polar cusp bearing on wave-particle interactions is that taken by OGO 5. Unfortunately, these data were gathered during one of the most intense geomagnetic storms of the solar cycle. For this reason, it is questionable whether similar phenomena involving plasma wave generation and subsequent wave-particle interactions would be present in quiet-time polar cusps. So far, no other spacecraft carrying a full complement of plasma wave detectors and other particles and fields sensors has been put into an orbit appropriate to study the polar cusp, with the exception of HEOS 2. However, the very low data rate available on that spacecraft may hinder the performance of an adequate experimental study of wave-particle interactions in the cusp.

7. Other Wave-Particle Interactions

Since, as stated in the introduction, this review is restricted to wave-particle inter-actions and their geophysical significance in the outer magnetosphere, which we have arbitrarily defined as that region above several thousand kilometers, we have essentially excluded a whole important low-altitude region. As is well-known, such phenomena as auroral electrojets, auroral arcs, F-layer irregularities and scintillation regions, equatorial electrojets, artificial gas clouds in the ionosphere, etc., all involve wave-particle interactions in perhaps basic and important ways. It is not my intent to imply that the absence of a review of these phenomena here signifies any belief on my part of their unimportance; quite to the contrary, to adequately review these important low-altitude features, covering experimental and theoretical aspects, would almost double the size of this review. Thus, it is left to another to round out the picture by reviewing the areas omitted here.

Also, the reader may note that another area of wave-particle interactions for which a vast body of literature exists, has been omitted here. Precisely, I refer to the phenomena of high-altitude triggered emissions (VLF), micropulsations (ULF) and artificial modification of high-altitude ambient populations by particle releases or strong wave injection. Also neglected was the area of ionospheric modification by powerful ground-based antennas, a phenomenon involving wave-wave interactions, linear mode conversion and wave-particle interactions. These omissions were made for a variety of reasons, but one reason common to all was that of final size of the manuscript.

Another area not reviewed is the large literature on the fielding-line merging, or reconnection problem. My feeling after having read the many theoretical papers on this problem is that it remains basically unsolved. Every available treatment can be criticized on the basis of over-idealization, and lack of inclusion of sufficient microscopic physics in the formulation. Thus, I felt that to make a negative, and non-constructive, review of this body of work would accomplish no useful purpose. Continuing attacks on the reconnection problem may yield new and exciting results in the next few years; at that time, a comprehensive review would be more appropriate.

8. Summary

In summarizing the importance of wave-particle interactions to the geophysical understanding of phenomena in the outer magnetosphere, one can point to the successes and deficiencies of theoretical explanations based on such interactions. A fair assessment of this sort is attempted below.

8.1. BOW SHOCK

Some features of bow shock structures seem readily and acceptably explained by wave-particle interactions. For example, electron heating at least qualitatively is understood to result from irreversible phase-space scattering by self-consistent waves generated by current-driven microscopic instabilities in the shock's macroscopic structure. However, to data, proton thermalization is not understood well, probably because plasma wave data are insufficient to provide the right clues.

8.2. MAGNETOPAUSE

Almost nothing seems really understood in magnetopause layers, including their formation macroscopically. Few plasma waves have been identified in the magneto-pause layer experimentally.

8.3. INNER ELECTRON BELT STRUCTURE

The occurrence of plasmaspheric ELF hiss throughout the plasmasphere seems ex-perimentally well established. Coupled pitch-angle and radial diffusion theory with an *assumed* wave turbulence distribution has been carried out for electrons, and if obliquely propagating ELF hiss wave turbulence is assumed, the general features of the inner zone radial profiles, including the slot location and dependence on electron energy, have been theoretically reproduced. The next step would be to attempt to solve the coupled pitch-angle, radial diffusion, and wave kinetic equations to obtain particle distribution function and wave turbulence spectrum self-con-sistently. It appears to me that this refinement, while it may give more accurate detailed fits to equilibrium flux profiles within the plasmasphere, would add very little to our almost complete qualitative understanding of the equilibrium trapped electron flux profiles. Thus, inner zone electron fluxes, at least during quiet times, appear adequately explained on the basis of wave-particle interactions leading to diffusion.

Transient changes in these fluxes are less well understood, but the basic wave-particle interactions seem to apply, so that the remaining problems are likely to be associated with dynamics of the particle-injection characteristics, which may well differ from one disturbance to another.

8.4. OUTER ZONE BELT AND RING CURRENT STRUCTURE

Beyond the plasmapause, flux profiles depend to a greater degree on magnetic ac-tivity, which in turn implies enhanced convection due to solar wind driving mech-anisms. The outer zone during enhanced convection apparently develops electron and proton pitch-angle distributions capable of producing both electrostatic and electromagnetic waves in the ELF and VLF range. How convection can produce the distorted pitch-angle distributions required theoretically to explain the observed electrostatic $(n+\frac{1}{2}) f_{ce}$ wave turbulence is unknown. However, the character of velocity-space diffusion of electrons due to these waves seems understood fairly well. Similarly, proton distributions must be distorted sufficiently to produce strong diffusion in this region if low-altitude precipitation patterns are to be understood.

Ring current protons decay by wave-particle interactions which put them on strong diffusion over a wide range of energies. The observation of isotropic precipitation at low altitudes outside the plasmapause indicates some wave-particle interaction other than that due to ion cyclotron waves must operate in the ring current. It may be an electrostatic loss cone mode. However, experimental evidence is lacking so no theory presently available can claim experimental verification. Some evidence now appears to be available to support the ion cyclotron theory of ring current proton precipitation in a thin zone just inside the plasmapause during the recovery phase of magnetic storms. This still fails to explain the bulk of the ring current precipitation beyond the plasmapause.

Evidence for electrostatic drift waves at mHz frequencies in ring current injection clouds at synchronous orbit is strong. However, such waves are not expected to produce pitch-angle or energy diffusion of significant magnitude. On the other hand, they can aid radial diffusion. Their theoretical importance as yet is not established.

8.5. CURRENT-DRIVEN INSTABILITIES AND ANOMALOUS RESISTIVITY

High-altitude field-aligned currents have been observed on both nightside ($L \sim 5$ to 6) plasma sheet and on dayside polar cusp boundaries. These currents appear to generate intense electrostatic wave turbulence. This turbulence, if present on a significant fraction of the flux tube carrying the current, implies anomalous resistance due to wave-particle collisions. Crude estimates of anomalous resistivity based on observed wave levels indicate that the $J_{\parallel}/\sigma$ parallel electric fields imply potential drops of kilovolts over dimensions of an Earth radius. Full implications of such parallel fields to magnetospheric dynamics are still not understood, but could be important to the explanation of auroral ionospheric phenomena such as arcs and electrojets. Whether there are implications to the dynamics of substorms is unknown.

Acknowledgements

I am grateful to many colleagues with whom I have discussed over a long period facets of the magnetospheric topics reviewed here. Some of these are F. L. Scarf, C. F. Kennel, F. V. Coroniti, D. J. Williams, W. Bernstein, W. Heikkila, L. R. Lyons, M. Schulz, L. A. Frank, S. DeForest, E. W. Greenstadt, V. Formisano, A. Eviatar, M. Ashour-Abdalla, T. S.-T. Young, D. A. Gurnett and C. T. Russell. It is also fitting to point out that although his name does not appear in my references, the late Neil Brice was among the first of the space scientists to emphasize wave-particle interactions and their importance to geophysical phenomena.

References

Alfvén, H.: 1968, *J. Geophys. Res.* **73**, 4379.

Anderson, R. R. and Gurnett, D. A.: 1973, *J. Geophys. Res.* **78**, 4786.

Armstrong, J. C. and Zmuda, A. J.: 1970, *J. Geophys. Res.* **75**, 7122.

Ashour-Abdalla, M. and Cowley, S. W. H.: 1974, in B. M. McCormac (ed.), *Magnetospheric Physics*, D. Reidel Publishing Company, Dordrecht, p. 241.

Aubry, M. P., Russell, C. T., and Kivelson, M. G.: 1970, *J. Geophys. Res.* **75**, 7018.

Aubry, M. P., Kivelson, M. G., McPherron, R. L., Russell, C. T., and Colburn, D. S.: 1972, *J. Geophys. Res.* **77**, 5487.

Axford, W. I. and Hines, C. O.: 1961, *Can. J. Phys.* **39**, 1433.

Axford, W. I.: 1964, *Planetary Space Sci.* **12**, 45.

Baxter, D. and LaQuey, R.: 1973, *J. Geophys. Res.* **78**, 6798.

Bernstein, W., Fredricks, R. W., and Scarf, F. L.: 1964, *J. Geophys. Res.* **69**, 1201.

Bernstein, W., Hultqvist, B., and Berg, H.: 1974, *Planetary Space Sci.*, in press.

Birmingham, T. J.: 1969, *J. Geophys. Res.* **74**, 2169.

Burch, J. L.: 1972, *J. Geophys. Res.* **77**, 6696.

Burtis, W. J. and Helliwell, R. A.: 1969, *J. Geophys. Res.* **74**, 3002.

Burton, R. K. and Holzer, R. E.: 1974, *J. Geophys. Res.* **79**, 1014.

Carpenter, D. L.: 1963, *J. Geophys. Res.* **68**, 1675.

Cole, K. D.: 1965, *J. Geophys. Res.* **70**, 1689.

Coleman, P. J., Jr. and McPherron, R. L.: 1970, in B. M. McCormac (ed.), *Particles and Fields in the Magnetosphere*, D. Reidel Publishing Company, Dordrecht, Holland, p. 171.

Cornwall, J. M.: 1964, *J. Geophys. Res.* **69**, 1251.

Cornwall, J. M.: 1965, *J. Geophys. Res.* **70**, 61.

Cornwall, J. M.: 1966, *J. Geophys. Res.* **71**, 2185.

Cornwall, J. M.: 1968, *Radio Sci.* **3**, 740.

Cornwall, J. M.: 1972, *J. Geophys. Res.* **77**, 1756.

Cornwall, J. M., Coroniti, F. V., and Thorne, R. M.: 1970, *J. Geophys. Res.* **75**, 4699.

Cornwall, J. M., Coroniti, F. V., and Thorne, R. M.: 1971, *J. Geophys. Res.* **76**, 4428.

Cornwall, J. M. and Schulz, M.: 1971, *J. Geophys. Res.* **76**, 7791.

Coroniti, F. V., Fredricks, R. W., Kennel, C. F., and Scarf, F. L.: 1971, *J. Geophys. Res.* **76**, 2366.

Coroniti, F. V. and Kennel, C. F.: 1972, *J. Geophys. Res.* **77**, 2835.

Coroniti, F. V., Fredricks, R. W., and White, R.: 1972, *J. Geophys. Res.* **77**, 6243.

Cuperman, S., Salu, Y., Bernstein, W., and Williams, D. J.: 1973, *J. Geophys. Res.* **78**, 7372.

D'Angelo, N.: 1973, *J. Geophys. Res.* **78**, 1206.

Davidson, R. C.: 1972, *Methods of Non-Linear Plasma Theory*, Academic Press, New York.

DeForest, S. and McIlwain, C.: 1971, *J. Geophys. Res.* **76**, 3587.

Dory, R. A., Guest, G. E., and Harris, E. G.: 1965, *Phys. Rev. Letters* **14**, 131.

Dungey, J. W.: 1958, *Cosmic Electrodynamics*, Univ. Press, Cambridge.

Dungey, J. W.: 1961, *Phys. Rev. Letters* **6**, 47.

Dungey, J. W.: 1963, *Planetary Space Sci.* **11**, 591.

Dungey, J. W.: 1965, in *Plasma Physics*, published by International Atomic Energy Agency, Vienna, pp. 349–372.

Eather, R. H. and Carovillano, R. L.: 1971, *Cosmic Electrodyn.* **2**, 105.

Etcheto, J., Gendrin, R., Solomon, J., and Roux, A.: 1973, *J. Geophys. Res.* **78**, 8150.

Eviatar, A. and Wolf, R. A.: 1968, *J. Geophys. Res.* **73**, 5561.

Fairfield, D. H.: 1969, *J. Geophys. Res.* **74**, 3541.

Fairfield, D. H.: 1973, *J. Geophys. Res.* **78**, 1553.

Fälthammar, C. G.: 1965, *J. Geophys. Res.* **70**, 2503.

Frank, L. A.: 1967, *J. Geophys. Res.* **72**, 3753.

Frank, L. A.: 1970, *J. Geophys. Res.* **75**, 1263.

Frank, L. A. and Ackerson, K. L.: 1971d, *J. Geophys. Res.* **76**, 3612.

Frank, L. A.: 1971a, *J. Geophys. Res.* **76**, 2512.

Frank, L. A.: 1971b, *J. Geophys. Res.* **76**, 5602.

Fredricks, R. W.: 1971, *J. Geophys. Res.* **76**, 5344.

Fredricks, R. W., Scarf, F. L., and Frank, L. A.: 1971, *J. Geophys. Res.* **76**, 6691.

Fredricks, R. W., Scarf, F. L., and Green, I. M.: 1972, *J. Geophys. Res.* **77**, 1300.

Fredricks, R. W. and Scarf, F. L.: 1973, *J. Geophys. Res.* **78**, 310.

Fredricks, R. W., Scarf, F. L., and Russell, C. T.: 1973, *J. Geophys. Res.* **78**, 2133.

Fredricks, R. W. and Russell, C. T.: 1973, *J. Geophys. Res.* **78**, 2917.

Gendrin, R.: 1972a, in E. R. Dyer, (ed.), *Solar-Terrestrial Physics/1970*, D. Reidel Publishing Co., Dordrecht-Holland, p. 236.

Gendrin, R.: 1972b, in B. M. McCormac (ed.), *Earth's Magnetospheric Processes*, D. Reidel Publ. Co., Dordrecht-Holland, p. 311.

Greenstadt, E. W., Inouye, G. T., Green, I. M., and Judge, D. L.: 1967, *J. Geophys. Res.* **72**, 3855.

Greenstadt, E. W. and Fredricks, R. W.: 1974, in B. M. McCormac (ed.), *Magnetospheric Physics*, D. Reidel Publishing Company, Dordrecht, Holland, p. 355.

Gurnett, D. A. and Frank, L. A.: 1973, *EOS Trans. Am. Geophys. Union* (abstract SM-111) **54**, 433.

Haerendel, G.: 1970, in B. M. McCormac (ed.), *Particles and Fields in the Magnetosphere*, D. Reidel Publ. Co., Dordrecht-Holland, p. 416.

Haerendel, G., Hedgecock, P. C., and Akasofu, S.-I.: 1971, *J. Geophys. Res.* **76**, 2382.

Heikkila, W. J. and Winningham, J. D.: 1971, *J. Geophys. Res.* **76**, 883.

Holzer, R. E., Farley, T. A., Burton, R. K., and Chapman, M. C.: 1974, *J. Geophys. Res.* **79**, 1007.

Hultqvist, B., Borg, H., Christophersen, P., Riedler, W., and Bernstein, W.: NOAA ERL SEL Report (unpublished).

Kadomtsev, B. B.: 1965, *Plasma Turbulence*, Academic Press, London and New York.

Kellogg, P. J.: 1959, *Nature* **183**, 1295.

Kennel, C. F. and Engelmann, F.: 1966, *Phys. Fluids* **9**, 2377.

Kennel, C. F. and Petschek, H.: 1966, *J. Geophys. Res.* **71**, 1.

Kennel, C. F., Scarf, F. L., Fredricks, R. W., McGehee, J. H., and Coroniti, F. V.: 1970, *J. Geophys. Res.* **75**, 6136.

Kindel, J. M. and Kennel, C. F.: 1971, *J. Geophys. Res.* **76**, 3055.

Kivelson, M. G., Russell, C. T., Neugebauer, M., Scarf, F. L., and Fredricks, R. W.: 1973, *J. Geophys. Res.* **78**, 3761.

Kitamura, T. and Jacobs, J. A.: 1968, *Planetary Space Sci.* **16**, 863.

Koons, H. C., Vampola, A. L., and McPherson, D. A.: 1972, *J. Geophys. Res.* **77**, 1771.

LaQuey, R. E.: 1973, *Phys. Fluids* **16**, 550.

Lerche, I.: 1966, *J. Geophys. Res.* **71**, 2365.

Lerche, I.: 1967, *J. Geophys. Res.* **72**, 5295.

Levy, R. H., Petscheck, H. E., and Siscoe, G. L.: 1964, *AIAA J.* **2**, 2065.

Liemohn, H. B.: 1967, *J. Geophys. Res.* **72**, 39.

Lyons, L. R.: 1971, in K. Folkestad (ed.), *Magnetospheric-Ionospheric Interactions*, University Press, Oslo, p. 47.

Lyons, L. R.: 1973, *J. Geophys. Res.* **78**, 6793.

Lyons, L. R.: 1974, *J. Geophys. Res.* **79**, 575.

Lyons, L. R., Thorne, R. M., and Kennel, C. F.: 1971, *J. Plasma Phys.* **6**, 589.

Lyons, L. R., Thorne, R. M., and Kennell, C. F.: 1972, *J. Geophys. Res.* **77**, 3455.

Lyons, L. R. and Thorne, R. M.: 1973, *J. Geophys. Res.* **78**, 2142.

Men'shutina, I. N. and Pudovkin, M. I.: 1971, *Cosmic Res.* **73**, 2355.

Nambu, M.: 1973, *J. Geophys. Res.* **78**, 1203.

Oliven, M. N. and Gurnett, D. A.: 1968, *J. Geophys. Res.* **73**, 2355.

Ossakow, S. L.: 1968, *J. Geophys. Res.* **73**, 6366.

Oya, H.: 1970, *J. Geophys. Res.* **75**, 4279.

Oya, H.: 1971, *Phys. Fluids* **14**, 2487.

Oya, H.: 1972, *J. Geophys. Res.* **77**, 3483.

Parker, E. N.: 1967, *J. Geophys. Res.* **72**, 2315.

Parker, E. N.: 1967, *J. Geophys. Res.* **73**, 4365.

Piddington, J. H.: 1960, *J. Geophys. Res.* **65**, 93.

Roberts, C. S.: 1969, *Rev. Geophys. Space Phys.* **7**, 305.

Rosenberg, T. J., Helliwell, R. A., and Katsufrakis, J. P.: 1971, *J. Geophys. Res.* **76**, 8445.

Rosenbluth, M. N.: 1965, in *Plasma Physics*, published by the International Atomic Energy Agency, Vienna, p. 501.

Russell, C. T. and Thorne, R. M.: 1970, *Cosmic Electrodyn.* **1**, 67.

Russell, C. T., Childers, D. D., and Coleman, P. J.: 1971, *J. Geophys. Res.* **76**, 845.

Russell, C. T., Chappell, C. R., Montgomery, M. D., Neugebauer, M., and Scarf, F. L.: 1971, *J. Geophys. Res.* **76**, 6743.

Sagdeev, R. Z. and Galeev, A. A.: 1966, Tech. Rpt. IC/66/64, Int'l. Centre for Theor. Phys., Trieste, Italy.

Scarf, F. L., Fredricks, R. W., Frank, L. A., Russell, C. T., Coleman, P. J., Jr., and Neugebauer, M.: 1970a, *J. Geophys. Res.* **75**, 7316.

Scarf, F. L., Fredricks, R. W., Green, I. M., and Neugebauer, M.: 1970b, *J. Geophys. Res.* **75**, 3735.

Scarf, F. L., Fredricks, R. W., Frank, L. A., and Neugebauer, M.: 1971, *J. Geophys. Res.* **76**, 5162.

Scarf, F. L., Fredricks, R. W., Green, I. M., and Russell, C. T.: 1972, *J. Geophys. Res.* **77**, 2274.

Scarf, F. L., Fredricks, R. W., Russell, C. T., Kivelson, M., Neugebauer, M., and Chappell, C. R.: 1973, *J. Geophys. Res.* **78**, 2150.

Scarf, F. L., Fredricks, R. W., Kennel, C. F., and Coroniti, F. V.: 1973, *J. Geophys. Res.* **78**, 3119.

Scarf, F. L., Fredricks, R. W., Neugebauer, M., and Russell, C. T.: 1974, *J. Geophys. Res.* **79**, 511.

Sen, A. K.: 1965, *Planetary Space Sci.* **13**, 131.

Shelly, E. G., Johnson, R. G., and Sharp, R. D.: 1972, *J. Geophys. Res.* **77**, 6104.

Southwood, D. J.: 1968, *Planetary Space Sci.* **16**, 587.

Spitzer, L., Jr. and Harm, R.: 1953, *Phys. Rev.* **89**, 977.
Spitzer, L., Jr.: 1956, 'Physics of Fully ionized Gases', in R. Marshack (ed.), *Interscience Tracts on Physics and Astronomy*, No. 3, Interscience Publishers, Inc., New York.
Taylor, H. A., Brinton, H. C., and Pharo, M. W., III: 1968, *J. Geophys. Res.* **73**, 961.
Thorne, R. M., Smith, E. J., Burton, R. K., and Holzer, R. E.: 1973, *J. Geophys. Res.* **78**, 1581.
Tsuda, T.: 1967, *J. Geophys. Res.* **72**, 6013.
Tsurutani, B. T. and Smith, E. J.: 1974, *J. Geophys. Res.* **79**, 118.
West, H. I., Jr., Buck, R. M., and Walton, J. R.: 1973, *J. Geophys. Res.* **78**, 1064.
Whalen, B. A. and McDiarmid, I. B.: 1973, *J. Geophys. Res.* **78**, 1608.
Williams, D. J. and Lyons, L. R.: 1974, *EOS Trans. Am. Geophys. Union* **55**, 394.
Williams, D. J. and Lyons, L. R.: 1974, *J. Geophys. Res.*, in press.
Young, T. S.-T.: 1971, Ph.D. Thesis, Report LPP-1, Dept. of Aeronautics and Astronautics, Laboratory for Plasma Physics and Space Sciences, Massachusetts Institute of Technology, Cambridge, Mass., 1971.

DRIFT OF PARTICLES AND WAVE-PARTICLE INTERACTIONS

JACQUES SOLOMON

Groupe de Recherches Ionosphériques, Centre National d'Etudes des Télécommunications, 3 Avenue de la République, 92131 – Issy-les-Moulineaux, France

1. Introduction

To improve our comprehension of the Earth's magnetosphere it is necessary to couple the dynamic of the particles and the plasma instabilities (see for example Ashour-Abdalla and Cowley, 1974). We will study here the effect on the ion-cyclotron interaction of the sudden appearance of hot protons on the nightside of the outer zone magnetosphere during substorms (Figure 1).

First, we will consider the dispersion relation of the electromagnetic ion-cyclotron waves (propagating parallel to the static magnetic field) in a finite β plasma and the respective effects of the cold and hot particle densities. Second, we will examine the effects of the gradient-curvature magnetic drift on the distribution function and on the ion-cyclotron interaction. Last, we will mention, as a conclusion, the radial drift due to the electric field.

2. Ion-Cyclotron Instability

The ion-cyclotron instability in a finite β plasma has been studied in numerous recent works (Cornwall and Schulz, 1971; Perraut and Roux, 1974), but essentially with the numerical computations (also notice the recent work of Cuperman and Landau on the electron-cyclotron instability). In this paragraph we try to give a brief analytical answer to the effects of a finite β plasma and of the cold plasma on the ion-cyclotron instability.

We write the total distribution function as the sum of a cold and hot plasma distribution function (e: electrons, i: protons):

$$F(v_\perp, v_\parallel) = n_0 \left[\frac{2\pi}{v_\perp} \delta(v_\perp) \, \delta(v_\parallel) + f_{i, e}(v_\perp, v_\parallel) \right] \tag{1}$$

and we choose for the hot protons a loss cone distribution function

$$f_i(v_\perp, v_\parallel) = \frac{n_1/n_0}{\pi^{3/2} v_0^3} \frac{1}{\Gamma(m+1)} \left(\frac{v_\perp}{v_0} \right)^{2m} \exp - \left(\frac{v_\perp^2 + v_\parallel^2}{v_0^2} \right), \tag{2}$$

where n_0 is the cold plasma density and n_1 is the hot plasma density.

The anisotropy (as defined in Kennel and Petschek's work, 1966, for example) is in this case

$$\mathscr{A} = m.$$

The particle velocities $v_\perp$, $v_\parallel$, v, and the pitch angle α are defined as usual with re-

V. Formisano (ed.), The Magnetospheres of the Earth and Jupiter, 153–159. All Rights Reserved
Copyright © 1975 by D. Reidel Publishing Company, Dordrecht-Holland

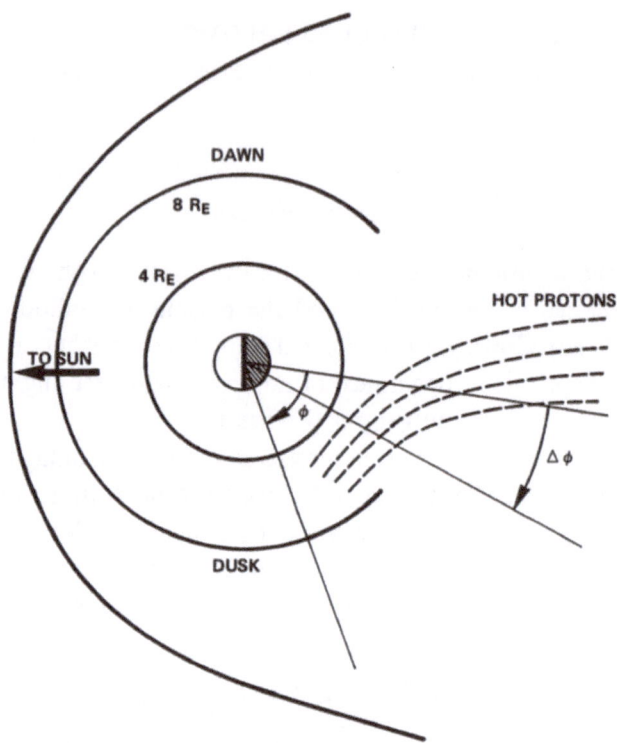

Fig. 1.   Schematic view of the trajectories, in the equatorial plane, of the hot protons which contribute to
the ring current. Definition of ϕ and $\Delta\phi$ as used in Section 3.

ference to the static magnetic field $\mathbf{B}_0$. The hot electrons, whose distribution function
is not specified, do not take an important part in the ion-cyclotron interaction as
the relativistic electrons only can interact, however they ensure the electric neutrality.

Then we can write the dispersion relation with the Fried and Conte function in
the following form:

$$D(\omega, k) = \frac{c^2 k^2}{\omega^2} - \frac{\omega_p^2}{\omega_{ci}(\omega_{ci} - \omega)} \times$$

$$\times \left\{ 1 + \frac{n_1}{n_0 + n_1} \frac{\omega_{ci}(\omega_{ci} - \omega)}{\omega^2} \left[1 + \frac{V_R}{U_{\parallel}} Z\left(\frac{V_R}{U_{\parallel}}\right) \right] \left(m - \frac{\omega}{\omega_{ci} - \omega} \right) \right\}$$

(3)

in which,

ω_p: ion plasma frequency calculated with the total density $n_0 + n_1$
c: light velocity
k: wave number ($\mathbf{k} \parallel \mathbf{B}_0$)
ω_{ci}: ion-cyclotron frequency
$f = \omega/2\pi$: wave frequency
$V_R = (\omega - \omega_{ci})/k$: resonant velocity
$U_{\parallel}^2 = v_0^2$.

The Fried and Conte function (Fried and Conte, 1961) is defined as:

$$Z(z) = \frac{1}{\sqrt{\pi}} \int\limits_{-\infty}^{+\infty} \frac{\exp - v^2}{v - z} \, dv.$$

From this dispersion relation, we can deduce the imaginary part of the wave frequency ω_i $(\omega = \omega_r + i\omega_i)$. We obtain schematically the curve vs x represented on Figure 2, where $x = \omega/\omega_{ci}$ is the reduced frequency. We get an amplification of the waves when ω_i is positive.

Now, due to the term

$$\left(m - \frac{\omega}{\omega_{ci} - \omega} \right)$$

which appear both in (3) and in the formula for ω_i (see below, formula (5)), there is a cut-off frequency x_c. For not too large anisotropies, i.e. $m \lesssim 1$, the cut-off frequency $x_c \lesssim 0.5$. Then it is possible to show that $V_R/U_{\|} \gtrsim 1$ in the amplification region $(0 \leqslant x \leqslant x_c)$.

This is sufficient to simplify $D(\omega, k)$ by using the well-known asymptotic development of the Fried and Conte function:

$$Z(V_R/U_{\|}) \simeq -(U_{\|}/V_R).$$

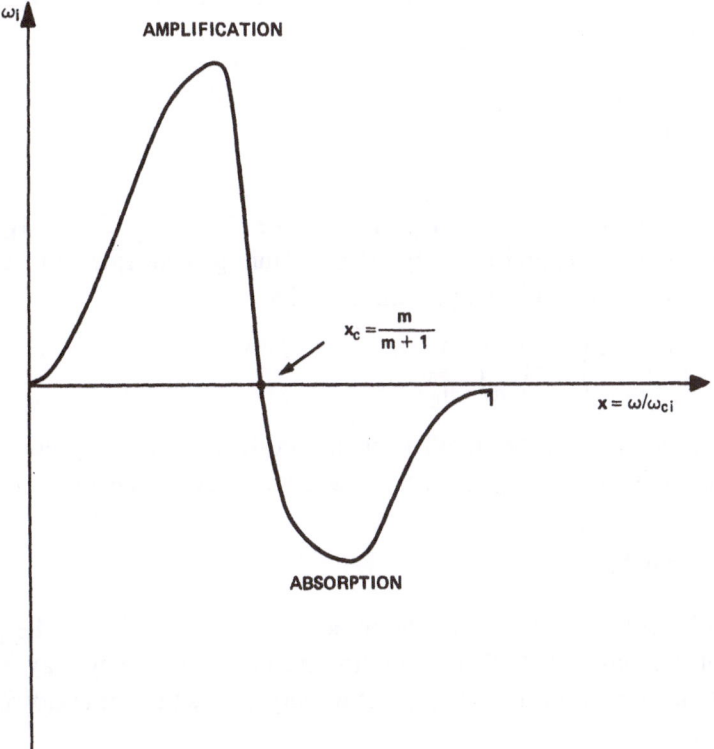

Fig. 2. Schematic plot of the growth rate of the ion-cyclotron wave vs the reduced frequency x.

Thus we will obtain all the usual expressions in the same form as when n_1/n_0 is much less than one, provided that we use the total density $(n_0 + n_1)$ instead of n_0 in all the quantities where it appears.

For instance we will get the classical expression for the resonant velocity (Kennel and Petschek, 1966):

$$V_R^2 \simeq V_A^2 \frac{(1-x)^3}{x^2},\tag{4}$$

where V_A is the ion Alfvén velocity computed with $n_0 + n_1$.

One important thing is to note: this can also be demonstrated for any distribution function, and thus it will be useful when the distribution function will be deformed by the gradient-curvature magnetic field.

The other point to be discussed is the importance of the cold plasma. We can write the growth rate of the ion-cyclotron interactions for the same loss cone distribution function, in the following form:

$$\omega_i(x)/\omega_{ci} \simeq \frac{\sqrt{\pi}}{2} \frac{n_1}{n_0 + n_1} \frac{V_R^2}{v_0^2} \frac{v_0}{V_A} \times$$

$$\times \left[\exp - \left(\frac{V_R^2}{v_0^2} \right) \right] \left(m - \frac{x}{1-x} \right)\tag{5}$$

and the maximum of this growth rate is obtained for a frequency x_0 satisfying the following equation:

$$\frac{V_R^2}{v_0^2} \simeq 1 + \frac{x_0}{2(1-x_0)} \frac{1}{m - \dfrac{x_0}{1-x_0}}.\tag{6}$$

But this maximum depends on V_A, through the resonant velocity V_R, thus on the cold plasma density n_0 and the 'largest' maximum growth rate will be obtained for an optimum value of n_0 given approximately by

$$\left(\frac{n_0 + n_1}{n_1} \right)_{opt} \approx \left(\frac{2}{m} \right)^2 \frac{1}{\beta_{\parallel}},\tag{7}$$

where $\beta_{\parallel}$ is the ratio of the parallel kinetic pressure to the magnetic pressure. The corresponding frequency x_{max} depends practically only on the anisotropy:

$$x_{max} \simeq \frac{m}{m+2}.\tag{8}$$

Formula (7), with n_0, emphasizes the importance of the bulge of the plasmapause. As the hot protons drift both in azimuth and in L-value, they may encounter this bulge and the increase of the cold plasma density allows the ion-cyclotron interaction to occur.

In the preceding discussion we have considered a given distribution function for the hot protons, but in fact these protons are drifting and we must include this drift in our computations.

3. Effects of the Gradient-Curvature Magnetic Drift of the Hot Protons on the Distribution Function

We assume that, at time $t=0$, the distribution function of the hot protons on a given L-shell is of the following form:

$$f(v_\perp, v_\parallel) \exp -\frac{\phi^2}{(\Delta\phi)^2}, \tag{9}$$

where ϕ is the azimuth angle for the drifting protons from a given initial position and $\Delta\phi$ a given initial width of injection (Figure 1).

We assume a dipole magnetic field. The bounce average drift velocity V_D is then (Hamlin *et al.*, 1961):

$$V_D = A r_0 v^2 (0.7 + 0.3 \sin\alpha) \quad \text{with} \quad r_0 = L R_E \quad \text{and} \quad A = \frac{3}{2} \frac{m_i}{q} \frac{1}{B_0 r_0^2} \tag{10}$$

L: McIlwain parameter
R_E: Earth radius
m_i: proton mass
q: electric proton charge
B_0: equatorial magnetic field at $r = r_0$.

Then at any positive time t and azimuthal angle ϕ, we obtain the new distribution function in the equatorial plane:

$$f(v_\perp, v_\parallel, \phi, t) = f(v_\perp, v_\parallel)$$
$$\times \exp -\frac{1}{(\Delta\phi)^2} [\phi - A v^2 t (0.7 + 0.3 \sin\alpha)]^2. \tag{11}$$

From this expression we can compute useful quantities such as the density, the thermal energy, the anisotropy.

For instance, for the density, we obtain (with $m = 1$)

$$n_1(\phi, t) \simeq 1.3 n_1(t=0) \frac{\Delta\phi}{\phi} \left(\frac{\phi}{0.85 A v_0^2 t}\right)^{5/2} \exp -\frac{\phi}{0.85 A v_0^2 t}, \tag{12}$$

where $n_1(t=0)$ is the initial value of the hot protons density.

We have plotted these quantities vs time for a given ϕ and given initial values on Figure 3. We have chosen $\Delta\phi = 30°$, a value which seems to be reasonable for the width of injection during substorms (Pfitzer and Winckler, 1969). All these quantities, after a rapid increase, decrease more slowly. By using the remark on the simplification of the dispersion relation, in Section 2, we can also compute the growth rate $\omega_i(x, t, \phi)$. We will not give here its analytical expression which is complicated, but the consequences for the emission: we determine the frequency for which it is maximum

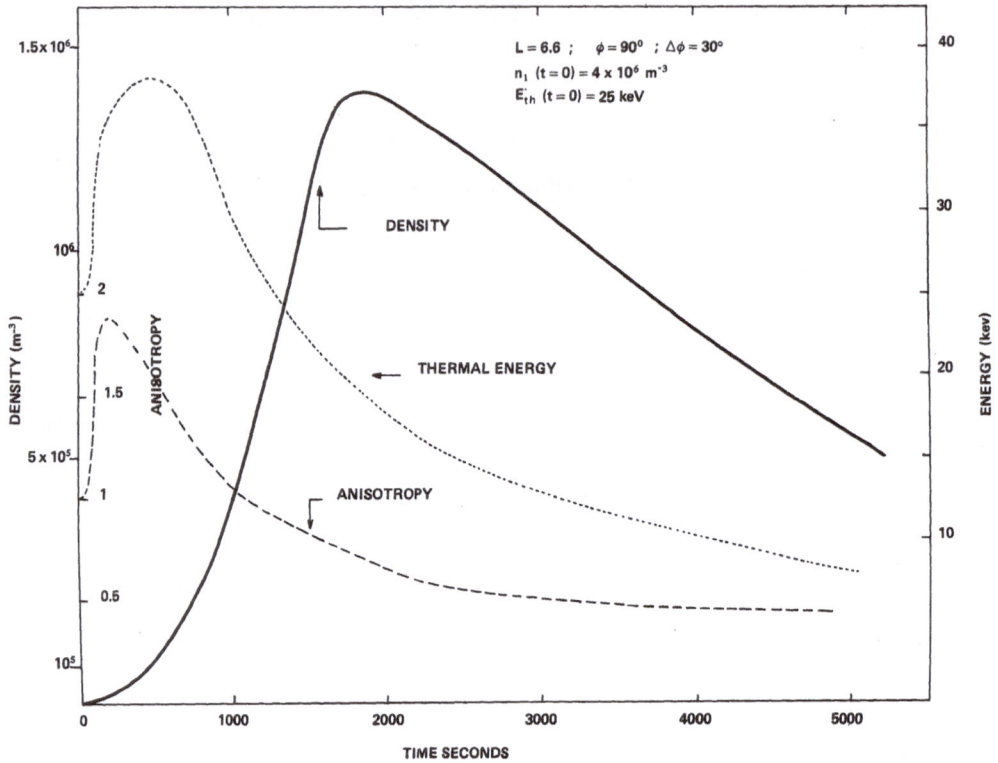

Fig. 3.   Plot of the density, the anisotropy and the thermal velocity of the hot protons vs time, for a given ϕ. For the anisotropy, we have fixed $V_R^2 = 2v_0^2$.

at any time for a given ϕ and the corresponding values of this growth rate (Figure 4).

We can see on Figure 4 that the emitted frequency for a given ϕ does not change very much as time elapses: about 15% if we consider the frequency range in which the largest growth rate changes by an order of magnitude. There is also a weak dependence of the frequencies with the azimuth angle ϕ. These results seem in agreement with measurements of the electromagnetic noise in the ULF range obtained on the ATS-1 satellite (McPherron *et al.*, 1972). We also find that the largest growth rates for different values of ϕ are obtained for $t \simeq 1100$ s ($\phi = 90°$) and $t \simeq 1500$ s ($\phi = 120°$). These values correspond to the same emitted frequency given by the following expression:

$$x \simeq \frac{2}{3(1 + v_0/V_A)}. \tag{13}$$

We will conclude this paper with some comments on this last formula.

4. Conclusion

From (13), as v_0 and V_A depend on L, the emitted frequencies will be different on each L-shell; and in fact, we should also take into account the radial drift of the

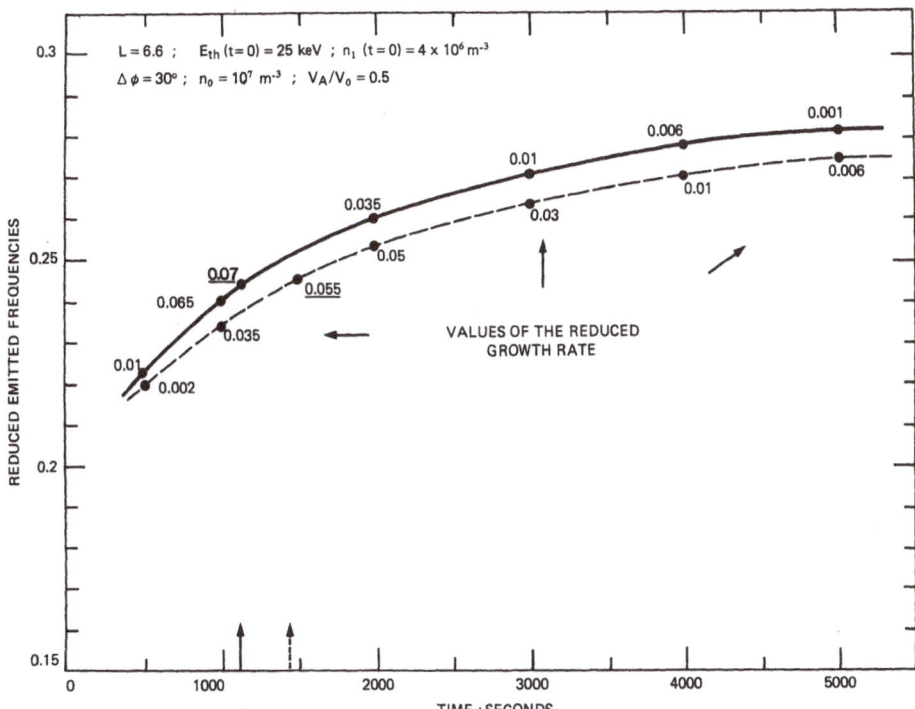

Fig. 4.   Plot of the emitted frequencies for which the growth rate is maximum vs time, for $\phi = 90°$ (solid line) and $\phi = 120°$ (dashed line), and given initial parameters (on top of the curves). The corresponding values of the reduced growth rate ω_i/ω_{ci} are also indicated on the curves.

protons. We hope that we will improve, by this way, the theory of the ion-cyclotron interaction in the magnetosphere and its applications to emission such as IPDP or Pc1, and further on the possible effects on the ring current.

Acknowledgement

We would like to thank Dr R. Pellat, from the Theoretical Center of 'Ecole Polytechnique', Paris, who has given us the basic idea of this work.

References

Ashour-Abdalla, M. and Cowley, S. W. H.: 1974, in B. M. McCormac (ed.), *Magnetospheric Physics*, D. Reidel Publ. Co., Dordrecht-Holland, p. 241.
Cornwall, J. M. and Schulz, M.: 1971, *J. Geophys. Res.* **76**, 7791.
Cuperman, S. and Landau, R. W.: 1974, *J. Geophys. Res.* **79**, 128.
Fried, B. D. and Conte, S. D.: 1961, *The Plasma Dispersion Function*, Academic Press, New-York.
Hamlin, D. A., Karplus, R., Vik, R. C., and Watson, K. M.: 1961, *J. Geophys. Res.* **66**, 1.
Kennell, C. F. and Petschek, H. E.: 1966, *J. Geophys. Res.* **71**, 1.
McPherron, R. L., Russell, C. T., and Coleman, P. J., Jr.: 1972, *Space Sci. Rev.* **13**, 411.
Perraut, S. and Roux, A.: 1974, 'Respective Role of the Cold and Warm Plasma Densities on the Generation Mechanism of ULF Waves in the Magnetosphere', to be published, *J. Atmospheric Terrest. Phys.*
Pfitzer, K. A. and Winckler, J. R.: 1969, *J. Geophys. Res.* **74**, 5005.

HOT-COLD PLASMA INTERACTIONS IN THE EARTH'S MAGNETOSPHERE

DONALD J. WILLIAMS

NOAA Space Environment Laboratory, Boulder, Colo., U.S.A.

1. Introduction

We present here a synopsis of observations and results obtained from the Explorer-45 satellite which pertain directly to the interaction of the magnetospheric hot and cold plasma populations. (Williams and Lyons, 1974a, b). This subject has not only been of considerable theoretical interest for the past several years but has also been used to conceptually explain a variety of magnetospheric phenomena (Kennel and Petschek, 1966; Cornwall, 1966; Brice and Lucas, 1971; Cornwall *et al.*, 1970, 1971; Cuperman *et al.*, 1973). Results from such studies have also been applied directly in early attempts to describe the magnetospheric environment of Jupiter (see for example Brice and Ioannidis, 1970; Ioannidis and Brice, 1971; Brice and McDonough, 1973; Coroniti *et al.*, 1973; Coroniti, 1974). In addition a number of active magnetospheric experiments have been described designed to trigger instabilities based on these previous hot-cold plasma interaction studies (Brice, 1970, 1971; Cornwall and Schultz, 1971; Williams, 1972; Cuperman and Landau, 1974).

Of particular interest are the results of Cornwall *et al.* (1970) who predict the stimulation of ion-cyclotron waves as the hot ring current plasma interacts with the cold plasmaspheric plasma. Wave amplification is expected to be sufficient to drive the hot plasma into strong pitch angle diffusion thereby resulting in an important loss process for the hot particles. Such a process should be most readily evident during the recovery phase of a geomagnetic storm since the overwhelming ring current source term, present during main phase, is absent.

Analysis of experimental results from Explorer 45 (Williams and Lyons, 1974a, b) yields basic agreement with these concepts with the priviso that the resulting pitch angle diffusion is not strong (full loss cone) but moderate (fractionally full loss cone). An additional interesting finding is that in regions of low cold plasma density ($\lesssim 0.1$ cm^{-3}) the hot ring current plasma is stably trapped with little or no losses due to wave-particle processes. Charge exchange and Coulomb losses operate continually and the extent to which they affect the hot plasma distribution at a particular spatial location is determined by the relative magnitudes of charge exchange lifetimes, Coulomb lifetimes and plasmaspheric cold plasma refilling times.

Explorer 45 (S^3) was launched on November 15, 1971 into an elliptical orbit having an apogee of 5.24 R_E, a perigee of 220 km, a period of 7.82 h and an inclination of 3.5°. An electrostatic analyzer-channeltron instrument plus three separate solid state detector systems measure proton intensities from 0.73 keV to 872 keV in 28

V. Formisano (ed.), The Magnetospheres of the Earth and Jupiter, 161–177. All Rights Reserved

energy steps and electron intensities from 0.73 keV to 560 keV in 20 steps. This report considers only the proton data.

The satellite spin axis is maintained in the plane of the orbit and pitch angle information is obtained by sectoring the satellite spin (8.451 s) into 32 segments. All pitch angles presented here are measured pitch angles using simultaneous data from the onboard 3-axis fluxgate magnetometer.

Figure 1 shows the Explorer-45 orbit for the period of interest. The outbound portion of the orbit intersects 1800 LT at $\sim 3.4\ R_E$ and the inbound portion intersects local midnight at $\sim 3.3\ R_E$. Apogee for the orbit shown is at ~ 2100 LT and there is an $\sim 12°$ per month precession towards dusk. The labeled regions refer to data to be displayed in subsequent figures.

2. Data and Results

2.1. HOT-COLD PLASMA CONSIDERATIONS

Figure 2 shows the hot ring current plasma distribution obtained during the recovery phase of the geomagnetic storm having sudden commencement and main phase on December 17, 1971. The data shown are for orbit 103 and give a clear example of the basic recovery phase patterns observed.

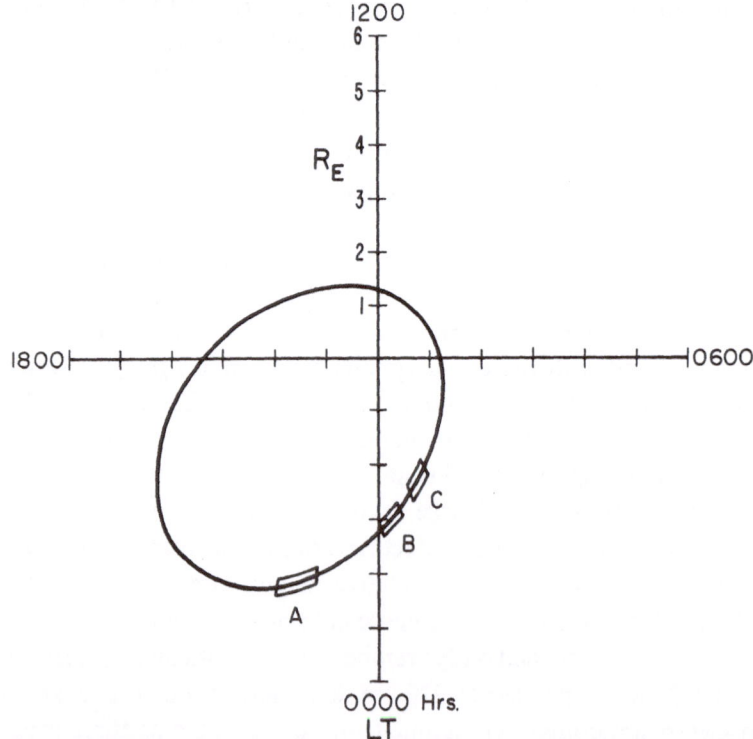

Fig. 1. Explorer-45 orbit shown in polar diagram of local time (hours) and altitude (Earth radii) for period of interest December 17, 18, 1971. Apogee precesses towards 1800 hours LT at rate of $\sim 12°$ per month. Regions A, B, and C refer to main phase data shown in Figure 7.

EXPLORER 45   ORBIT 103   INBOUND

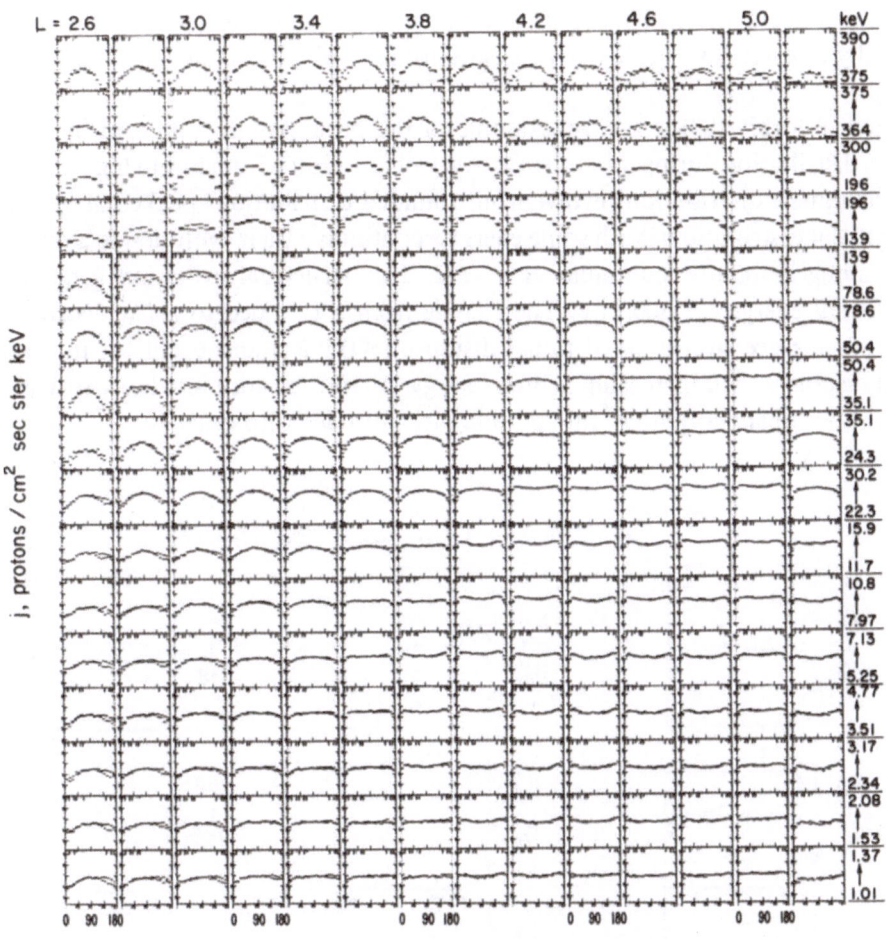

Measured Local Pitch Angle (Degrees)

Fig. 2.   Proton ring current (hot plasma population) snapshot, December 18, 1971. Storm recovery phase. Note transformation of flat-top (concave-top) pitch angle distributions to rounded pitch angle distributions peaked at 90° with lower energies transforming at lower altitudes. When pitch angle scans in the flat-top distributions reach the loss cone region, intensity decreases are seen implying an empty loss cone. Each individual plot shows $\log_{10}$ differential flux vs measured local pitch angle for a specific energy and L value. Plots for sixteen energies covering 1–390 keV are stacked at a given L value and shown every 0.2 R_E. This is a subset of the full display covering 1–872 keV and every 0.1 R_E which was used for analysis. No data editing has been done. The region where the solid state detector (24.3–300 keV) often suffers a saturation problem (Williams *et al.*, 1973) occurs for $L \lesssim 3.2$. This saturation problem appears as unusual depressions in intensities for pitch angles 90° ± 45°. Contamination of the channeltron instrument (1–30.3 keV) by reflected sunlight at near background count rates can be seen during the 90°–180° pitch angle sweep. This problem is related to spin axis orientation, is easily identified, and in the present data does not exist in the 0°–90° pitch angle sweep. Neither of the above effects has any influence on the present study. Data dropouts and telemetry noise effects can also easily be identified.

As no editing of the data has been performed, regions of contaminated data are indicated in the figure caption. These suspect data have no impact on the present results. The individual plots in each figure show the observed flux vs measured local pitch angle for a particular energy and L value. Plots for all energies are stacked vertically at each L value. Sixteen energies from 1 keV to 390 keV are shown every $0.2\ R_E$.

Figure 2 (orbit 103) is a clear example of the basic pattern observed in the hot proton distribution during the entire recovery phase. Going from low to high altitudes, the pitch angle distributions evolve from rounded distributions peaked at 90° pitch angle to flat distributions, with some energies continuing their evolution to a distribution having a shallow minimum at $\alpha = \pi/2$. This structural evolution generally is seen for all energies $\lesssim 300$ keV. The change from a rounded to a flat distribution often takes place over a small radial distance ($\lesssim 0.1\ R_E$) and occurs at increasing radial distances with increasing proton energy. A study of additional recovery phase orbits shows that this pattern moves to higher altitudes with time.

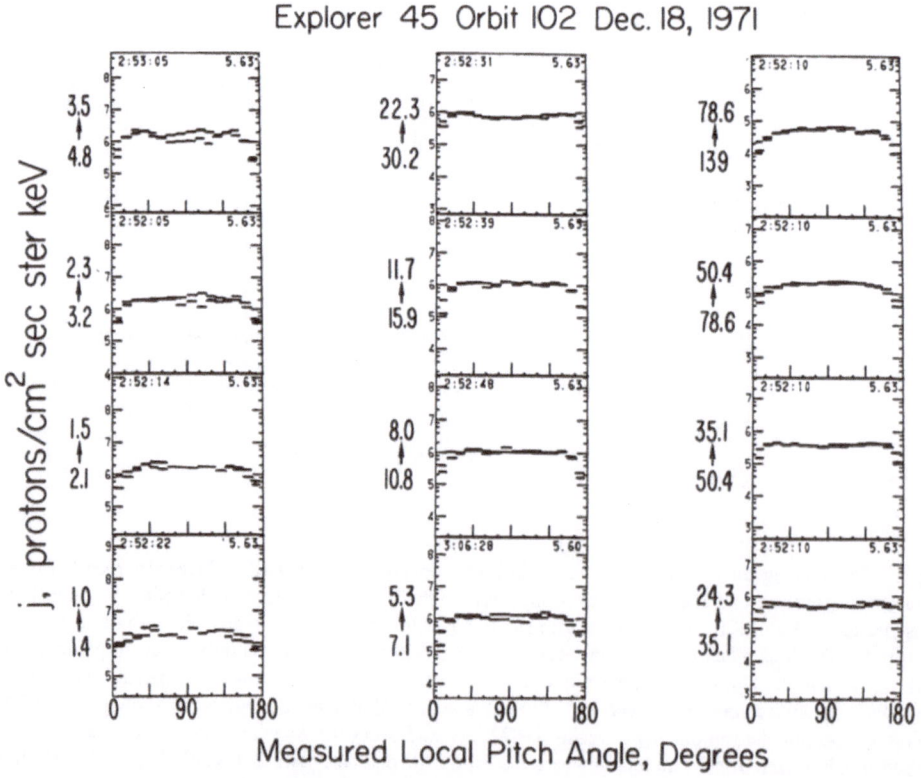

Fig. 3. Example of proton observations indicating an empty loss cone at $5.6\ R_E$. Log_{10} proton differential flux vs measured pitch angle shown for the range 1–139 keV. Intensity decreases in region of expected loss cone are clearly evident. These decreases indicate an empty loss cone since the detector angular aperture plus pitch angle scan is significantly larger than the theoretical loss cone. All recovery phase orbits show this effect whenever the pitch angle scan extends to very small angles.

A close inspection of our data further shows that for all recovery phase orbits in which adequate pitch angle scans exist, the loss cone is nearly empty in the region of the flat pitch angle distributions. An example of this is shown in Figure 3 where a pronounced intensity drop is evident in the region of the loss cone. Since the detector angular aperture (11°) plus the angular scan (11.25°) is larger than the theoretical loss cone, a drop in intensity as seen in Figure 3 implies a nearly empty loss cone.

Thus, in the spatial evolution of the pitch angle distribution during recovery phase discussed earlier, the high altitude flat distributions are actually flat-top distributions having a nearly empty loss cone.

Before leaving the data, we wish to show our initial attempts at displaying the hot plasma data in a format more convenient for theoretical analysis. In Figure 4 we show the spatial evolution of the distribution function in the velocity plane for both main phase (orbit 101) and recovery phase (orbit 103). Contours of constant phase space density (divided by $1/2 M_{prot} = 3(10)^{23}$), $f = j/E$, are plotted in the $v_\perp$, $v_\parallel$ plane where the velocity components, $v_\perp$ and $v_\parallel$ are measured relative to the local magnetic field direction, E is the proton kinetic energy and j is the measured proton flux. These plots are computor generated from the same data used to construct displays as shown in Figure 2. We hope that the additional perspective afforded by the distribution function displays will make it more convenient to identify and follow the evolution of various processes acting on the hot plasma.

The preceding discussion and Figures 2 and 3 yield the following observational facts concerning the hot ring current plasma during recovery phase (Williams and Lyons, 1974a, b):

(1) At high Explorer-45 altitudes, pitch angle distributions are isotropic except for a nearly empty loss cone (flat-top distributions).

(2) A transition to a rounded pitch angle distribution peaked at $\alpha = \pi/2$ occurs at lower altitudes with the transition altitude decreasing with decreasing energy.

(3) The above pattern moves toward higher altitudes during recovery phase.

We consider the existence of flat-top pitch angle distributions having nearly empty loss cones indicative of a stably trapped particle population in a region of negligible losses due to pitch angle diffusion.

The transition to rounded pitch angle distributions appears to result from a depletion of particles. Since the energy and spatial dependence and the abruptness of the transition are inconsistent with charge exchange effects, we have tested the hypothesis of Cornwall et al. (1970) concerning the stimulation of ion-cyclotron waves at the hot-cold plasma interface of the plasmasphere and the ring current.

The resonant energy equation describing the cyclotron resonance condition for protons and the amplified em waves is (Kennel and Petschek, 1966)

$$E_{\parallel,\,\mathrm{res}} = \frac{B^2}{8\pi N}\, A^{-2}(1+A)^{-1} \equiv \frac{B^2}{8\pi N}\, F(A),$$

where B is the local magnetic field magnitude and N is the total plasma density. A is a particle anisotropy factor obtained from a combination of integrals over the

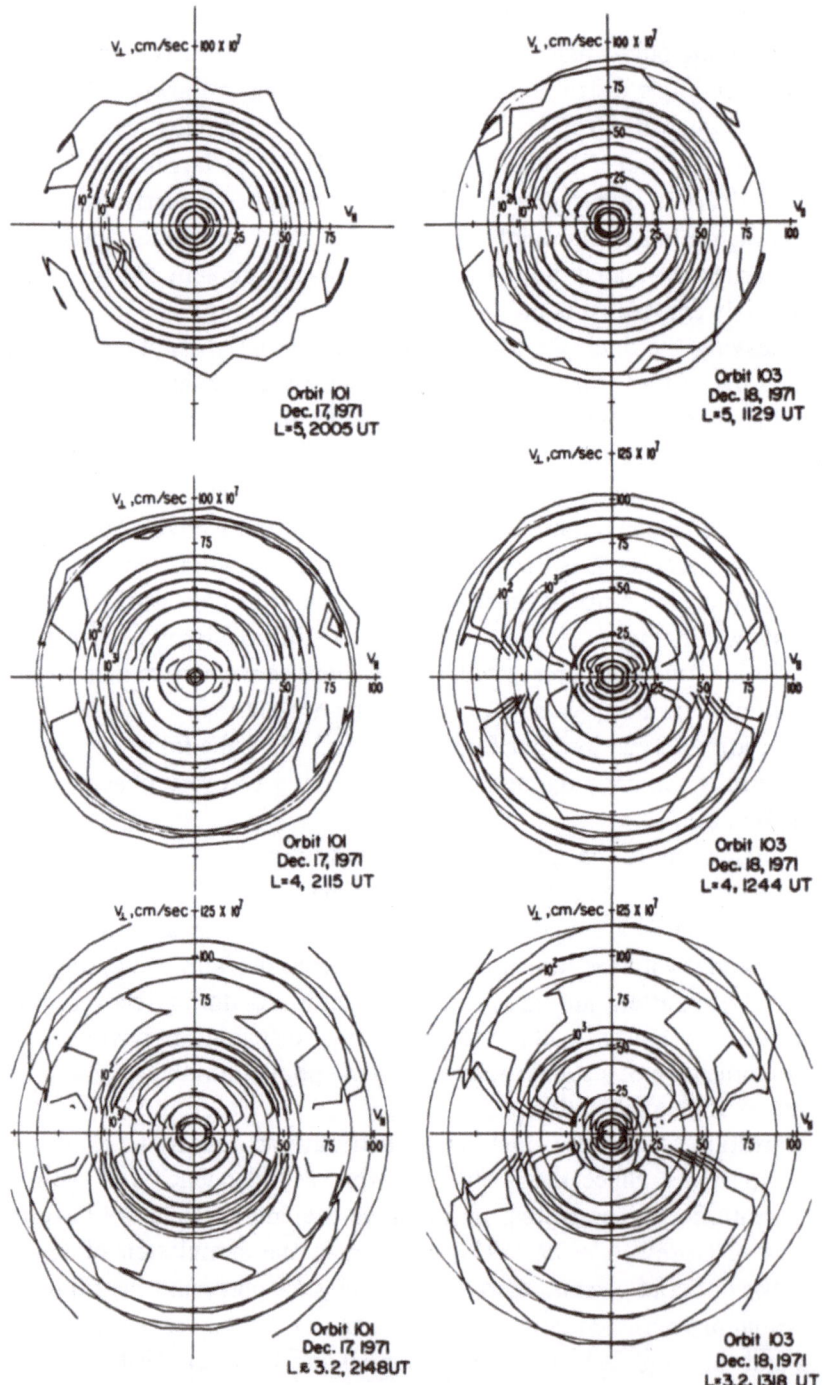

Fig. 4. Contours of constant phase space density, $f = j/E$, plotted in $v_\perp$, $v_\parallel$ plane. Circles centered at $v_\perp = v_\parallel = 0$ (thin lines) added for reference. Contours generated every 0.5 units in $\log_{10} f$ by computer with no smoothing applied to the data. Contours generated for $v_\perp > 0$ and $v_\perp < 0$ separately and region of no available data separates the two sets of contours. Spatial evolution of distribution function shown during main phase (orbit 101) and recovery phase (orbit 103). Count rate statistics are the cause of jagged nature of several contours. Break in orbit 101, $L = 5$ contours is a telemetry problem.

distribution function and its first derivative in velocity space. Pending a numerical integration through the distribution function, we estimate $F(A)$ to be of order 1, $\pm$ a factor of four, within the regions of rounded pitch angle distributions.

We estimate $E_{\parallel,\,\text{res}}$ as a function of altitude by observing the altitude at which the various differential energy channels begin their transition from flat-top to rounded pitch angle distributions. This is defined as that altitude where the intensity in the pitch angle scan immediately adjacent to the loss cone scan suddenly decreases towards the loss cone value.

This measurement of $E_{\parallel,\,\text{res}}$ combined with the measured value of B allows us to plot as a function of altitude a normalized plasma density

$$\frac{N}{F(A)} = \frac{B^2}{8\pi E_{\parallel,\,\text{res}}}\;(\text{cm})^{-3}.$$

This result is shown in Figure 5a for orbit 103 where we also include an *in situ* estimate of plasma density obtained from the saturation of an onboard DC electric field

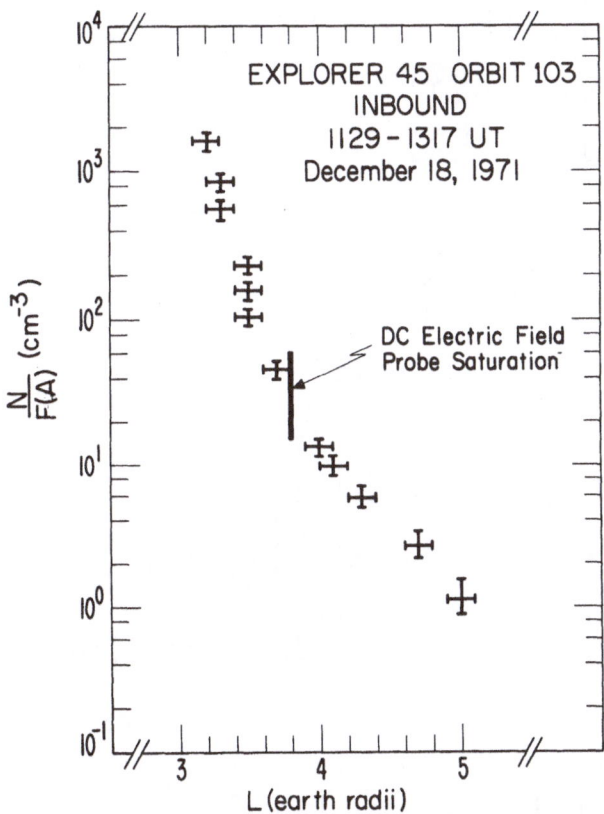

Fig. 5a. Plot of $N/F(A)$ vs altitude. Resonant energy equation used to obtain $N/F(A)=B^2/8\pi E_{\parallel,\,\text{res}}$. Altitude determined as that point where flat top distributions begin their transformation to a rounded distribution. Altitude and density estimate of DC electric field probe saturation shown. This analysis strongly indicates that the moderate pitch angle diffusion responsible for the rounded pitch angle distributions is due to the amplification of ion-cyclotron waves as the hot ring current plasma intereacts with the cold plasmaspheric plasma in the region of the plasmapause.

probe (Maynard and Cauffman, 1973; Maynard, personal communication). The agreement is good and the inferred plasmapause shape is quite reasonable.

To further test the ion-cyclotron hypothesis we have performed the same analysis on additional orbits to investigate the time behavior of the $N/F(A)$ boundary. These results are shown in Figure 5b. The apparent outward motion is consistent with average plasmaspheric refilling rates of $\sim 2(10)^8$ ions cm^{-2} s^{-1} at $L = 3.5$ to $\sim 2.5(10)^7$ ions cm^{-2} s^{-1} at $L = 4.5$. These refilling rate estimates are in agreement with earlier observations by Chappell *et al.* (1970) and Park (1970) and with the theoretical expectations of Banks (1972).

We thus make the following conclusions relevant to the hot ring current plasma

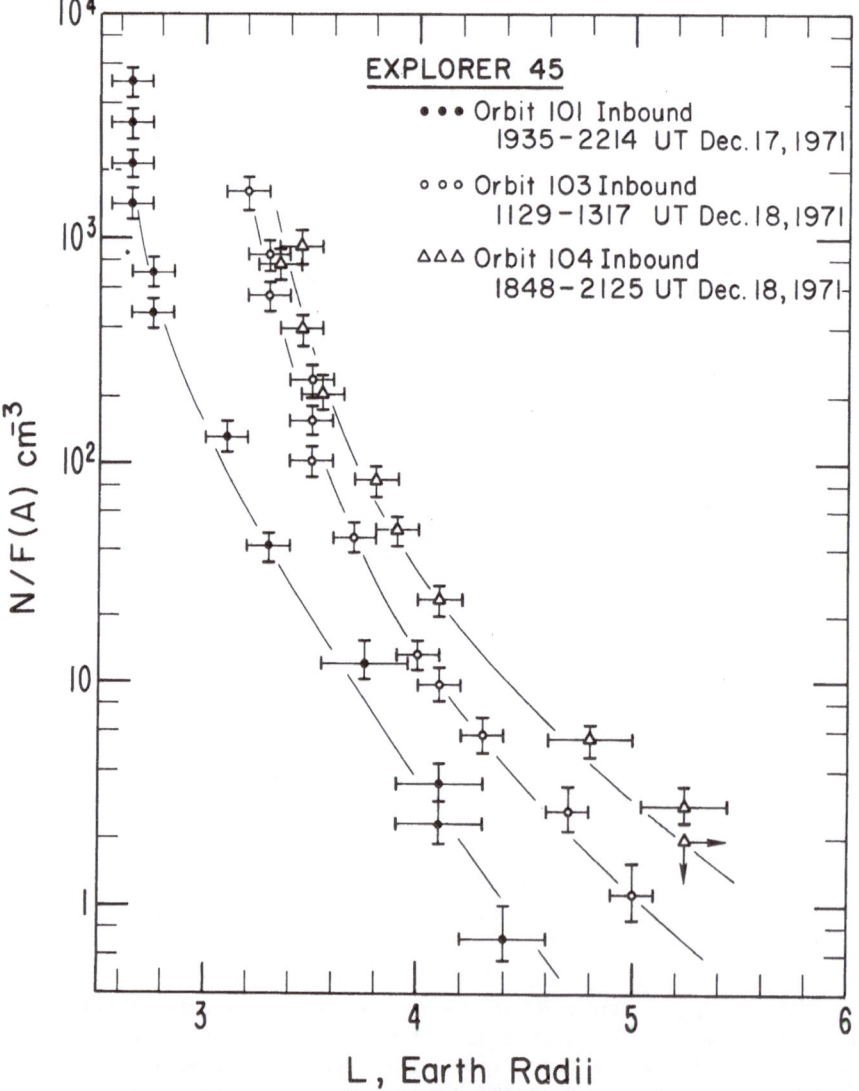

Fig. 5b.   Same as Figure 5a with orbits 101 and 104 added. Outward motion of $N/F(A)$ boundary is clearly seen and is consistent with expected plasmaspheric cold plasma refilling rates.

during the recovery of magnetic storms (Williams and Lyons, 1974a):

(1) At altitudes immediately above the plasmapause region, the hot ring current plasma is stably trapped with negligible losses due to pitch angle scattering.

(2) In the region of the plasmapause (cold plasma density gradient) the hot ring current plasma experiences moderate pitch angle diffusion.

(3) The spatial, energy, and temporal dependence of the initiation of the pitch angle diffusion, analysis of the data via the ion-cyclotron resonant energy equation, and a comparison of these results with an *in-situ* estimate of the plasma density strongly indicate that the pitch angle diffusion process is due to the amplification of ion-cyclotron waves as the hot ring current plasma interacts with the cold plasmaspheric plasma.

These results are schematically summarized in Figure 6. A variety of low altitude observations (Amundsen *et al.*, 1972; Kleckner and Hock, 1973; Bernstein *et al.*, 1974; Mizera, 1974) have been shown to be consistent with this picture (Williams and Lyons, 1974a). In particular low altitude observations showing a loss cone full of precipitating protons at geomagnetic latitudes above the plasmapause fit into Figure 6 by noting that such observations would correspond to an equatorial altitude during recovery phase above that reached by Explorer 45.

The relative sizes and locations of the regions shown in Figure 6 are dependent on past and existing magnetic and electric field activity. During main phase and after prolonged periods of magnetic quiet, the plasmapause region and the inner edge of the plasma sheet may nearly coincide with little if anything left of the stable region. While Explorer-45 data cannot help in the case of prolonged magnetic quiet since the plasmapause region is above satellite apogee, main phase data can be useful in further delineating the relative positions of the regions shown in Figure 6.

We have studied (Williams and Lyons, 1974b) Explorer-45 data from the main phase of the December 17, 1971 magnetic storm and found two regions above $L = 3.6$ and outside the plasmapause where the geomagnetic field line orientation allowed small ($\lesssim 10°$) pitch angles to be sampled. In both these cases, the pitch angle distributions were consistent with isotropy. In addition, the inbound portion of orbit 101 which occurred during the latter half of main phase, showed the characteristic recovery phase flat-top pitch angle distributions with empty loss cone and the transition to the rounded pitch angle distributions.

Data from portions of orbit 101, as indicated in Figure 1, are shown in Figure 7 illustrating the possibility that Explorer 45 sampled all three of the regions depicted in Figure 6 during the main phase of the December 17, 1971 storm. This indicates significant penetration of the turbulent region to lower altitudes during the main phase of the storm. However we were unable to determine if the turbulent region and plasmapause region coincided at any time during main phase because of inadequate pitch angle scans resulting from geomagnetic field distortions.

2.2. CHARGE EXCHANGE CONSIDERATIONS

As seen in Figure 2 and noted by Williams and Lyons (1974a), the fluxes near $\alpha \sim \pi/2$

Schematic Model of Recovery Phase

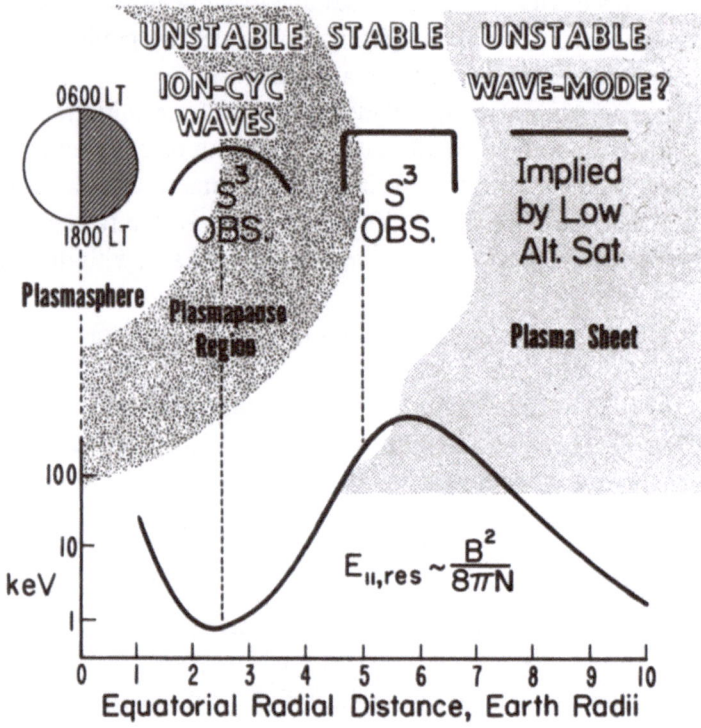

Fig. 6. Schematic picture of geomagnetic storm recovery phase showing hot ring current plasma behavior in the geomagnetic field and its interaction with the cold plasmaspheric plasma. A semi-quantitative resonant energy vs altitude plot which uses Explorer-45 results is shown for reference. Moderate pitch angle diffusion occurs for the hot ring current plasma in the plasmapause region due to the amplification of ion-cyclotron waves. Each energy of the ring current plasma begins its interaction in the plasmapause region at the appropriate value of $B^2/8\pi N$. Above the plasmapause region the hot ring current plasma is stably trapped with negligible or no losses due to pitch angle scattering. Above this region, and above Explorer-45 (S^3) apogee during recovery phase, there is the plasma sheet region exhibiting strong turbulence and a full loss cone as implied by low altitude measurements.

do not show as sudden an intensity decrease in the transition region as do the lower pitch angle intensities. This is consistent with the inefficiency of the ion-cyclotron interaction near $\alpha \sim \pi/2$ ($v_{\parallel} \sim 0$) and indicates that other loss mechanisms may be required to explain the behavior of the $\alpha \sim \pi/2$ protons.

Charge exchange is a loss mechanism which operates throughout the trapping regions. Although the three dimensional density profile for neutral hydrogen to high altitudes around the earth is not well known, results exist showing the importance of charge exchange effects on storm and substorm associated protons (Swisher and Frank, 1968; McIlwain, 1972; P. Smith, personal communication).

We show in Figure 8 the time histories of protons locally mirroring at $\alpha \sim \pi/2$ during the December 1971 storm period. Three altitudes and three energies are shown along with charge exchange lifetimes from Swisher and Frank (1968) and Liemohn

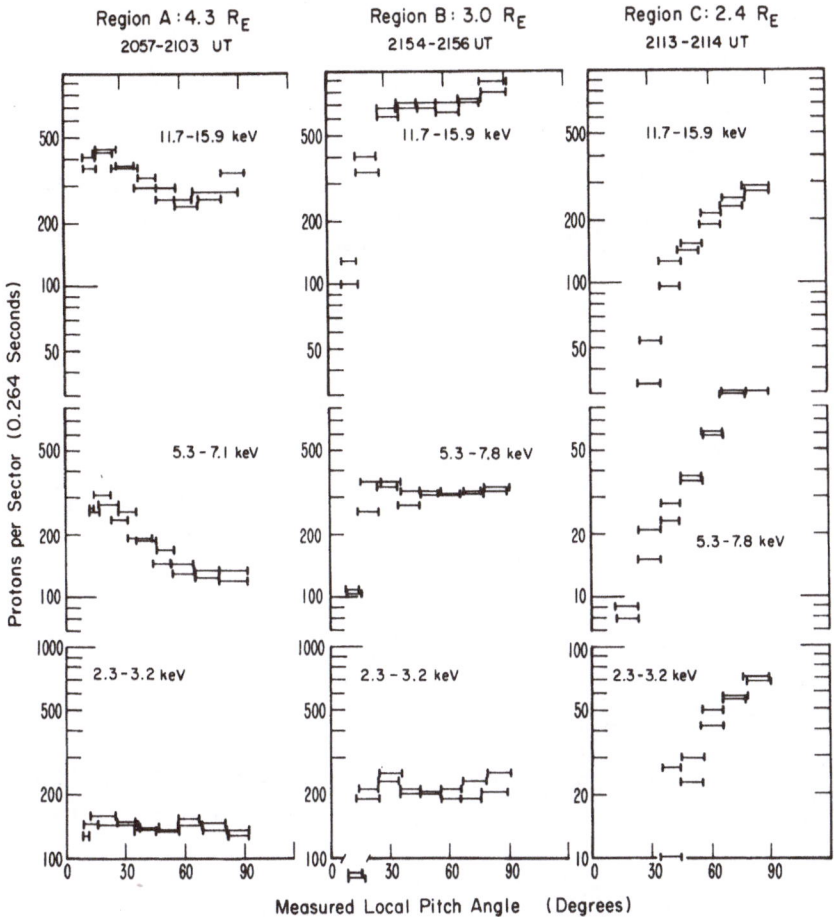

Fig. 7. Sample of proton pitch angle distributions obtained during main phase (Explorer-45 orbit 101) of December 17, 1971 magnetic storm. Regions A, B, and C refer to regions shown in Figure 1. The data at small pitch angles near the loss cone show that Explorer 45 probably sampled all the regions shown in Figure 6; full loss cone in Region A indicating turbulent hot plasma, empty loss cone in Region B indicating stable hot plasma and rounded pitch angle distributions in Region C indicating occurrence of moderate pitch angle diffusion in plasmapause region.

(1961). There is a tendency for the $\alpha \sim \pi/2$ protons generally to follow the charge exchange lifetimes during recovery phase. Some of the departures from this tendency may be due to substorm effects. The effect of the plasmapause-ring current interaction discussed in the previous section is most noticeable at smaller pitch angles and may occur at a different altitude for some of the energies and times shown in Figure 8. Thus it may not be visible at all in such a time-intensity plot.

While we see the $\alpha \sim \pi/2$ protons generally following the charge exchange lifetimes, a proper evaluation of charge exchange requires a detailed study of the temporal and spatial evolution of the pitch angle distributions at all energies. Only then can a quantitative comparison with the effects of ion-cyclotron resonance and other plasma effects be undertaken.

In spite of this uncertainty, we can qualitatively illustrate a recovery phase picture

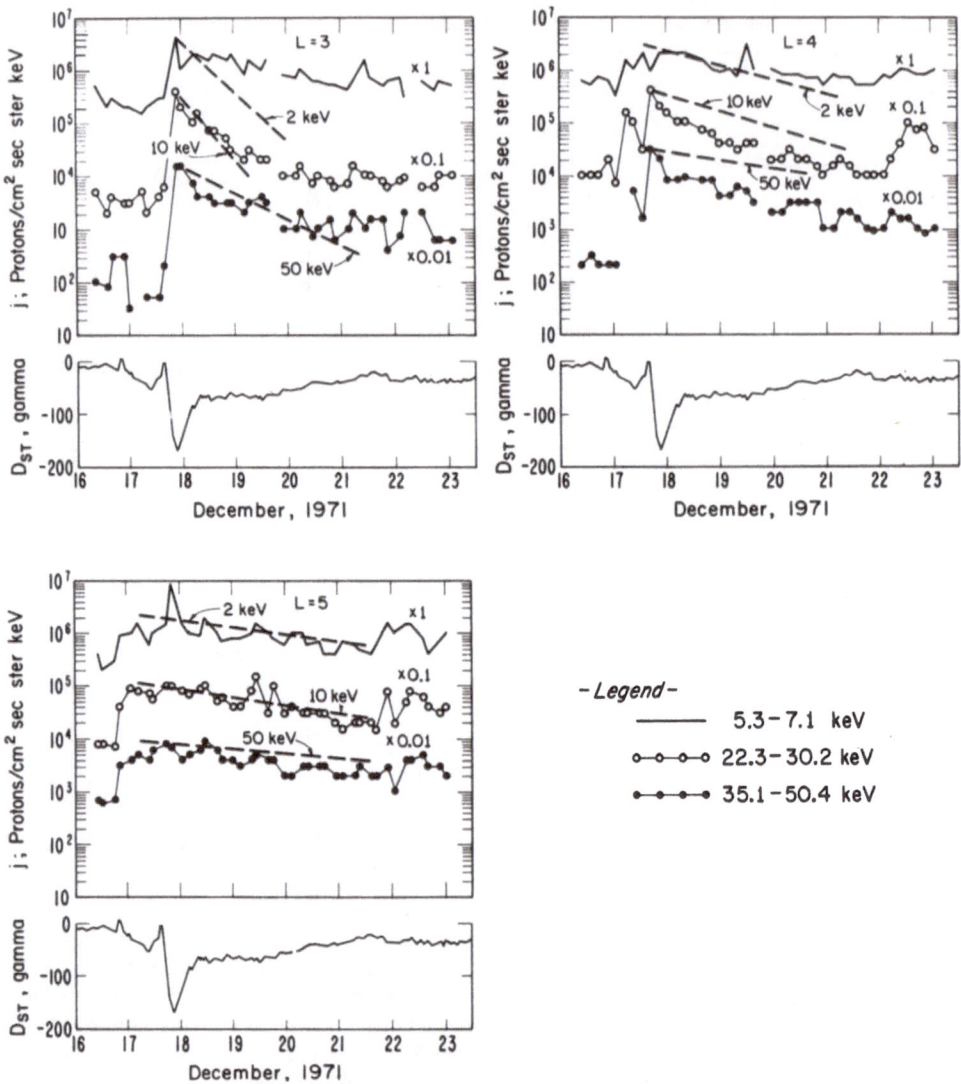

Fig. 8.   Proton differential flux vs time during December 1971 geomagnetic storm events. Only local pitch angles of $\alpha \sim \pi/2$ are shown. Proton energies indicated by symbols and expected charge exchange lifetime curves shown as dashed lines. General tendency to follow charge exchange is evident although discrepancies can be seen. Ion-cyclotron resonance losses will not show on a time history plot such as this.

including both ion-cyclotron processes and charge exchange effects. This is shown in Figure 9. The horizontal bar qualitatively shows, in relation to the plasmapause region, what the relative strengths of charge exchange (CE) and ion-cyclotron resonance (IC) losses are. The bar shows the situation for only one energy and pitch angle and will move up and down the figure for other energy-pitch angle combinations. The IC region is extended to overlap the CE region because of possible parasitic precipitation effects which may compete with charge exchange time scales (Lyons and Thorne, 1972). The shaded IC region moves left and right in accordance with the time varying plasma density profile in the plasmapause region. No other losses are included.

It is apparent that the hot plasma distribution function is dominated by ion-cyclotron resonance effects in the plasmapause region. However the specific effects to be observed at altitudes above the plasmapause region will depend critically on the relative values of charge exchange lifetimes and plasmasphere refilling times.

Well inside the plasmapause region, the remainder of the hot plasma distribution should be dominated by charge exchange and Coulomb losses. A significant hot plasma intensity exists inside the plasmapause region during recovery phase presumably because the ion-cyclotron resonance initiates only a moderate pitch angle diffusion mode and is relatively inefficient at removing particles near $\alpha \sim \pi/2$. The arrest of the instability at a moderate diffusion level may be due to absorption of wave energy by cold electrons (Cornwall et al., 1971) and by quenching effects due to attainment of large ratios of cold plasma density to hot plasma density (Cuperman and Landau, 1974; Cuperman et al., 1973).

2.3. FURTHER LOSS CONSIDERATIONS

The data which has been discussed show quantitatively the role of ion-cyclotron resonance and qualitatively the role of charge exchange as losses operating on the hot plasma population during recovery phase. However in addition to the need of

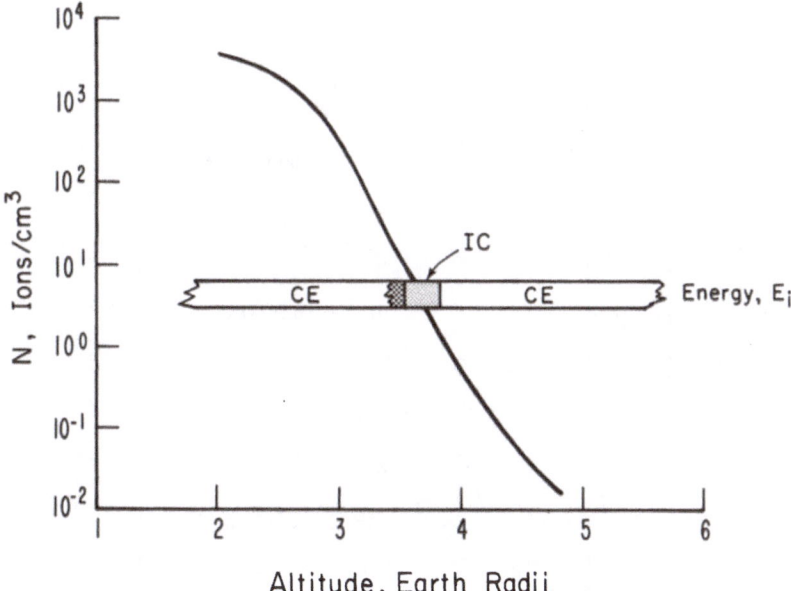

Fig. 9.   Schematic diagram illustrating regions where charge exchange (CE) and ion-cyclotron resonance (IC) losses are expected to dominate. Situation illustrated for one specific energy, E_i, and pitch angle. Horizontal bar moves up and down (up for lower energies) for other energy-pitch angle combinations. As cold plasma refilling occurs, plasmapause region and shaded IC region move towards higher altitudes. Overlap region illustrates extension of IC region through possible parasitic precipitation effects. No other losses are considered here.

quantifying the charge exchange process, the effects of radial diffusion, convection, and enhanced ion-cyclotron resonance leading to strong pitch angle diffusion have to be ascertained to determine their role in the recovery phase loss picture.

We show in Figures 10 and 11 a simple example of recovery phase loss which requires the assessment listed above. Figure 10 shows low altitude segments of the inbound portions of orbits 101 (main phase), 102 (1st recovery phase orbit), and 103 (2nd recovery phase orbit). The differential flux at $\alpha \sim \pi/2$ is shown for four energies. An anomalously large decrease is seen from orbit 101 to orbit 102. The changes from orbit 102 to 103 are qualitatively consistent with charge exchange if the geomagnetic latitude of the observations are taken into account.

To quantify these results, we have corrected for geomagnetic latitude and plotted the ratio of orbit 102 flux to orbit 101 flux as a function of altitude for several energies. These results are shown in Figure 11 along with solid and dashed lines showing the expected ratio at the energies indicated based on charge exchange losses. The

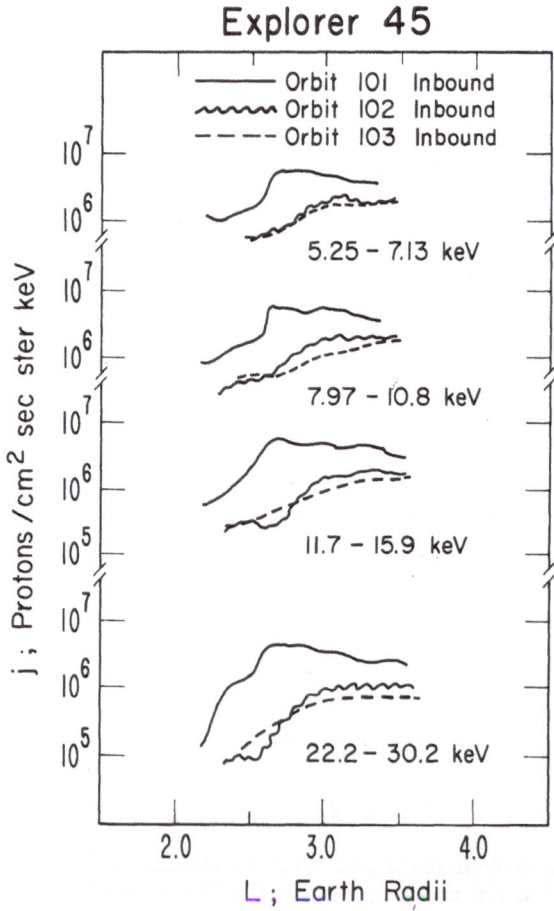

Fig. 10. Differential flux vs altitude for segments of inbound portions of orbits 101, 102, and 103. These data show the large intensity decrease occurring between main phase (orbit 101) and the beginning of recovery phase (orbit 102).

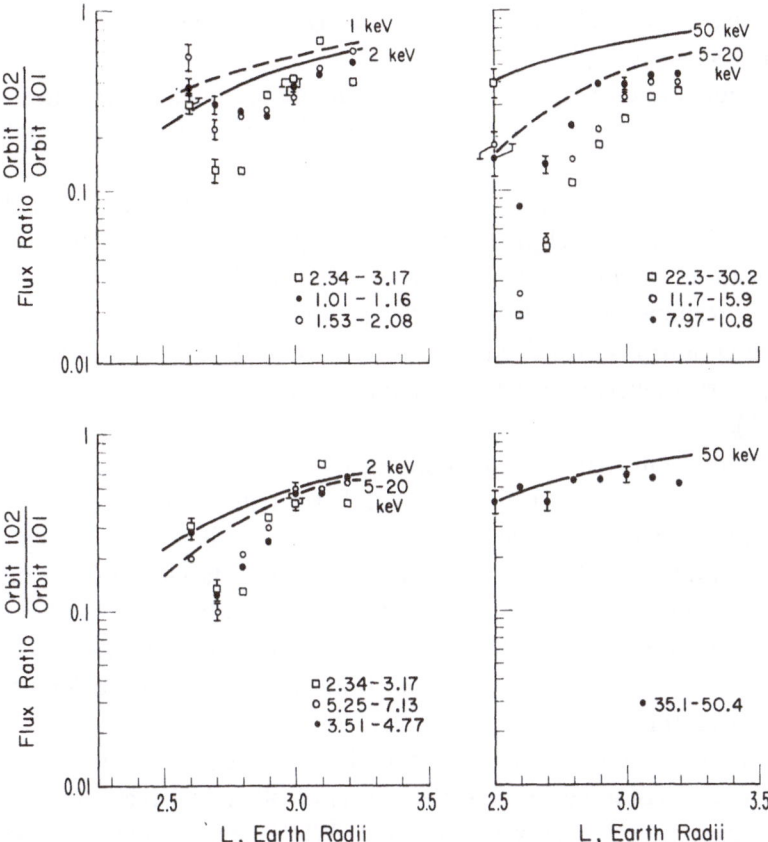

Fig. 11.   Plot of ratio of orbit 102 flux to orbit 101 flux vs altitude for several energies. Solid lines show expected ratio at indicated energies assuming only charge exchange losses. Losses faster than those are seen at all energies with the 35.1–50.4 keV losses exceeding charge exchange losses for $L \gtrsim 3.1$.

regions of loss rates faster than charge exchange generally are apparent. However for the 35.1–50.4 keV channel, the losses become faster than charge exchange for $L \gtrsim 3.1$. There appears to be a slight tendency for the region of high losses to move toward higher altitude at higher energies.

This example is shown to illustrate the importance of knowing the effects of all the processes listed above.

3. Summary

We have presented a discussion of recent results pertaining to the interaction of the Earth's magnetospheric hot and cold plasmas. The specific case discussed was the interaction of the hot ring current plasma with the cold plasmaspheric plasma in the evening to early morning local time sector during magnetic storm recovery phase.

The basic features found were:

(1) Above the plasmapause region, the hot ring current plasma is stable with negligible losses due to pitch angle diffusion. Thus, in the absence of cold plasma,

a hot plasma can be stable in a trapping configuration on a scale size comparable to the Earth's magnetosphere.

(2) The hot ring current plasma enters a moderate pitch angle diffusion regime in the plasmapause region. Therefore the addition of cold plasma destabilizes the hot plasma.

(3) The energy, spatial, and temporal dependence of the above destabilization, an analysis via the ion-cyclotron resonant energy equation and comparison of this analysis with an *in situ* estimate of the plasma density strongly indicate that the mechanism responsible for the destabilization of the hot plasma is the amplification of ion-cyclotron waves due to the interaction of the cold plasmaspheric plasma with the hot ring current plasma in a manner similar to that discussed by Cornwall *et al.* (1970).

These results, obtained from observations made by Explorer 45, are schematically summarized in Figure 6 where magnetospheric regions of hot plasma stability, moderate turbulence, and strong turbulence are shown during recovery phase. Data were shown indicating that the region of hot plasma stability became destabilized in a manner consistent with expected cold plasma refilling rates into the formerly depleted cold plasma region. Data were also shown indicating that Explorer 45 sampled all the regions shown in Figure 6 during the main (injection) phase of a geomagnetic storm.

To further understand the magnetospheric hot-cold plasma interaction, a semi-quantitative discussion concerning the effects of other loss mechanisms, with emphasis on charge exchange, was presented. While charge exchange does operate throughout the magnetospheric trapping regions, a detailed analysis of its effects on proton energy, pitch angle, and spatial distributions as a function of time is required before quantitative comparisons can be made with other losses at a given time and point in space. This will require much better knowledge of the three dimensional neutral atmospheric density distribution to high altitudes ($\sim 10\ R_{\rm E}$).

A qualitative picture of the complimentary effects of charge exchange and ion-cyclotron losses was given for a particular energy E_i and pitch angle in Figure 9. In the region of depleted cold plasma (hot plasma stable region) the effects on the hot plasma distribution function will depend on the relative values of charge exchange lifetimes and cold plasma refilling times.

While much is now known about the interaction between the magnetospheric hot and cold plasmas, it is clear that there is much to learn. All the considerations in this paper, for example, concern plasmas for which $\beta \leqslant 1$. We have not yet experimentally studied in detail the situation where $\beta > 1$, a situation which often exists in magnetospheric trapping regions particularly during magnetic storm main phase.

Results such as these have clear applications to other planetary systems and in particular Jupiter. It is only through the past years of magnetospheric study that it is possible to obtain such sophisticated data analyses and results as are now obtained from one or two flyby missions past the distant planets.

Acknowledgements

It is a pleasure to acknowledge the work of Dr L. R. Lyons, a co-researcher in the studies described in this report. I wish to further acknowledge several helpful discussions with Drs P. Smith and T. A. Fritz and to acknowledge the Explorer 45 experiment principals, R. Hoffman, T. Fritz, L. Cahill, and N. Maynard without whose cooperation this study would not have been possible. Figures 2, 3, 4, 5a, and 6 are from Williams and Lyons, 1974a. Figure 5b is from Williams and Lyons, 1974b. These studies were partially funded under NASA Contract Number S-50028.

References

Amundsen, R., Søraas, F., Lindalen, H. R., and Aarsnes, K.: 1972, *J. Geophys. Res.* **77**, 556.
Banks, P. M.: 1972, in E. R. Dyer (ed.), *Critical Problems of Magnetospheric Physics*, Proceedings of the Joint COSPAR/IAGA/URSI Symposium Madrid, Spain, May 11–13, Pub. IUCSTP Secretariat November 1972.
Bernstein, W., Hultqvist, B., and Borg, H.: 1974, *Planetary Space Sci.*, **22**, in press.
Brice, N.: 1970, *J. Geophys. Res.* **75**, 4890.
Brice, N.: 1971, *J. Geophys. Res.* **76**, 4698.
Brice, N. M. and Ioannidis, G. A.: 1970, *Icarus* **13**, 173.
Brice, N. and Lucas, C.: 1971, *J. Geophys. Res.* **76**, 900.
Brice, N. M. and McDonough, T. R.: 1973, *Icarus* **18**, 206.
Chappell, C. R., Harris, K. K., and Sharp, G. W.: 1970, *J. Geophys. Res.* **75**, 3848.
Cornwall, J. M.: 1966, *J. Geophys. Res.* **71**, 2185.
Cornwall, J. M., Coroniti, F. V., and Thorne, R. M.: 1970, *J. Geophys. Res.* **75**, 4699.
Cornwall, J. M., Coroniti, F. V., and Thorne, R. M.: 1971, *J. Geophys. Res.* **76**, 4428.
Cornwall, J. M. and Schulz, M.: 1971, *J. Geophys. Res.* **76**, 7791.
Coroniti, F. V.: 1974, *Astrophys. J. Suppl. Ser.* **27**, 261.
Coroniti, F. V., Kennel, C. F., and Thorne, R. M.: 1974, *Astrophys. J.* **189**, 383.
Cuperman, S. and Landau, R. W.: 1974, *J. Geophys. Res.* **79**, 128.
Cuperman, S., Salu, Y., Bernstein, W., and Williams, D. J.: 1973, *J. Geophys. Res.* **78**, 7372.
Ioannidis, G. and Brice, N.: 1971, *Icarus* **14**, 360.
Kennel, C. F. and Petscheck, H. E.: 1966, *J. Geophys. Res.* **71**, 1.
Kleckner, F. W. and Hock, R. J.: 1973, *J. Geophys. Res.* **78**, 1187.
Liemohn, H.: 1961, *J. Geophys. Res.* **66**, 3593.
Lyons, L. R. and Thorne, R. M.: 1972, *J. Geophys. Res.* **77**, 5608.
Maynard, N. C. and Cauffman, D. P.: 1973, *J. Geophys. Res.* **78**, 4745.
McIlwain, C. E.: 1972, in B. M. McCormac (ed.), *Earth's Magnetospheric Processes*, D. Reidel, Dordrecht, Holland, p. 268.
Mizera, P. F.: 1974, *J. Geophys. Res.* **79**, 581.
Park, C. G.: 1970, *J. Geophys. Res.* **75**, 4249.
Swisher, R. L. and Frank, L. A.: 1968, *J. Geophys. Res.* **73**, 5665.
Williams, D. J.: 1971, NOAA Tech. Memo ERL SEL-19, Boulder, Colorado.
Williams, D. J. and Lyons, L. R.: 1974a, *J. Geophys. Res.* **79**, 4195.
Williams, D. J. and Lyons, L. R.: 1974b, *J. Geophys. Res.* **79**, 4791.
Williams, D. J., Fritz, T. A., and Konradi, A.: 1973, *J. Geophys. Res.* **79**, 4751.

CONCEPTS OF MAGNETOSPHERIC CONVECTION

VYTENIS M. VASYLIUNAS

Dept. of Physics and Center for Space Research, Massachusetts Institute of Technology, Cambridge, Mass. 02139, U.S.A.

Magnetospheric physics, which grew out of attempts to understand the space environment of the Earth, is becoming increasingly applicable to other systems in the Universe. Among the planets, in addition to the Earth, Jupiter, Mercury, Mars, and (in a somewhat different way) Venus are now known to have magnetospheres. The magnetospheres of pulsars have long been regarded as an essential part of the pulsar phenomenon. Other astrophysical systems, such as supernova remnant shells or magnetic stars and binary star systems, may be describable as magnetospheres. The major concepts of magnetospheric physics thus need to be formulated in a general way not restricted to the geophysical context in which they may have originated.

Magnetospheric convection has been one of the most important and fruitful concepts in the study of the Earth's magnetosphere. The purpose of this paper is to describe the basic theoretical notions of convection in a manner applicable to magnetospheres generally and to discuss the relative importance of convective and corotational motions, with particular reference to the comparison of the Earth and Jupiter.

As illustrated schematically in Figure 1, a magnetosphere contains a central object I (planet, star, or pulsar) with a magnetic field, surrounded by a region of space II in which the configuration and dynamics are largely influenced by the central object; outside of the magnetosphere lies the external medium III, which may be the solar wind (in the case of planets), the interstellar medium (magnetic stars), the stellar wind from one of the components (binary star systems), or the shell of swept-up interstellar material (supernova remnants). The magnetosphere shown in Figure 1 has an 'open' magnetic topology, i.e., magnetic field lines from the polar regions of the central object connect to magnetic field lines in the external medium. Observationally, the only magnetosphere for which the magnetic topology has been definitely established so far, that of the Earth, is open; theoretically, from many points of view, a closed magnetosphere is a very singular case; thus, in my opinion, any magnetosphere should be assumed to be open unless there is compelling evidence to the contrary. Accordingly, the rest of this paper is phrased in terms of an open magnetosphere (although the results can be transcribed for the case of a closed magnetosphere by substituting 'boundary layer' for 'external medium').

Rather different physical principles and equations apply in each of the three regions. The central object can usually be treated, to a good approximation, as a moving ohmic conductor; hence in region I the electric field **E** and the current

V. Formisano (ed.), The Magnetospheres of the Earth and Jupiter, 179–188. All Rights Reserved
Copyright © 1975 by D. Reidel Publishing Company, Dordrecht-Holland

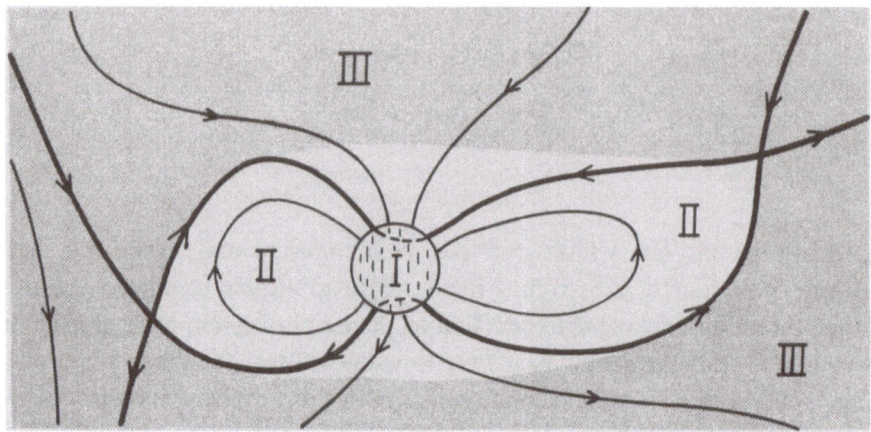

Fig. 1. A schematic representation of magnetic field lines in an open magnetosphere. The heavy lines represent the separatrix, a surface that separates open from closed magnetic field lines. The shading represents the external medium. In this and all succeeding figures, the orientation of all the vectors is that appropriate for the case of the Earth: magnetic dipole moment pointing down and angular momentum pointing up; the external medium flows from left to right.

density $\mathbf{J}$ are related by the equation (in Gaussian units)

$$\mathbf{E}+\frac{1}{c}\mathbf{V_s}\times\mathbf{B}=\boldsymbol{\eta}\cdot\mathbf{J} \qquad (1)$$

where $\boldsymbol{\eta}$ is the effective resistivity (including the Hall term) and $\mathbf{V_s}$ is the bulk flow velocity of the resistive medium. In the case of a planet, $\mathbf{V_s}$ is usually the velocity of the neutral atmosphere at ionospheric heights; in the case of a star, $\mathbf{V_s}$ is the velocity of the plasma in the outer (surface) layers; for a pulsar, $\mathbf{V_s}$ represents the rotational motion. In general, $\mathbf{V_s}$ is governed by the internal dynamics of the central object (or its atmosphere or outer layers) and to first approximation can be considered as given when discussing magnetospheric convection (for planetary atmospheres, however, the influence of convection itself on $\mathbf{V_s}$ is not always negligible).

In region II, on the other hand, $\mathbf{E}$ nd $\mathbf{J}$ are governed by the equations

$$\mathbf{E}+\frac{1}{c}\mathbf{V}\times\mathbf{B}=0 \qquad (2)$$

$$\frac{1}{c}\mathbf{J}\times\mathbf{B}=\varrho\left(\frac{\partial}{\partial t}+\mathbf{V}\cdot\nabla\right)\mathbf{V}+\nabla\cdot\mathbf{P}-\varrho\mathbf{g}, \qquad (3)$$

where $\mathbf{V}$ is the bulk flow velocity of the plasma and the right-hand side of Equation (3) represents the mechanical stresses acting on the plasma. In region III, the same Equations (2) and (3) apply, but the flow velocity $\mathbf{V}$ is now governed by the dynamics of the external medium and its interaction with the magnetosphere and to first approximation is little affected by magnetospheric convection. Finally, the three regions are coupled by the equations of current continuity and Faraday's law, which apply everywhere:

$$\nabla \cdot \mathbf{J} = 0 \tag{4}$$

$$\nabla \times \mathbf{E} + \frac{1}{c}\frac{\partial}{\partial t}\mathbf{B} = 0. \tag{5}$$

The simplest motion within the magnetosphere would correspond to plasma just comoving (which means, in most practical cases, corotating) with the central object. This implies that, at the interface between regions I and II, $\mathbf{V} = \mathbf{V}_s$; Equation (2), together with the continuity of the tangential components of $\mathbf{E}$ required by Equation (5), then implies that the left-hand side of Equation (1) is zero and hence $\mathbf{J} = 0$, from which together with Equation (4) it follows that there are no currents flowing between the central object and the magnetosphere. Comotion thus occurs only if the mechanical stresses are such that the currents implied by Equation (3) close entirely within the magnetosphere; as long as the stresses satisfy this condition, the plasma in the magnetosphere will comove with the central object even if the conductivity of the latter is not 'very high'. Conversely, departures from comotion require mechanical stresses such that Equations (3) and (4) imply current flow between the magnetosphere and the central object. As a very simple example, assume that plasma in the polar regions of the magnetosphere rotates uniformly but somewhat more slowly than the central object. In the frame of reference of the rotating object, there exists then an electric field $\mathbf{E}^*$ given by $c\mathbf{E}^* = (\mathbf{V}_s - \mathbf{V}) \times \mathbf{B}$, which vanishes at the pole and increases with decreasing latitude; the horizontal current then has a non-zero divergence which implies vertical currents flowing down along the magnetic field lines near the pole and up at lower latitudes, as illustrated in Figure 2. (We have implicitly assumed that the currents within the central object flow on a thin spherical shell, but the same qualitative results are obtained for distributed currents by solving the appropriate equation for the electric potential within the central object with $\mathbf{E}^*$ as the boundary condition.) These vertical currents must close somewhere within the magnetosphere or the external medium. It is obvious (although not always appreciated) that the closing currents must flow across field lines (no matter how the field lines might be twisted) and hence are associated with $\mathbf{J} \times \mathbf{B}$ forces which must be balanced by mechanical stresses. These stresses, it is easily verified, are in the sense required to oppose the rotation of the central object, and the difference in rotational speed, i.e., the 'slippage', between the magnetospheric plasma and the object is proportional to the external mechanical force divided by the integrated conductivity of the object. These concepts can be used to discuss the slowing down of pulsars (see, e.g., Cohen *et al.*, 1973).

In the case of planetary magnetospheres, however, the motions of interest are not pure corotation or simple slippage but rather the motions that result from the flow of the external medium (the solar wind) past the magnetosphere. Consider the 'open' magnetic field lines that connect the external medium with the central object (the part of the surface crossed by them is usually termed the *polar cap*). By virtue of Equation (2), the flow of the external medium is associated with an electric field

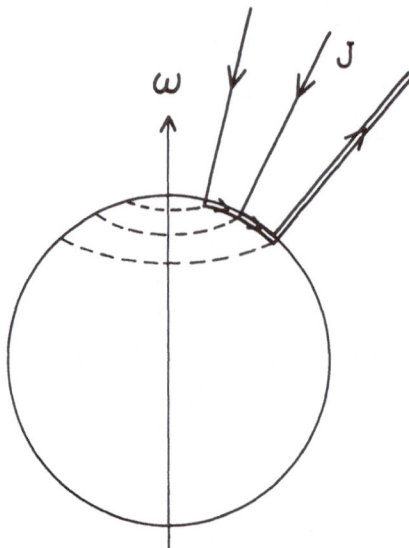

Fig. 2.   The pattern of currents when the polar region of the magnetosphere rotates more slowly than the central object. Only one hemisphere is shown. The pattern has axial symmetry about the rotation axis.

E; but Equation (2) also implies that (in a steady state) magnetic field lines are equipotentials and hence **E** is mapped down to the polar cap, as illustrated in Figure 3a. The tangential component of **E** must be continuous across the edge of the polar cap; thus **E** cannot be confined to the polar cap but must exist over the entire surface of the central object, having the qualitative pattern sketched in Figure 3a. Within the closed field line region, **E** is then mapped from the surface out into the entire magnetosphere. Equation (2) now implies a pattern of magnetospheric motions whose projection just above the surface is shown in Figure 3b. (We are neglecting rotation of the central object, whose effects we will consider later.) The motion of the external medium is directly impressed on the region traversed by the open field lines and continuity then requires a return flow within the closed field line region of the magnetosphere, giving rise to a convective circulation pattern first described by Axford and Hines (1961). (Axford and Hines dealt with a closed magnetosphere and the role of the polar cap was played by field lines crossing a boundary layer to which the motion of the external medium was assumed communicated by viscous drag; convection in an open magnetosphere was first described by Dungey (1961).)

The electric field associated with the convection also exists within the outer layers of the central object (or its ionosphere), where it drives currents in accordance with Equation (1). The horizontal divergence of the currents driven by the **E** pattern of Figure 3a requires currents along the magnetic field lines, down on one side of the polar cap and up on the other, as indicated schematically in Figure 3a. These field-aligned (Birkeland) currents have been termed the *driving currents* of magnetospheric convection (Vasyliunas, 1970); as far as the ionosphere is concerned, the convection electric field may be viewed as determined by the requirement of closing these currents. The existence of the driving currents in the Earth's magnetosphere has long

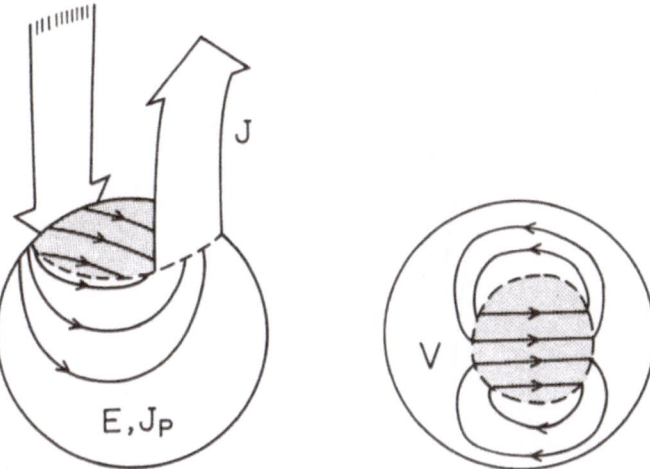

Fig. 3. (a) (*left*) Configuration of the electric field associated with convection and the required currents in and out of the central object. (The current density in the direction of **E** is the Pedersen current, in ionospheric terms.) (b) (*right*) The streamlines of the corresponding flow of plasma, which are also equipotentials of the electric field. Shading represents the polar cap. The pattern shown here results if no field-aligned currents other than the driving currents exist; it is, thus a sketch of the quantitative models obtained by Iwasaki and Nishida (1967), Vasyliunas (1970), and Wolf (1970).

been obvious on theoretical grounds (Vasyliunas, 1968; Schield *et al.*, 1969) and direct observational evidence for them has recently been obtained (Zmuda and Armstrong, 1974).

In addition to the driving currents, other sets of field-aligned currents between the magnetosphere and the central object may exist, particularly as a result of the plasma pressure gradients in the closed field line region of the magnetosphere. Each such set of Birkeland currents must close through the central object by means of suitable electric fields and hence modifies the convection pattern; since the pressure gradients themselves depend on the convection, the behavior of the entire system is governed by a closed self-consistent chain of equations that was described by Vasyliunas (1970) following the earlier work of Fejer (1964) and Swift (1967). Quantitative models of pressure gradient effects on the convection in the Earth's magnetosphere have been extensively described in the literature (see, e.g., Block, 1966; Swift, 1967; Karlson, 1971; Vasyliunas, 1972; Wolf, 1974; review by Boström, 1974) and will not be discussed here. Instead, I will consider the driving currents themselves, a topic little discussed in most of the previous work in which the detailed configuration of the driving currents (or, equivalently, the electric potential at the edge of the polar cap) has been simply postulated.

The driving currents are expected to close within the external medium, since the open field line region of the magnetosphere generally is unable to support the stresses required by Equation (3). As illustrated in Figure 4, the $\mathbf{J} \times \mathbf{B}$ force associated with the closure of the driving currents is always opposite to the direction of plasma flow in the external medium and hence should lead to deceleration of the external

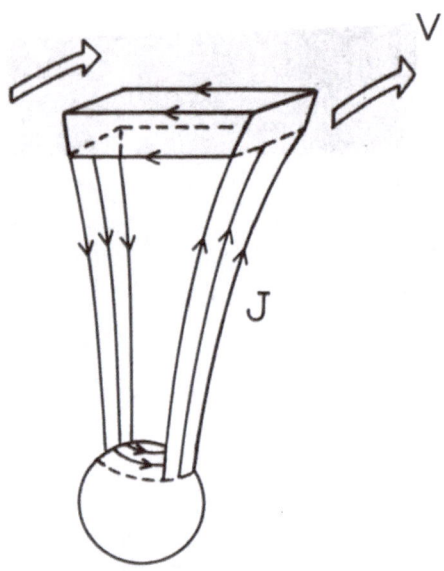

Fig. 4. Schematic illustration of the closure of the driving currents in the flowing external medium (shading above).

flow; the fractional decrease in speed required to accomplish the closure may be roughly estimated from order-of-magnitude approximations to Equations (1) and (3) as

$$\frac{\Delta V}{V} \sim \frac{4\pi \Sigma V_{\mathrm{A}}}{c^2},$$ (6)

where Σ is the integrated conductivity of the central object and V_{A} is the Alfvén speed in the external medium. Equivalently, estimate (6) may be obtained by requiring all the energy dissipated as Joule heat in the central object to be supplied by the decrease of mechanical energy in the external flow. (For the case of the Earth, $\Sigma \sim 1$ mho $= 100c$ and $V_{\mathrm{A}} \sim 100$ km s^{-1}, hence $\Delta V/V \ll 1$; from this point of view the Earth's ionosphere is a relatively poor conductor.) The slowing down of the external flow may be viewed as being produced by a drag force resulting from 'line-tying' to the resistive central object and the whole process is entirely analogous to the functioning of a magnetohydrodynamic generator, the external medium playing the role of flowing conducting fluid and the central object that of the resistive load. There is thus nothing mysterious about the driving currents, and in particular they do not involve in any crucial way the magnetic field line merging or reconnection process that is as yet little understood (see, e.g., review by Vasyliunas, 1975): the merging process is required in order to have an open magnetosphere, but given the fact that the magnetosphere *is* open, the configuration of the driving currents should be obtainable from the application of familiar MHD concepts.

So far we have discussed the convective circulation of the magnetosphere without considering the rotation or other intrinsic motions of the central object. Convection over the polar cap represents a mapping of the external flow and thus is not directly affected by the rotation. Within the closed field line region, the electric field (such as that sketched in Figure 3) was obtained by considering currents described by Equation (1) with $V_s = 0$ and thus is the electric field in a frame of reference moving with the central object; to obtain the electric field in a fixed frame of reference, we need to add the term $V_s \times B/c$ (suitably mapped out along the magnetic field lines). Now the superposition of a two-cell convective motion and a corotational motion leads to the flow topology illustrated in Figure 5a, where the heavy dot represents the stagnation point in the flow, at which the two motions cancel. This flow topology was independently proposed by Nishida (1966) and Brice (1967), who both pointed out that the inner region of closed circulating flow lines was likely to have a high plasma density and hence should be identified with the observed plasmasphere (the flow topology was also suggested earlier by Carpenter, 1962). However, this topology exists only as long as the stagnation point lies within the closed magnetic field line region; if the computed stagnation point should come out to be within the polar cap region, where the superposition of convective and corota-

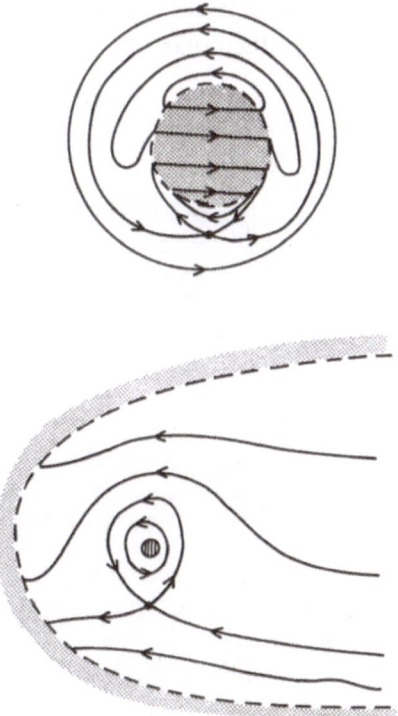

Fig. 5a.   Streamlines of the convective flow, including the effects of rotation, looking down on the surface of the central object (*top*) and projected along magnetic field lines to the equatorial plane of the magnetosphere (*bottom*), for the case when convection dominates over corotation. Shading represents the polar cap (*top*) and the equatorial projection of the part of the polar cap near its edge (*bottom*).

tional motions does not apply, the flow topology is instead that illustrated in Figure 5b: an essentially one-cell circulatory pattern. Brice and Ioannidis (1970) suggested, on the basis of a simple scaling of the Earth's convection electric field, that the topology of Figure 5b should exist in Jupiter's magnetosphere.

Which flow topology applies in any given case is obviously determined by whether the stagnation point computed from simple superposition lies within or outside the polar cap region. This leads to the following quantitative criterion: if Φ_0 is the potential difference across the polar cap due to the convection electric field alone and Φ_{CR} is the potential difference across half the polar cap due to the corotation electric field alone, then (omitting an unimportant numerical factor of order unity) the two-cell flow topology of Figure 5a applies if $\Phi_0 > \Phi_{CR}$ and the one-cell flow topology of Figure 5b applies if $\Phi_0 < \Phi_{CR}$. Assume for simplicity that the polar cap is a circle of colatitude θ around the coincident magnetic and rotation axis; then, approximately for small θ,

$$c\Phi_{CR} \approx \omega a^2 \theta^2 B_p / 2, \tag{7}$$

where a is the radius of the central object, ω the rotation frequency, and B_p the magnetic field at the pole. To estimate Φ_0, imagine following the magnetic field lines from the polar cap far out into the external medium, where the flow and the field are uniform; then the projection of the polar cap onto a plane containing **V** and **E**

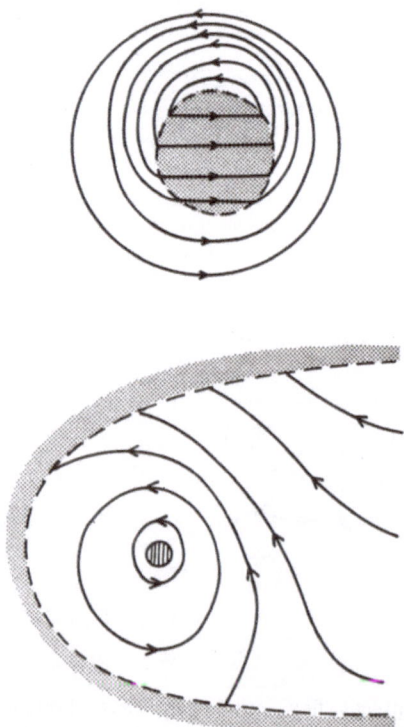

Fig. 5b.   Same as (5a) but for the case when corotation dominates over convection.

appears as sketched in Figure 6. In terms of the dimensions defined in the figure, we have

$$c\Phi_0 = lV_0B_0 \sin\theta_{VB}, \tag{8}$$

where V_0 and B_0 are the uniform magnitudes of **V** and **B** and θ_{VB} the angle between them. Also, from conservation of magnetic flux,

$$lLB_0 \sin\theta_{VB} \approx \pi a^2\theta^2 B_p. \tag{9}$$

Combining Equations (7), (8), and (9) we obtain

$$\frac{\Phi_{CR}}{\Phi_0} \approx \frac{\omega L}{2\pi V_0} \tag{10}$$

i.e., the ratio of corotation to convection potential is simply the distance L (roughly the length of the magnetospheric tail) divided by the distance the external medium flows in one rotation period of the central object. Assume for V_0 the solar wind speed of 400 km s^{-1}. Then, for the Earth $2\pi V_0/\omega = 5400\ R_E$, which is somewhat longer than the empirically estimated value of $L \approx 1000\ R_E$ (Dungey, 1965); hence the flow topology of Figure 5a should apply to the Earth's magnetosphere. For Jupiter, on the other hand, $2\pi V_0/\omega = 200$ Jovian radii; the value of L is not known but is expected to be much larger than 200 Jovian radii, since the latter is roughly the diameter of the front part of Jupiter's magnetosphere (Smith *et al.*, 1974; Wolfe *et al.*, 1974) and the magnetospheric tail is typically much longer than the dimension of the front part. Hence, in agreement with the suggestion of Brice and Ioannidis, the convective flow in Jupiter's magnetosphere should have the topology of Figure 5b, with corotation being the dominant effect rather than, as in the case of the Earth, a small (albeit significant) modification. (It should be noted that this calculation, unlike that of Brice and Ioannidis, deals only with electric fields in the ionosphere and does not assume a dipolar magnetic field in the outer regions of Jupiter's magnetosphere. On the other hand, we have simply assumed that Jupiter's atmosphere at ionospheric heights corotates with the planet and have not considered the question raised by Kennel and Coroniti (1975) whether the dynamical effects of convection may be large enough to modify greatly the motions of the upper atmosphere.)

In summary, the theory of magnetospheric convection is in essence the unraveling of the electric fields and currents that are required by the principles of magneto-

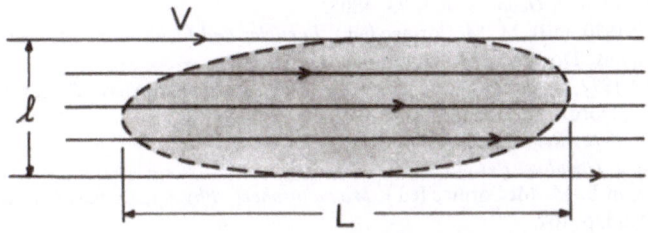

Fig. 6.   Schematic representation of the polar cap (shaded) projected along magnetic field lines into the external medium.

hydrodynamics when a moving plasma is connected by magnetic field lines to a resistive object. The theory provides a unified description of plasma motions in magnetospheres as diverse as those of the Earth and Jupiter and can also be applied to a variety of astrophysical systems. The basic physical principles of the theory appear to be well understood; what is needed is, on the one hand, a further development of detailed quantitative models (including in particular the hitherto neglected description of convection on open field lines) and, on the other hand, a better understanding of important related topics such as the magnetic field merging process, the configuration of open magnetospheres, and the dynamical interaction between magnetospheric convection and the motions of neutral atmospheres.

Acknowledgements

This work was supported by the National Aeronautics and Space Administration under grant NGL 22-009-015. I am grateful to the Laboratorio Plasma nello Spazio and the Consiglio Nazionale delle Richerche of Italy for providing travel support to attend the Neil Brice Memorial Symposium, and to Dr W. I. Axford and Dr G. Haerendel for useful comments on the manuscript.

References

Axford, W. I. and Hines, C. O.: 1961, *Can. J. Phys.* **39**, 1433.

Block, L. P.: 1966, *J. Geophys. Res.* **71**, 855.

Boström, R.: 1974, in B. M. McCormac (ed.), *Magnetospheric Physics*, D. Reidel Publishing Company, Dordrecht-Holland, p. 45.

Brice, N. M.: 1967, *J. Geophys. Res.* **72**, 5193.

Brice, N. M. and Ioannidis, G. A.: 1970, *Icarus* **13**, 173.

Carpenter, D. L.: 1962, Thesis, Stanford University, Palo Alto, California.

Cohen, R. H., Coppi, B., and Treves, A.: 1973, *Astrophys. J.* **179**, 269.

Dungey, J. W.: 1961, *J. Geophys. Res.* **70**, 1753.

Fejer, J. A.: 1964, *J. Geophys. Res.* **69**, 123.

Iwasaki, N. and Nishida, A.: 1967, *Rept. Ionosph. Space Res. Japan* **21**, 17.

Karlson, E.: 1971, *Cosmic Electrodyn.* **1**, 474.

Kennel, C. F. and F. V. Coroniti: 1975, this volume, p. 451.

Nishida, A.: 1966, *J. Geophys. Res.* **72**, 5669.

Schield, M. A., Freeman, J. W., and Dessler, A. J.: 1969, *J. Geophys. Res.* **74**, 247.

Smith, E. J., Davis, L., Jr., Jones, D. E., Colburn, D. S., Coleman, P. J., Jr., Dyal, P., and Sonett, C. P.: 1974, *Science* **183**, 305.

Swift, D. W.: 1967, *Planetary Space Sci.* **15**, 835.

Vasyliunas, V. M.: 1968, *J. Geophys. Res.* **73**, 5805.

Vasyliunas, V. M.: 1970, in B. M. McCormac (ed.), *Particles and Fields in the Magnetosphere*, D. Reidel Publishing Company, Dordrecht-Holland, p. 60.

Vasyliunas, V. M.: 1972, in B. M. McCormac (ed.), *Earth's Magnetospheric Processes*, D. Reidel Publishing Company, Dordrecht-Holland, p. 29.

Vasyliunas, V. M.: 1975, *Rev. Geophys. Space Phys.* **13**, No. 1.

Wolf, R. A.: 1970, *J. Geophys. Res.* **75**, 4677.

Wolf, R. A.: 1974, in B. M. McCormac (ed.), *Magnetospheric Physics*, D. Reidel Publishing Company, Dordrecht-Holland, p. 167.

Wolfe, J. H., Collard, H. R., Mihalov, J. P., and Intriligator, D. S.: 1974, *Science* **183**, 303.

Zmuda, A. J. and Armstrong, J. C.: 1974, *J. Geophys. Res.* **79**, 4611.

ELECTRIC FIELD EFFECTS ON THE ACCESS OF MAGNETOSHEATH PLASMA TO THE DISTANT MAGNETOTAIL

M. K. BIRD

Institut für Astrophysik und Extraterrestrische Forschung, Universität Bonn, Auf dem Hügel 71, 53 Bonn, F.R.G.

1. Introduction

The low-energy background plasma populating the plasma sheet of the magneto-tail has arrived there either via the polar wind (Banks and Holzer, 1968) or via the solar wind through magnetic 'soft spots' on the magnetopause (Stevenson and Comstock, 1968). The conjugate neutral points on the dayside magnetopause have been suggested by Frank (1971) as logical access zones whereby solar wind plasma convects into the nightside plasma sheet from the observed polar cusp regions (Heikkila *et al.*, 1972). Eviatar and Wolf (1968), Beard *et al.*, (1970), and other authors have proposed a more direct entry of plasma to the plasma sheet through the lateral flanks of the extended cylindrical magnetotail. Trajectory calculations by Bird (1974) in a simple analytical model of the magnetotail field/magnetosheath interface sup-port the flank access theory. Solar wind protons impinging on the dawn magneto-tail boundary are caught up in the tail field provided their entry point is close enough to the quasi-neutral plane.

The field model used in the trajectory calculations of Bird (1974) was based on a self-consistent calculation of the tail magnetic field under very quiet conditions (Bird and Beard, 1972). Only drift and magnetization currents in the plasma sheet and on the tail boundary were considered sources of the observed magnetic field. Recent trajectory calculations by Pudovkin and Tsyganenko (1973) yielded results qualitatively similar to those of Bird (1974), but no analysis of particle access from the magnetosheath was made. Both models assumed the electric field to be zero. A more realistic model, in view of the highly variable solar wind flux, would be to include the possibility of non-vanishing electric fields. The effects of a constant dawn-to-dusk electric field of arbitrary strength on the penetration of magnetosheath protons along the dawn magnetotail flank is investigated in this paper. A complete description of the self-consistent magnetic configuration and its analytical repre-sentation are presented in Bird (1974) and will not be repeated here.

2. Proton Trajectories in a Model Magnetotail

2.1. PARTICLE ACCESS TO MAGNETOTAIL WITH $E = 0$

The coordinate system used here is an idealized solar magnetospheric system with its origin at the center of the Earth, the x-axis directed along the magnetotail, and

V. Formisano (ed.), The Magnetospheres of the Earth and Jupiter, 189–196. All Rights Reserved

the z-axis aligned perpendicular to the quasi-neutral sheet, which is centered on the xy-plane. The zero-divergence magnetic field and corresponding current density of the magnetotail in this coordinate system is taken to be

$$B_x = B_0 \left(\frac{x_0}{x}\right)^\alpha \tanh\left(\frac{z}{z_0}\right) \tag{1}$$

$$B_z = \eta B_0 \left(\frac{x_0}{x}\right)^{\alpha+1} \left\{1 - C_0 \ln \cosh\left(\frac{z}{z_0}\right)\right\} \tag{2}$$

$$j_y \simeq \frac{c}{4\pi} \frac{B_0}{z_0} \left(\frac{x_0}{x}\right)^\alpha \operatorname{sech}^2\left(\frac{z}{z_0}\right) \tag{3}$$

where, for the region of validity of these equations ($-100\ R_E < x < -30\ R_E$), the following parametric values are appropriate

$$
\begin{aligned}
&B_0 = 10\ \gamma & &\alpha = 0.3 \\
&x_0 = -60\ R_E & &\eta = 0.1 \\
&z_0 = 4\ R_E & &C_0 = 0.2.
\end{aligned}
\tag{4}
$$

The field line configuration of the magnetotail field represented by (1) and (2) with the parameters (4) is displayed in Figure 1. The normal component B_z is always positive, and the y-component of $\mathbf{B}$ is taken in this approximation to be zero. The parameter z_0, which corresponds roughly with the half-thickness of the plasma sheet,

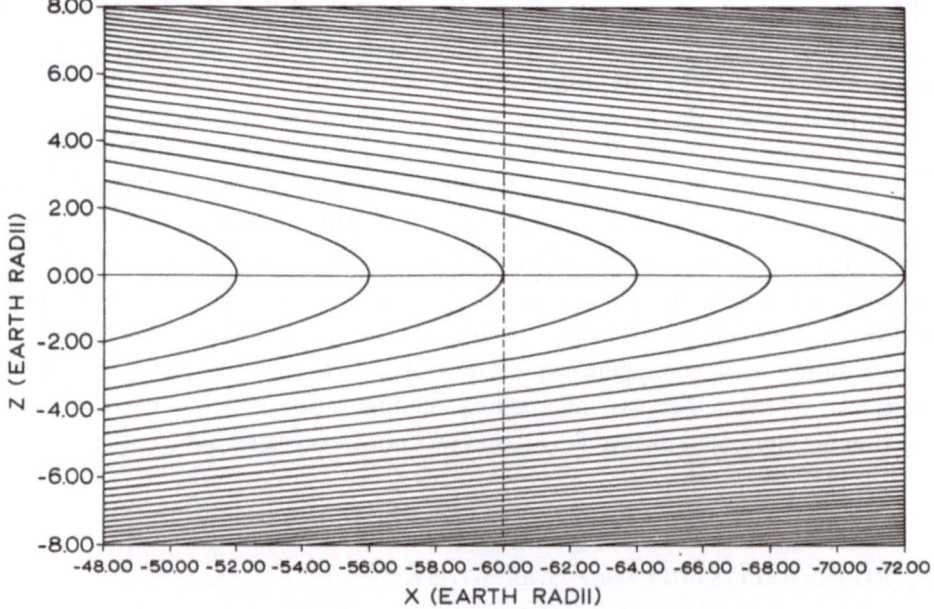

Fig. 1.   Model magnetic field lines in distant magnetotail. The above field line structure is that of the field given by Equations (1) and (2) in text. Only field lines that intersect the xy-plane (quasi-neutral sheet) at $x = -4n$, where n is an integer, have been drawn.

is quite compatible with the data of Meng and Mihalov (1972) and of Nishida and Lyon (1972) who report plasma sheets 8–10 R_E thick even at the lunar distance. Behannon (1968) first suggested the inverse power dependence of the main tail field component and his analysis yielded $\alpha = 0.3 \pm 0.2$ as a most probable value for the fall-off exponent.

The proton trajectories are followed in the magnetic field (1) and (2) by numerically integrating the equation of motion

$$m\frac{d\mathbf{v}}{dt} = e\left\{\mathbf{E} + \frac{1}{c}\,\mathbf{v} \times \mathbf{B}\right\}. \tag{5}$$

The protons are assumed to impinge at time $t=0$ upon the magnetopause, which along the dawn flank of the magnetotail is approximated by the plane $y=y_0$. The incoming protons have a given energy ε_i (which remains constant when $\mathbf{E}=0$), a given impact point on the magnetopause (x_i, z_i), and an initial angle of incidence specified by the angles (θ_i, ϕ_i) defined as follows:

$$\theta_i = \cos^{-1}\left(\frac{v_{iy}}{v_0}\right) \tag{6}$$

$$\phi_i = \tan^{-1}\left(\frac{v_{iz}}{v_{ix}}\right) \tag{7}$$

with (v_{ix}, v_{iy}, v_{iz}) the initial velocity vector in the plane $y=y_0$ at $t=0$, and $v_0 = \sqrt{v_{ix}^2 + v_{iy}^2 + v_{iz}^2}$. Specification of a particle's initial energy, impact point, and angles (θ_i, ϕ_i) completely determine the trajectory, which is followed until either (a) the particle crosses from the magnetotail back into the magnetosheath (y becomes less than y_0) at which point it is assumed 'repelled' by the magnetotail field, or (b) the particle progresses deep enough into the tail that it is indefinitely trapped. This second possibility was determined inductively to occur whenever $y > y_0 + 2a_{max}$, where a_{max} is the maximum proton gyroradius in the magnetotail field at the point $(x, z) = (x_i, 0)$

$$a_{max} = \frac{mcv_0}{e\eta B_0}. \tag{8}$$

It was found that all protons penetrating at least $2a_{max}$ into the magnetotail would drift steadily toward the dusk flank while executing bounce motion about the quasi-neutral sheet. The drift motion proceeded with velocities typically two orders of magnitude lower than the particle's kinetic velocity, resulting nonetheless in a sizeable sheet current that seems capable of supporting the entire observed magnetotail magnetic field.

For incident magnetosheath protons of energy $\varepsilon_i = 1$ keV and initial impact point $(x_i, z_i) = (-60 R_E, 0)$, only certain values of the angles (θ_i, ϕ_i) will result in a proton being trapped in the tail under condition (b) above. Figure 2 is a velocity space diagram with θ_i as radial coordinate and ϕ_i as the azimuthal coordinate that shows the combinations (θ_i, ϕ_i) that result in indefinite proton penetration at the dawn

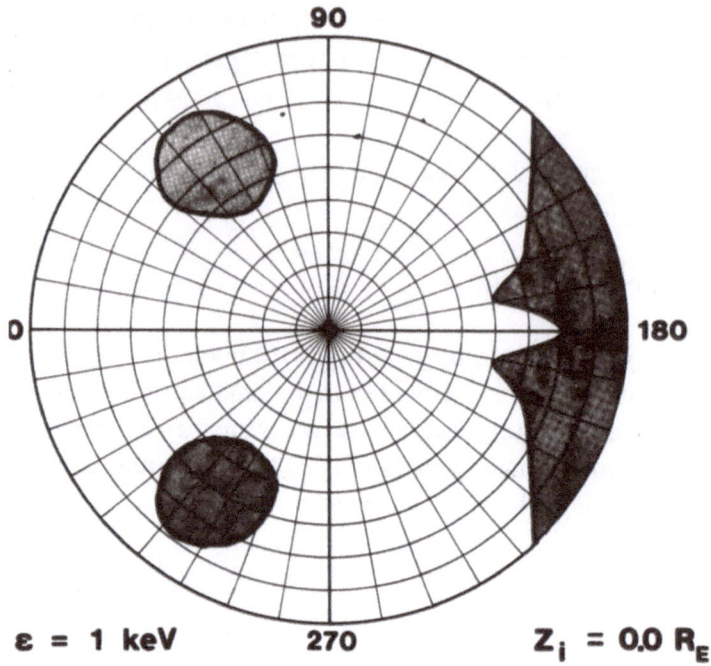

Fig. 2.   Proton penetration at dawn magnetotail flank for $\Phi_0 = 0$. This velocity-space diagram uses θ_i as the polar coordinate (from 0 at center to 90° at outer circle) and ϕ_i as the azimuthal coordinate as in-dicated. See text for definition of (θ_i, ϕ_i). Shaded regions are those values of θ_i and ϕ_i for which 1 keV protons impacting at the point $(x_i, z_i) = (-60\ R_E, 0)$ penetrate indefinitely into the tail. Protons with initial velocity vectors lying in non-shaded regions of the diagrams will be repelled from the tail.

flank (shaded region). It is seen that incident velocities around $(\theta_i, \phi_i) = (\pi/2, \pi)$ are the most susceptible to being trapped. These particles have a velocity in the mag-netosheath very near that of the undisturbed solar wind (flowing in the $-x$ direction), and it is thus possible that a large fraction of the solar wind protons impacting in the distant magnetotail will gain access to the plasma sheet. The shaded regions of indefinite penetration are larger for higher incident energy, and they tend to shrink rapidly with increasing distance of the initial impact point z_i from the quasi-neutral sheet.

2.2. PARTICLE ACCESS TO MAGNETOTAIL WITH $\mathbf{E} \neq 0$

Models of the geomagnetic tail for which an inherent dawn-to-dusk electric field is present have been developed by Speiser (1965, 1967), Cowley (1971, 1973), and other authors. Annihilation and reconnection of oppositely directed field lines at the neutral sheet together with considerable energizing of the ambient plasma are salient fea-tures of the models, which appears to help explain the observed plasma fluxes during geomagnetic storms (Burke and Reasoner, 1973). The role of the electric field in the quiescent magnetotail remains a widely debated topic, and it is appropriate to ex-amine the possible effects such a field would have on the penetration of magnetosheath protons to the plasma sheet of the distant magnetotail. Only a constant uni-directional electric field $(\mathbf{E} = E_0 \hat{\mathbf{y}})$ of varying strength will be superimposed on the magnetic field (1) and (2), and the trajectories will be computed according to the scheme outlined in the previous subsection. The particle energy will no longer remain a constant, but rather vary according to

$$\tfrac{1}{2}mv^2 + eE_0 y = \text{const} \tag{9}$$

and the particles will also no longer be tied to a given 'drift shell', since they now experience a non-current producing drift given by

$$\mathbf{v}_D = \frac{c}{B^2} E_0 \hat{\mathbf{y}} \times \mathbf{B}. \tag{10}$$

Alfvén (1968) first derived a compact formula for the expected self-consistent dawn-to-dusk potential drop for a simple neutral sheet with no normal magnetic field component. He obtained

$$\Phi_A = E_0 d = \frac{B_0^2}{4\pi e N_0} \tag{11}$$

where d is the total magnetotail width (taken to be 40 R_E), and N_0 is the plasma number density at the center of the plasma sheet. If one further requires, as the observations appear to verify, that the magnetic pressure in the main magnetotail lobe be equal to the particle pressure in the center of the plasma sheet at the same value of x, then

$$\frac{B_0^2}{8\pi} = p = \tfrac{2}{3} N_0 \langle \varepsilon \rangle \tag{12}$$

where $\langle \varepsilon \rangle$ is the mean thermal energy of the ambient plasma. Combining (11) with (12) one obtains

$$E_0 = \frac{\Phi_A}{d} = \frac{4 \langle \varepsilon \rangle}{3ed} \tag{13}$$

which for $\langle \varepsilon \rangle \simeq 1$ keV, results in a cross-tail potential $\Phi_A \simeq 1.33$ kV. This value is considerably lower than that assumed by Speiser (1967) or by Alfvén (1968).

Trajectory calculations performed with an electric field corresponding to the cross-tail potential Φ_A showed virtually no change from the zero potential trajectories with respect to particle penetration at the dawn flank. The velocity-space diagram for a cross-tail potential $\Phi_0 = \Phi_A$ and impact point $(x_i, z_i) = (-60\ R_E, 0)$ is almost indistinguishable from Figure 2. If the potential Φ_0 is, however, arbitrarily raised by a factor of 10 or 100, the velocity diagram at the same impact point with the same initial energy changes its form to that displayed in Figure 3. The electric field is directed such that it enhances the penetrability of positively charged particles at the dawn flank, and the shaded region of indefinite penetration grows rapidly for increasing cross-tail potentials between 10 and 100 kV.

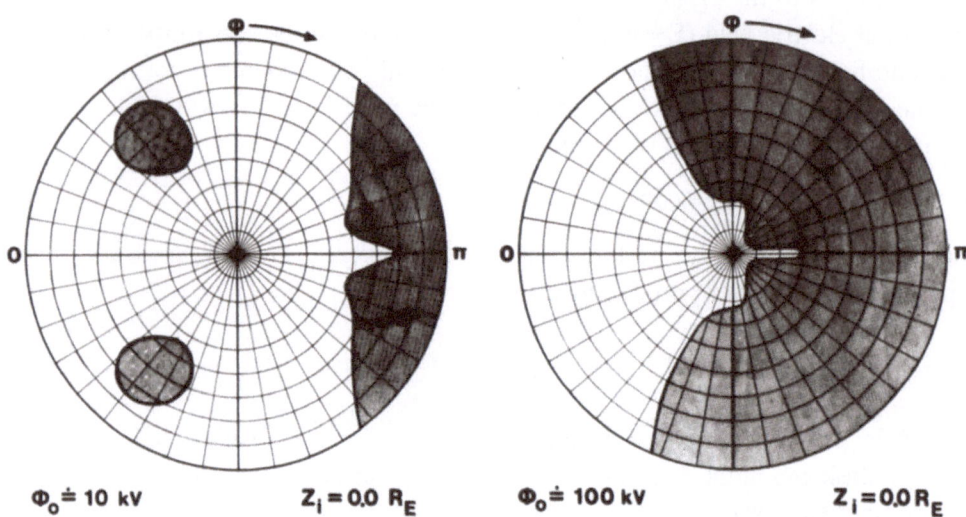

$\Phi_0 \doteq 10$ kV $Z_i = 0.0\ R_E$ $\Phi_0 \doteq 100$ kV $Z_i = 0.0\ R_E$

Fig. 3.   Proton penetration at dawn magnetotail flank for $\Phi_0 \neq 0$. Refer to Figure 2 for explanation of the coordinates (θ_i, ϕ_i) used here. The inclusion of a constant dawn-to-dusk electric field results in enhanced proton penetration (shaded regions increase in size for increasing electric field strength). Compare these diagrams with Figure 2, for which $\Phi_0 = 0$. Incident particle energy and impact point on the dawn magnetopause are kept a constant $\varepsilon_i = 1$ keV and $(x_i, z_i) = (-60\ R_E, 0)$ respectively.

The accessibility of magnetosheath protons with incident velocity vectors in the $-x$ direction are of particular importance since the bulk of the magnetosheath plasma is expected to flow in this approximate direction along the flanks of the distant tail. For incidence angles $(\theta_i, \phi_i) = (\pi/2, \pi)$ of protons impacting at $x_i = -60\ R_E$ there exists for each proton energy and each cross-tail potential a definite maximum value of the impact point $z_{i\ max}$, above which all incident particles are repelled (i.e. only protons with $|z_i| < z_{i\ max}$ can penetrate into the plasma sheet). The variation of $z_{i\ max}$ as a function of the cross-tail potential Φ_0 for different values of the initial proton energy ε_i is shown in Figure 4. The width of the penetration 'window' on the dawn flank is $2z_{i\ max}$ and is about 4 R_E thick for 1 keV protons in tail models with weak electric fields $(\Phi_0 \simeq \Phi_A)$. This width increases with increasing Φ_0 at all particle energies, although the penetration of 100 keV particles is not significantly enhanced by higher Φ_0. Indeed, for cross-tail potentials $\Phi_0 \simeq 100\ \Phi_A$, the penetration window $2z_{i\ max}$ for 1 or 10 keV particles is actually larger than for 100 keV protons. At lower potentials the value of $z_{i\ max}$ is generally greater for higher particle energies since the $+y$ drift velocity is, to a first approximation, linearly proportional to the proton's energy.

Eastwood (1972), considering the effects of the normal component B_z on the required electric field, determined that the self-consistency of the model would be retained provided the cross-tail potential were raised according to

$$\Phi_0 \simeq \Phi_A \frac{d}{2a_{max}} \tag{14}$$

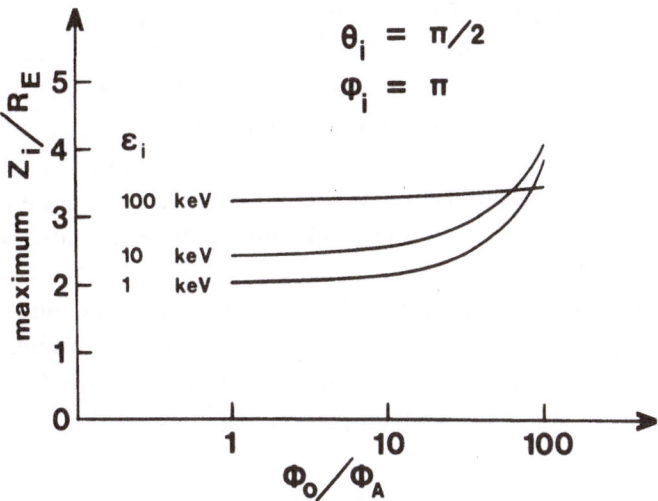

Fig. 4.   Maximum incident value of z_i for penetration of undisturbed solar wind particles. The angles of incidence are fixed here to represent protons flowing in the $-x$ direction ($\theta_i = 90°$, $\phi_i = 180°$). The impact point of the protons is $(x_i, z_i) = (-60\ R_E, z_i)$ and only incoming particles with $z_i < z_{i\ max}$ can gain access to the plasma sheet in the distant magnetotail. Φ_0 is measured in units of Φ_A as defined in the text $(\Phi_A = 1.33\ \text{kV})$.

which for $a_{max} = 0.72\ R_E\ (\varepsilon_i = 1\ \text{keV})$, yields $\Phi_0 \simeq 37$ kV, a value which would produce a velocity-space diagram intermediate to the two in Figure 3.

Whereas only constant single-component electric fields are considered here, it should be pointed out that Cowley (1971) has shown that the potential gradient across the tail is highest toward the dusk flank. The electric field E_y in the Cowley model reaches its minimum at the dawn flank and would thus be expected to be of lesser significance there for particle penetration than along the dusk flank, where solar wind electrons can be trapped upon entering the tail from the magnetosheath.

3. Summary

Magnetosheath protons incident on a model magnetopause on the dawn flank of the distant geomagnetic tail are found to be trapped within the magnetic field for certain combinations of impact energy, angle of incidence, and impact point. The trapped protons propagate slowly from the dawn to the dusk side of the tail and produce an appreciable drift current, which is an important source of the tail magnetic field. Solar wind electrons enter the tail on the dusk side and produce a similar although much weaker drift current. Solar wind protons moving along the tail in the $-x$ direction are caught up in the tail's plasma sheet at $x_i = -60\ R_E$ as long as the impact point $|z_i| < 2\ R_E$. The width of the window for particle penetration $2z_{i\ max}$ increases for higher energies.

The inclusion of a constant dawn-to-dusk electric field in the trajectory calculations produces no noticeable effect on the regions of penetration (shaded regions in the incident velocity-space diagrams) for cross tail potentials $\Phi_0 < 5$ kV. For potentials $\Phi_0 > 5$ kV a definite enhancement in the penetrability of incident protons is realized both in the velocity-space diagrams at $z_i = 0$ and in the penetration window width $2z_{i\,max}$ for particles with incident velocities in the $-x$ direction $(\theta_i = \pi/2, \phi_i = \pi)$.

In view of the oversimplification of the structure of the dawn flank magnetopause and the magnetotail magnetic field assumed in these calculations, a direct application to the real magnetotail should be made only with reservation. Nevertheless, it is felt that the model demonstrates the penetration capability of low-energy protons entering the distant magnetotail within a few Earth radii of the quasi-neutral sheet. A substantial dawn-to-dusk electric field increases the susceptibility of incident protons to indefinite penetration and supports further the hypothesis that a large fraction of the magnetotail background plasma originates in the solar wind.

References

Alfvén, H.: 1968, *J. Geophys. Res.* **73**, 4379.
Banks. P. M. and Holzer, T. E.: 1968, *J. Geophys. Res.* **73**, 6846.
Beard, D. B., Bird, M., and Huang, Y. H.: 1970, *Planetary Space Sci.* **18**, 1349.
Behannon, K. W.: 1968, *J. Geophys. Res.* **73**, 907.
Bird, M. K. and Beard, D. B.: 1972, *Planetary Space Sci.* **20**, 2057.
Bird, M. K.: 1974, *Planetary Space Sci.* (in press).
Burke, W. J. and Reasoner, D. L.: 1973, *J. Geophys. Res.* **78**, 6790.
Cowley, S. W. H.: 1971, *Cosmic Electrodyn.* **2**, 90.
Cowley, S. W. H.: 1973, *Cosmic Electrodyn.* **3**, 448.
Eastwood, J. W.: 1972, *Planetary Space Sci.* **20**, 1555.
Eviatar, A. and Wolf, R. A.: 1968, *J. Geophys. Res.* **73**, 5561.
Frank, L. A.: 1971, *J. Geophys. Res.* **76**, 2512.
Heikkila, W. J., Winningham, J. D., Eather, R. H., and Akasofu, S.-I.: 1972, *J. Geophys. Res.* **77**, 4100.
Meng, C. I. and Mihalov, J. D.: 1972, *J. Geophys. Res.* **77**, 4661.
Nishida, A. and Lyon, E. F.: 1972, *J. Geophys. Res.* **77**, 4086.
Pudovkin, M. I. and Tsyganenko, N. A.: 1973, *Planetary Space Sci.* **21**, 2027.
Speiser, T. W.: 1965, *J. Geophys. Res.* **70**, 4219.
Speiser, T. W.: 1967, *J. Geophys. Res.* **72**, 3919.
Stevenson, T. E. and Comstock, C.: 1968, *J. Geophys. Res.* **73**, 175.

NEUTRAL SHEETS

J. W. DUNGEY

Imperial College, Physics Dept., London SW7, England

1. Introduction

As the theory has progressed, the topic of neutral sheets has developed into an extensive subject area. This statement is not based on neutral sheets occurring frequently in nature, though they well may, but on the complexity of the neutral sheet in the geomagnetic tail, which is the best observed and towards which most theoretical work is aimed. It will be found that the structure and behaviour of a neutral sheet depends on the charged particle population in neighbouring regions, so that other cases would require modified models. 'Neutral sheet' generally means a thin layer separating regions in which the fields are approximately uniform, but substantially different, and to justify the name the field strength should drop at least an order of magnitude below the strength on either side. A natural simplification, appropriate here, is to assume symmetry so that the fields on the two sides are equal and opposite, except for any small component normal to the sheet. The thinness of the sheet is the one simplifying feature of the problem, but also the reason why the guiding centre approach is invalid, with the consequence that neutral sheets remain a relatively poorly understood part of plasma physics.

Although the model of the magnetic field, outlined above, is justified for the tail by direct observation, the topology of the magnetosphere should be mentioned, though only the usual, simple possibility need be described. The closed field lines are taken to occupy a single region and the tail field lines occupy two regions, north and south, though they must connect somewhere, normally outside the bow shock. The surfaces separating different regions must be parallel to the field, since they separate field lines of different type. These surfaces intersect in a closed curve, which mathematicians would call a separatrix, and this must also be parallel to the field. This curve runs through the magnetopause between the clefts and through the neutral sheet. In a meridian section the point of intersection of the separatrix looks like a magnetic neutral point and in two dimensional models of the tail it becomes a neutral line. Interest in the topology arises because the quantity of flux in the tail field lines is an important parameter for the state of the magnetosphere. For instance the distance of the magnetopause depends primarily on the solar wind pressure, but significantly also on the tail flux and the configuration on the night side is more sensitive to the tail flux. Now there is a rigorous relation, derived directly from Faraday's law, which states that the rate of change of the tail flux equals the EMF round the separatrix. This is the reason why determination of the electric field on the 'neutral line' is a major objective. Taking the sheet to be a plane normal to the z-direction (northwards), the simple topology gives a region of closed field lines with

V. Formisano (ed.), The Magnetospheres of the Earth and Jupiter, 197–204. All Rights Reserved
Copyright © 1975 by D. Reidel Publishing Company, Dordrecht-Holland

$B_z > 0$ separated by a single 'neutral line', which may be curved, from the region of interplanetary lines with $B_z < 0$.

The rate of change of B_z, and hence also motion of the neutral line, is given by $(\text{curl}\,\mathbf{E})_z$. In magnetospheric coordinates the cross-tail electric field is E_y and $(\text{curl}\,\mathbf{E})_z$ may be approximately $\partial E_y / \partial x$.

It will be seen that, while considerable progress can be made with two dimensional models, the complete problem has some essentially three-dimensional features. It is to be hoped that existing knowledge of two-dimensional models will enable the building of an adequate three-dimensional model. One observed feature demonstrating the inadequacy of two-dimensional models is the Harang discontinuity which is most commonly observed near midnight. However near the auroral oval the discontinuity is near 2100 local time (Heppner, 1972) and probably maps out to the dusk edge of the neutral sheet (Cowley, 1973b).

2. Simple Models

The basic results concerning particle trajectories will first be summarised. The field is dominated by the component $B_x(z)$, an odd function of z, and, in the y, z-plane, trajectories crossing the plane $z = 0$ are shown in Figure 1.

The average y-velocity can have either sign, but there is a preference for motion agreeing with the current, and hence giving a gain in energy when there is an E_y corresponding to some effective resistivity. The oscillation of z can be seen mathematically from the component $v_y B_x$, if v_y is approximately constant and B_x approximately proportional to z. With the same conditions the WKB approximation is valid and the amplitude of oscillation of v_z is proportional to $v_y^{1/4}$.

For any static model the electric field has a potential ϕ and the energy $\frac{1}{2}mv^2 + q\phi$ is an integral of the motion. For sufficiently simple models integrals of canonical momentum are useful. The only restriction made throughout this paper is $B_y = 0$ and the general momentum equations are

$$m(v_x) = q\left\{\int B_z \, dy + \int E_x \, dt\right\} \tag{1}$$

$$m(v_y) = q\left\{\int B_x \, dz - \int B_z \, dx + \int E_y \, dt\right\} \tag{2}$$

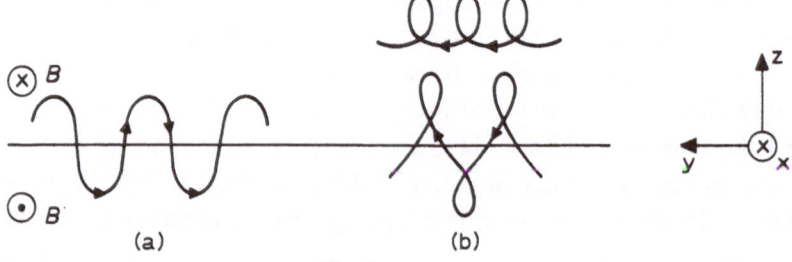

(a)   (b)

Fig. 1.

the integrals following the trajectory. Schindler and Soop (1968) have given a general treatment for distribution functions, which are functions only of the energy and y-component of canonical momentum. Self-consistent solutions are found and the tearing instability is also treated. They cannot however allow any quantity to vary with y and they can only guess their distribution function, and it will be seen that these difficulties are related.

When E_y is included, particles on both sides of the neutral sheet drift into the sheet and subsequently move along trajectories like those shown in Figure 1, except that now E_y ensures that they all eventually move in such a way as to gain energy. A further complication of the model is needed to prevent all particles from gaining energy indefinitely. Inclusion of B_z as first discussed by Speiser (1965) will be deferred to Section 5. Alfvén's (1968) approach was simply to make the model finite in the y-direction, as the real tail is, but the resulting problem is not as simple as it looks. Alfvén did however obtain a value Φ_A for our crucial quantity the cross-tail potential and he used a short cut which remains very valuable in tackling more complicated models. After considering trajectories in a similar way to the above description, Alfvén concluded that all the electrons drifting into the neutral sheet would go out at the dawn edge and all the protons at the dusk edge. This gave him a relation between the electric current and the cross-tail potential which he combined with consistency between the current and magnetic field to give

$$\Phi_A = B_x^2/4\pi\, ne, \tag{3}$$

where B_x and n refer to the uniform regions on either side of the sheet. For $B_x = 15\ \gamma$, $\Phi_A \sim n^{-1}$ kV and with $n \sim 10^{-1}$ cm^{-3}, the order of magnitude fits with various estimates based indirectly on observation, though it seems likely that large values are confined to expansion phases. The discussion of the next section will suggest that the cross-tail potential is substantially less than Φ_A unless noise causes anomalous resistivity.

3. Developments Based on Alfvén's Model

Cowley (1971) pointed out that the assumption of a uniform electric field in Alfvén's model is not self-consistent, because space charge occurs. Because of their different masses, protons would spend longer in the neutral sheet than electrons and a large positive space charge would result. Cowley (1973a) generalised the model to allow any electrostatic field and solved for the self-consistent potential. He found that the plasma approximation is valid, and that there is indeed positive charge in the sheet such that all the equipotentials go nearly to the dusk edge of the sheet as shown in Figure 2. He found that the dimensions of the region where the potential varies rapidly near the dusk edge are both of the order of the 'proton plasma wavelength' $(m_p c^2/4\pi\, ne^2)^{1/2}$, but the structure of this region has not been studied. It may be affected by particles coming from the magnetosheath, which will be discussed next. On either side of the neutral sheet in Figure 2 all charged particles drift approximately

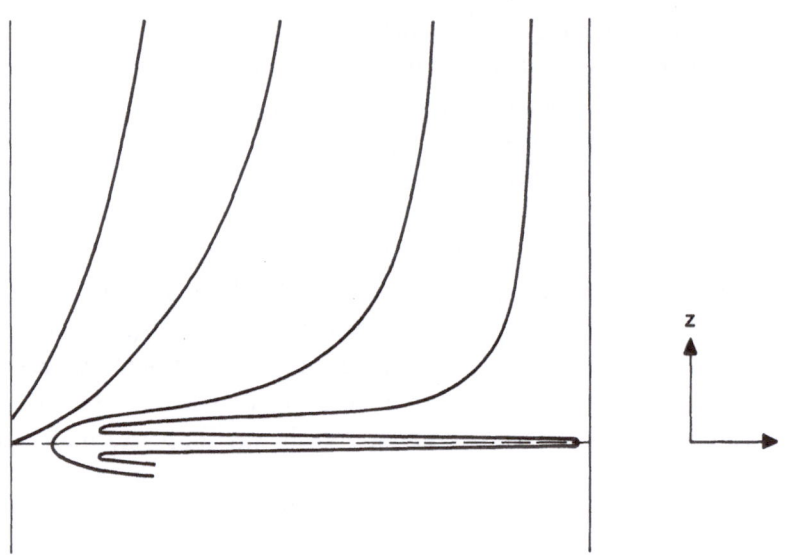

Fig. 2.

along the equipotentials and the self-consistency problem is equivalent to an aero-dynamic problem with a sink at the dusk edge of the neutral sheet.

Even in the absence of any electric field the neutral sheet would be accessible to particles from the magnetosheath entering at the edges. They would carry an electric current and this has been discussed previously in terms of their pressure (Dungey, 1972). It seems probable that these particles carry the greater part but not the whole of the total current, and it now seems desirable to modify Cowley's model to include such particles. The shape of the equipotentials may be unchanged, but, applying Alfvén's argument, the cross-tail potential Φ should be $\Phi_A(1 - J_{mag}/J_{tot})$ where J_{tot} and J_{mag} are the total current and the contribution from magnetosheath particles. It will be assumed that $e\Phi$ is still several times larger than the temperature of either species in the magnetosheath and can therefore have an appreciable effect on particles entering from the sheath. A further modification should be considered next.

In Cowley's model the electrons drift almost to the dusk edge before reaching $z=0$, after which they move dawnwards in a layer which must be much thinner than the layer of positive charge. This electron layer must have negative charge and the potential at $z=0$ should be intermediate between the dawn and dusk values and could be determined from self-consistency in a way similar to Cowley's method. We assume that the potential change is concentrated at the two edges of the sheet and well away from the edges E_y is negligible and all quantities are approximately independent of y. It may be noted that the y-component of canonical momentum (Equation (2)) is then useful, particularly for relating the distribution function at small values of z to that at $z=0$. It is now necessary to construct distribution functions by deciding for each region of velocity space and each species whether the particles come from the lobes, the dawn sheath or the dusk sheath. This has not been done rigorously but it may be expected that the majority of protons come from the dawn sheath, gaining some energy on entry, move duskwards and are accelerated into the sheath at the dusk edge. Similarly electrons from the dusk sheath are important, but lobe electrons should be included as they are important in the early Cowley model. Simple considerations show that lobe electrons arrive with less energy than those from the dusk sheath and that both generally have v_z small compared to v_y. However this result uses the fact that electrons in the lobe have low energy and it would be preferable to map the different regions of velocity space by computing trajectories backward. Furthermore there are duskward moving electrons from the dawn sheath which are reflected by the dusk potential jump and then occupy a low energy dawnward moving region. The problem for protons is similar, but may be less important.

Consideration of the Cowley potential or E_D, required for self-consistency, led to a natural explanation of field-aligned current layers, whose magnetic perturbation is easily and frequently observed on satellites (Zmuda *et al.*, 1970; Aubry *et al.*, 1972; Fairfield, 1973). If the potential is mapped along field lines, it gives a north-south electric field over a thin section of the ionosphere. Questions arise concerning parallel electric fields near the neutral sheet and anomalous resistivity near the ionosphere, but it is reasonable to expect an intense double layer of current. Cowley's model refers to the first open field lines and would give a downward current on the equator ward side of the double layer, and hence an eastward magnetic spike in the northern hemisphere. It is encouraging that the magnetic spikes observed far out are often associated with the edge of the plasma sheet and at low altitudes with auroras. However similar electric fields may exist on the last closed field lines, so that the theoretical model is as yet incomplete, but the observation of field-aligned currents will be a useful test.

4. Stability of Cowley Models

It would be unwise to pursue the building of self-consistent models too far without considering their stability. On the other hand stability studies require self-consistent models for their unperturbed states. The overall impression of observations is of remarkable quiet in the tail and even in the neutral sheet most of the time. However the onset of the expansion phase in substorms seems to call for an explosive instability leading to anomalous resistivity somewhere in the neutral sheet. This is the most exciting problem: to discover the nature of the instability and the critical condition. Explosivity can be envisaged if, with a given tail flux, stability requires the cross-tail potential Φ to be less than some critical value, because then anomalous resistivity, which increases Φ, makes for less stability.

In a Cowley model it is natural to look for something like a double stream instability and electrons from the dusk sheath may be expected to provide a second stream. Bowers (1973a) studied the Vlasov formulation for the central electron layer and waves varying like $\exp i(\omega t + k_y y)$ and later (1973b) added electrons from the sheath. The exact integrals are complicated by the closely spaced resonances occurring when the average frequency seen by a particle is a multiple of the frequency of its oscillation in z. Bowers studied the real part of the dispersion equation using the short wave-length approximation equivalent to assuming quasiuniform plasma. His solutions correspond to ion acoustic waves with phase varying more rapidly with z than with y. He claims that these waves cannot escape from the neutral sheet, because of a singularity when the ion plasma frequency equals the wave frequency, but this is based on a questionable approximation. However it is reasonably certain that some such waves are evanescent outside the sheet, though there may be an eigenvalue condition relating ω and k_y.

The Landau energy exchange involves the resonances, but it appears that the few lowest order resonances dominate and the sign, determining stability or instability, can be discussed qualitatitively. In a frame moving in the y-direction, at a speed such that the wave is static in that frame, the electric field has a potential and the energy of any particle is a constant. This restricts the combinations of derivatives of f which can occur in the Vlasov theory, and can be used to predict the phase speed of waves that are likely to grow. Bowers (1973b) includes electrons from the dusk sheath, which have a minimum energy corresponding to the dusk potential jump, and finds that the phase speed of the first waves to grow is near the electron speed corresponding to this minimum energy. The existence of a minimum energy must contribute to the instability of any wave simply because the distribution function steps up at this speed, so that any diffusion of the step is towards lower energy. The total energy exchange is given by an integral over all velocities and leads to a generalised Penrose criterion. Bowers (1973b) results suggest that instability sets in for a cross-tail potential much less than Alfvén's value. If this is correct, most of the potential inferred indirectly for the expansion phase is probably due to anomalous resistivity resulting from the instability. One of the remaining problems concerns the location of the anomalous resistivity. If waves are amplified as they travel dawn-wards, the intensity of turbulence should increase towards the dawn edge. A further problem is then the propagation of the nonlinear effect of the turbulence duskwards in the form of surface waves.

5. Models with Appreciable B_z

Observation shows a sudden reversal of B_x, even though B_z is appreciable and this implies that a field line turns through a very sharp hairpin bend. It is assumed that such field lines occupy the plasma sheet and consideration of anisotropic stress is then important. The component $B_x B_z/8\pi$ of magnetic stress reverses on crossing the sheet, but the sheet is so thin, that the corresponding component of total stress can not change appreciably. It must then be concluded that the total stress in the plasma sheet is isotropic. The condition $p_\parallel - p_\perp = B^2/4\pi$ (critical stability for the firehose mode) requires a cigar shaped pitch angle anisotropy, which may be achieved by plasma streaming parallel to the field. It should be noted that sharp reversals of B_x are observed as close as 10 R_E from the Earth (Thomas and Hedgecock, 1975).

Most satellite crossings show detectable values of B_z and for our purposes 'neg-ligible' means much less than 'detectable', but at the same time B_z is usually much less than the field strength outside the sheet. Speiser (1975) found a simple description of trajectories in models with small but appreciable B_z. Provided the amplitude of oscillation in z is not too great, it can be separated from the x, y-motion, which depends on B_z and E_y. (Inclusion of E_x is now desirable, but not too difficult). The electric field can be removed at least locally by a frame transformation and the significant quantity is the gyroradius in the B_z field. If the gyroradius for protons as well as electrons is much smaller than the width of the tail, the difficulty explained in Section 2 disappears, and the dimensionality of the problem can be reduced. The x-component of canonical momentum given by (1) is useful. If the particles mirror between the half-gyrations found by Speiser, (1) can be used to estimate the cross-tail drift speed. Assuming B_z to be larger over the mirroring part of the trajectory than in the neutral sheet part, the resulting drift speed is similar to the dipole formula with the actual neutral sheet B_z in the denominator. It is still not clear how this compares to the drift due to electric field, and their relative importance is significant in determining where the particles in the neutral sheet come from.

Eastwood (1972) obtained self-consistent models assuming that the particles came from the polar wind. The current is carried mainly by the protons, which are less sensitive to E_z than electrons and iteration for the current density profile converged rapidly. For the simplest model the sheet was very thin and modifications including hot electrons and consequent E_z corresponding to positive charge gave thicknesses of the order observed. In the simple model an electron is in the sheet for only a very short time, but Eastwood points out that electrons can be trapped by the positive potential, allowing time for them to be heated by noise, and also notes that they then contribute more to the current. He concludes that convection away from the neutral line is important for these trapped electrons and further progress requires consideration of the intermediate region, where the tail width is much bigger than the electron gyroradius, but not the proton gyroradius.

Eastwood has also considered stability and expects electron double streaming to generate electrostatic noise in the neutral sheet. Electromagnetic noise in the plasma sheet could originate by propagation from this or could be generated locally by the cigar shaped pitch angle anisotropy.

References

Alfvén, H.: 1968, *J. Geophys. Res.* **73**, 4379.
Aubry, M. P., Kivelson, M. G., McPherron, R. L. Russell, C. T., and Colburn, D. S.: 1972, *J. Geophys. Res.* **77**, 5487.
Bowers, E. C.: 1973a, *Astrophys. Space Sci.* **21**, 399.
Bowers, E. C.: 1973b, *Astrophys. Space Sci.* **24**, 349.
Cowley, S. W. H.: 1971, *Cosmic Electrodyn.* **2**, 90.
Cowley, S. W. H.: 1973a, *Cosmic Electrodyn.* **3**, 448.
Cowley, S. W. H.: 1973b, *Astrophys. Space Sci.* **20**, 491.
Dungey, J. W.: 1972, in B. M. McCormac (ed.), *Earth's Magnetospheric Processes*, D. Reidel Publ. Co., Dordrecht-Holland, p. 210.
Eastwood, J. W.: 1972, *Planetary Space Sci.* **20**, 1555.
Fairfield, D. H.: 1973, *J. Geophys. Res.* **78**, 1553.

Heppner, J. P.: 1972, in Holtet and Egeland (eds.), *Geofys. Publ. Minneskrift for Professor L. Harang*, Universitetsforlaget, Oslo.
Schindler, K. and Soop, M.: 1968, *Phys. Fluids* **11**, 1192.
Speiser, T. W.: 1965, *J. Geophys. Res.* **70**, 4219.
Thomas, B. T. and Hedgecock, P. C.: 1975, this volume, p. 55.
Zmuda, A. J., Armstrong, J. C., and Heuring, F. T.: 1970, *J. Geophys. Res.* **75**, 4757.

EVIDENCE FOR MAGNETIC FIELD LINE
RECONNECTION IN THE SOLAR WIND

V. FORMISANO

Laboratorio Plasma nello Spazio, C.P.27, Frascati, Italy

and

E. AMATA

Physics Dept., Imperial College, London S.W.7, England

1. Introduction

The reconnection of magnetic field lines of force has been invoked as a key process for astrophysical and geophysical problems such as solar flares, geomagnetic storms, strong radio sources and star formations. The problem was first investigated by Dungey (1953), then by Sweet (1958) and Parker (1963). The main objective in most of these and other papers was to investigate a mechanism (magnetic field annihilation) capable of accelerating to high energies the large number of particles observed in large solar flares. The mathematical models developed therefore used mainly the conditions of the plasma in the solar atmosphere (see Petschek, 1963; Sonnerup, 1970; Yeh and Axford, 1970; Stevenson, 1972). Until recently, however, no experimental study of the reconnection problem was available.

Recently a laboratory experiment has been developed to demonstrate that magnetic field reconnection does occur in a collisional plasma (see Bratenahl and Yeates, 1970). In space physics only a tentative identification of the process has been published by Unti *et al.* (1972) while the recent work by Formisano *et al.* (1974) was mainly concerned with instabilities (such as the tearing mode) present in neutral sheets where reconnection was probably occurring. On the other hand, conventional explanations of the geomagnetic substorm are centered on magnetic field reconnection both at the magnetopause and in the geomagnetic tail (cf. Russell and McPherron, 1973 and references therein).

Since the plasma conditions in space are very different from laboratory experiments and may change appreciably with time we think that an experimental study of the reconnection problem over a wide range of parameters may considerably improve the understanding of this important process. Data from HEOS 1 and from OGO 5 were therefore analysed, first to find positive evidence for the reconnection process and then to perform a quantitative study of the phenomenon.

In this report we discuss a few observations that provide evidence for the reconnection process in the solar wind.

2. What do we Expect to Observe?

The theoretical models published in the literature have always assumed the recon-

V. Formisano (ed.), The Magnetospheres of the Earth and Jupiter, 205–217. All Rights Reserved

nection problem to be two dimensional. Therefore the first difficulty in studying reconnection in the solar wind with a satellite is that we explore in only one dimension, while the structure may be three- or at least two-dimensional. Another difficulty arises from the large differences in the plasma parameters between the solar corona considered in most studies, and the solar wind, where our data were taken. The difference in the plasma parameters is probably very important for the scale lengths on which the reconnection occurs. It should be noted that there is no predetermined value for the scale length of the reconnection process in the solar wind so we may expect, on occasion, both small and large size phenomena. This fact is important in that a small size structure carried at the solar wind speed can only be observed in the magnetic field data or in other data with very high time resolution, while large size structures can be observed also in the plasma data (which usually has low time resolution).

Another important difficulty we shall deal with when comparing data with theoretical calculations is that nature does not always chose simple situations. Therefore, in general, we would expect the magnetic field on sides 1 and 2 of the reconnection boundary not to be antiparallel as is usually assumed in the models. Furthermore the magnetic field intensities on the two sides may be different, while they were assumed to be equal in the models. While it is possible to have magnetic field line reconnection with $B_1 \neq B_2$ and with directions that are not antiparallel, the effect of this change on the model predictions is not clear.

Figure 1 reproduces the usuall scheme of Petschek's models for low (a) and high (b) reconnection rate. The geometry is such that a satellite would most likely cross the structure with some inclination at a distance from the region where the reconnection actually occurs (called the diffusion region). The plot at the top of Figure 1 shows how the structure would appear in the magnetic field data taken by a satellite crossing the geometry along the thick line shown in the figure. The structure would appear as a decrease in the magnetic field intensity bounded by two discontinuities, slow shocks in Petschek's model.

The main differences in the observations between slow (a) and fast (b) reconnection would be in the perturbation of the magnetic field outside the neutral sheet in the fast reconnection. In the latter case, a slow decrease followed by a small increase of magnetic field intensity would appear before the first slow shock (the reverse after the second slow shock).

A few criticisms have been made of this model mainly due to the requirement for an extremely fast reconnection rate to explain solar flare energy release through annihilation of magnetic field energy. In this model, indeed, the field lines would move toward the null point at the Alfvén speed. Therefore, a limit to the reconnection is

$$U = \frac{\pi}{4} \frac{V_A}{\lg(\mu\sigma L V_A)},$$

where L is the scale length, σ the electric conductivity, V_A the Alfvén speed and μ is the magnetic permeability, i.e. the reconnection rate could take place at not more than $\frac{1}{10}$ the Alfvén speed.

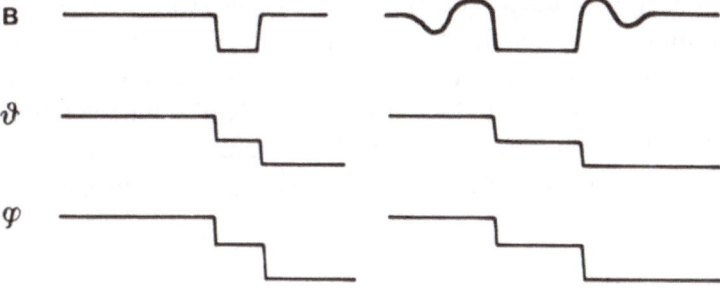

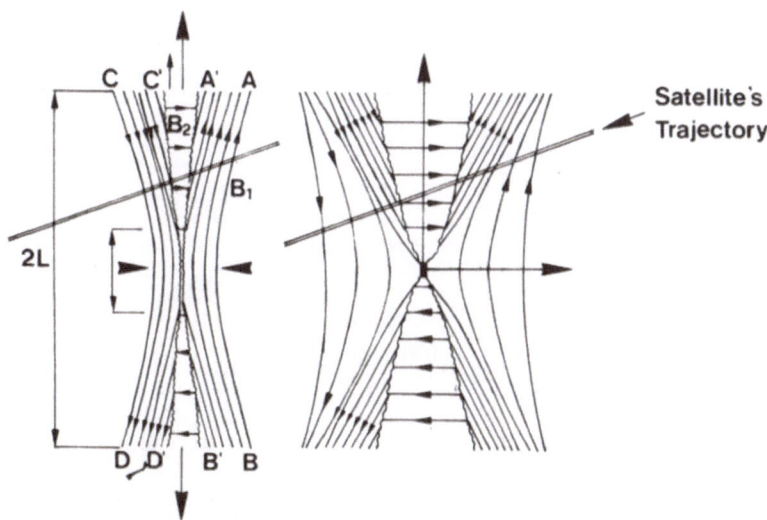

Fig. 1.   Structure for magnetic field line reconnection according to Petscheck's model, (a) for low reconnection rate, (b) for high reconnection rate. The plots at the top show how the structure should appear in
the magnetic field data of a satellite crossing the structure along the trajectory indicated.

Uncertainty also stems from the assumption of an incompressible fluid made by .
the authors in all the published models. In Petschek's model there are standing
Alfvén waves in the incompressible fluid and these waves may become slow or intermediate shocks (or a mixture of both) in the compressible fluid. Another criticism
concerns the requirement of a uniform field to produce a slow reconnection rate, while
such regions appear to be excluded in more accurate calculations. The general
conclusion, therefore, is that Petschek derived only an approximate solution of the
MHD equations.

Since we do not know how good this approximation is for solar wind conditions,
we have considered this model for comparison with data.

Another model using an incompressible fluid was developed by Yeh and Axford

(1970) and Sonnerup (1970) independently. A sketch of the geometry of the reconnection region according to this model is given in Figure 2 (a) and (b) for slow and fast reconnection rates respectively. The figure shows the geometry only for the non-singular solution of the MHD equations, out of a family of solutions found by Yeh and Axford (1970). In the incompressible fluid there are now two sets of standing Alfvén waves instead of one and, as in Petschek's model, these Alfvén waves should

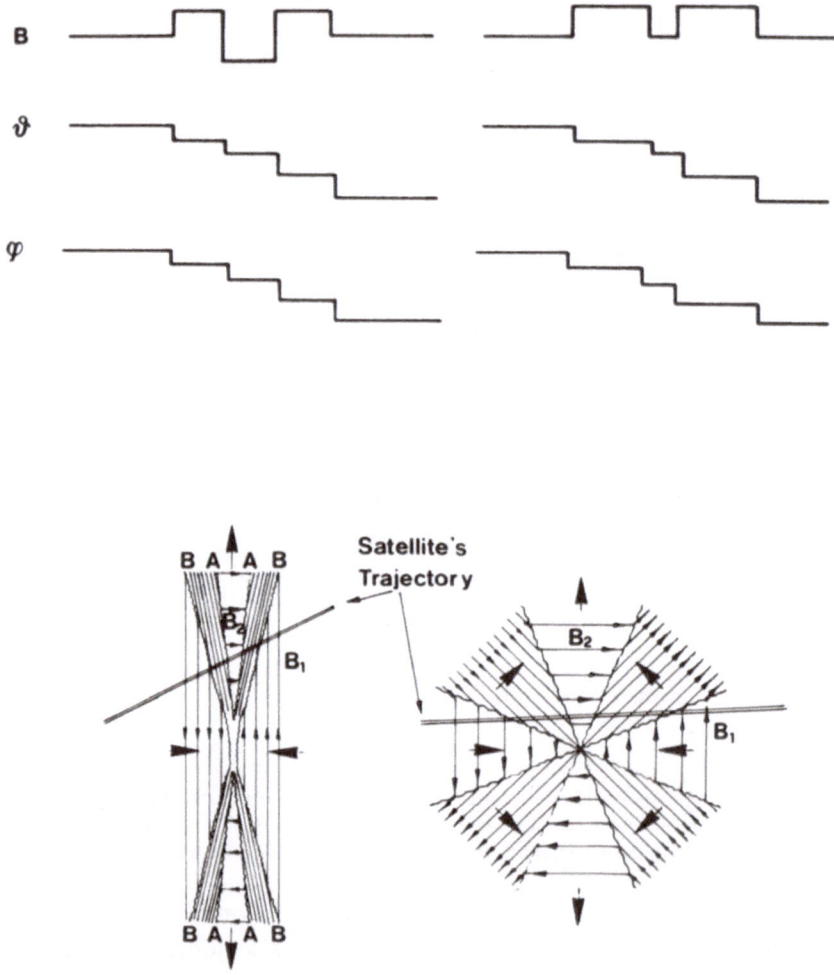

Fig. 2.   As Figure 1, but for the model by Sonnerup (1970) and Yeh and Axford (1970).

appear as slow shocks in a compressible fluid. Sonnerup (1972) argued that in a compressible fluid these two new slow shocks may not appear at all as they may be regarded as a mathematical approximation of the MHD interactions. Yeh and Dryer (1973) studied this problem for a compressible fluid and found that the two discontinuities close to the low B region are slow shocks while, in contrast, the two more distant discontinuities are diminutive expansion waves. The maximum reconnection rate would now be $U \lesssim (1 + \sqrt{2}) \, V_A$. This is independent of both the conductivity and the scale length L.

It should be noted that in both models the larger the reconnection rate the larger is the angle between the slow shocks. If this is true also for a compressible fluid we should more easily observe the signature of magnetic field reconnection when the latter occurs at a slow rate, i.e. when the dimensions are large.

If we accept Sonnerup's results as valid for a compressible fluid, the magnetic field data from a satellite should reveal structures similar to those shown at the top of Figure 2. Between the two regions with opposite magnetic field directions there

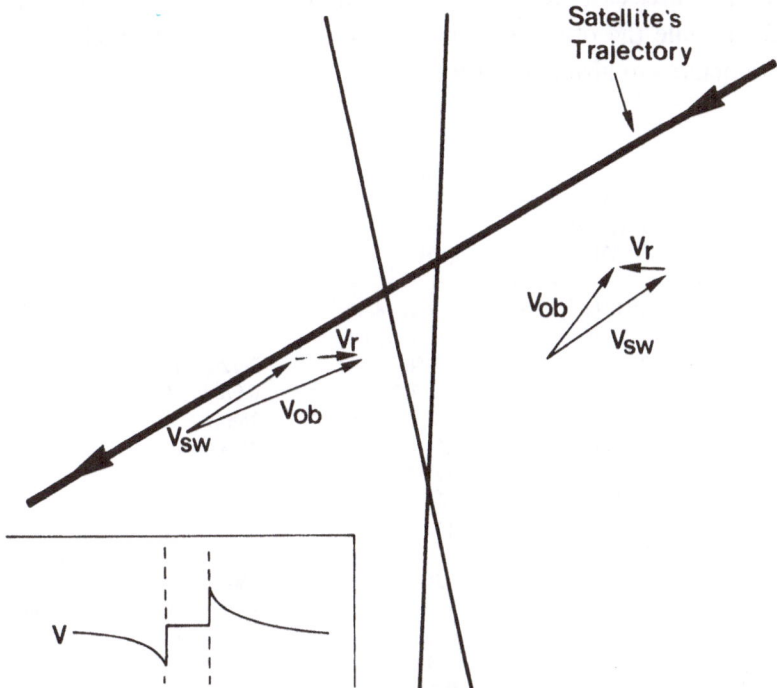

Fig. 3.   Solar wind velocity modification close to a magnetic field line reconnection structure.

should be three regions, two with enhanced magnetic intensity and one with much lower magnetic field intensity. The five regions should be separated by four slow shocks.

It should be noted that both models predict that the particles in the regions with opposite fields should move toward the diffusion region at about the Alfvén speed (maximum $\approx 2 \, V_A$). Therefore if a pair of discontinuities carried by the solar wind fall on either side of a neutral sheet where reconnection occurs a satellite will observe different solar wind directions before and after the pair, the magnitude of the velocity vector being lower than the unperturbated solar wind before the discontinuity pair and higher after because of finite electrical conductivity (see Figure 3).

It should finally be remembered that Sonnerup (1972) has suggested that the magnitude of the ion gyro radius may limit the reconnection rate. Furthermore a turbulent diffusion of particles close to the null point may increase the reconnection rate only if β (the ratio of particle pressure to magnetic pressure) is sufficiently small. Sonnerup actually concluded by emphasizing the importance of β since, for low β, laminar and turbulent resistivity effects would permit large reconnection rates (small dimensions) while, for high β, very small reconnection rates should result (large dimensions).

3. Observations

We present here only four observations of structures with large changes of magnetic field directions. Three cases (Day 260, Day 36 and Day 306, 1969) were found in the HEOS-1 data while the other (Day 64, 1969) was found in the OGO-5 data. The plasma parameters are given in Table I.

TABLE I

Plasma parameters

Time (UT)	Day 260 (1552)	Day 64 (1513)	Day 306 (0100)	Day 36 (0212)
* N_1 P cm^{-3}	10.8/11.9/12.4	5.9/6.4/8.4	8.3/9.6/11/11.3	4.8/4.8/4.7
* V_1 km s^{-1}	412/390/396	383/382/386	289/283/282/284	465/461/459
* W_1^1 km s^{-1}	36/47/49	47/48/62	24/26/31/33	34/35/37
* N_2 P cm^{-3}	14.98/5.6/6.9	7.2/7.0/7.6	6.9/6.9/8.5	5.8^a
* V_2^2 km s^{-1}	393/452/416	382/395/393	288/284/277	460^a
* W_2 km s^{-1}	68.7/33.6/33.1	55/49/51	35/33/44	32^a
B_1	13	12	5.5	3
B_{min}	10.5	11	4	1.5
B_2	13.5	11	7	4
1	295°	–	80°	60°
2	11°	125°	260°	
1	0°	–	−75°	−80°
2	55°	–	30°	55°
B_1 B_2	125°	100°		
1	1.00	0.55	1.2	3.7
2	0.27	0.96	2.1	2.4
** V_1 km s^{-1}	22	3.5	7	6
** V_2 km s^{-1}	36	3.5	11	
CA_1 CA_2 km s^{-1}	77 117	102 86	34 51	28 34
M	0.29 0.30	0.034 0.04	0.2 0.21	0.21 –
B_{min}/B_2	0.77	1	0.57	0.375
B_{min}/B_1	0.81	0.91	0.73	0.5
*** L_B km	5.3×10^3	5.4×10^3	5.0×10^3	5.5×10^3
*** L_P km	10^5	–	0.83×10^5	1.1×10^5

^{a} No more data available.
* Measurements, in time sequence, made immediately before (or immediately after) the discontinuity pair.
** Variation of measured bulk speed in proximity of the discontinuity pair.
*** Dimension of the structure from the magnetic field (L_B) or from the plasma (L_p).

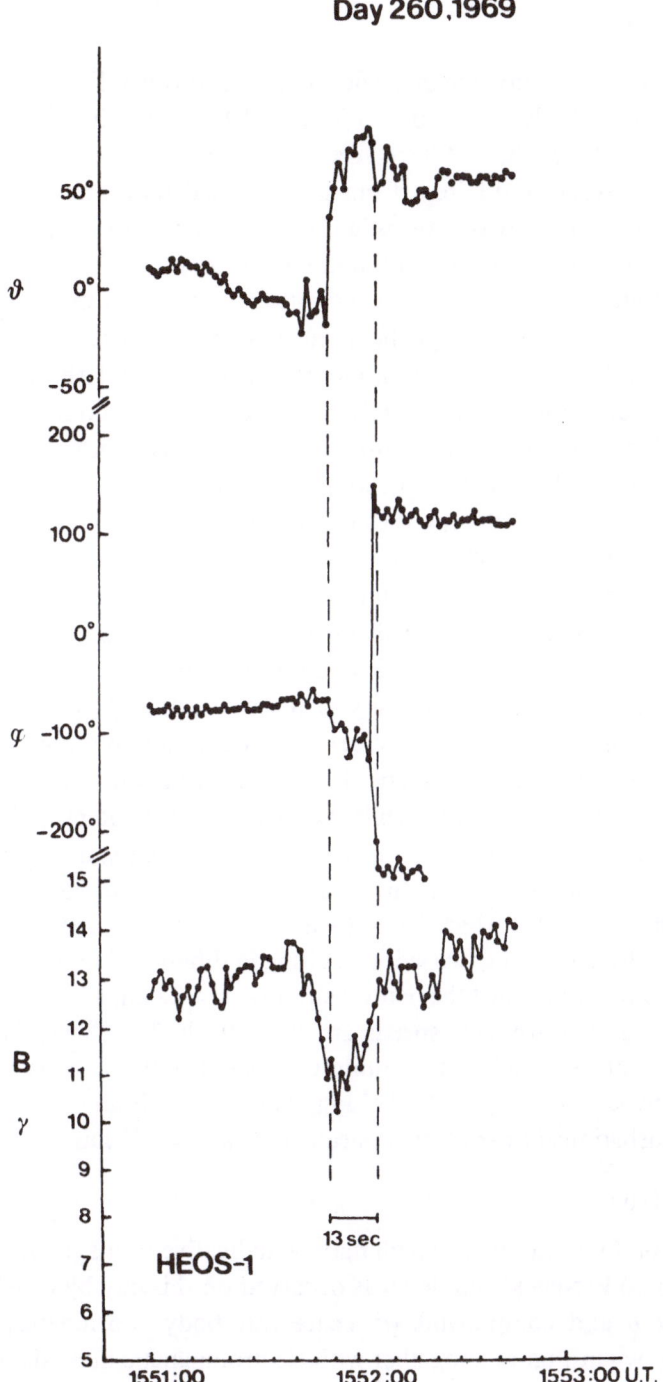

Fig. 4. High time resolution magnetic field data from HEOS 1 on day 260, 1969. Note the two distinct discontinuities in the magnetic field direction and the small gradual change of θ and ϕ on either side.

3.1. DAY 260, 1969

HEOS-1 high time resolution magnetic field data are shown in Figure 4. The magnetic field intensity showed a decrease from 13.5 γ to ≈ 10.5 γ for ≈ 20 seconds with a shape similar to the so called D sheet of Burlaga and Ness (1969) but the magnetic field direction indicates a structure different from a broadened tangential discontinuity (D-sheets). There are, indeed, two changes in the direction on either side of the minimum in the magnetic field, these two discontinuities being separated by 13 seconds. We note also the small gradual change in magnetic field direction as we approach the two discontinuities from the high magnetic intensity regions, suggesting that the lines of force are smoothly distorted while approaching the neutral sheet. The magnetic field changes direction by approximately 100° at each discontinuity.

The general features of this structure appear to be very similar to the predictions of Petschek's model when the reconnection rate is slow (top panel in Figure 1a). The presence of hydromagnetic fluctuations throughout this region resembles the similar observation by Unti *et al.*, of two discontinuities that were interpreted as possible evidence of field reconnection according to Petschek's model.

The plasma parameters given in Table I show the three measurements close to the discontinuity pair. We see that as the discontinuities are approached from the distant unperturbed regions the number density increases towards the discontinuity on the upstream side while the density decreases towards the discontinuity on the downstream side. The observed plasma speed decreases from the upstream side but increases from the downstream side, while the thermal speed increases from upstream and seems to be constant when we approach from downstream. Although the bulk speed and thermal speed are roughly equal in the unperturbed upstream plasma (side 1) and in the unperturbed downstream plasma (side 2), the number density appears to be different on the two sides resulting in different β for sides 1 and 2. Note that the plasma moves toward the neutral sheet with ≈ 22 km s^{-1} on side 1 and ≈ 36 km s^{-1} on side 2. The Alfvénic speeds are respectively 77 and 117 km s^{-1} so that $\Delta V/C_A$ is $= 0.3$ on both sides. The minimum scale length of the structure derived from the magnetic field is $L_B = 5.3 \times 10^3$ km while the maximum scale length derived from the perturbations in the plasma parameters is $L_p = 10^5$ km.

3.2. DAY 64, 1969

Another pair of discontinuities (which may be indicative of magnetic line reconnection according to Petschek's model) was observed on this day by OGO 5. Magnetic field intensity B and components (in spacecraft body coordinates) observed by OGO 5 are shown in Figure 5 together with electromagnetic noise data. The plasma parameters are displayed in Table I. The magnetic field intensity shows a small decrease between 15 13:05 and 15 13:19 UT with two small increases on either side. Two discontinuities are also indicated by the magnetic field components on either side of the field decrease. A third directional discontinuity observed 20 seconds earlier might be related to the pair but the association cannot be determined at

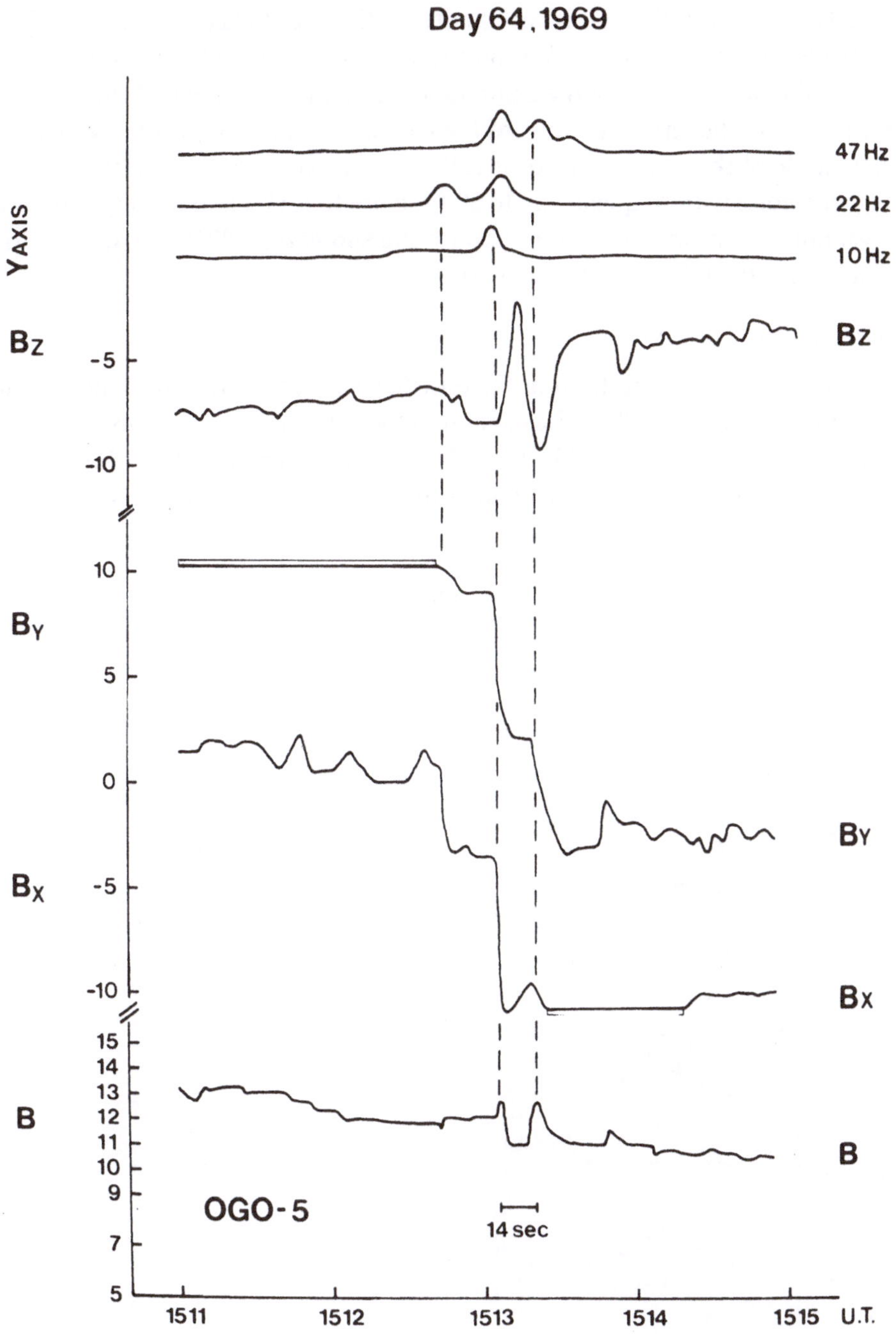

Day 64, 1969

Fig. 5.   OGO-5 magnetic field data for Day 64, 1513 UT, 1969. Note again the two directional discontinuities separated by a region of lower magnetic field intensity and, at the top, the presence of electromagnetic noise up to 50 Hz across the discontinuities.

present. This discontinuity pair is very similar to the one studied by Unti *et al.* (1972) and indicates that magnetic reconnection is occurring, although this example is not as clear as the previous one. Note again that the number density and thermal speed appear to be modified as we go toward the reconnection region, while the change in the magnitude of the velocity vector is rather small. Note also the presence of electromagnetic noise with frequency up to 50 Hz close to each discontinuity. This noise has been noticed on other occasions (see Formisano *et al.*, 1974) and appears to be associated with the presence of large currents.

3.3. DAY 36 AND 306, 1969

Two similar observations noticed in the HEOS-1 data are shown in Figures 6 and 7 for Day 36 and Day 306, 1969. We again find the same features described above for the Day 260 event: a slow modification of the ambient magnetic field direction and plasma parameters as we approach a neutral sheet characterised by the decrease in

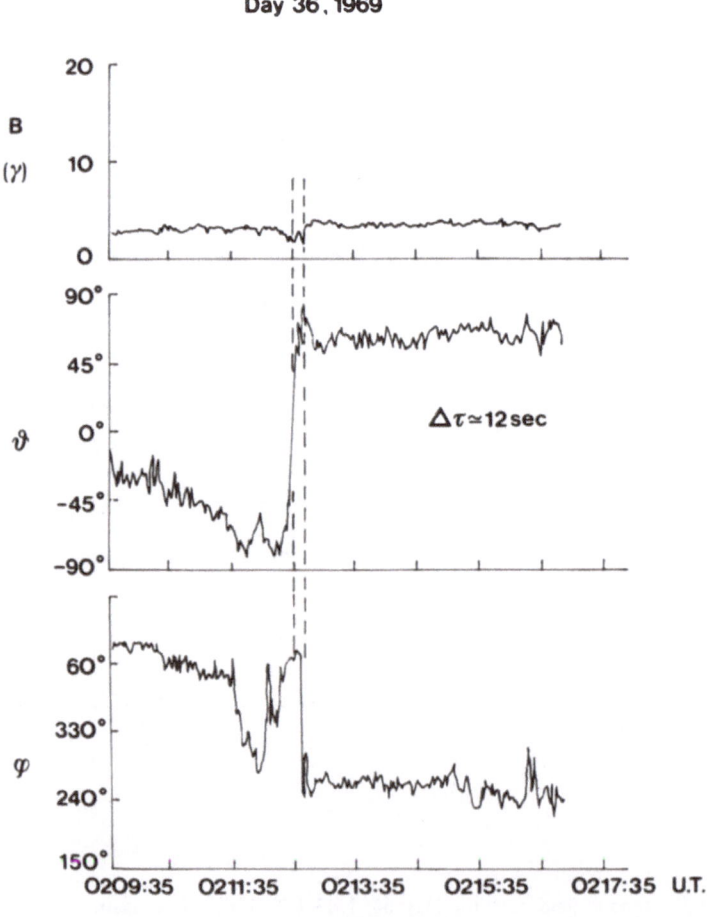

Fig. 6. HEOS-1 magnetic field data for Day 36, 1969.

Day 306, 1969

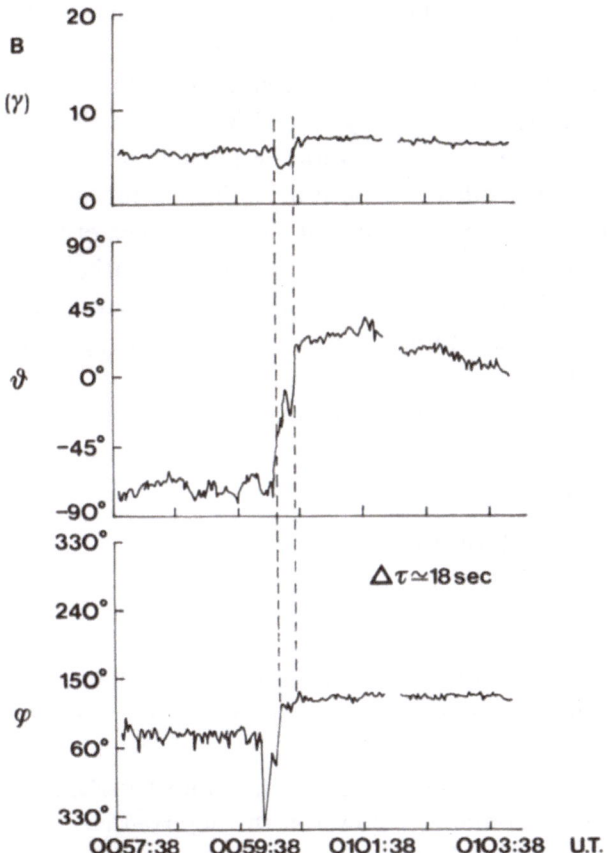

Fig. 7.   HEOS-1 magnetic field data for Day 306, 1969.

the magnetic field intensity bounded by two directional discontinuities. All the comments made for the Day 260 neutral sheet hold again for these observations.

4. Conclusions

Four neutral sheet observations have been presented. All four may be interpreted as indicating magnetic line reconnection rather than as D-sheets because all of them show two distinct discontinuities which bound a lower magnetic field intensity region. This interpretation of the data although partially supported by the plasma data, has not been definitely proved in this paper, being material of a future study. From the plasma data we have that the solar wind speed, indeed, slows down before the structure crosses the satellite and speeds up afterwards suggesting that the lines of force move toward the neutral sheet from the unperturbed regions. The geometry of B in the reconnection region appears to be similar to that described by Petschek

(1963) since only two discontinuities are observed. It is not clear whether these discontinuities are slow shocks due to the poor time resolution of the plasma data.

The dimensions of these structures are consistently between 5000 km (distance between the discontinuities) and 10^5 km/region over which the plasma parameters are distorted by the flow toward the neutral sheet. All these cases had a high $\beta(\beta>1)$ and should be examples of low reconnection rate.

If the reconnection rate is defined by $M = \Delta V/C_A$, M is smaller than $\frac{1}{10}$ of the Alfvén speed only on Day 64 in agreement with the possible reconnection rate predicted by Petschek. In the three other cases M is 0.3 or 0.2, i.e. much larger than Petschek's limit. We note that M is consistently equal on both sides of the neutral sheet also when the parameters are different (see Day 260 and Day 306) because the same amount of magnetic lines of force have to be reconnected in the neutral sheet. The reconnection rates observed agree with Sonnerup's limit $M < (1 + \sqrt{2})$ rather than with Petschek's limit $M < 0.1$.

It appears also that reconnection is dominated by laminar resistivity (see Sonnerup, 1972) for which

$$M^2 < \frac{4}{\pi^2} \frac{\beta_1}{1 + (T_{ei}/T_{i1})}$$

$$\beta_1 < \frac{4}{\pi^2} \left(1 + \frac{T_{ei}}{T_{i1}}\right)$$

These relationships are satisfied if we use the average condition $T_e/T_i = 2$. In this case, indeed the inequalities become $M < 0.27$, $\beta_1 < 1$ which are approximately valid in our examples.

In conclusion the reconnection rates reached with Petschek model appear to be higher than previously estimated, however, the observed geometry agrees with this model and the scale length in the solar wind is 5×10^3–10^5 km. Structures similar to those predicted by Sonnerup and Yeh and Axford are not observed, however, we do not know whether the scale lengths should be very much different according to the two models.

Acknowledgements

We are pleased to acknowledge Dr C. T. Russell and Dr P. C. Hedgecock for providing us with unpublished magnetic field data and for useful discussions.

This research has been supported by the Consiglio Nazionale delle Ricerche of Italy.

The magnetic field experiment on board HEOS 1 was supported by the British Science Research Council.

The OGO-5 magnetometer and search coil experiment (University of California, Los Angeles, U.S.A.) were supported by NASA grant NGR 05-007-004.

References

Bratenahl, A. and Yeates, C. M.: 1970, *Phys. Fluids* **13**, 2696.

Dungey, J. W.: 1953, *Phil. Mag.* **44** (7), 725.

Formisano, V., Hedgecock, P. C., Russell, C. T., and Means, J.: 1974, in *Proc. of the Third Solar Wind Conference*, Asilomar, 1974.

Parker, E. N.: 1963, *Astrophys. J. Suppl.* **8**, 177 (1963).

Petscheck, H. E.: 1964, in *Proc. of the AAS-NASA Symposium on the Physics of Solar Flares*, Greenbelt, Md., Oct. 1963, NASA SP-50, p. 425.

Sonnerup, B. U. O.: 1970, *J. Plasma Phys.* **4**, 161.

Sonnerup, B. U. O.: 1972, in *High Energy Phenomena on the Sun*, Proceedings 1972 symposium NASA-G.S.F.C. X693-73-193, preprint.

Stevenson, J. C.: 1972, *J. Plasma Phys.* **7**, 293.

Sweet, P. A.: 1958, in B. Lehnert (ed.), 'Electromagnetic Phenomena in Cosmical Physics', *IAU Symp.* **6**, 123.

Tyan, Yeh and Axford, W. I.: 1970, *J. Plasma Phys.* **4**, 207.

Unti, T. W. J., Atkinson, G., Wu, C. S., and Neugebauer, M.: 1972, *J. Geophys. Res.* **77**, 2250.

Yeh, T. and Dryer, M.: 1973, *Astrophys. J.* **182**, 301 (1973).

PART II

THE MAGNETOSPHERE OF JUPITER

THE RADIOASTRONOMY OF JUPITER

D. STANNARD

University of Manchester, Nuffield Radio Astronomy Laboratories, Jodrell Bank, Macclesfield,
Cheshire, U.K.

1. Introduction

Jupiter is of particular interest to radio astronomers because of the richness and
variety of its radio emission, and the evident relationship of this emission to the
magnetic field structure of the planet. The two main types of radio emission are
the long wavelength decametric 'burst' emission and the shorter wavelength decimetric
'synchrotron' emission. In addition, at very short wavelengths, there is also thermal
radio emission from the planetary disk.

An excellent review of the magnetosphere of Jupiter has been given by Carr and
Gulkis (1969), so that only a brief summary of the basic features of the Jovian radio
emission will suffice here (Sections 2 and 3). Evidence for departures of the magnetic
field from that of a simple dipole is discussed in Section 4, and two models which
have been proposed to account for the observed irregularities are described. A crucial
observational test of these models is provided by measurements of the centroid of
the decimetric emission relative to the optical disk (Section 5), with the evidence
appearing to favour the centred dipole model with a localized field anomaly. The
Pioneer 10 magnetometer results are described in Section 6, and in Section 7 recent
observations of the decimetric emission are reviewed, and some unexplained features
of the radiation discussed.

2. Radio Emission from Jupiter

2.1. THERMAL EMISSION

Thermal radiation from the planetary disk dominates the high frequency spectrum
of the Jovian radio emission (Figure 1), with the form of the spectrum in good agreement with a black body curve corresponding to the brightness temperature of $\sim 130\,\mathrm{K}$
measured in the infrared range. Dickel et al. (1970) have presented evidence for a
decrease in the thermal emission between 18 and 28 GHz which they suggest may
be attributed to absorption from ammonia inversion transitions. At lower frequencies
(2.9 and 1.4 GHz) observations (Berge, 1966; Branson, 1968) show that the brightness
temperature of the planet rises to approximately $250\,\mathrm{K}$, presumably because the
transparency of the atmosphere is such that the radiation at these frequencies originates from a lower level, where the temperature is higher.

2.2. DECIMETRIC EMISSION

Although the terminology strictly only refers to radio emission at wavelengths of
the order of ten centimetres, it has come to describe the broad continuum of emission

V. Formisano (ed.), The Magnetospheres of the Earth and Jupiter, 221–236. All Rights Reserved
Copyright © 1975 by D. Reidel Publishing Company, Dordrecht-Holland

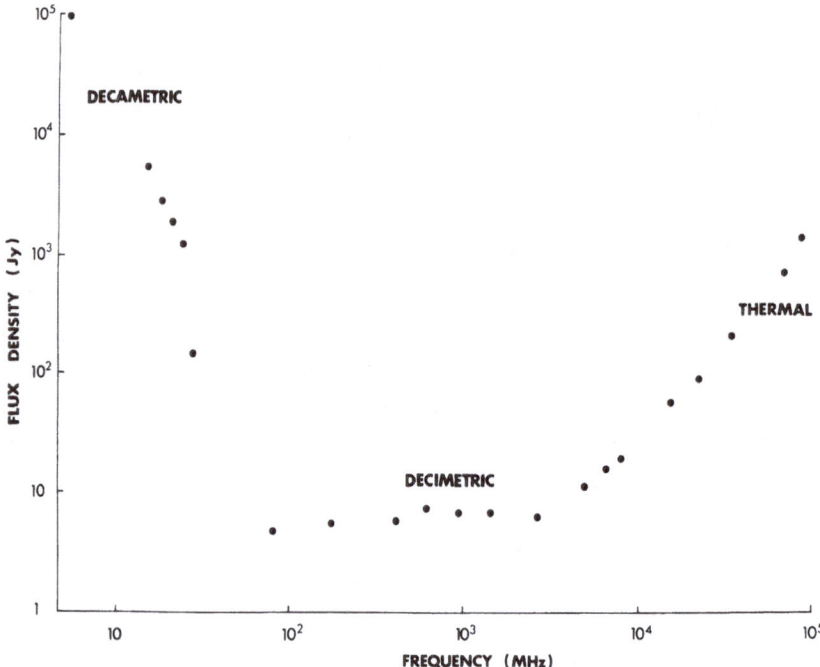

Fig. 1.   The mean spectrum of the Jovian radio emission.

which extends in wavelength from a few centimetres to over a metre. The spectrum is relatively flat in this range, with a slight maximum near 800 MHz, and a mean intensity of approximately 7 Jy at the standard distance of 4.04 AU (1 Jy$= 10^{-26}$ W m^{-2} Hz^{-1}).

The main features of the decimetric radiation can be satisfactorily accounted for by synchrotron emission from relativistic electrons which are trapped in radiation belts of a basically dipolar magnetic field, with the axis of the magnetic dipole inclined by about 9° to the rotation axis of the planet. The radiating electrons have an aniso-tropic distribution of pitch angles, with the majority moving in relatively flat helical orbits and mirroring well within one planetary radius of the magnetic equator. The resulting radiation is strongly 'beamed' into the plane of the magnetic equator, and is linearly polarized at about the 20% level with the E-vector perpendicular to the projected direction of the magnetic field lines.

As the planet rotates the changing aspect of the magnetic equator, as viewed from the Earth, causes variations in the observed intensity and polarization (Figure 2). The total intensity goes through two maxima per revolution, when the magnetic equator is viewed edge on, and two minima, when the equatorial plane is viewed from the extreme positive and negative magnetic latitudes. (The minima are in general unequal as the Jovicentric declination of the Earth (D_E) may lie up to $\pm 3°$ from the plane of planet's rotational equator). The position angle of the linear polarization rocks back and forth in a near sinusoidal manner, with the amplitude of the varia-tion giving the inclination between the magnetic and rotation axes. The degree of

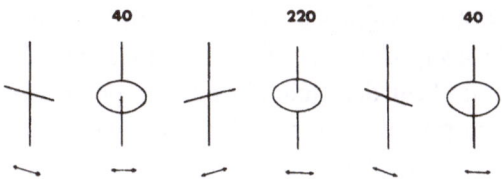

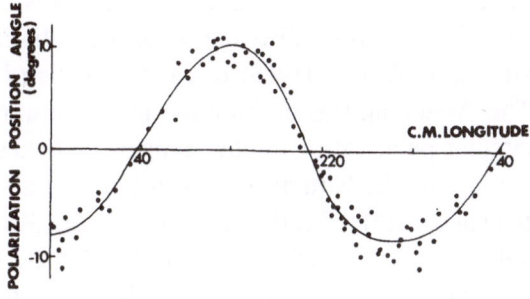

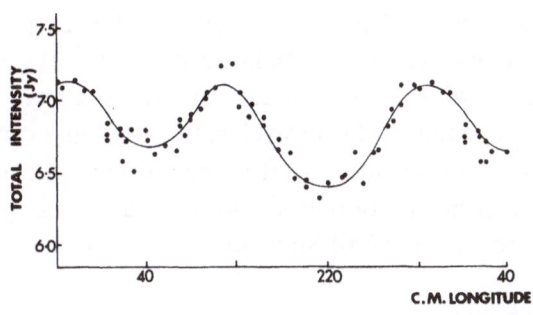

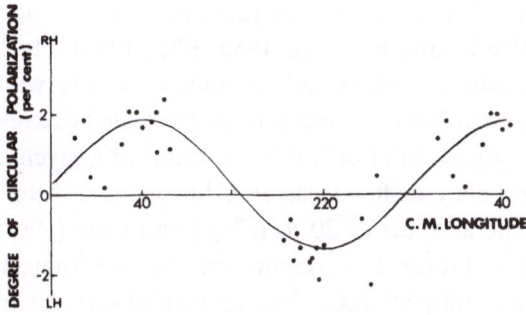

Fig. 2. Variation of the position angle of linear polarization, the total intensity, and the degree of circular polarization of the decimetric emission with c.m. longitude. The orientation of the magnetic equatorial plane and the direction of the polarization *E*-vector are sketched at the top of the figure.

linear polarization (not shown) has a variation with longitude similar to that of the total intensity, whilst away from the magnetic equatorial plane a small amount of circular polarization has been detected. The circular polarization is left-handed when Jupiter is viewed from positive magnetic latitudes, which shows that the magnetic field of Jupiter is opposite in polarity to that of the Earth, with the pole in the northern hemisphere a north magnetic pole. The pole is tipped towards the Earth when the central meridian (c.m.) longitude is near 220°.

Quantitatively, the variation of intensity and polarization with magnetic latitude may be used to deduce the pitch angle distribution of the radiating electrons. Using a thin shell approximation, Thorne (1965) calculated a distribution of the form $N(\alpha) = \sin^2 \alpha + 2 \sin^{40} \alpha$. Assuming this distribution and a power law spectrum, the observed degree of circular polarization implies a magnetic field strength of between 0.4 and 1.9 G in the radiation belts (that is, at a distance of about 2 Jovian radii). The corresponding field at the surface of the planet is between 3 and 15 G (see Komesaroff *et al.*, 1970; Gleeson *et al.*, 1969; Clarke, 1970; Berge, 1974; Stannard and Conway, 1974). With this field strength, the shape of the integrated spectrum implies a typical electron energy of 13 MeV, with a total number of 10^{28} radiating electrons (Barber and Gower, 1965; Carr and Gulkis, 1969).

The structure of the radiation belts has been examined by Berge (1966) and Branson (1968). Interferometric observations by the latter at wavelengths of 21 and 74 cm suggest that the emission at longer wavelengths may originate from slightly further out in the radiation belts than that at short wavelengths. The lunar occultations of Gulkis (1970) support this conclusion, and the effect may account for departures from the predicted frequency dependence of both the linear and circular polarization (Dickel *et al.*, 1970; Komesaroff *et al.*, 1970; Stannard and Conway, 1974).

2.3. DECAMETRIC EMISSION

At frequencies below 40 MHz the Jovian radio emission is characterized by strong bursts of radiation (Burke and Franklin, 1955; Ellis, 1965). The bursts are sporadic, but occur predominantly during periods of intense activity which may last from a few minutes up to several hours. These periods are called decametric 'noise storms'. A typical burst has a duration of only a few seconds at a given frequency, but often exhibits a rapid frequency drift. Occasional bursts have been observed of much shorter duration, some as short as 20 μs (Flagg and Carr (1967)). The spectrum of the emission shown in Figure 1 represents the average intensity over both active and quiescent periods, and peak intensities may be observed which are 3 or 4 mag. stronger (Carr *et al.*, 1964).

The probability of occurrence of the decametric emission varies as a function of both c.m. longitude and frequency (Figure 3). There are three distinct peaks or 'sources' visible in this figure. A fourth source, characterized by a distinctive narrow band spectrum (Dulk, 1965), is believed to exist in the zone of low emission between c.m. longitudes 20° and 80°. It is generally assumed that the decametric emission occurs at or near the local electron gyrofrequency (Wu *et al.*, 1973; Warwick, 1967), so that

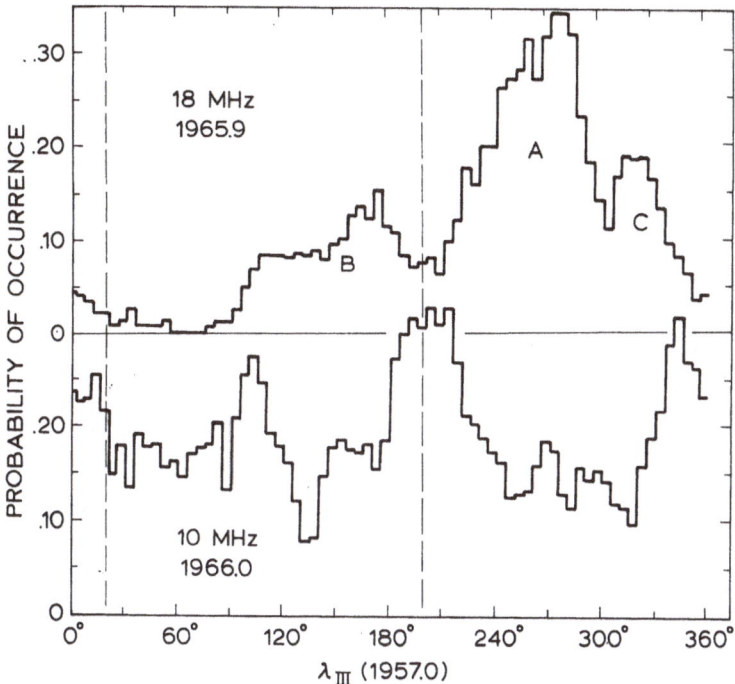

Fig. 3.   Histograms of the occurrence probability of the decametric emission as a function of c.m. longitude (λ_{III}). The longitudes of the poles are indicated by vertical lines. (From Carr and Gulkis, 1969.)

the cut off in emission above 39 MHz implies a maximum field strength of some 10–15 G, possibly close to the surface of the planet.

Frequency time spectrograms of the emission also show a systematic variation which is repeatable with longitude (Warwick, 1963). These 'dynamic spectra' show that source B and the fourth source tend to drift upwards in frequency with time, whilst sources A and C have drifts which are predominantly downwards in frequency. The stability of these patterns suggest that the mechanism responsible may be linked to the magnetic field structure of the planet (Warwick, 1967).

One of the most important developments in the investigation of the decametric radiation was the discovery of a modulation effect by the satellite Io (Bigg, 1964) (Figure 4). Events associated with source B are almost entirely dependent on the position of Io, whereas there is a component of source A which is largely Io independent. Io independent events become fewer relative to Io dependent ones as the frequency is increased (McCulloch and Ellis, 1966).

Interferometric studies (Dulk, 1970; Lynch et al., 1972) of the structure of individual decametric events show that the majority are produced in regions smaller than 400 km.

The decametric radiation has a high degree of elliptical polarization. At frequencies greater than 20 MHz the polarization is mainly right-handed (Sherrill, 1965; Warwick and Dulk, 1964). As the frequency is decreased left-handed components become

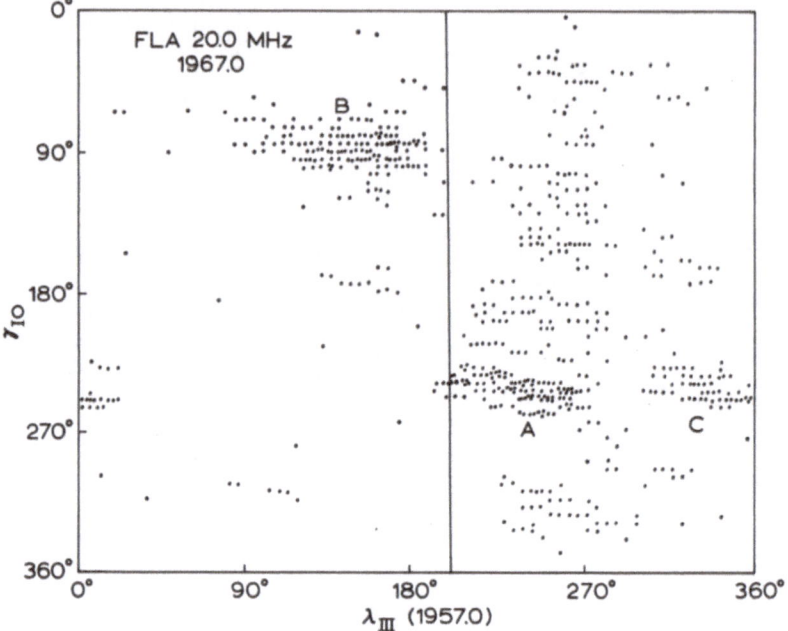

Fig. 4. Distribution of the occurrence probability of the decametric emission (proportional to the density of dots) as a function of the position of Io from superior conjunction and the c.m. longitude of Jupiter (λ_{III}). The vertical line indicates the longitude of the pole in the northern hemisphere. (From Carr and Gulkis, 1969.)

stronger, but the polarization remains right-handed near the longitude of the pole in the northern hemisphere. This enabled the sense of the Jovian magnetic field to be deduced (Warwick, 1963) in advance of the circular polarization measurements of the decimetric radiation. The hand of polarization is opposite to that of the deci-metric emission because the radiation originates from near the pole of the magnetic dipole, where the field lines run in the opposite direction to those near the dipole equator.

The observed properties of the decametric radiation are complex, and a wide variety of emission mechanism have been proposed to account for them (see for example the reviews given by Carr and Gulkis (1969), Warwick (1967), and Warwick (1970)). In a co-rotating plasma disk model, Gledhill (1967) has suggested that the emission is generated near Io, and that the geometrical position of the satellite with-in the magnetosphere then determines the observed emission probabilities and fre-quencies. Alternative theories suggest that the radiation is produced close to the surface of the planet, but is in some way coupled to the motion of Io. This coupling may be via electromagnetic waves (Ellis, 1965), by low frequency hydromagnetic waves (Warwick, 1967), by magnetic coupling (either with Io magnetized or highly conducting (Piddington and Drake, 1968), or acting as a unipolar inductor (Gold-reich and Lynden-Bell, 1969), or just as an inert disturbing body (Shatten and Ness, 1971)), by tidal effects, or by ring currents (Bigg, 1966).

3. Radio Rotation Period

An important link between the decimetric and decametric emissions comes from the observed radio rotation periods of the planet. Early observations of the decametric emission led to the IAU adopting in 1962 a formal system III radio rotation period of $9^h55^m29.37^s$. Subsequent observations of both the decametric and decimetric emissions suggest that this period is in error, being too short by some 0.4 s. This causes a drift in the apparent c.m. longitude of Jovian radio features of approximately 3.2° per year.

Recently derived values for the radio rotation period of Jupiter are given in Table I. Determination of the decametric period is complicated by a cyclic drift in the apparent source positions due to the changing aspect of the planet during its orbital period of 11.86 yr about the Sun. Averages over a complete orbital cycle are however in good agreement with the decimetric period, suggesting that both sets of radio data measure the same fundamental quantity, presumably the rotation period of the planet's magnetic field.

TABLE I

Radio rotation period of Jupiter

IAU System III	$9^h55^m29.37^s$
Decametric period	$9^h55^m29.73 \pm 0.04^s$ (Donivan and Carr, 1969)
	$9^h55^m29.76 \pm 0.04^s$ (Carr, 1971)
Decimetric period	$9^h55^m29.71 \pm 0.07^s$ (Carr, 1971)
	$9^h55^m29.74 \pm 0.03^s$ (Stannard and Conway, 1974)

4. Asymmetries in the Emission

4.1. DECAMETRIC EMISSION

Many of the features of the decametric emission suggest an underlying poloidal character to the Jovian magnetic field, but anomalies remain, both in the predominance of a single sense of circular polarization, and in the location of the 'discrete sources' seen at higher frequencies. Although no single theory is generally accepted for the origin of the decametric emission, two particular theories are widely invoked to account for these asymmetries.

In the theory of Ellis and McCulloch (1963) the emission is produced by amplified cyclotron radiation from streams of near relativistic electrons at high latitudes in the Jovian magnetosphere. Rotation of the inclined dipole field then produces the two main sources observed near the longitudes of the poles at low frequencies. The additional sources seen at higher frequencies are explained by local anomalies in the angle of magnetic dip. In this model Io control is supposed to be exercised by electromagnetic coupling in the whistler mode (McCulloch, 1971).

An alternative theory has been proposed by Warwick (1963a, b) who attributes the asymmetries in the decametric emission to a dipole field which is considerably offset from the centre of the planet. Warwick suggests that the dipole centre is located $0.75^{+0.1}_{-0.4}$ Jovian radii south of the centre of the planetary disk, at a distance from the rotation axis of 0.2 ± 0.1 radii towards the c.m. longitude plane of $232^{+20°}_{-10°}$. The theory has been revised (Warwick, 1967) to account for the influence of Io in terms of hydromagnetic waves and particle disturbances.

As we shall see in Section 4.3, the same two types of explanation have also been advanced to account for the asymmetries present in the decimetric emission.

4.2. OBSERVED ASYMMETRIES IN THE DECIMETRIC EMISSION

Although the gross features of the decimetric emission are well accounted for by the inclined dipole model outlined in Section 2.2, irregularities remain in the shape of the polarization position angle curve, in the brightness distribution of the radiation belts, and in the beaming of the radiation about the magnetic equatorial plane.

4.2.1. *The Shape of the Polarization Position Angle Curve*

If the Jovian magnetic field were that of a perfect centred dipole, then the polarization position angle curve shown in Figure 2 would be sinusoidal. In practice the curve is grossly distorted, with the slope at the downward crossover near 220° c.m. longitude much steeper than that at the upward crossover near 40° c.m. longitude (Roberts and Komesaroff, 1964). The shape of the curve is generally represented by observers (Roberts and Komesaroff, 1965; Komesaroff and McCulloch, 1967; Whiteoak *et al.*, 1969) as a three harmonic series of the form

$$P(l) = A_0 + A_1 \sin(l - l_1) + A_2 \sin 2(l - l_2) + A_3 \sin 3(l - l_3),$$

where the amplitudes A_1, A_2 and A_3 have typical values of 9.5, 1.0 and 0.5° respectively. The amplitude of the fundamental (A_1) is taken to be the inclination between the magnetic and rotation axes, whilst the phase (l_1) gives the longitude of the magnetic poles. The detailed shape of the position angle curve appears to be independent of both frequency (Roberts and Komesaroff, 1965) and epoch of measurement (Berge, 1974; Stannard and Conway, 1974; Komesaroff and McCulloch, 1967).

4.2.2. *The Brightness Distribution of the Radiation Belts*

In 1967 Branson used the Cambridge one-mile synthesis radio telescope to obtain maps of the radiation belts, smoothed over three longitude ranges (Figure 5). The distributions show the expected 'double structure', but contain an anomaly in the form of a localized 'hot spot' of emission near c.m. longitude 200°. This feature is behind the planet on the first map, and moves from the left hand side of the second map to the right hand side of the third.

4.2.3. *The Beaming of the Radiation about the Magnetic Equator*

Because of the relatively flat pitch angle distribution of the radiating electrons, the decimetric emission is strongly beamed into the plane of the magnetic equator

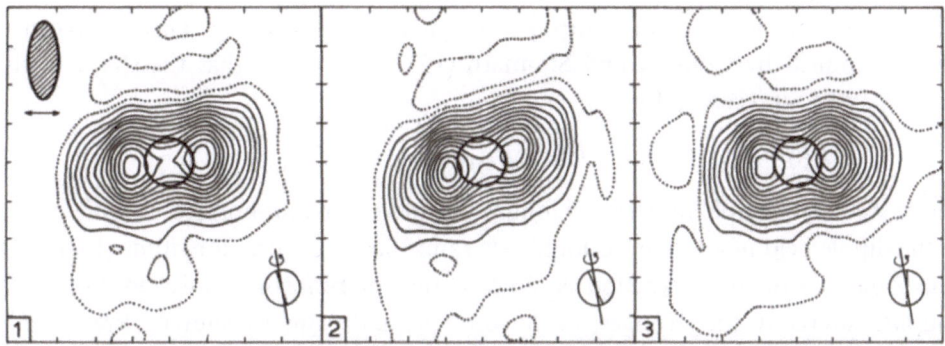

Fig. 5.   Maps of the radio emission from Jupiter at $\lambda 21$ cm for c.m. longitudes of 15° (map 1), 135° (map 2) and 255° (map 3). In each map north is at the top and east to the left. The half power synthesized beamwidth and the direction of received polarization are shown to the left of map 1, with a sketch of the magnetic and rotation axes of the planet given below each map. The dark circles represent the optical limb of the planet. (From Branson, 1968.)

(Thorne, 1965). To first order the beaming can be represented by a symmetric fall off in intensity with magnetic latitude (D_{ME}) of the form

$$S \approx S_0 \cos^n D_{\mathrm{ME}}$$

Measurements of the intensity at epoch 1963–4 (Roberts and Komesaroff, 1965; Barber, 1966) showed however a considerable asymmetry in the latitude dependence of the beaming, with northern magnetic latitudes having a beaming index n of ~ 4, whilst southern latitudes had a much steeper index of nearer 10. In these calculations the magnetic latitude is obtained assuming a 9 or 10° inclination between the magnetic and rotation axes (as deduced from the polarization position angle curve). It was pointed out by Roberts and Ekers (1968) that the asymmetry could be removed if the adopted inclination angle between the axes were increased to approximately 15°.

4.3. POSSIBLE EXPLANATIONS FOR THE DECIMETRIC ASYMMETRIES

Warwick (1967, 1970) has shown that his off-centred dipole model, which was originally proposed to account for the decametric asymmetries, also provides a ready explanation for the asymmetries in the decimetric emission. Shadowing of parts of the radiation belts accounts well for the observed shape of the polarization position angle curve, and the displacement of the dipole towards 230° c.m. longitude would produce an apparent enhanced region in the brightness distribution of the emission with the relatively low equatorial resolution of Branson's synthesis maps. Warwick points out that the asymmetry in the beaming is removed if the radiation is not beamed about the dipole magnetic equator, but symmetrically about a magnetic latitude of $+1.2°$. He interprets this effect as evidence for a quadrupole term in the Jovian magnetic field, with the southern pole having a stronger field as inferred from the decametric data.

Most observers prefer to account for the distorted shape of the polarization position angle curve in terms of a localized magnetic field anomaly. This has recently
been quantified by Conway and Stannard (1972) who show that the shape of the
curve may be represented by

$$P(l) = A° + A^1 \sin(l - l^1) + \Delta PA,$$

where ΔPA is a simple function of longitude, and can be interpreted as a distortion
of the dipole field in a restricted longitude range near 220° c.m. longitude (Figure 6).
The longitude of this distortion is close to that of Branson's emission 'hot spot',
and also source A, the strongest of the anomalous decametric sources. Presumably
all three features are intimately connected. As pointed out by Ellis and McCulloch
(1963) and Conway and Stannard (1972) the proposed Jovian field distortion is
similar to a distortion in the Earth's own magnetic field, the so-called 'Atlantic
anomaly'. It is intriguing that the only two planets in the solar system whose magnetic
fields are known in any detail both seem to possess the same type of field anomaly.

On the localized anomaly theory the apparent asymmetry in the decimetric beaming at epoch 1963–4 follows as a natural consequence of the occultation of the hot
spot by the planetary disk. The occultation occurs when Jupiter is viewed from the

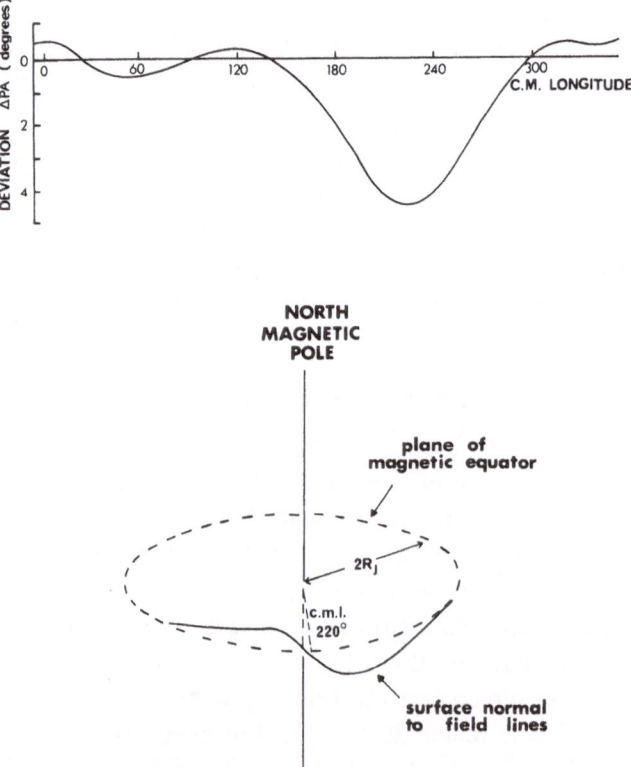

Fig. 6.   (a) ΔPA, the deviation of the position angle curve from a pure sinusoid. (b) A sketch of the surface
normal to the Jovian field lines showing the proposed magnetic anomaly.

greatest southern magnetic latitudes, which is just the latitude range where the deficit in emission is observed.

Quantitatively both Warwick's theory of an off-centred dipole, and the alternative theory of a centred dipole with a localized field anomaly, have considerable success in accounting for the observed asymmetries in the decametric and decimetric emissions. Fortunately there is an observational test which can distinguish between the two theories – a measurement of the centroid of emission of the radiation belts. If Warwick's theory is correct, the centroid should be displaced well to the south of the planet, whereas the localized anomaly theory predicts at most only a small displacement.

5. Measurements of the Centroid of the Decimetric Emission

Accurate determinations of the position of the centroid of the decimetric emission relative to the optical disk have been made by Roberts and Ekers (1966), McCulloch and Komesaroff (1973), Berge (1974), and Stannard and Conway (1974). All of these measurements rule out the large southerly displacement of 0.7 radii required by Warwick's off-centred dipole model (Warwick, 1970), and indicate that the Jovian magnetic field is quite well centred and reasonably symmetric. An early indirect measurement of the radio centroid by Berge and Morris (1964), which was taken as evidence of a highly displaced field, has been reinterpreted by Berge (1965) as indicating the presence of circular polarization in the decimetric emission and should be discounted.

Although the radio centroid determinations show that there is no large departure from a centred field, there is some suggestion in the data of a slight displacement to the north of the planet by 0.1 ± 0.2 radii (Figure 7). There is also evidence of a sinusoidal variation with rotation in the equatorial direction (Table II). This variation is barely significant in the individual determinations, but is present in approximately the same phase in all four sets of data and is thus thought to be real. An apparent variation of this form may either be interpreted in terms of a symmetric

TABLE II

Equatorial displacement of the centroid of the decimetric emission

Observer	Equatorial displacement	
	(radii)	(c.m. longitude plane)
Roberts and Ekers (1966)	0.07 ± 0.05[a]	238°[b]
McCulloch and Komesaroff (1973)	0.02 ± 0.05	218°[b]
Berge (1974)	0.08 ± 0.05[c]	$228 \pm 25°$
Stannard and Conway (1974)	0.11 ± 0.06[d]	$221 \pm 15°$

[a] Estimated error.
[b] Longitude adjusted to epoch 1970. No error assigned.
[c] Measurement at 21 cm, other measurements at 11 cm.
[d] Measurement with the radiation belts partially resolved.

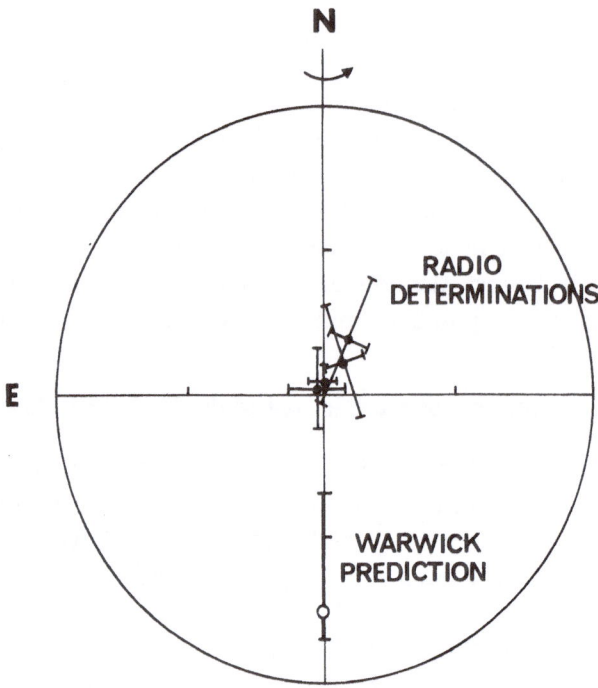

Fig. 7.   Mean displacement of the radio centroid of the decimetric emission relative to the optical disk.
The data are averages over 360° of longitude. The position predicted by Warwick (1970) is
shown for comparison.

dipole whose centre is slightly displaced from the rotation axis, or again as a centred dipole with a localized anomaly. It should be noted that on the latter explanation the longitude of the inferred anomaly is the same as that of the hot spot of emission in Branson's maps. If the hot spot is located with the rest of the emission at a distance of about 2 Jovian radii, then the intensity of the hot spot obtained from the centroid measurement agrees well with the intensity obtained from Branson's maps. The measurements of the radio centroid of the decimetric emission are thus in excellent agreement with the localized field anomaly theory outlined earlier.

6. The Location of the Jovian Dipole from the Pioneer 10 Magnetometer Results

Observations of the Jovian magnetic field with the vector helium magnetometer on board Pioneer 10 have been used to produce a model for the planetary dipole (Smith *et al.*, 1974). The data were obtained over the radial range 2.8 to 6 radii between c.m. longitude of 180° and 320° and over the magnetic latitude range $-13°$ to $+13°$. A least squares fit to the data gives a magnetic dipole offset by 0.12 ± 0.03 radii to the north of the planet, and 0.20 ± 0.03 radii in the equatorial plane towards 168° c.m. longitude. The dipole is inclined by approximately 15° to the rotation axis, and lies in the meridian plane near 230° c.m. longitude.

The small offset of the dipole to the north of the planet agrees well with the radio centroid determinations. (In making the comparison the radio displacements at 11 cm should be increased by a factor of ~ 1.4 to give a measurement of the centroid of the radiation belts alone, free from the diluting effect due to the thermal brightness of the planetary disk.) The longitude of the north magnetic pole is also close to that predicted from the radio observations (Berge, 1974; Stannard and Conway, 1974), but there appears to be serious disagreement over the inclination between the magnetic and rotation axes, and the equatorial displacement of the dipole.

One possible explanation for the discrepancy is that the Pioneer 10 data were obtained over a longitude range which is effectively centred on the region containing the localized hot spot of emission and proposed field anomaly. The detailed structure of the magnetic field in this region is liable to be complex, and may well not be typical of the dipole field as a whole. The spacecraft data should be carefully scrutinized for evidence of such a feature, since the data may well be providing valuable information about the localized field distortion to complement the radio data.

7. Recent Observations of the Intensity and Beaming of the Decimetric Emission

Although the dipole field model with a localized distortion has some success in accounting for the observed asymmetries in the decametric and decimetric emissions, recent observations of the total intensity and beaming of the decimetric radiation by Gulkis *et al.* (1973) and Berge (1974) have revealed further irregularities.

7.1. Variation of the total intensity with epoch

There is conflicting evidence as to whether short term variations of the decimetric emission are correlated with solar activity (Gulkis *et al.*, 1973; Roberts and Hugvenin, 1963; Gerard, 1970). The observations of Gulkis *et al.* (1973) and Berge (1974) show however that the overall emission from the Jovian radiation belts has decreased significantly in recent years (Figure 8). The data show no evidence for an eleven or twelve year periodicity which might be expected from a correlation with either the solar activity cycle or Jupiter's orbital position about the Sun. There is as yet no convincing explanation for the decline, although Berge (1974) has speculated that it may be associated with a decrease of the planet's magnetic moment.

7.2. Beaming of the decimetric emission

At epoch 1969 (with $D_E = -2.2°$) Gulkis and Gary (1971) remeasured the latitude dependence of the beaming of the decimetric emission. They found a symmetric decrease of intensity about the magnetic equator, with no evidence of the latitude asymmetry present in the earlier measurements at epoch 1963–4 (when D_E was $+3.3°$). As these authors comment, the absence of an asymmetry is difficult to explain if Branson's hot spot of emission is a real feature which is occulted by the planetary disk.

One theory advanced to resolve this difficulty (Stannard, 1972) was that the radiation from the hot spot is more strongly beamed than the emission from the rest of

D. STANNARD

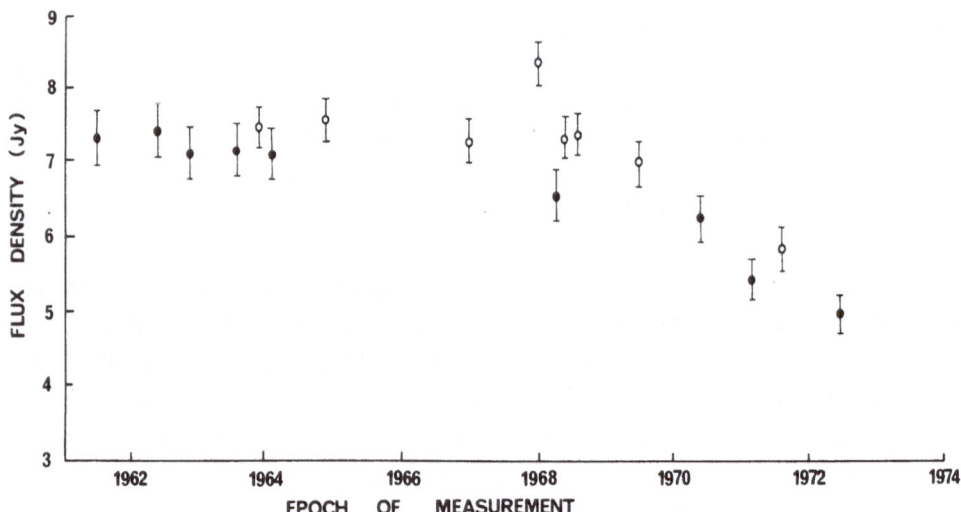

Fig. 8. Decrease of the decimetric emission with time. Data at $\lambda 11$–13 cm cited by Gulkis *et al.* (1973)
are shown as open circles. Data at $\lambda 21$ cm by Berge (1974) are shown as filled circles.

the radiation belts. At the early epoch the occultation of the hot spot would occur
at approximately $-6°$ magnetic latitude, whilst at the later epoch this latitude is
nearer $-12°$. Thus if a beam from the hot spot was not observable from $-12°$ magnetic latitude the occultation would produce no noticeable asymmetry in the beaming
curve. Other properties of the hot spot such as a small north-south extent inferred
from the high resolution centroid measurements (Stannard and Conway, 1974), and
possibly a flat radio spectrum and enhanced polarization, were shown to be consistent with this hypothesis.

However, measurements at epoch 1970 ($D_E = -3.1$) and 1971 ($D_E = -3.0$) by Berge
(1974) and Gulkis *et al.* (1973) showed that the beaming curve had changed still
further, with an asymmetry again present but in the opposite sense to that at epoch
1963–4 (that is with the steepest fall off now to the north of the magnetic equator).
The interpretation of this effect is by no means clear. It could be caused by a temporal
dependence of a north-south asymmetry (cf. Warwick, 1967), or it might suggest an
underestimate of the inclination angle between the magnetic and rotation axes
(Berge, 1974; Roberts and Ekers, 1966). This is interesting in the context of the Pioneer
results, but it presents many difficulties in the interpretation of the polarization
position angle data. A more intriguing possibility has been suggested by Berge (1974),
namely that the effect can be explained by assuming that D_E applicable to the radio
measurements is only about half the value obtained by direct optical observation.

The observations of Berge (1974) also suggest that the two maxima in intensity
per revolution are unequal, which could indicate that the beaming is a function of
longitude as well as latitude. Such an effect may also account for the apparent change
in the beaming curve with epoch. There is no evidence to indicate whether this
property can be ascribed to the hot spot alone, or whether it must be a function
of the entire radiation belt.

Continued observations of the decimetric emission are required to investigate these irregularities in the beaming and total intensity. Observations during the next few years would be particularly valuable, being a complete orbital period after the comprehensive investigations at epoch 1963–4. Theoretical consideration of the possible magnetic field geometry in the distorted region containing the hot spot may be of value for the interpretation of the Pioneer 10 magnetometer results.

Acknowledgements

It is a pleasure to thank R. G. Conway for many stimulating discussions on the radio astronomy of Jupiter, and G. L. Berge for permission to quote material in advance of publication. The author is grateful to colleagues at Jodrell Bank for helpful comments on this manuscript, and to the U.K. Science Research Council for financial support in the form of a Research Fellowship.

References

Barber, D.: 1966, *Monthly Notices Roy. Astron. Soc.* **133**, 285.
Barber, D. and Gower, J. F. R.: 1965, *Planetary Space Sci.* **13**, 889.
Berge, G. L.: 1965, *Astrophys. J.* **142**, 1688.
Berge, G. L.: 1966, *Astrophys. J.* **146**, 767.
Berge, G. L.: 1974, *Astrophys. J.* **191**, 775.
Berge, G. L. and Morris, D.: 1964, *Astrophys. J.* **140**, 1330.
Bigg. E. K.: 1964, *Nature* **203**, 1008.
Bigg, E. K.: 1966, *Planetary Space Sci.* **14**, 741.
Branson, N. J. B. A.: 1968, *Monthly Notices Roy. Astron. Soc.* **139**, 155.
Burke, B. F. and Franklin, K. L.: 1955, *Nature* **175**, 1074.
Carr, T. D.: 1971, *Astrophys. Letters* **7**, 157.
Carr, T. D. and Gulkis, S.: 1969, *Ann. Rev. Astron. Astrophys.* **7**, 577.
Carr, T. D., Brown, G. W., Smith, A. G., Higgins, C. S., Bollhagen, H., May, J., and Levy, J.: 1964, *Astrophys. J.* **140**, 778.
Clarke, J. N.: 1970, *Radio Sci.* **5**, 529.
Conway, R. G. and Stannard, D.: 1972, *Nature Phys. Sci.* **239**, 142.
Dickel, J. R., Degioanni, J. J., and Goodman, G. C.: 1970, *Radio Sci.* **5**, 517.
Donivan, F. F. and Carr, T. D.: 1969, *Astrophys. J.* **157**, L65.
Dulk, G. A.: 1965, *Science* **148**, 1585.
Dulk, G. A.: 1970, *Astrophys. J.* **159**, 671;
Ellis, G. R. A.: 1965, *Radio Sci.* **69D**, 1513.
Ellis, G. R. A. and McCulloch, P. M.: 1963, *Australian J. Phys.* **16**, 380.
Flagg, R. S. and Carr, T. D.: 1967, *Astrophys. Letters* **1**, 47.
Gerard, E.: 1970, *Radio Sci.* **5**, 513.
Gledhill, J. A.: 1967, *Nature* **214**, 155.
Gleeson, L. J., Legg, M. P. C., and Westfold, K. C.: 1969, *Proc. Astron. Soc. Australia* **1**, 274.
Goldreich, P. and Lynden-Bell, D.: 1969, *Astrophys. J.* **156**, 59.
Gulkis, S.: 1970, *Radio Sci.* **5**, 505.
Gulkis, S. and Gary, B.: 1971, *Astron. J.* **76**, 12.
Gulkis, S., Gary, B., Klein, M., and Stelzried, C.: 1973, *Icarus* **18**, 181.
Komesaroff, M. M. and McCulloch, P. M.: 1967, *Astrophys. Letters* **1**, 39.
Komesaroff, M. M., Morris, D., and Roberts, J. A.: 1970, *Astrophys. Letters* **7**, 31.
Lynch, M. A., Carr, T. D., May, J., Bluck, W. F., Robinson, V. M., and Six, N. F.: 1972, *Astrophys. Letters* **10**, 153.

McCulloch, P. M.: 1971, *Planetary Space Sci.* **19**, 1297.

McCulloch, P. M. and Ellis, G. R. A.: 1966, *Planetary Space Sci.* **14**, 347.

McCulloch, P. M. and Komesaroff, M. M.: 1973, *Icarus* **19**, 83.

Piddington, J. H. and Drake, J. F.: 1968, *Nature* **217**, 935.

Roberts, M. S. and Huguenin, G. R.: 1963, *Mem. Soc. Roy. Sci. Liège* **7**, 569.

Roberts, J. A. and Ekers, R. D.: 1966, *Icarus* **5**, 149.

Roberts, J. A. and Ekers, R. D.: 1968, *Icarus* **8**, 160.

Roberts, J. A. and Komesaroff, M. M.: 1964, *Nature* **203**, 827.

Roberts, J. A. and Komesaroff, M. M.: 1965, *Icarus* **4**, 127.

Schatten, K. H. and Ness, N. F.: 1971, *Astrophys. J.* **165**, 621.

Sherrill, W. M.: 1965, *Astrophys. J.* **142**, 1171.

Smith, E. J., Davis, L., Jones, D. E., Colburn, D. S., Coleman, P. J., Dyal, P., and Sonett, C. P.: 1974, *Science* **183**, 305.

Stannard, D.: 1972, Ph.D. thesis, University of Manchester.

Stannard, D. and Conway, R. G.: 1974, in preparation.

Thorne, K. S.: 1965, *Radio Sci.* **69D**, 1557.

Warwick, J. W.: 1963a, *Astrophys. J.* **137**, 41.

Warwick, J. W.: 1963b, *Astrophys. J.* **137**, 1317.

Warwick, J. W.: 1967, *Space Sci. Rev.* **6**, 841.

Warwick, J. W.: 1970, NASA CR-1685.

Warwick, J. W. and Dulk, G. A.: 1964, *Science* **145**, 380.

Whiteoak, J. B., Gardner, F. F., and Morris, D.: 1969, *Astrophys. Letters* **3**, 81.

Wu, C. S., Smith, R. A., and Zmuidzmas, J. S.: 1973, *Icarus* **18**, 192.

LONG TERM VARIATIONS OF THE DECIMETRIC RADIO EMISSION OF JUPITER (AND SATURN?)

E. GERARD

Observatoire de Meudon, Dept. de Radioastronomie, 92 Meudon, France

1. Long Term Variations of the Decimetric Radio Emission of Jupiter

The Jovian total flux density at decimeter wavelengths is known to possess a short modulation due to the beaming of the synchrotron radiation of the relativistic electrons trapped in the dipole magnetic field of the planet. The maximum flux (later called I_{max}) occurs twice per rotation when the Earth is at zero magnetic declination. It has been a matter of dispute for many years to know whether I_{max} was variable over long periods of time i.e. months and years. After more than a decade of observations it is clear that the flux density of Jupiter is variable (Roberts and Huguenin, 1963; Gerard, 1970; Klein *et al.*, 1972; Berge, 1974; Gerard, 1974, in preparation) at 21, 11 and 6 cm wavelength. At 21 cm we find that I_{max} measured with the E-vector along the magnetic equator is 40% lower in 1973 than what Roberts and Komesaroff (1965) found in 1962. A similar variation is observed by Berge. At 13 cm Klein *et al.* note a decrease of 20% for I_{max} between 1964 and 1971. Finally at 6 cm we find more than 10% decrease for I_{max} measured in 1968 (Whiteoak *et al.*, 1969) and 1973. The smaller percentage decrease towards shorter wavelengths is mostly due to the increasing thermal radiation from the disk which contributes about 10, 30 and 60% of the total flux at 21, 11 and 6 cm respectively. The non-thermal radiation (the synchrotron emission) thus suffered a 40% decrease during the last decade i.e. solar cycle No. 20 and will be much weaker during the 1974 solar minimum than it was during the 1964 solar minimum. However the observations that we made at 11 cm (Gerard, 1970) between December 1967 and August 1968 indicate that I_{max} underwent variations as large as 30% during the maximum of the solar cycle and rose well above its 1964 value in January 1968. A correlation with solar activity as measured by the 10.7 cm solar flux density was also suggested with phase lags lying between 3 and 9 days. In 1971 Klein *et al.* (1972) failed to detect variations of I_{max} larger than 9% but the Sun was considerably less active than in 1967–1968.

In conclusion the observations show yearly variations of I_{max} and suggest monthly (perhaps weekly) variations during the maximum of the solar cycle. The time required for a relativistic electron to radiate half its energy is

$$T_{1/2} (\text{days}) = 6.75 \times 10^{-3} \ B^{-3/2} (\gamma_{max})^{-1/2} \qquad (1)$$

(Ginzburg and Syrovatskii, 1965) where B is the magnetic field strength expressed in gauss and γ_{max} the initial frequency of maximum emission expressed in megahertz. Taking $B = 4$ G for the equatorial surface magnetic field (Smith *et al.*, 1974) and assuming a critical frequency of $\sim 3 \times 10^3$ Mhz (Roberts, 1965) i.e. $\gamma_{max} = 850$

V. Formisano (ed.), The Magnetospheres of the Earth and Jupiter, 237–239. All Rights Reserved

Mhz, the electron lifetime is one year at $L = 1.8$, in the heart of the radiation belt. An acceleration and/or injection mechakism is thus continuously at work throughout the solar cycle compensating for the synchrotron losses and other losses that may be present.

In addition one could suggest a variation within the solar cycle along the following lines: during an intense solar maximum (like cycle No. 19 from 1957 to 1959 and cycle No. 20 from 1967 to 1968) magnetic and electric field perturbations driven by the solar wind produce energetic particle diffusion and acceleration throughout the Van Allen belts of Jupiter. Once the solar maximum is over the radiation belt electrons decay with their own synchrotron lifetime to quiescent conditions where only in-house acceleration and/or injection can operate. At $L = 2.5$, Equation (1) yields $T_{1/2} = 5$ yr hence a stationary regime would hardly be reached within a solar cycle. This may explain part of the 40% decrease of I_{max} since the 1957 solar maximum was unusually intense. At $L = 1.2$, $T_{1/2} = 2$ months and perhaps only disturbed conditions can produce and maintain significant electron fluxes for a limited time.

2. Possible Non-Thermal Decimetric Radio Emission from Saturn

Recent attempts at detecting non-thermal emission from Saturn at decimeter wavelengths have failed so far although there are hints of a possible modulation of the total flux density with longitude (Gerard, 1969; Berge and Muhleman, 1973). The equatorial surface magnetic field can be estimated by assuming that the ratio between the magnetic dipole moment and the angular momentum is constant for all planets. This empirical relationship has recently been proven to be roughly correct for the terrestrial planets. Adopting 0.3 G for the equatorial surface field of the Earth one finds respectively 14.8, 4.8, 1.8 and 1.6 G for Jupiter, Saturn, Uranus and Neptune. If one rather scales the Jovian planets magnetic field dipoles to that recently measured for Jupiter by Pioneer 10 the actual surface field of Saturn is $\frac{1}{3}$ of that of Jupiter i.e. ~ 1.3 G. One can likewise predict relativistic electron energies smaller than those attained on Jupiter by a factor of $3^{1/2}$ (see for instance Brice and Mc-Donough, 1973) hence the synchrotron losses proportional to $B^2 \times E^2$ could be smaller by at least a factor of 30. The Saturn rings and the first satellites form an almost continuous screen up to $L = 3$ and should seriously reduce the particle fluxes in the inner belt: the effect of the Galilean satellites on the Jovian radiation belt has been calculated by Hess et al., 1973 and observed with Pioneer 10. One can then reasonably expect the radiation belt of Saturn to emit less than one hundredth of that of Jupiter in which case the detection of the decimeter non-thermal emission is marginally possible with modern equipment. The circumstances will also be more favourable when the rings are closed if the synchrotron emission is highly beamed (Gerard and Kazès, 1973).

In the meantime a more accurate assessment of the magnetic dipole field will be possible when Pioneer 11 flies by Saturn around 1979.

References

Berge, G. L.: 1974, *Astrophys. J.*, in press.
Berge, G. L. and Muhleman, D. O.: 1973, *Astrophys. J.* **185**, 373.
Brice, N. and McDonough, T. R.: 1973, *Icarus* **18**, 206.
Gerard, E.: 1969, *Astron. Astrophys.* **2**, 246.
Gerard, E.: 1970, *Astron. Astrophys.* **8**, 181.
Gerard, E. and Kazès, I.: 1973, *Astrophys. Letters* **13**, 181.
Ginzburg, V. L. and Syrovatskii, S. I.: 1965, *Annual Rev. Astron. Astrophys.* **3**, 297.
Hess, W. N., Birmingham, T. J., and Mead, G. D.: 1973, *Science* **182**, 1021.
Klein, M. J., Gulkis, S., and Stelzried, C. T.: 1972, *Astrophys. J.* **176**, L85.
Roberts, J. A.: 1965, *Radio Sci.* **69 D**, 1543.
Roberts, J. A. and Komesaroff, M. M.: 1965, *Icarus* **4**, 127.
Smith, E. J., Davis, L., Jones, D. E., Colburn, D. S., Coleman, P. J., Dyal, P., and Sonett, C. P.: 1974, *Science* **183**, 305.

UPPER LIMITS TO JOVIAN X-RAY EMISSION
FROM THE UHURU SATELLITE

KEVIN C. HURLEY

Centre d'Etude Spatiales des Rayonnements, B.P. 4057, 31-Toulouse, France

1. Introduction

In the past decade, the study of celestial X-radiation has led to a deeper understanding of physical processes at work both in the Cosmos in general, and in our own solar system in particular. As magnetospheres seem to be universal astrophysical phenomena, it is interesting to search for rather general scaling laws which might govern their large scale energy balance. The study of terrestrial auroral zone X-rays, created by the precipitation of energetic electrons into the upper atmosphere under a wide variety of geomagnetic and solar wind conditions, has given us considerable information about the dynamics of our magnetosphere; in principle, a study of Jovian X-ray emission would do the same for Jupiter. Moreover, if Jovian X-ray activity were detectable at Earth – from satellite or balloon observations, for example – we could acquire a long term picture of the Jovian magnetosphere which might be as revealing as the information gathered by flyby experiments.

X-ray activity on the Earth's dayside exhibits a wide range of temporal structures, from 'microburst' events lasting less than a second (Anderson and Milton, 1964) to storms lasting hours; studies of Jovian decametric emission suggest that Jupiter's electron precipitation might also occur on this range of time scales. A variety of models can be constructed to estimate the energy in Jovian precipitating electron fluxes and the resulting X-ray fluxes which reach the Earth; three models will be outlined here, the details concerning the assumed electron spectra, cross sections, etc. having appeared elsehwere (Hurley, 1972).

Model IA: this model considers the solar wind energy input to the terrestrial and Jovian magnetospheres, and scales the energy in Jovian precipitating electron fluxes by the ratio of the two intercepted energies.

Model IB: this model scales the energy in precipitating fluxes by the ratio of the magnetic field energy density at Jupiter to that at the Earth; the basis for this idea is that, in the Earth's auroral zones, precipitating electron fluxes are observed which originate in a highly diamagnetic plasma sheet.

Model II: in this model, the electron fluxes are found by assuming that Jovian decametric emission is created by electron gyroradiation (Warwick, 1970).

The different X-ray fluxes which these models give, and the question of their de-

V. Formisano (ed.), The Magnetospheres of the Earth and Jupiter, 241–244. All Rights Reserved

tectability, will be discussed below, following a description of the experiment and data reduction procedures.

2. Experiment and Data Reduction

The X-ray satellite UHURU, launched in December, 1970, carried aboard two large area proportional counter arrays (840 cm^2 each), sensitive to X-rays in the 1.7–18 keV energy range, oriented perpendicular to the satellite spin axis, and collimated to $0.52° \times 5.2°$ (referred to as side 1) and $5.2° \times 5.2°$ (referred to as side 2). A detailed description of the satellite has appeared elsewhere (Giacconi *et al.* 1971). Uhuru's spin axis orientation is changed approximately once a day in order to scan different regions of the sky at a spin rate of about 1 revolution/720 s; during 1971, Jupiter was scanned on 26 different days.

A partial analysis of 22 days of data has now been done: X-ray data from sides 1 and 2, in the 2–6 keV energy range, have been examined for evidence of Jovian X-ray activity on three different time scales:

(1) On a daily basis. This is done by superposing all the Jupiter scans from a given day; in this way, one obtains 22 observations of Jupiter with a typical limiting sensitivity of about 5×10^{-3} photons cm^{-2} s^{-1} keV^{-1} at Earth, after correction for instrumental effects.

(2) On a long term basis. This is done by superposing all the Jupiter scans from 22 days into one single observation with a limiting sensitivity of about 5×10^{-4} photons cm^{-2} s^{-1} keV^{-1}.

(3) On a short term basis. Here, each individual scan of Jupiter is examined for X-ray activity. About 500 scans from side 1 were analyzed this way; each lasts approximately 2 s and has a sensitivity of 2×10^{-2} photons cm^{-2} s^{-1} keV^{-1}.

None of the three analyses described above gave any statistically significant evidence for Jovian X-ray activity; three sigma upper limits from (1) and (2) are shown in Figure 1, along with the upper limits from previous experiments. An approximate conversion between the Uhuru fluxes at Earth and the X-ray energy at Jupiter is 1 photon cm^{-2} sec^{-1} $keV^{-1} = 10^{20}$ erg s^{-1} at Jupiter.

3. Discussion

According to Model IA, the Jovian magnetosphere should intercept at least 100 times more solar wind energy than the Earth.

This leads to 2–6 keV X-ray fluxes of the order of 10^{-7} photons cm^{-2} s^{-1} keV^{-1} at Earth. Even by increasing the solar wind energy input by a factor of 10^3, to take into account conditions during the largest solar storms, the fluxes remain at the very limit of detectability on a daily basis. The Model IB fluxes are on the order of 5×10^{-4} photons cm^{-2} s^{-1} keV^{-1}; thus, if they were present over long periods of time, they would be detectable.

The most encouraging results come from Model II, which gives fluxes around

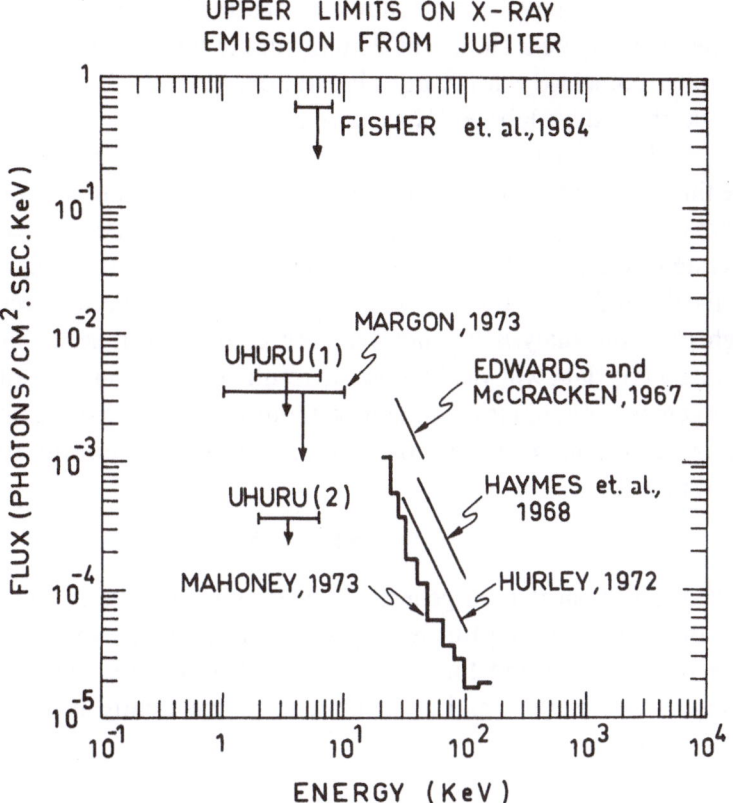

Fig. 1. After Mahoney (1973). Upper limits to Jovian X-ray emission from various experiments. Uhuru (1) is a typical upper limit from one day's data; Uhuru (2) is the result of combining 22 days of data. Uhuru upper limits to the short term X-ray activity (not shown), on a time scale of seconds, are around 2×10^{-2} photons $\text{cm}^{-2} \text{s}^{-1} \text{keV}^{-1}$.

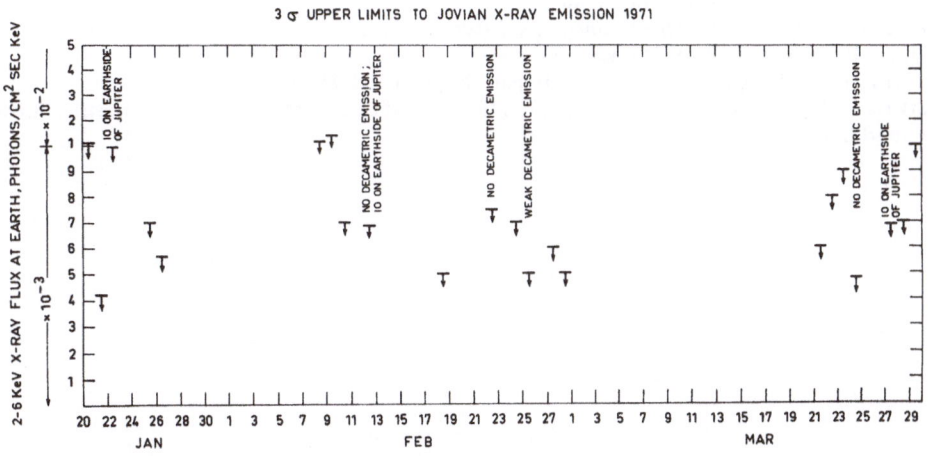

Fig. 2. Daily upper limits to Jovian X-ray activity in early 1971. On four days, marked either 'No Decametric Emission' or 'Weak Decametric Emission', radio coverage of Jupiter was coincident with X-ray coverage. On three days, the satellite Io was on the Earth side of Jupiter at the same time as the X-ray observations, and presumably, decametric emission was being generated.

10^{-2} photons cm^{-2} s^{-1} keV^{-1}; these should be observed when decametric radiation is being generated. Figure 2 shows the 22 daily upper limits; decametric radio coverage of Jupiter coincided with X-ray observations on four days (Warwick *et al.*, 1973). On the three days labeled 'No Decametric Emission', no radio fluxes were reported; a weak decametric event, as indicated, was present on one day. Finally, there were three days when the satellite Io was on the Earthward side of Jupiter for 80% of more of the X-ray observation period; presumably, decametric emission was being generated during these periods (although not neccessarily being beamed towards the Earth), and associated X-ray emission should have been detected.

Although the data analysis has not yet been completed, it seems unlikely that further work on this data will yield a positive result. The intriguing question of how Jupiter can generate decametric emission without associated X-radiation must remain, at least for the moment (and perhaps only until the following paper is presented), a mystery.

Acknowledgements

I am indebted to R. Giacconi for generously offering me the use of Uhuru data and data reduction facilities, to S. Murray for his help with the data analysis, and to the staff of American Science and Engineering. I thank F. Cambou and G. Vedrenne for encouraging this project, and I am grateful to the Centre National d'Études Spatiales for a fellowship.

References

Anderson, K. A. and Milton, D. W.: 1964, *J. Geophys. Res.* **69**, 4457.
Edwards, P. J. and McCracken, K. G.: 1967, *J. Geophys. Res.* **72**, 1809.
Fisher, P. C., Clark, D. B., Meyerott, A. J., and Smith, K. L.: 1964, *Nature* **204**, 983.
Giacconi, R., Kellogg, E., Gorenstein, P., Gursky, H., and Tananbaum, H.: 1971, *Astrophys. J.* **165**, L27.
Haymes, R. C., Ellis, D. V., and Fishman, G. J.: 1968, *J. Geophys. Res.* **73**, 867.
Hurley, K.: 1972, *J. Geophys. Res.* **77**, 46.
Mahoney, W.: 1973, Ph.D. Thesis, Univ. of California, Berkeley.
Margon, B.: 1974, private communication.
Warwick, J. W.: 1970, NASA Contractor Report, NASA CR-1685.
Warwick, J. W., Dulk, G. A., and Swann, D. G.: 1973, World Data Center A for Solar Terrestrial Physics, Report UAG-25.

UPPER LIMITS FOR X-RAY EMISSION FROM JUPITER AS MEASURED FROM THE COPERNICUS SATELLITE

JOHN F. VESECKY

Astronomy Dept., The University, Leicester, England

and

J. L. CULHANE and F. J. HAWKINS

Physics Dept., Mullard Space Science Laboratory, University College, London, England

1. Introduction

Since Jupiter is known to have an extensive and active magnetosphere in some ways analogous to that of the Earth, a number of observations have been made in the hopes of detecting X-ray emission from Jupiter analogous to X-rays from terrestrial aurorae. If one assumes an X-ray flux inversely proportional to photon energy (as with terrestrial aurorae), then observations at relatively low energies ($\lesssim 10$ keV) from satellites provide an excellent opportunity to detect any Jovian X-ray emission. The Uhuru and Copernicus (OAO-3) satellites provide just such an opportunity. Unfortunately observations reported to date have not been successful in either case and provide only upper limits – for Uhuru results see Hurley (1975).

In this paper we report upper limit X-ray fluxes from three Copernicus detectors covering the 0.6 to 7.5 keV energy range. First the detectors are briefly described and the times at which observations were made are given. A portion of the observational period fortunately coincided with a period of decametric radio activity on Jupiter and upper limit fluxes for this time are noted separately. Photon number spectra proportional to E^{-2}, E^{-3}, and E^{-4} ($E=$ photon energy in keV) are fitted to the observed upper limit counting rates in each detector and the resulting upper limit spectra are compared with previous X-ray observations of Jupiter. Finally the upper limit X-ray fluxes are discussed in relation to magnetospheric activity on Jupiter.

2. Copernicus X-Ray Telescope System

On board the third orbiting astronomical observatory (OAO-3), in addition to the prime ultraviolet telescope experiment, an X-ray telescope package was also included. The relevant parameters for the X-ray detectors of interest here are listed in Table I and further details are given by Bowles *et al.* (1974) and Sanford and Hawkins (1973). Although these detectors have rather small collecting areas, they can be pointed at a given object for relatively long times, thus obtaining a reasonable sensitivity. The background counting rates in the three detectors are determined by the ambient trapped particle environment and thus vary with time. The guard counters, which

V. Formisano (ed.), The Magnetospheres of the Earth and Jupiter, 245–251. All Rights Reserved
Copyright © 1975 by D. Reidel Publishing Company, Dordrecht-Holland

TABLE I

Copernicus detector parameters during Jupiter observations

Optics[a]	Energy range	Effective area	Mean efficiency	Field of view
Collimating tube	2.5–7.5 keV	17.8 cm^2	0.21	3.3° FWHM[b]
Grazing incidence paraboloid mirror	1.0–3.1 keV	3.7 cm^2	0.33	12′ FWHM
	0.6–1.9 keV	12.3 cm^2	0.14	13′ FWHM

[a] All detectors used proportional counters.
[b] Full angular width at half maximum response

operate in anticoincidence with the X-ray detectors, give a measure of the trapped particle flux affecting the detectors. By correlating the X-ray detector count rate with the guard count rate when the X-ray telescope is not pointed at an X-ray source one can relate the guard count rate to the background count rate in the X-ray detectors. X-ray background count rates determined in this manner were used in the results reported below.

3. Observations

The Copernicus X-ray telescope package was pointed nearly or directly toward Jupiter from 1906 UT, May 2, 1973 until 2010 UT, May 3, 1973. Usable data were collected during the time intervals shown in Figure 1. The total observing time amounted to some 3.3 hours in all. During the observations Copernicus was directed toward regions of sky near Jupiter and toward the star Theta Capricornus as well as toward Jupiter. As a result Jupiter was at times not in the relatively small ($\sim 10'$) fields of view of the lower energy detectors while remaining within the relatively

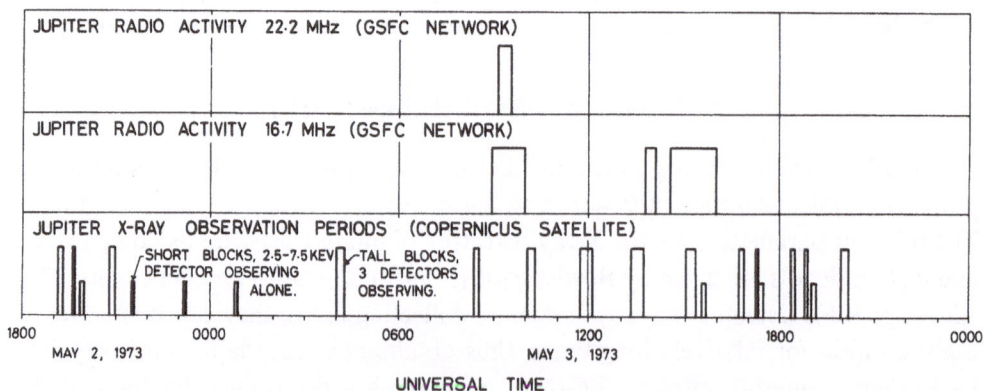

Fig. 1. Periods of X-ray observation of Jupiter by the Copernicus satellite and their relationship to periods of decametric radio activity on Jupiter. Tall blocks indicate intervals of observation by all three Copernicus detectors and short blocks indicate intervals of observation by the 2.5–7.5 detector alone.

wide (3.3°) field of view of the 2.5–7.5 keV detector. This difference is noted in Figure 1 by the use of tall blocks to indicate intervals of observation by all three detectors and short blocks to indicate intervals of observation by the 2.5–7.5 keV detector alone.

Io related decametric radio bursts from Jupiter were observed in the interval 0855 to 1600 UT (May 3) by the Goddard Space Flight Centre Jupiter radio network. The precise time intervals and frequencies are indicated in Figure 1 with a time resolution of 5 min. Fortunately some X-ray observations coincided with the decametric radio activity near 1530 UT on May 3rd.

4. Upper Limit Results and Comments

The background counting rate was determined for each detector using the guard count rates as discussed above. The observed count rates did not exceed the background rate by more than three standard deviations (3σ) over the entire 3.3 h period (a shorter period applies to the lower energy detectors as noted on Figure 1). Hence no source was detected and the $+3\sigma$ level background count rate was taken as the upper limit. Since the efficiencies of the detectors vary with energy, a given count rate in a detector can correspond to a variety of inputs depending on how the photon flux varies with photon energy. Accordingly we fitted photon flux spectra proportional to E^{-2}, E^{-3}, and E^{-4} to the upper limit count rates. The real spectrum is, of course, unknown; but power law spectra are common in X-ray sources and have been observed for auroral X-rays in the 2–10 keV energy range by Wilson et al. (1969). The photon flux spectra consistent with the observed upper limit count rates are given in Table II below. The figures in parentheses are upper limits for the observations made during decametric radio activity near 1530 UT, May 3rd. The other figures are for the total observing time.

TABLE II

Copernicus upper limits for X-ray emission from Jupiter

Assumed spectral shape	Upper limit flux spectra for detectors in three energy ranges[a]		
	0.6–1.9 keV	1.0–3.1 keV	2.5–7.5 keV
E^{-2}	$8 \times 10^{-3} E^{-2}$	$3 \times 10^{-2} E^{-2}$	$5 \times 10^{-2} E^{-2}$
	$(3 \times 10^{-2} E^{-2})$[b]	$(9 \times 10^{-2} E^{-2})$	$(2 \times 10^{-1} E^{-2})$
E^{-3}	$1 \times 10^{-2} E^{-3}$	$6 \times 10^{-2} E^{-3}$	$2 \times 10^{-1} E^{-3}$
	$(3 \times 10^{-2} E^{-3})$	$(2 \times 10^{-1} E^{-3})$	$(9 \times 10^{-1} E^{-3})$
E^{-4}	$1 \times 10^{-2} E^{-4}$	$1 \times 10^{-1} E^{-4}$	$1 E^{-4}$
	$(4 \times 10^{-2} E^{-4})$	$(4 \times 10^{-1} E^{-4})$	$(4 E^{-4})$

[a] All upper limit flux spectra are in photons cm^{-2} s^{-1} keV^{-1} at Earth orbit. E = photon energy in keV.

[b] Spectra in parentheses refer to observations during decametric radio activity from Jupiter. Spectra not in parentheses refer to the total observing time, 3.3 h.

The spectra given in Table II can be integrated to obtain the photon flux in a given detector's energy range. In the case of an assumed E^{-2} spectral shape the photon fluxes (photons cm^{-2} s^{-1}) are as follows: 9×10^{-3} (3×10^{-2}) in the 0.6–1.9 keV detector, 2×10^{-2} (6×10^{-2}) in the 1.0–3.1 keV detector and 1×10^{-2} (6×10^{-2}) in the 2.5 to 7.5 keV detector; fluxes in parentheses are for upper limits obtained during Jovian decametric radio activity.

In Figure 2 we compare the results given in Table II for E^{-2} spectra with a sum-

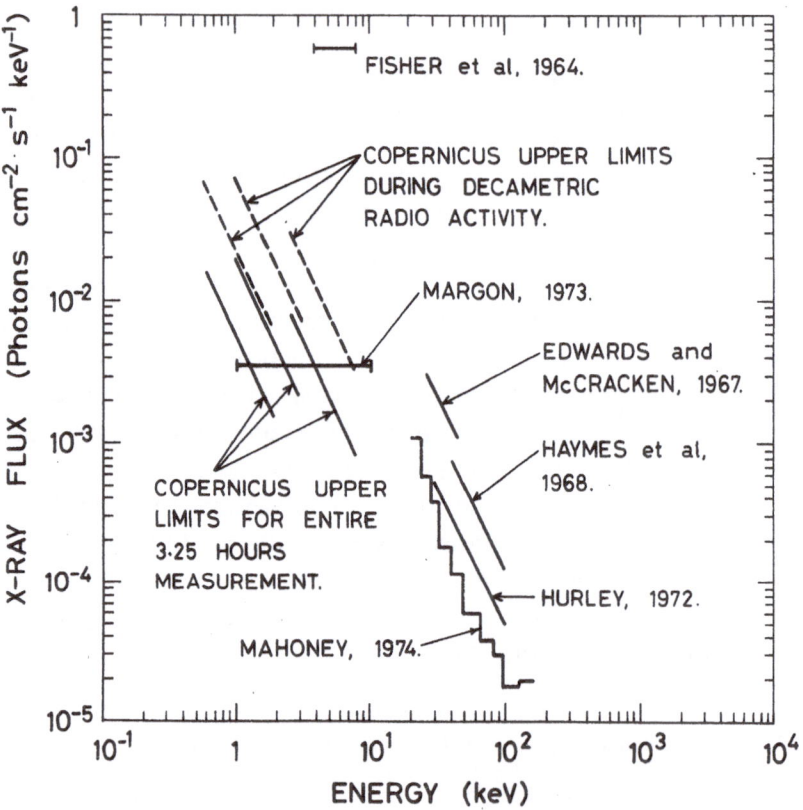

Fig. 2. Upper limits on X-ray emission from Jupiter. The summary of other X-ray upper limits was compiled by Mahoney (1974).

mary of other upper limits compiled by Mahoney (1974). The particular significance of the Copernicus upper limits results from their being made at low energies and during a period of considerable Jovian decametric radio activity. The figure makes clear the importance of low energy observations since any X-ray emission from Jupiter is likely to decrease with increasing photon energy. This point is especially noteworthy if indeed Jovian X-ray emission is similar in spectral shape to emission from a stable auroral arc reported by Wilson *et al.* (1969). They measured a photon number spectrum proportional to $E^{-3.4}$ in the 0.2 to 10 keV range. Even if one assumes an E^{-2} spectrum, the 0.6–1.9 keV measurement by Copernicus is clearly the lowest of the upper limits shown in Figure 2.

Decametre radio noise bursts are the best evidence presently available regarding magnetospheric activity on Jupiter (excluding, of course, flyby spacecraft such as Pioneer 10). During periods of magnetospheric activity on Earth X-rays are often observed in the auroral zone. Akasofu (1968) describes this phenomenon as the X-ray substorm. If one extrapolates the terrestrial phenomenon to Jupiter, then the Copernicus observations made during the decametric activity are particularly significant in that they were made during a period when X-ray bursts were likely to occur. The decametric radio noise during the X-ray observations was in fact relatively strong. Although only two intervals of X-ray observation actually coincided with decametre bursts, Figure 1 shows that the whole period of Io related (Earth-Jupiter-Io angle near 90°) radio noise (0855–1600 UT) was sampled by X-ray observations at frequent intervals. The observations thus suggest that if X-ray bursts lasting an hour or more are regularly associated with decametre radio activity on Jupiter, they are likely to be weaker than the appropriate Copernicus upper limits in Table II and Figure 2.

5. Discussion

A better appreciation of the above results can be obtained by comparing the upper limit fluxes of Table II with an extrapolation of terrestrial auroral X-ray emission to Jupiter. Wilson *et al.* (1969) measured the X-ray emission from a stable auroral arc by looking down on the arc from a rocket at about 800 km altitude. The X-ray spectrum for the most intense region was a very close fit to $3 \times 10^2 \, E^{-3.4}$ photons cm^{-2} s^{-1} keV^{-1} sr^{-1} in the 2 to 10 keV range. If we assume that 10% of Jupiter's visible disc is covered by auroral arcs of the same intensity, we have a source solid angle of about 1.5×10^{-7} sr. The X-ray flux at Earth orbit from such a source is then about $4 \times 10^{-5} \, E^{-3.4}$ photons cm^{-2} s^{-1} keV^{-1}. Comparing this spectrum with the upper limits for E^{-3} spectra in the 2.5–7.5 keV Copernicus detector (Table II) we find that this hypothetical Jovian aurora is weaker than the total time upper limit by a factor of about 4×10^3 and weaker than the upper limit during decametric activity by a factor of about 2×10^4. Thus precipitating electron fluxes on Jupiter could be about 10^4 times more intense than those required to produce the stable auroral arc studied by Wilson *et al.* (1969) and still be consistent with the upper limits reported here. Clearly more observations to reduce the present upper limits would be worthwhile.

Following a method used by Warwick (1970) we can obtain a rough upper limit for the precipitating electron flux in a given small energy range. Chamberlain (1961) gives the following expression for ΔE, the total energy radiated as bremsstrahlung by an electron of kinetic energy E_1, $\Delta E/E_1 = 6.4 \times 10^{-4}(E_1/E_0)$, where $E_0 = 511$ keV, the electron rest energy. ΔE is radiated over a spectrum from $h\nu \sim E_1$ to $h\nu \sim 0.1 \, E_1$. We shall take E_1 equal to the average energy within a given detector's energy range and assume that the energy ΔE is radiated isotropically and within the given detector's energy range. Taking the area over which the precipitation occurs as 10% of the hemisphere facing the Earth, i.e. 3×10^{19} cm^2, we obtain the upper limits for precipitating electron fluxes given in Table III below. These upper limits are about

TABLE III

Upper limits for electron precipitation on Jupiter

	X-ray detector energy range		
	0.6–1.9 keV	1.0–3.1 keV	2.5–7.5 keV
Upper limit on photon flux[a]	9×10^{-3} (3×10^{-2})[c]	2×10^{-2} (6×10^{-2})	10^{-2} (6×10^{-2})
Average energy (E_1)	1.3 keV	2.5 keV	5.3 keV
Upper limit on electron precipitation[b]	10^{13} (4×10^{13})	10^{13} (4×10^{13})	3×10^{12} (2×10^{13})

[a] All photon fluxes are in photons cm^{-2} s^{-1} in the given energy range.
[b] Electron precipitation fluxes are in electrons cm^{-2} s^{-1} near energy E_1.
[c] Parentheses indicate upper limits during decametric radio activity.

five orders of magnitude greater than typical precipitating electron fluxes in the Earth's auroral zone (Hess, 1968).

The comparisons made above suggest that precipitating electron fluxes (~ 1 to 10 keV) on Jupiter are not more than 10^4 to 10^5 times as intense as the fluxes associated with terrestrial aurorae. It has been suggested that precipitating electron fluxes on Jupiter may be as low as 10 times those associated with terrestrial aurorae. If one accepts these estimates, an X-ray experiment of considerably greater sensitivity will be required to detect X-ray emission from Jupiter and gather the information such observations can provide, i.e. precipitating particle flux intensity and spectral shape.

Acknowledgements

The authors would like to thank all the staff from Mullard Space Science Laboratory and Goddard Space Flight Centre who assisted in making the X-ray observations from Copernicus. and Philip Charles of M.S.S.L. for help in the data analysis. We are also grateful to Dr R. G. Stone, Dr. J. K. Alexander, and Mrs Susan Vaughan for kindly providing the decametre radio observations of Jupiter. Financial assistance from the Science Research Council, the Royal Society, and NASA Goddard Space Flight Centre is gratefully acknowledged.

References

Akasofu, S. I.: 1968, *Polar and Magnetic Substorms*, Springer-Verlag, New York.
Bowles, J. A., Patrick, T. J., Sheather, P. H., and Eiband, A. M.: 1974, *J. Phys.* E7, 183.

Chamberlain, J. W.: 1961, *Physics of the Aurora and Airglow*, Academic Press, New York.

Edwards, P. J. and McCracken, K. G.: 1967, *J. Geophys. Res.* **72**, 1809.

Fisher, P. C., Clark, D. B., Meyerott, A. J., and Smith, K. L.: 1964, *Nature* **204**, 982.

Haymes, R. C., Ellis, D. V., and Fishman, G. J.: 1968, *J. Geophys. Res.* **73**, 867.

Hess, W. H.: 1968, *The Radiation Belt and the Magnetosphere*, Blaisdell, Waltham, Massachusetts.

Hurley, K. C.: 1972, *J. Geophys. Res.* **77**, 6558.

Hurley, K. C.: 1975, this volume, p. 241.

Mahoney, W. A., Ph.D. Thesis, University of California at Berkeley, Space Sciences Laboratory, 1974.

Margon, B.: Results from 1969 rocket flight from Brazil, quoted by Mahoney (1974).

Sanford, P. W. and Hawkins, F. J.: 1973, *Annals of N.Y. Academy of Sciences* **224**, 285.

Warwick, J. W.: 1970, NASA Contractor Report NASA CR-1685.

Wilson, B. G., Baxter, A. J., and Green, D. W.: 1969, *Can. J. Phys.* **47**, 2427.

JUPITER AND SATURN*

R. HIDE

Geophysical Fluid Dynamics Laboratory, Meteorological Office, Bracknell, Berkshire, England

Jupiter, the largest planet, and Saturn, the second largest, contain nine-tenths of the material of the solar system outside the Sun and most of the angular momentum of the solar system is associated with their orbital motion. Both planets rotate very rapidly (rotation periods ~ 10 h) and possess rich satellite systems. Owing to their strong gravitational fields and low surface temperatures, Jupiter and Saturn may, unlike the 'terrestrial' planets, be fairly close in chemical composition to the primordial material out of which the solar system originally formed; they consist mainly of hydrogen, much of which is compressed to a metallic form. Jupiter is the only planet other than Earth showing evidence of a general magnetic field.

Absorption of incident solar energy accounts for less than one-half the estimated total thermal (infrared) radiation emitted by Jupiter and Saturn. The balance is probably due to internal heat sources and could be accounted for in terms of a gravitational contraction at about 0.1 cm yr^{-1}. The outward flow of heat should maintain the atmospheric temperature gradients close to their adiabatic values, which is a significant result for theories of atmospheric motions (see Appendix A). These theories are largely concerned with explaining the rough alinement of clouds in bands parallel to the equator, the presence of strong eastward equatorial currents, the occurrence of transient spots and other irregular markings and, in the case of Jupiter, the nature of the enigmatic Great Red Spot.

Jupiter, unlike Saturn, is a strong emitter of non-thermal radio noise on decametre and decimetre wavelengths. Plausible theories of this radio emission invoke a strong Jovian dipole magnetic field and an associated system of van Allen-type 'radiation' belts of electrically-charged particles extending beyond and interacting with the first Galilean satellite Io. The most likely source of the Jovian magnetic field – which theories of Jupiter's internal constitution must now take properly into account – is a hydromagnetic dynamo associated with fluid motions in the electrically-conducting parts of Jupiter's interior. The absence of a non-thermal component in Saturn's radio spectrum implies that radiation belts cannot form around that planet, possibly because Saturn is non-magnetic or, if it is magnetic, because charged particles in the vicinity of Saturn are rapidly removed through interactions with Saturn's rings.

Modern research on Jupiter and Saturn is based on a rich variety of data, soon

* R. Hide's invited lecture entitled 'The Origin of Jupiter's Magnetic Field' dealt with the dynamics, internal constitution and magnetic fields of the Earth, Jupiter and Saturn and the theory of hydromagnetic dynamos in rotating fluids. As a partial record of this contribution to the Neil Brice Memorial Symposium we present here, with the kind permission of the Royal Society of London, the abstract of an article originally published in the *Proceedings of the Copernicus Quincentenary Celebrations*, held by the Royal Society in January 1973 (*Proc. Roy. Soc.* **A336**, 63–84, 1974).

to be augmented by observations from space-craft (notably Pioneer 10 and Pioneer 11). Future progress with the theoretical interpretation of these data in terms of improved models of the structure and evolution of the giant planets will involve not only the further application of a wide range of established knowledge but also the development of new ideas in several areas of basic science.

The paper ends with two appendices, on the dynamics of rapidly rotating non-homogeneous fluids and on hydromagnetic dynamos.

PHOTOCHEMISTRY OF HYDROCARBONS IN THE JOVIAN UPPER ATMOSPHERE

S. S. PRASAD, L. A. CAPONE*, and L. J. SCHNECK

Dept. of Physics and Astronomy, University of Florida, Gainesville, Fla. 32611, U.S.A.

1. Introduction

The recent remarkable success of Pioneer 10 has greatly expanded the arena of planetary aeronomy. New and invigorating incentives and justifications now exist for more thorough studies in Jovian upper atmospheric physics and chemistry. One important problem in this area is the problem of the photochemistry of the hydrocarbons. Apart from being interesting in its own right, knowledge of the distributions of the various hydrocarbons is needed in many other studies, too (e.g., the thermal structure). The recent occultation of the multiple star system β-Scorpii by Jupiter, which occurred on March 13, 1971, provided another opportunity to probe the Jovian atmospheric conditions. Utilizing this event and the advancements in technology that have taken place since the time of the last occurrence of a similar occultation studied by Baum and Code (1953), Hubbard *et al.* (1972) and Veverka *et al.* (1974) have reported the existence of several smaller scale stratified thermal structures. In analogy with the heating produced by methane absorption in the 3.3 μ band (Gillet *et al.*, 1969), it has been suggested by Hubbard (1969) that these smaller scale layers may be the result of a similar absorption process in hydrocarbon or other trace constituents. Some hydrocarbons are good infrared radiators also, and could influence the thermal structure on the manner discussed by McGovern and Burk (1972), and Strobel and Smith (1973). Another important area of study where a knowledge of the height profile of the hydrocarbons could be extremely useful, relates to Jovian ionospheric structure and conductivities. It may be noted that CH_3 has the lowest ionization potential compared to all the other major constituents in the Jovian atmosphere. It could, therefore, play a role analogous to that of NO in the terrestrial ionosphere. Depending on the altitude where the concentration of CH_3 maximizes, it may also be a candidate worth investigating in the context of the stratified ionization detected by the S-band occultation experiment on board Pioneer 10 (Kliore *et al.*, 1974).

Guided by these considerations, we report here a study of the hydrocarbon photochemistry in the Jovian atmosphere.

2. Chemical Reactions and Their Rate Coefficients

Given a model for the height profile of temperature, altitude distribution of molecular hydrogen and helium, CH_4 mixing ratio in the lower region, and the solar UV

* Now at NASA's Ames Research Center, Moffett Field, California, 94035

V. Formisano (ed.), The Magnetospheres of the Earth and Jupiter, 255–268. All Rights Reserved
Copyright © 1975 by D. Reidel Publishing Company, Dordrecht-Holland

fluxes, all of which are to be described in the subsequent sections, the concentrations of the various hydrocarbons are controlled by chemical reactions and eddy and molecular diffusion. The chemical reactions used in our work are given in Table I

TABLE I

Chemical reactions in the Jovian atmosphere and their rate constants[a]

Reaction	Reaction rate	Reference
(1) $CH + CH_4 \rightarrow C_2H_4 + H$	$2.5\,(-12)$	Braun et al. (1970)
(2) $CH + H_2 + M \rightarrow CH_3 + M$	$1.0\,(-30)\,[M]$	Braun et al. (1970)
(3) $CH + CH + M \rightarrow C_2H_2 + M$	$6.0\,(-29)\,[M]$	Strobel (1969)
(4a) $^1CH_2 + CH_4 \rightarrow C_2H_6^* \rightarrow 2CH_3$	$1.9\,(-12)$	Braun et al. (1970)
(4b) $^1CH_2 + CH_4' \rightarrow C_2H_4 + H_2$	$1.2\,(-13)$	Braun et al. (1970)
(4c) $^1CH_2 + CH_4 \rightarrow {}^3CH_2 + CH_4$	$3.0\,(-13)$	Braun et al. (1970)
(5a) $^1CH_2 + H_2 \rightarrow CH_3 + H$	$1.0\,(-12)$	Braun et al. (1970)
(5b) $^1CH_2 + H_2 \rightarrow {}^3CH_2 + H_2$	$5.0\,(-12)$	Braun et al. (1970)
(6) $^1CH_2 + He \rightarrow {}^3CH_2 + He$	$3.0\,(-13)$	Braun et al. (1970)
(7) $^3CH_2 + CH_4 \rightarrow 2CH_3$	$5.0\,(-14)$	Braun et al. (1970)
(8) $^3CH_2 + H_2 \rightarrow CH_3 + H$	$5.0\,(-14)$	Braun et al. (1970)
(9) $^3CH_2 + {}^3CH_2 \rightarrow C_2H_2 + H_2$	$5.0\,(-11)$	Braun et al. (1970)
(10) $CH_3 + H + M \rightarrow CH_4 + M$	$8.5\,(-29)\,[M]$	Barker et al. (1970)
(11) $CH_3 + CH_3 + M \rightarrow C_2H_6 + M$	$1.4\,(-29)\,[M]\,T^{1/2}$	See Text
(12) $C_2H + H_2 \rightarrow C_2H_2 + H$	$5.0\,(-11)$	Estimated
(13) $C_2H_2 + H \rightarrow C_2H_3^*$	$1.5\,(-13)\exp\left(-\dfrac{750}{T}\right)$	Schofield (1967)
(14) $C_2H_2 + C_2H_3 \rightarrow C_4H_5^* \rightarrow C_4H_4 + H$	$1.0\,(-14)$	Estimated
(15a) $C_2H_2^* \rightarrow C_2H_2 + h\nu$	$\tau = 5.0\,(-6)\,\text{s}$	Estimated
(15b) $C_2H_2^* \rightarrow C_2H + H$	$\tau = 5.0\,(-6)\,\text{s}$	Estimated
(16) $C_2H_2^* + H_2 \rightarrow C_2H_3 + H$	$8.0\,(-11)$	Becker et al. (1971)
(17) $C_2H_3 + H_2 \rightarrow C_2H_4 + H$	$5\,(-12)\exp\left(-\dfrac{3200}{T}\right)$	Benson and Haugen (1967)
(18) $C_2H_3 + H \rightarrow C_2H_2 + H_2$	$7.0\,(-12)$	Volpi and Zocchi (1966)
(19) $C_2H_3^* \rightarrow C_2H_2 + H$	$\tau = 1.0\,(-7)$	Volpi and Zocchi (1966)
(20) $C_2H_3^* + H_2 \rightarrow C_2H_4 + H$	$6.4\,(-10)$	Estimated
(21) $C_2H_4 + H + M \rightarrow C_2H_5 + M$	$5.9\,(-29)\,[M]\exp\left(-\dfrac{750}{T}\right)$	Benson and Haugen (1967)
(22) $C_2H_5 + H \rightarrow C_2H_6^* \rightarrow 2CH_3$	$6.0\,(-11)$	Kurylo et al. (1970)
(23) $H + H + H_2 \rightarrow 2H_2$	$8.3\,(-33)\,[H_2]\left(\dfrac{T}{300}\right)^{-0.6}$	Trainor et al. (1973)

[a] Reaction rate coefficients are in $cm^3\,s^{-1}$, $2.5\,(-12) \equiv 2.5 \times 10^{-12}$, $[M]$ is the number density of species M, and τ is the lifetime.

along with their reaction rate coefficients. In choosing the chemical reactions to be used in the Jovian atmospheric studies, the low temperature ($\simeq 140\,K$) prevailing there must be kept in mind. This very low temperature in the Jovian atmosphere precludes all such reactions that have any sizeable activation energy. Even then Table I is long, and it is desirable to discuss here some of the implications of the reaction scheme being used.

In the context of the problem at hand, this discussion is, perhaps, best initiated by mentioning the following two reactions:

$$C_2H_2 + H + M \rightarrow C_2H_3 + M \tag{1}$$
$$C_2H_3 + H \quad\quad \rightarrow C_2H_2 + H_2 \tag{2}$$

which have been used by Strobel (1973a) in his most recent study of the hydrogen and hydrocarbon distribution in Jupiter's upper atmosphere. The set of reactions (1) and (2) could form a circuitous path and severely deplete atomic hydrogen. Reaction (1), however, is a simplified description of the following:

$$C_2H_2 + H \rightarrow C_2H_3^* \tag{3}$$
$$C_2H_3^* \quad \rightarrow \quad C_2H_2 + H, \quad \text{or} \quad C_2H_3 + h\nu \tag{4}$$
$$C_2H_3^* + M \rightarrow C_2H_3 + M \quad\quad (M \neq H_2) \tag{5}$$
$$C_2H_3^* + M (= H_2) \rightarrow C_2H_3 + H_2 \tag{6a}$$
$$\rightarrow C_2H_4 + H \tag{6b}$$

H_2 being more abundant than any other third body, most of the collisions of $C_2H_3^*$ will be with H_2. If the result is the path (6b), then the circuitous depletion of hydrogen, as implied by the set (1) and (2) is non-existent. In this context, it may be worthwhile to add here that the results of Takita et al. (1969) have shown that path (6b) is indeed preferred over the path (6a). Assuming, for the sake of discussion only, that reactions (3) through (6) are replaceable by (1), there is yet another reaction which is very important as a regulator of the catalysis and should have been included in any correct description of the role of C_2H_2. This reaction, as pointed out by Prasad et al. (1974), is

$$C_2H_3 + C_2H_2 \rightarrow C_4H_4 + H \tag{7}$$

The implication of this reaction is that as soon as the hydrogen density falls below that of C_2H_2 (as it does in Figure 3 of Strobel, 1973a), C_2H_3 would begin to react with C_2H_2 and the catalytic action would be broken until the hydrogen density increases again. Thus the reaction (7) appears to switch on and off the catalytic action of C_2H_2, if at all that catalytic process exists.

Failure to properly incorporate the temperature dependence of the various reaction rates is another pitfall which might give rise to spurious results. An excellent example is again the case of the reaction of C_2H_2 with H. The primary reaction is the reaction (3) for which the laboratory experiments, as summarized by Schofield (1967), have established a temperature dependent rate coefficient of $1.5 \times 10^{-13} \exp(-750/T)$. The high pressure limiting value of the rate coefficient of the reaction (1) should correspond to the rate coefficient of (3). Strobel (1973a) adopted a value of 4.0×10^{-14} cm^3 s^{-1} for the high pressure limit, based on experimental measurements of Volpi and Zocchi (1966) made at about 400 K. This value is within a factor of 2 of the value determined from the temperature dependent expression given by Schofield (1967), but about two orders of magnitude larger than that appropriate for the Jovian atmospheric temperature, if the temperature dependence observed in the laboratory is allowed to persist down to the low Jovian temperatures. Since there is no evidence to the contrary, preserving the temperature dependence is preferable to completely ignoring it. This realization is extremely important inasmuch as a temperature dependent reaction rate will make the catalytic action very unimportant indeed, even if it survived the difficulties associated with chemistry explained earlier. Ignoring the temperature dependence, on the other hand, will introduce false conclusions about the role of C_2H_2 in depressing hydrogen concentration. We have used a temperature dependent rate coefficient for the reaction $H + C_2H_4 + M \rightarrow C_2H_5 + M$ also, in contrast to the constant rate pertaining to 300 K as used by Strobel (1973a). Our choice of reaction rate for the three body association of CH_3 resulting in the C_2H_6 formation needs some explanation. The starting point was the same as that used by Strobel (1973a) which was based upon experiments of Kistiakowsky and Roberts (1953); however, we modified it to incorporate a $\sqrt{T}$ dependence of the rate and the fact that the rotating sector method could be overestimating the rate coefficient by a factor of two (Shepp, 1956). These modifications resulted in a rate coefficient 3.3 times lower than that used by Strobel (1973a), and should show its effect on the C_2H_6 concentration.

3. The Jovian Model Atmosphere and the Solar Fluxes

Several authors (Owen and Mason, 1968; Belton, 1969) have evaluated the abundances of hydrogen, methane, and ammonia in the atmosphere of Jupiter. These measurements have been reviewed by McElroy (1969). Also, the brightness temperature of Jupiter has been measured at a number of IR wavelengths between 2 and 20 μ by Gillett et al. (1969). These measurements indicate that the brightness temperature of Jupiter in the atmospheric window near 5 μ may be as high as 230 K, and may be characteristic of the temperature at the top of the thick clouds. Furthermore, a temperature inversion may be present in the middle atmosphere due probably to the solar heating in the 3020 cm^{-1} CH_4 bands (Gillett et al., 1969).

Using these new sets of observational data as separate constraints which must be satisfied by any acceptable model, Hogan et al. (1969) have constructed five models for the thermal structure in the Jovian atmosphere above the level of the dense clouds. These models are consistent also with the recent microwave observations of Wrixon and Welch (1970). We have, therefore, adopted Hogan et al.'s (1969) model #3 in our studies. In this model, the mesopheric temperature reaches a value of about 140 K. Thus, in our model atmosphere, we assumed a constant temperature of 140 K from the level of $n(H_2) = 10^{17}$ to the level of the mesopause, viz., that of $n(H_2) = 3 \times 10^{14}$. Beyond this level the temperature profile obtained by Shimizu (1971) for low solar activity and for a solar zenith-angle of 60° was used. For the mixing ratio of He we used a value of 0.26, based upon a recent study by Veverka et al. (1974). The mixing ratio for CH_4 was assumed to be 0.001; and the role of ammonia was neglected. The neglect of ammonia should be a very good approximation except in the lowest portions of the altitude range under study here (Strobel, 1973a, b).

It may be worthwhile to point out that the models for the Jovian atmosphere are by no means settled. Different values for many of the parameters mentioned above are quite possible; these differences will not, however, in all probability, affect the problem being discussed here in any appreciable way.

Our calculations were intended to model the diurnal and global averages; we assumed, therefore, a solar flux reduced by a factor of $\frac{1}{2}$ to take into account diurnal averages. The solar zenith angle was taken to be 60° to simulate a global average. The solar fluxes used were those of Hinteregger et al. (1965) reduced by a factor of 27 to account for the Sun-Jupiter distance. Recently, Hinteregger (1970) has given a revised tabulation of the solar EUV flux, but this tabulation has been doubted by several investigators (e.g., Roble and Dickinson, 1973). We therefore used the earlier values for the fluxes (i.e., Hinteregger et al., 1965) as indicated above.

The choice of eddy diffusion coefficient to be used in our calculations was rendered difficult largely because the physical mechanisms of the production and propagation of eddy turbulence are not yet well known (even for the terrestrial atmospheric environment). On the basis of their analysis of the Lα albedo as measured by Moos and Rottman (1972), Wallace and Hunten (1973) have proposed an eddy diffusion coefficient of 5×10^5 cm^2 s^{-1} for the region around the turbopause and above. This led Strobel (1973a) to adopt a value of 10^5 and 5×10^5 cm^2 s^{-1} for the eddy diffusion coefficient in his studies. The Lα flux measurement from the Earth could, however, be contaminated by the possible contribution to the observed flux from scattering in a toroid of gas suspected to exist near Io's orbit (McElroy et al., 1974). In fact, based upon the preliminary analysis of the data from the UV photometer on board Pioneer 10, Carlson and Judge (1974) have suggested an eddy diffusion coefficient of the order of 10^8 cm^2 s^{-1}. We have, therefore, considered the eddy diffusion coefficient as an unknown and adjustable parameter to be varied in this study, and have presented three cases for this coefficient. In case 1, studied previously by Prasad et al. (1974), the eddy diffusion coefficient had an altitude independent value of 10^5 cm^2 s^{-1}. This was similar to the model of Strobel (1973a). In case 2 the coefficient increased exponentially with altitude from its value of 10^5 cm^2 s^{-1} at 80 km to 10^6 cm^2 s^{-1} at 100 km; beyond this altitude the value was held constant. In case 3, the maximum value of the eddy diffusion coefficient was 10^7 cm^2 s^{-1}, which was reached at 150 km, and varied below this height as in case 2.

4. Details of the Photolysis

In this section we shall discuss the details of the photochemistry of CH$_4$ beginning with the discussion of the relevant photoabsorption and photodissociation processes. Much of the laboratory data on the photolysis have already been reviewed by the various authors. In view of this we have presented here only an abridged version for the sake of completeness and continuity of the paper. Laboratory studies of Rebbert and Ausloos (1972) have shown that the quantum yields of CH at 1236 Å and 1048–1067 Å are 0.06 and 0.23 respectively. Also the quantum yield of ^{1}CH$_2$ at 1236 Å was found to be 0.91. Based upon these and other experimental evidence, such as those of Gorden and Ausloos (1967a, b), Strobel (1973a) has suggested the following photodissociation scheme for CH_4:

$$CH_4 + hv \rightarrow {}^1CH_2 + H_2 \qquad 92\% \qquad (8a)$$
$$(L\alpha) \rightarrow CH \;\; + H_2 \qquad 8\% \qquad (8b)$$
$$CH_4 + hv \rightarrow {}^1CH_2 + H_2 \qquad\qquad (9)$$
(all other wavelengths)

For the photolysis of other hydrocarbons of interest, C_2H_4 and C_2H_6, we have used:

$$C_2H_4 + hv \rightarrow C_2H_2 + H_2 \qquad 25\% \qquad (10a)$$
$$(L\alpha) \rightarrow C_2H_2 + 2H \qquad 75\% \qquad (10b)$$
$$C_2H_4 + hv \rightarrow C_2H_2 + H_2 \qquad\qquad (11)$$
(all other wavelengths)
$$C_2H_6 + hv \rightarrow C_2H_2 + 2H_2 \qquad 45\% \qquad (12a)$$
$$(L\alpha) \rightarrow C_2H_4 + 2H \qquad 30\% \qquad (12b)$$
$$\rightarrow C_2H_4 + {}^1CH_2 \qquad 25\% \qquad (12c)$$
$$C_2H_6 + hv \rightarrow C_2H_2 + 2H_2 \qquad 90\% \qquad (13a)$$
(all other
wavelengths) $\rightarrow C_2H_4 + 2H \qquad 10\% \qquad (13b)$

based upon the studies of Gorden and Ausloos (1967a, b), Back and Griffiths (1967) for C_2H_4, and of Lias *et al.* (1970) for C_2H_6.

Absorption of UV radiation by acetylene yields excited C_2H_2. Studies of Steif *et al.* (1965) and Becker *et al.* (1971) suggest that $C_2H_2^*$ has a lifetime of a few microseconds. In the less dense region of the atmosphere ($[M] < 10^{15}$ cm^{-3}), therefore, we would have:

$$C_2H_2 + hv \rightarrow C_2H_2^* \qquad\qquad (14)$$
$$C_2H_2^* \qquad \rightarrow C_2H_2 + hv \qquad (15)$$
$$C_2H_2^* \qquad \rightarrow C_2H + H \qquad (16)$$
$$C_2H + H_2 \rightarrow C_2H_2 + H. \qquad (17)$$

There are indications that $k_{15} \simeq 1.5\, k_{16}$ (Takita *et al.*, 1968, 1969). Thus about 40% of C_2H_2 will be available to go through the reaction chain (14), (16), and (17), which form a circuitous path through which absorption of UV results in decomposition of H_2. The extent to which the above reaction chain could become a catalytic source of H will depend upon how much of C_2H_2 is available to take part in it. In the lower region of the atmosphere, where $C_2H_2^*$ can undergo many collisions during its lifetime, the following reactions could occur:

$$C_2H_2^* + M (\neq H_2) \rightarrow C_2H_2 + M \qquad (18)$$
$$C_2H_2^* \qquad\qquad \rightarrow C_2H_2 + H_2 \qquad (19)$$
$$C_2H_2^* + H_2 \qquad \rightarrow C_2H_3 + H. \qquad (20)$$

Becker *et al.* (1971) have measured the quenching coefficient of $C_2H_2^*$ by H_2 as

8×10^{-11} cm^3 s^{-1}, but it was not determined whether the process was (19) or (20). Assuming, for a moment, that reaction (20) is the process involved above, we see that if C_2H_3 reacts with H according to

$$C_2H_3 + H \rightarrow C_2H_2 + H_2 \tag{21}$$

then the initial photoabsorption by C_2H_2 does not produce any net chemical change. However, in the region where $[H] \ll [C_2H_2]$, C_2H_3 may react with C_2H_2 producing C_4H_4 and H. In this case the C_2H_2 is not recovered (at least easily), and the photo-absorption by C_2H_2 amounts to the product of two H atoms per absorbed photon, a conclusion quite different from that reached by Strobel (1973a).

4. The Continuity Equation

The concentration of any given species is controlled by its continuity equation involving production, loss, and transport via eddy and molecular diffusion:

$$\frac{\partial n_i}{\partial t} = Q_i - L_i - \frac{\partial \phi_i}{\partial z} \tag{22}$$

with

$$\phi_i = \phi_i + \psi_i \tag{23}$$

$$\phi_i = -D_i \left[\frac{\partial n_i}{\partial z} + \frac{n_i}{H_i} + \frac{n_i(1 + \alpha_t^i)}{T} \frac{\partial T}{\partial z} \right] \tag{24}$$

$$\psi_i = -K_e \left[\frac{\partial n_i}{\partial z} + \frac{n_i}{H_{av}} + \frac{n_i}{T} \frac{\partial T}{\partial z} \right], \tag{25}$$

where

ϕ_i = total flux,
ϕ_i = molecular diffusion flux,
ψ_i = eddy diffusion flux,
D_i = binary diffusion coefficient of the ith species in H_2,
K_e = eddy diffusion coefficient,
α_t^i = thermal diffusion coefficient,
n_i = number density of the ith constituent,
T = temperature,
H_i and H_{av} = scale height of the given species i, and the average scale height, respectively,
z and t = height and time variables.
It is useful to rewrite the total diffusion flux as:

$$\phi_i = -\alpha_i \frac{\partial n_i}{\partial z} - \beta_i n_i \tag{26}$$

with

$$\alpha_i = D_i + K_e \tag{27}$$

$$\beta_i = \frac{D_i}{H_i} + \frac{K_e}{H_{av}} + \frac{\alpha_i}{T} \frac{\partial T}{\partial z} (1 + \alpha_t^i). \tag{28}$$

The continuity Equation (31) can now be written as:

$$\frac{\partial n_i}{\partial t} = \gamma_i + \alpha_i \frac{\partial^2 n_i}{\partial z^2} + \left[\frac{\partial \alpha_i}{\partial z} + \beta_i \right] \frac{\partial n_i}{\partial z} + \frac{\partial \beta_i}{\partial z} n_i \tag{29}$$

with

$$\gamma_i = Q_i - L_i. \tag{30}$$

There were fourteen coupled continuity equations, one for each species, viz: CH_4, H, CH_3, CH_2, CH, C_2H_2, C_2H_4, C_2H_6, $C_2H_2^*$, $C_2H_3^*$, C_2H_5, $^3CH_2$, C_2H, and C_2H_3. It was immediately obvious that this set of fourteen continuity equations must be solved numerically; a huge task indeed.

Some relief was, however, possible inasmuch as C_2H, CH, $C_2H_2^*$, and $C_2H_3^*$ were always in chemical equilibrium. The divergence of the flux term in their continuity equations can, therefore, be omitted.

6. Numerical Method of Solving the Continuity Equation

Starting from an initial ($t=0$) profile, the continuity equations were solved in time until the equilibrium conditions were reached. Thus both the time and height derivatives had to be expressed numerically. For this purpose, the height and time domains were divided into intervals which yielded a set of grid points in the $z-t$ plane. The grid points were equally spaced in the time domain, but had unequal intervals in the height domain. The height intervals used ranged from 0.25 km to 2 km, and were so arranged that the altitude grid points were closer together at the higher altitudes. Henceforth, we shall abbreviate any function $f(z, t)$ of $z = z_k$ and $t = l\Delta t$ as $f^l(k)$, where $k = 1, 2, 3, ..., k-1, K$, and $l = 0, 1, 2, 3, ...$

In the region of unequal height intervals, the first and second derivatives can be written as follows:

$$\left[\frac{dn}{dz} \right]_k = \frac{(n_{k+1} - n_k)(z_k - z_{k-1})}{(z_{k+1} - z_{k-1})(z_{k+1} - z_k)} + \frac{(n_k - n_{k-1})(z_k - z_{k+1})}{(z_{k+1} - z_{k-1})(z_{k-1} - z_k)} \tag{31}$$

$$\left[\frac{d^2n}{dz^2} \right]_k = \frac{(n_{k+1} - n_k)}{(z_{k+1} - z_{k-1})(z_{k+1} - z_k)} + \frac{(n_k - n_{k-1})}{(z_{k+1} - z_{k-1})(z_{k-1} - z_k)}, \tag{32}$$

where n_{k-1}, n_k, n_{k+1} refer to the densities of any given species at the corresponding altitude grid points. The time derivative was, however, expressed in a forward difference scheme.

A completely implicit method was used to solve our difference equation. This meant that the number densities of the ith species occurring in the loss term L_i were also represented at the $(t + \Delta t)$ time. Such a procedure, however, presented some complications in the case of atomic hydrogen, CH_3, C_2H_6, etc., due to the presence of the non-linear terms like $[n(H)]^2$ in the expressions for production and loss. This difficulty was surmounted by the usual linearization procedure, in which $[n^{l+1}(k)]^2$ can be replaced by $2n^{l+1}(k) n^l(k) - (n^l(k))^2$.

Thus we obtained

$$-A_i^l(k) \, n_i^{l+1}(k+1) + B_i^l(k) \, n_i^{l+1}(k) - C_i^l(k) \, n_i^{l+1}(k-1) = D_i^l(k) \tag{33}$$

for each species. In this equation, the coefficients A, B, C, and D involve known values of densities at the current lth time step, $t = l\Delta t$.

This equation can be written for k values between 2 and $K-1$. To solve these $(K-2)$ equations for K unknowns $n_i^{l+1}(1)$, $n_i^{l+1}(2), \ldots, n_i^{l+1}(K)$, we need two more equations; these are supplied by the upper and lower boundary values. These problems are discussed in the following two subsections.

6.1. THE BOUNDARY CONDITIONS

For all species except atomic hydrogen, the upper boundary condition may be taken to be that of a distribution with an effective scale height of

$$H_{\text{eff}, i} = \left(\frac{D_i}{H_i} + \frac{K_e}{H_{\text{av}}} \right) \bigg/ (D_i + K_e), \tag{34}$$

where $H_{\text{eff}, i}$ is the effective scale height of the ith constituent and the rest of the symbols have the same meaning as in the contexts Equations (31) through (36). The appropriate boundary condition for atomic hydrogen is a flux boundary condition obtained by integrating the continuity equation form the upper boundary z_k to ∞, in a manner described by Strobel *et al.* (1970) in connection with the nitric oxide problem in the terrestrial atmosphere. To minimize the difficulties associated with this boundary condition, the upper boundary was kept at sufficiently high altitudes so that the losses are minimal.

In the lower denser regions many species have chemical lifetimes much shorter than either the eddy or the molecular diffusion time constants. For these species photochemical equilibrium provides an appropriate lower boundary condition. However, C_2H_6 is practically inert with respect to the other species in the ambient temperatures prevailing in the Jovian atmosphere. Furthermore, those photons which could have dissociated C_2H_6 are already completely absorbed by CH_4; thus the Jovian atmosphere shields C_2H_6 against photodestruction also. For C_2H_6, therefore, a flux boundary condition must be applied at the lower boundary.

The flux at the lower boundary, however, is unknown. This makes the problem of the C_2H_6 distribution largely indeterminate. Progress is possible through model studies, using plausible flux conditions. One plausible condition is that of complete mixing or, in other words, zero flux through the lower boundary. Another plausible flux condition could be that of a stationary flux. All possible values of fluxes can be conveniently parameterized by an adjustable parameter α in the expression

$$\frac{1}{n_i}\left[\frac{dn}{dz}\right]_i = -\frac{1}{\alpha H_{av}} \tag{35}$$

with α ranging from $-\infty$ to $+\infty$. Zero flux through the lower boundary or complete mixing would be represented by $\alpha = 1$, whereas $\alpha = \infty$ would stand for stationary flux and an altitude independent number density at the lower boundary.

6.2. THE INITIAL CONDITIONS

Let us now consider the initial conditions. In this study our aim is to calculate the equilibrium profile. In such an attempt the choice of the initial profile is immaterial in the sense that after a sufficient number of iterations (i.e., after the lapse of sufficient time), the answers must converge to the true value. The amount of computer time involved, though, sets a practical restriction, and it is thus advisable to start with an initial profile which is as close to equilibrium as possible. For this purpose we have used an improved version of Strobel's (1969) approximate method of taking transport into account. The improvement consisted in taking

$$L_i = L_{i,\,\text{chem}} + \frac{1}{\tau_{\text{diff}}(i)} + \frac{1}{\tau_{\text{eddy}}(i)}, \tag{36}$$

where the meaning of the various symbols are the same as in Strobel (1969). Our improvement avoids the sharp discontinuity that may result from abrupt switching from one loss process to another, and from mixing to diffusive separation. Using this type of loss rate, it was a very easy matter to set up a set of simultaneous equations and to solve them by a Newton-Raphson method. The solutions so obtained were used as the initial profiles in a program which solves the continuity equations rigorously.

In our study the approximate solutions were also used for studying the relative importance of the various physicochemical factors that enter into consideration. This preliminary evaluation enabled us to select the most fruitful and interesting combination of possibilities for which to run the rigorous calculation. This advance planning was necessary in view of the computer costs involved.

7. Results and Discussion

Our results are shown in Figures 1 through 3. Figure 1, taken essentially from Prasad *et al.* (1974), shows our results for the case of an eddy diffusion coefficient of 10^5 cm^2 s^{-1}. The same results for the other values of the coefficient are presented in Figures 2 and 3. The concentrations of the excited species and of C_2H_5, C_2H_3, and C_2H were too small to be shown. For the convenience of comparison, we have also shown in Figure 1 the atomic hydrogen profile obtained by Strobel (1973a). A rigorous comparison is not possible due to differences in the values of the solar fluxes and the model atmosphere; however, these differences could not be responsible for the features

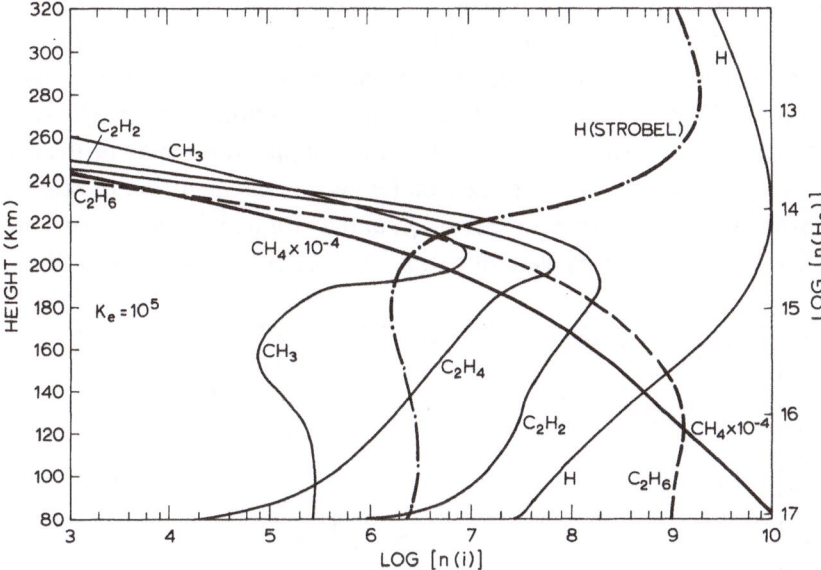

Fig. 1.   The altitude profile of the concentrations of H, CH_4, CH_3, C_2H_2, C_2H_4 and C_2H_6 in a model Jovian upper atmosphere. The altitude scale is arbitrary. The molecular hydrogen concentration at the various levels are indicated on the right hand side Y axis. The curves belonging to the various species have been labelled. For further clarity the C_2H_6 profile is shown by a dashed line and the CH_4 profile by a heavy continuous line. An altitude independent eddy. diffusion coefficient of 10^5 cm^2 s^{-1} was used to obtain the results presented in the figure. Also shown in the figure is the atomic hydrogen profile obtained by Strobel (1973).

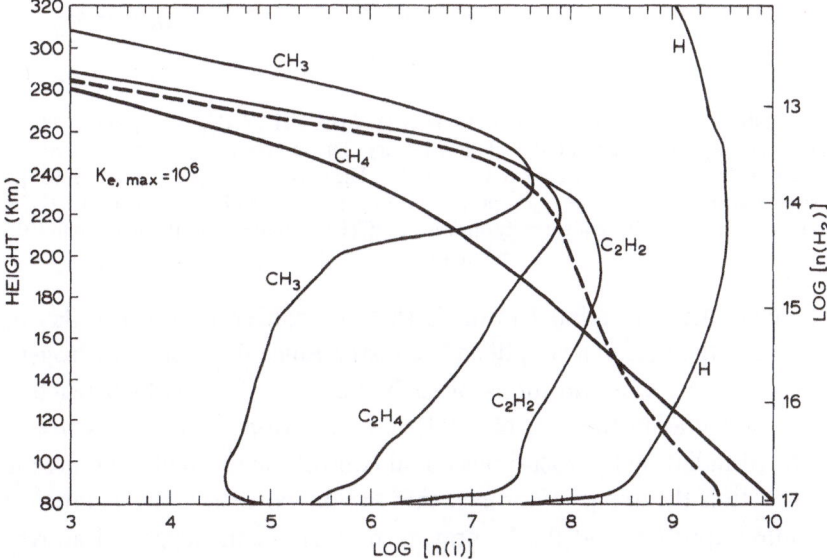

Fig. 2.   The altitude profile of the concentrations of H, CH_4, CH_3, C_2H_2, C_2H_4 and C_2H_6 in a model Jovian upper atmosphere. The altitude scale is arbitrary. The molecular hydrogen concentration at the various levels are indicated on the right hand side Y axis. The curves belonging to the various species have been labelled. For further clarity the C_2H_6 profile is shown by dashed line and the same for CH_4 by a heavy continuous line. An altitude dependent eddy diffusion coefficient ranging from 10^5 at 80 km to 10^6 at 100 km and above were used.

we shall be comparing and contrasting here, viz., the presence or absence of a dramatic decrease in atomic hydrogen in the lower region as reported by Strobel (1973a). This dramatic decrease was attributed to the catalytic action of C_2H_2 as embodied by Equations (1) and (2). When a more complete chemistry, Equations (3) through (7), is considered, the catalytic destruction disappears. This disappearance may be the result of the temperature dependent rate for reaction (3) and/or the inclusion of branch (6b). It is thus evident that for a correct description of the atomic hydrogen distribution, it is imperative that a laboratory determination of these two vital points in the chem-

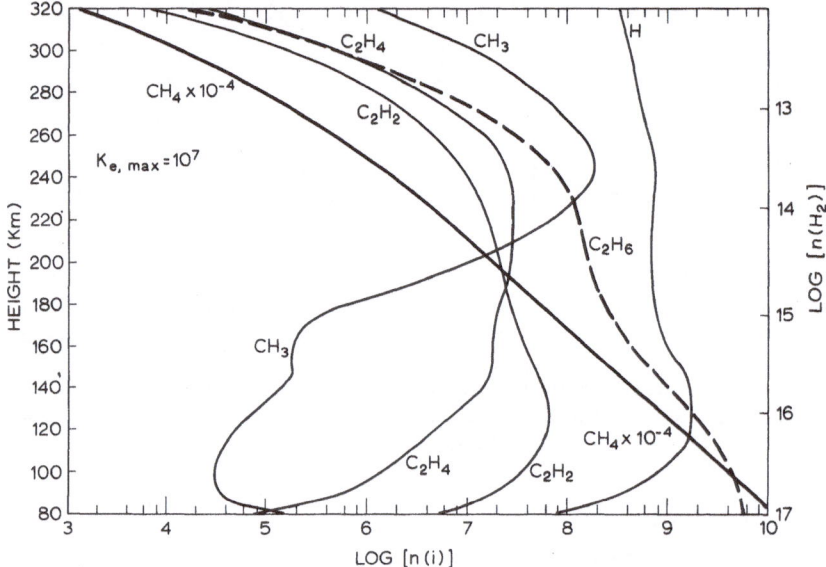

Fig. 3. The altitude profile of the concentrations of H, CH_4, CH_3, C_2H_2, C_2H_4 and C_2H_6 in a model Jovian upper atmosphere. The altitude scale is arbitrary. The molecular hydrogen concentration at the various levels are indicated on the right hand side Y axis. The curves belonging to the various species have been labeled. For further clarity the C_2H_6 profile is shown by a dashed line and that for CH_4 by a heavy continuous line. An altitude dependent eddy diffusion coefficient ranging from 10^5 at 80 km to 10^7 at 150 km and above were used.

istry be made. Another related result is that the concentration of C_2H_2 is not as abundant as in the case of a simplified chemistry. Since the atomic hydrogen density is not depleted, the concentrations of such species as CH_3 (which reacts with H) were not as large as in the case of catalytic destruction of atomic hydrogen. Since these infrared radiating hydrocarbons could control the thermal structure and ionosphere formation, the impact of the existence or non-existence of the catalytic destruction is quite important and the laboratory measurements suggested above, all the more desirable.

As expected, the results for the higher eddy diffusion coefficients are of the same qualitative nature. The higher the eddy diffusion coefficient, the higher is the range of altitudes over which the hydrocarbons predominate. Clearly, in any discussion of the ionosphere formation, the role of the hydrocarbon ion chemistry can not be neglected, if the eddy diffusion coefficient is as large as that being suspected on the basis of Pioneer 10 UV photometer data (Carlson and Judge, 1974).

Atomic hydrogen concentrations above the turbopause level decrease with an increase in the eddy diffusion coefficient. This is a result of the more effective downward transport of atomic hydrogen in the denser region, where it recombines to form molecular hydrogen. However, this decrease is offset by increased production from the photodissociation of CH_4. Thus, the atomic hydrogen concentration may not be a sensitive indicator of the eddy diffusion coefficient appropriate to the Jovian upper atmosphere. This is easily verified by comparing our $n(H)$ profiles in Figures (1) through (3). A two order of magnitude change in the eddy diffusion coefficient resulted in roughly only one order of magnitude change in hydrogen densities. On the other hand, the hydrocarbon distribution is quite sensitive to the eddy diffusion coefficient. Hydrocarbon column content above a suitably selected altitude can show a dramatic increase with the eddy diffusion coefficient. For example, the column densities of CH_4 above 273 km, i.e., above the $n(H_2) \simeq 10^{13}$ or the region of the maximum dayglow volume emission rates according to the calculations of Olivero *et al.* (1973), were 5.0×10^3 cm^2 and 2.0×10^{10} cm^2 for the eddy diffusion coefficient of 10^5 and 10^7 cm^2 s^{-1} respectively. One sensitive way of determining the eddy diffusion coefficients could, therefore, be the search for distinctive hydrocarbon emission features in the day glow of the Jovian atmosphere.

8. Concluding Remarks

To conclude our paper, we wish to stress the uncertainty in the suggested catalytic destruction of atomic hydrogen by C_2H_2. Several reaction rate coefficients and their temperature dependences are involved. These must be determined in the laboratory under temperatures similar to those in the Jovian atmosphere. Only then can we assess the reality of the catalytic action. Our study has also revealed that, for an eddy diffusion coefficient as high as those being suggested now, the hydrocarbons can exist in abundance up to the ionospheric altitudes. Their role, therefore, must be included in any realistic modelling of the ionospheric structure.

Acknowledgements

The research reported in this paper has been supported by the National Aeronautics and Space Administration through its grant NGL-10-005-008 to the University of Florida. The computations were partly supported by the Northeast Regional Data Center of the University of Florida.

References

Back, R. A. and Griffiths, D. W. L.: 1967, *J. Chem. Phys.* **46**, 4839.
Barker, J. R., Keil, D. G., Michael, J. V., and Osborne, D. T.: 1970, *J. Chem. Phys.* **52**, 2079.
Baum, W. A. and Code, A. D.: 1953, *Astron. J.* **58**, 108.

Becker, K. H., Haaks, D., and Schuergers, M.: 1971, *Z. Naturforsch.* **26a**, 1770.

Belton, M. J. S.: 1969, *Astrophys. J.* **157**, 469.

Benson, S. W. and Haugen, G. R.: 1967, *J. Phys. Chem.* **71**, 4404.

Braun, W., Bass, A. M., and Pilling, M.: 1970, *J. Chem. Phys.* **52**, 5131.

Carlson, R. W. and Judge, D. L.: 1974, *EOS, Trans. Am. Geophys. Union* **55**, 339.

Gillett, F. C., Low, F. J., and Stein, W. A.: 1969, *Astrophys. J.* **157**, 925.

Gordon, R., Jr. and Ausloos, P.: 1967a, *J. Chem. Phys.* **46**, 4823.

Gordon, R., Jr. and Ausloos, P.: 1967b, *J. Chem. Phys.* **47**, 1799.

Hinteregger, H. E.: 1970, *Ann. Geophys.* **26**, 547.

Hinteregger, H. E., Hall, L. A., and Schmidtke, G.: 1965, *Space Research* **5**, 1175.

Hogan, J. S., Rasool, S. I., and Encrenaz, T.: 1969, *J. Atmospheric Sci.* **26**, 898.

Hubbard, W. B.: 1969, *Astrophys. J.* **155**, 333.

Hubbard, W. B., Nather, R. E., Evans, D. S., Tull, R. G., Wells, D. C., Van Citters, G. W., Warner, B., and Varden Bout, P.: 1972, *Astron. J.* **77**, 41.

Kliore, A. J., Cain, D. L., Fjeldbo, G., Seidel, B. L., and Rasool, S. I.: 1974, *EOS, Trans. Am. Geophys. Union* **55**, 339.

Kistiakowsky, G. B. and Roberts, E. K.: 1953, *J. Chem. Phys.* **21**, 1637.

Kurylo, M. J., Peterson, N. C., and Braun, W.: 1970, *J. Chem. Phys.* **53**, 2776.

Lias, S. G., Collin, G. J., Rebbert, R. E., and Ausloos, P.: 1970, *J. Chem. Phys.* **52**, 1841.

McElroy, M. B.: 1969, *J. Atmospheric Sci.* **26**, 798.

McElroy, M. B., Yung, Y. L., and Brown, R. A.: 1974, *Astrophys. J.* **187**, L127.

McGovern, W. E. and Burk, S. D.: 1972, *J. Atmospheric Sci.* **29**, 179.

Moos, H. W. and Rottman, G. J.: 1972, *Bull. Am. Astron. Soc.* **4**, 360.

Oliviero, J. J., Bass, J. N., and Green, A. E. S.: 1973, *J. Geophys. Res.* **78**, 2812.

Owen, T. and Mason, H. P.: 1968, *Astrophys. J.* **154**, 317.

Prasad, S. S., Capone, L. A., and Schneck, L. J.: 1974, to be published.

Rebbert, R. E. and Ausloos, P.: 1972, *J. Photochem.* **1**, 167.

Roble, R. G. and Dickinson, R. E.: 1973, *J. Geophys. Res.* **78**, 249.

Schofield, K.: 1967, *Planetary Space Sci.* **15**, 643.

Shepp, A.: 1956, *J. Chem. Phys.* **24**, 939.

Shimizu, M.: 1971, *Icarus* **14**, 273.

Steif, L. J., Decerlo, V. J., and Mataloni, R. J.: 1965, *J. Chem. Phys.* **42**, 3113.

Strobel, D. F.: 1969, *J. Atmospheric Sci.* **26**, 906.

Strobel, D. F.: 1973a, *J. Atmospheric Sci.* **30**, 489.

Strobel, D. F.: 1973b, *J. Atmospheric Sci.* **30**, 1205.

Strobel, D. F. and Smith, G. R.: 1973, *J. Atmospheric Sci.* **30**, 718.

Strobel, D. F., Hunten, D. M., and McElroy, M. B.: 1970, *J. Geophys. Res.* **75**, 4307.

Takita, S., Mori, Y., and Tanaka, I.: 1968, *J. Phys. Chem.* **72**, 4360.

Takita, S., Mori, Y., and Tanaka, I.: 1969, *J. Phys. Chem.* **73**, 2929.

Trainor, D. W., Ham, D. O., and Kaufman, F.: 1973, *J. Chem. Phys.* **58**, 4599.

Veverka, J., Wasserman, L. H., Elliot, J., and Sagan, C.: 1974, *Astron. J.* **79**, 73.

Volpi, G. G. and Zocchi, F.: 1966, *J. Chem. Phys.* **44**, 4010.

Wallace, L. and Hunten, D. M.: 1973, *Astrophys. J.* **182**, 1013.

Wrixon, G. T. and Welch, W. J.: 1970, *Icarus* **13**, 163.

SOLAR WIND HEATING AT MID-HELIOSPHERIC
DISTANCES DUE TO SECONDARY IONS

H. J. FAHR

*Institut für Astrophysik und Extraterrestrische Forschung der Universität Bonn, Auf dem Hügel 71,
53 Bonn, F.R.G.*

1. Introduction

Amongst the various solar wind models that have been discussed in the past the
purely hydrodynamical ones proposed in their original form by Parker (1960) are
generally favoured up to the present. This is because the main features of the solar
wind observed near the Earth's orbit can more satisfactorily be described by this
kind of models without the need of too many *ad-hoc* assumptions. Special disad-
vantages that could not be overcome in exospheric treatments of the solar wind
expansion up to now are that exospheric models are very sensitive to boundary
conditions at the solar coronal base and that they do not include mixing of thermal
velocities of different orientations to the magnetic field and coupling between thermal velocities of electrons and protons.

Whereas Parker's original hydrodynamical model was a one fluid model regarding only the combined motion of electrons and ions, later more refined models of
Sturrok and Hartle (1966), Hartle and Sturrok (1968), and Hartle and Barnes (1969)
have considered the solar wind plasma as a two-fluid medium consisting of electrons
and protons which are essentially decoupled energetically.

The problem, however, that has remained for the hydrodynamical treatment is
the energy supply needed to make the solar wind expand with supersonic velocities
of about 300 km s^{-1} which is about half the escape velocity at the base of the solar
corona. The thermal energy available at this base can only cover one half of the
energy required to lift the solar wind plasma out of the solar gravitational field.
This clearly means that additional thermal energy of any hydrodynamically conceivable form must be supplied to the expanding plasma. A simple way to overcome
this problem is to use a polytropic energy law with a best-fitting *ad-hoc* exponent
smaller than the adiabatic one (Parker, 1960). The true nature of the energy supplied
to the expanding plasma has been considered to be thermal conduction, damping
of acoustic and hydromagnetic waves and viscous energy dissipation from discontinuous velocity structures. Thermal conduction due to Coulomb collisions has been
shown to be a coupling mechanism unable to tie electrons and protons energetically
together. (Hartle and Sturrok, 1968). To test the effect of an energy input due to
hydromagnetic wave absorption Hartle and Barnes (1969) have included into their
model a region of additional heating. The authors could show that the solar wind
velocities are increased, if the heating occurs inside 25 solar radii, whereas heating

V. Formisano (ed.), The Magnetospheres of the Earth and Jupiter, 269–278. All Rights Reserved

outside this region mainly enhances the temperatures of the distant solar wind $(r=r_E=1$ AU). This result though obtained for an unspecified artificial heat source is important for the discussion of any real source one might think of.

2. Solar Wind and Interstellar Matter as Interacting Media

Models mentioned in the foregoing section have only considered the solar wind as a medium that expands into a circumsolar vacuum. The circumsolar space, however, is filled with a neutral interstellar matter that advances towards the Sun with a velocity v_0 and thereby is subjected to the ionizing effect of the solar electromagnetic and corpuscular radiations. Due to the corpuscular interaction the distribution of interstellar matter in the solar system is at least partly determined by the dynamics of the ionizing solar wind. On the other hand, ions originating from ionizations of interstellar matter will be integrated into the ambient solar wind plasma. This incorporation of secondary ions into the primary plasma has the effect of changing energy and momentum of the primary plasma. Due to this coupling by secondary ions the joint system plasma-neutral medium could only be accurately treated by a simultaneous solution of the total set of dynamical equations of both the interacting media. This joint system of equations seems to be unsolvable up to now, however some support for mathematical simplifications is given by the fact that the motion and the density of neutral interstellar matter is unaffected by solar radiations outside $r=5\ r_E$, at least at the upwind side. On the other hand, the injection of energy and momentum by secondary ions does not affect the solar wind dynamics inside $1\ r_E$.

Therefore regions are existing in the solar system where the two interacting media are essentially decoupled and their dynamics can approximately be treated separately, i.e. inside $1\ r_E$ the solar wind expansion can be calculated without consideration of the effect of interacting interstellar gas, and outside $5\ r_E$ the interstellar gas can be taken to be undisturbed by the solar system. Recent calculations of Holzer (1973) and Wallis (1973) have considered the effect of interstellar gas on the solar wind expansion beyond $5\ r_E$ adopting an undisturbed interstellar gas flow which is a good approximation on the upwind, however, an indequate approximation on the downwind side of the solar system. They found a strong deceleration of the outwards moving solar wind, as it had already been predicted earlier from simpler calculations by Fahr (1971a, b). The effect of the interacting interstellar gas on the temperatures of the solar wind had only been inadequately considered in these papers and essentially represented the adiabatic heating due to the compression of the decelerated plasma.

The adequate mechanism how the solar wind should be heated by the interacting interstellar gas has been proposed by Fahr (1973) and later by Holzer and Leer (1973). This heating mechanism is due to suprathermal secondary ions that are injected into the primary solar wind plasma. This heat source that gives additional energy supply to the solar wind may be discussed in the following.

3. Heating by Secondary Ions

Ions originating from the ionization of interstellar neutrals immediately after their generation have velocities that are very different from the solar wind bulk velocity. Therefore the primary solar wind is sweeping over the secondary ions with a high velocity of the order of a few 100 km s^{-1}. This results in a reacceleration of the secondary ions by the frozen-in magnetic field of the solar wind plasma. This reacceleration process has been studied by Burns and Halpern (1968) and Fahr (1973). The secondary ion carries out a drift dependent on the angle between magnetic field and bulk velocity. Superimposed on this drift motion a Larmor precession is carried out that can be described analogous to a thermal velocity spread of the secondary ions corresponding to temperatures T. These temperatures T that are directly proportional to the mass of the secondary ion in consideration are given by Fahr (1973) and are shown to be of the order of 10^7 K for secondary protons and four times as high for secondary helium ions He$^+$.

The high thermal excess energies of secondary ions are transferred to the ambient solar wind plasma that has a temperature T. As transfer mechanisms, pitch angle scattering at magnetic irregularities (Feldman *et al.*, 1972), Coulomb collisions (Fahr, 1973) and hydromagnetic wave interaction (Wu and Davidson, 1972; Wu *et al.*, 1973) have been considered. In the paper of Wu *et al.* (1973) it had become evident that wave interaction is likely to be the dominant energy transfer mechanism that couples suprathermal ions to the ambient primary solar wind plasma with a time constant of 10^2 to 10^3 s. That means the excess energy of secondary ions can be taken as instantaneously being transferred to the primary plasma, since the characteristic expansion time of the solar wind plasma at $r = 1 \, r_E$ is of the order of 10^6 s (Hartle and Sturrok, 1968).

Therefore the production of secondary ions at some place in the solar system gives rise to a local heating of the primary solar wind plasma that can be easily included into the energy equation of the solar wind.

4. Nonthermal Heating of the Tri-Fluid Solar Wind

Since the ionization of the interstellar gas leads to the production of secondary protons and secondary helium ions that are four times as hot as protons and that are most likely to be mainly coupled to primary helium ions, we have used a tri-fluid solar wind model consisting of electrons, protons and helium ions. This enables us to study the different heating effect that secondary ions might have on the temperature of the different constituents of the solar wind plasma. In a simple estimate of the total heating possible by secondary ions Fahr (1973) has shown that heating of this kind only starts becoming active outside 1 r_E. That means secondary ions represent a heat source in the temperature-sensitive region of the solar wind, where additional energy supply has the effect of only increasing solar wind temperatures, whereas the solar wind velocities and densities are essentially unaffected. This is true as long as thermal energies of the heated solar wind plasma remain far enough below its kinetic energy so that thermal pressure gradients cannot influence the plasma motion. Thus, if this condition is fulfilled, we can adopt the solutions for the solar wind velocity and density as integrals of the equation of mass and momentum continuity and can introduce these into the equation of energy continuity. What remains to solve is a set of three energy equations for the three plasma species considered in this paper.

We start from the consistent solar wind model of Hartle and Sturrok (1968) and adopt their solutions for the solar wind velocity v_{sw} and density $n = n_e = n_p$ as integrals of the system of differential equations. We then enlarge their equations of energy continuity by introducing terms Δ_i for the local energy supply by secondary ions. With these terms the equations attain the following form:

$$\tfrac{3}{2}n_i k v_{sw}\frac{dT_i}{dr} - v_{sw}kT_i\frac{dn_i}{dr} - \frac{1}{r^2}\frac{d}{dr}\left(r^2\gamma_i\frac{dT_i}{dr}\right) = -\tfrac{3}{2}kn_i\sum_j v_{ij}(T_j - T_i) + \Delta_i, \quad (1)$$

where the suffixes i, j sign the various constituents, electrons $(i=1)$, protons $(i=2)$, and He-ions $(i=3)$, n_i means their densities, v_{sw} is taken to be constant beyond a distance r_0, T_i are the temperatures of the different species i, and γ_i are their heat conduction coefficients. v_{ij} are the frequencies of momentum transfer due to Coulomb collisions between the species i and j. The terms Δ_i represent the energy gain due to secondary ions characteristic for the species i. The term Δ_i is set equal to zero, since the suprathermal secondaries excite hydromagnetic waves of the ion cyclotron type that can only very badly be absorbed by the solar wind electrons. Thus only Δ_2 and Δ_3 are considered different from zero with the following form:

$$\Delta_2 = E_2(\alpha, v_{sw})\,P_2 + (1-q)\,\Delta_3/q \tag{2}$$

and

$$\Delta_3 = E(\alpha, v_{sw})\,P_3 q \tag{3}$$

P_2 and P_3 are the production rates of suprathermal protons and He-ions, respectively. $E_{2,3}(\alpha, v_{sw})$ is the specific energy presented per suprathermal proton and He-ion, respectively, for an inclination α of the magnetic field against the solar wind velocity v_{sw}. These energies are given by:

$$\begin{aligned} E_2 &= \tfrac{1}{2}m_p v_{sw}^2 \sin^2\alpha \\ E_3 &= 4E_2. \end{aligned} \tag{4}$$

In the energy gain Δ_3 given in formula (3) the direct absorption of proton cyclotron waves excited by suprathermal protons has been neglected due to the fact that the main heating of solar wind α-particles may most likely be due to the presence of suprathermal He-ions that are four times as hot as suprathermal protons.

The heat conduction coefficients and momentum transfer frequencies in Equation (1) have been determined with formulae analogous to those used by Hartle and Sturrok (1968). The quantity q takes account of the fact that some fraction of the excess energy of suprathermal He-ions might be transferred to solar wind protons rather than to solar wind α-particles. This quantity has been adopted here as an open parameter. Thermal conduction is one of the most important terms in the energy equation for species 1, whereas it is negligibly small for species 2 and 3 beyond $r_0 = 0.5$ AU.

Since in the energy equation of species 1 thermal convection and conduction largely dominates over terms of thermal Coulomb coupling to protons and α-particles, we can assume the temperature profile $T_{1,0}$ for the electrons given by Hartle and Sturrok (1968) not to be changed by the introduced sources $\Delta_{2,3}$ due to secondary ions.

We therefore start calculating first order temperature profiles $T_{i,1}$ for species 2 and 3 from Equation (1) introducing zeroth-order terms for thermal coupling calculated from the Hartle-and-Sturrok-solutions for T_1 and $T_2 = T_3$ which are taken as zero-order-approximations. In an integration procedure higher order solutions $T_{i,n}$ can be calculated taking the coupling terms from solutions $T_{j,n-1}$ of the preceding order. Our integration is started from $r_0 = 0.5$ AU where heating by suprathermal ions can be shown to be still ineffective (Fahr, 1973).

5. Discussion of Results

For an upwind direction facing the advance of interstellar matter the production rates $P_{2,3}$ have been calculated according to those given in the papers of Fahr (1973a) and Blum and Fahr (1970). In these calculations a constant solar wind speed of $v_{sw} = 400$ km s^{-1} has been assumed. The density of the solar wind protons and α-particles at 1 AU is adopted with 5 cm^{-3} and 0.2 cm^{-3}. The undefined parameter q has been varied between 1 and 0 according to the assumption of exclusive or vanishing coupling of suprathermal He-ions to solar wind α-particles.

In Figure 1 we have shown the results. The dashed-dotted line represents the solution of Hartle and Sturrok (1968), whereas the other curves show the increased temperatures T_2 and T_3 for different values of q, if suprathermal heating by secondary ions is included. It is evident that beyond $r = 1$ r_E the solar wind ion temperatures are noticeably enhanced. On the other hand it could be shown (Fahr, 1973) that the increased temperatures T_2 and T_3, and thereby the lowered thermal losses of solar wind electrons to solar wind ions, do not affect the adopted electron temperature profile T_1 taken from Hartle and Sturrok (1968).

In Figure 2 we show the obtained temperature gradients dT_2/dr and dT_3/dr, as well as ratio T_3/T_2 for q = 1 calculated both with and without thermal coupling between solar wind protons and α-particles due to Coulomb collisions. It can be seen that between 1.5 and 2.5 AU there is to be expected a crossover from negative to positive gradients, i.e. temperatures beyond this region again should increase with increasing solar distances r. This could be an important observable feature in the solar wind temperature profile that could be used to decide what energy sources are in fact active in the region outside 1 AU.

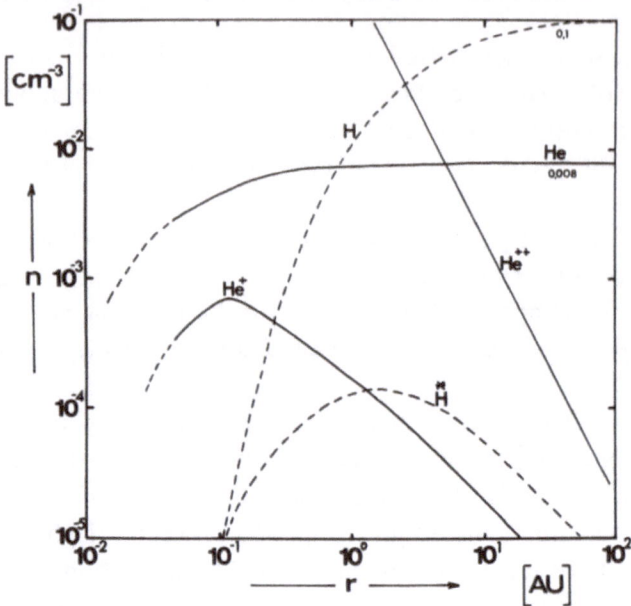

Fig. 1. The density profiles of interstellar H and He, of solar He^{++} and of the interaction products He$^+$ and fast neutral hydrogen atoms $\overset{*}{H}$ are given vs the distance r from the Sun for a direction inclined by an angle $\theta = 90°$ against the velocity v_0 of approach of interstellar matter. Densities of hydrogen and helium in the nearby interstellar space have been assumed to be 0.1 cm^{-3} and 0.008 cm^{-3}, respectively.

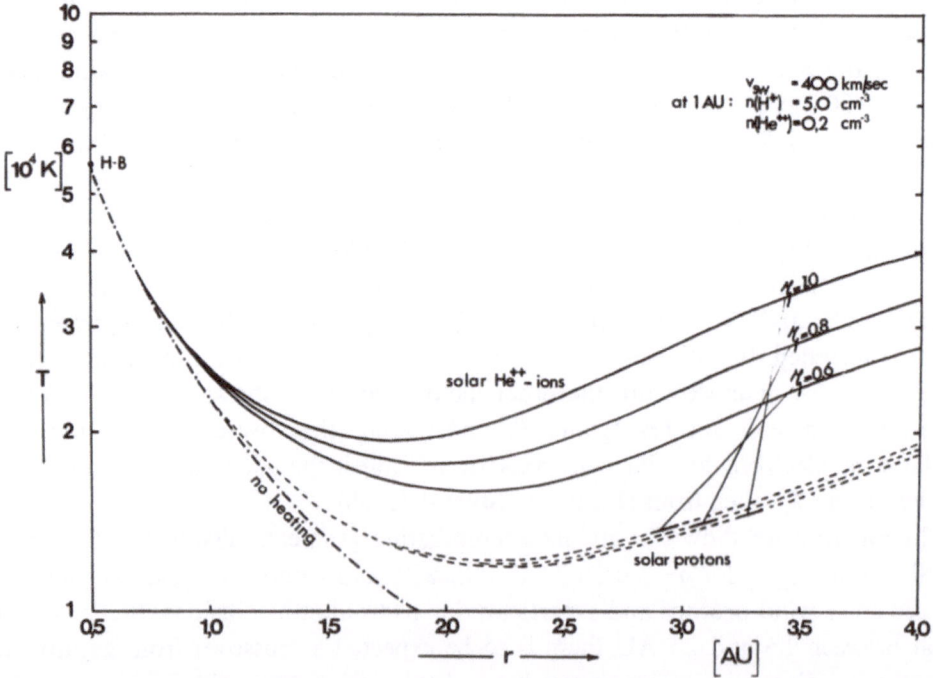

Fig. 2. The temperatures T_2, T_3 of solar wind protons and α-particles are given vs the solar distance r for the direction facing the approach of interstellar matter. The solutions inside of $r_0 = 0.5$ AU are taken from Hartle and Barnes, 1969. The parameter takes account of the fractional transfer of suprathermal He$^+$-energies to solar wind α-particles.

Comparison of the dashed and the dashed-dotted curves in Figure 2 reveals that the thermal energy coupling between protons and α-particles due to Coulomb collisions is still effective beyond $r=1$ AU contrary to what has been suggested by several authors.

The thermal behaviour of the solar wind ion plasma shown in Figures 2 and 3

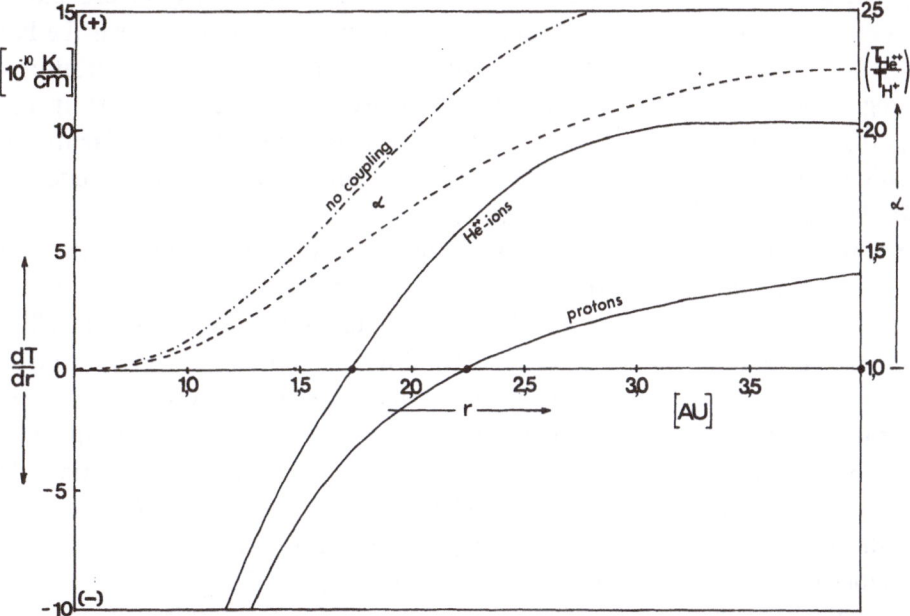

Fig. 3. Shown by the full lines are the gradients (left ordinate) of the temperatures T_2, T_3 of solar wind protons and α-particles. The dashed and the dashedKdotted lines show the ratio $\alpha=(T_3/T_2)$ (right ordinate) with and without consideration of thermal coupling between the ion species 2 and 3 due to Coulomb collisions.

should represent a permanent feature of the expanding solar wind that is due to the interaction of interstellar gases advancing towards the Sun. It may now be discussed whether or not this interstellar heat source introduced here can be considered to be dominant heat source for the solar wind plasma beyond $r=1$ AU, and what other distant energy sources might possibly compete with it. As far as we see there exists only one further energy source besides the interstellar one that could be made responsible for a solar wind heating beyond 1 AU, namely the conversion of kinetic into thermal energy of high velocity plasma flows penetrating the steady solar wind background plasma.

The problem of the expansion of non-linear high velocity plasma disturbances up to large solar distances has recently been investigated extensively by Hundhausen (1973a, b, c). The author investigates how a pressure pulse of $2\tau_0 = 100$ h duration propagates through the steady solar wind. We may use here this investigation in order to compare the heating effects of high velocity plasma flows and of interacting neutral gases. It has been shown by Hundhausen (1973a) that the maximum expansion velocity of a pulse superimposed on a steady solar wind of an asymptotic velocity of 320 km s^{-1} decreases from 500 km s^{-1} very slowly during the propagation of the pulse from a few solar radii to $r = 2.5$ AU. This means that only a very minor fraction of the kinetic energy of the pulse is irreversibly dissipated to the steady solar wind environment. This has also been confirmed by the solar wind data analysis of Burlaga and Ogilvie (1973) who investigated the $T_p - v_{sw}$-relation for phases of negative and positive velocity derivatives dv_{sw}/dt, i.e. immediately before and after the passage of a pressure pulse through the solar wind environment near the Earth. They found that at most a 15% change of the T_p-values in the $T_p - v_{sw}$-relation can be registered after the passage of a high velocity pulse, whereas the temperature T_{max} that is reached during the passage and that is purely due to adiabatic compression of the plasma within the pulse is more than one order of magnitude higher than the steady solar wind temperature T_p.

The peak temperature of the pulse used by Hundhausen (1973a) in his theoretical treatment of nonlinear disturbances, here referred to as T_{max}, is shown in Figure 4 vs the solar distance r. It does not show a minimum. Outside $r = 1$ AU the effect of high velocity streams on the quiet solar wind temperature T_p will even decrease in comparison to $r = r_E$ rather than increase, since the effectivity of the dissipating mechanisms should decrease outwards from the Earth's orbit leading thus to less than 15% changes of T_p. Then from the fact of a very inefficient dissipation of the kinetic energy of the pulse it becomes clear that the temperature T_p belonging to the steady solar wind phases outside regions of sharply increasing or decreasing velocities everywhere beyond $r = 1$ AU will only very slightly be enhanced by the energy dissipation of passing high velocity plasma flows. Therefore it seems to be obvious that heating due to high velocity flows is not able to give rise to an ascending temperature profile beyond 1 to 2 AU for quiet solar wind temperatures T_p as it has been found to occur, if the interstellar energy source is taken into account.

Another point is to consider mean solar wind temperatures T_p averaged over quiet and disturbed solar wind phases. As it can be seen in the paper of Hundhausen (1973a), there exist two reasons why the disturbed solar wind temperatures T_{max} within high velocity pulses gain an increasing statistical weight in the average temperature T_p for increasing solar distances r. One is the fact that the expansion velocity of the pulse is decreasing with r, the other that the radial and spatial extent of the temperature pulse which is due to adiabatic compression is increasing with r. Therefore the temporal presence of disturbed pulse temperatures T_{max} is appreciably enhanced at increasing solar distances r. Hundhausen (1973a) has given five consecutive expansion phases of an expanding pressure pulse with an initially triangular profile, the three latest phases showing the pulse at distances of 1.5, 2.0, 2.5 AU. Simplifying the strongly steepened temperature profiles of these three pulse phases with their orthogonal inner and outer temperature flanks into square-profiled pulses of equal area, one is able to determine the product of a mean temperature $\overline{T}_{max}$ during the passage of the pulse at 1.5, 2.0 and 2.5 AU and the mean duration of the pulse with the use of mean pulse expansion velocities $\overline{v}_{max}$ given by Hundhausen (1973a) for those distances.

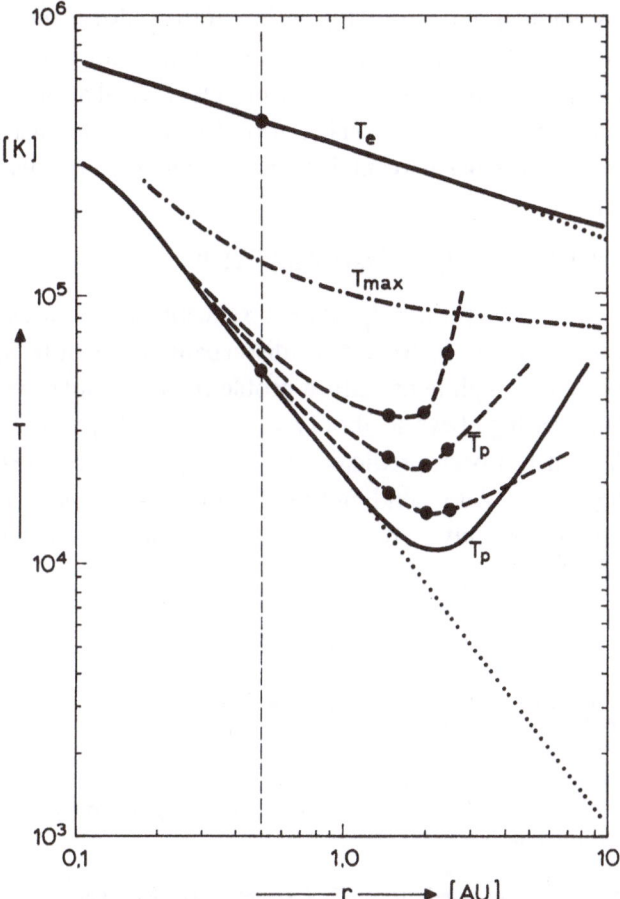

Fig. 4. The temperatures T_e and T_p are shown for quiet solar wind conditions ($v_{sw} = 400$ km s^{-1} and $n_p = 5$ cm^{-3} at $r = r_E$). Full lines give the profile for an interstellar heat source included, dotted lines give the profile of the Hartle and Barnes model. The dashed-dotted line gives the profile of the peak temperature T_{max} during the passage of a pressure pulse as it had been used by Hundhausen (1973a). The dashed lines give temperatures $\overline{T}_p$ averaged over quiet and disturbed solar wind conditions for $\varrho = 5\%$ (lowermost curve), $\varrho = 10\%$ and $\varrho = 20\%$ (uppermost curve) disturbances of the quiet solar wind at the coronal base.

For a width $D(r)$ of the pulse temperature profile and an expansion velocity $\bar{v}_{max}(r)$ of the pulse we obtain a duration $\tau(r)$ of the pulse given by:

$$\tau(r) = D(r)/\bar{v}_{max}(r).$$

Adopting now the triangular pressure pulse used by Hundhausen (1973a) as a typical one representative for a majority of disturbances at the solar corona, we can easily determine average temperatures $\overline{T}_p$ at different solar distances r assuming that the solar wind at the base of the corona may be characterized by ϱ percent disturbances of this kind of pulses. For this case we have to assume that typical pressure pulses are emitted from the solar corona with a frequency

$$f_0 = \frac{1}{2\tau_0(100/\varrho)}, \tag{6}$$

where $\tau_0 = 50$ h is one half of the initial duration of the pulse at the coronal base.

Due to the fact that all emitted pulses as long as they do not interfere have the same expansion characteristics, the frequency of pulse arrival at other solar distances is not changed, but remains $f(r) = f_0$. Therefore the average temperature $\bar{T}_p$ at r for a solar wind with a fraction of ϱ percent disturbances by pressure pulses at the coronal base can be given:

$$\bar{T}_p = \tau(r)\, \bar{T}_{\max}(r) - T_p(r)\, (2\tau_0 (100/\varrho) - \tau(r)). \tag{7}$$

The obtained average temperatures $\bar{T}_p$ for $r = 1.5$, 2.0 and 2.5 AU are shown in Figure 4 for various percentages $\varrho = 5$, 10, 20% of disturbances. It can be seen that due to the periodic passage of a high temperature profile the mean solar wind temperature T_p increases with increasing r beyond about $r = r_E$. This would give the same behaviour for the T_p profile as it has been predicted for the T_p profile due to the action of the interstellar energy source. An ascending branch of the quiet solar wind temperature profile T_p seems, however, only to be understandable with an active interstellar heat source as regarded in this paper.

References

Blum, P. W. and Fahr, H. J.: 1970, *Astron. Astrophys.* **4**, 280.
Burlaga, L. F. and Ogilvie, K. W.: 1973, *J. Geophys. Res.* **78**, 2028.
Burns, H. A. and Halpern, G.: 1968, *J. Geophys. Res.* **73**, 7377.
Fahr, H. J.: 1971a, *Planetary Space Sci.* **19**, 1121.
Fahr, H. J.: 1971b, *Cosmic Plasma Physics Conf. Frascati*, Rome, Sept. 1971, p. 81.
Fahr, H. J.: 1973a, *Space Research* **XIII**, 837.
Fahr, H. J.: 1973b, *Solar Phys.* **30**, 193.
Feldman, W. F., Lange, J. J., and Sherb, F.: 1972, *J. Geophys. Res.* **77**, 5389.
Hartle, R. E. and Sturrok, P. A.: 1968, *Astrophys. J.* **151**, 1155.
Hartle, R. E. and Barnes, A.: 1969, *J. Geophys. Res.* **75**, 6915.
Holzer, T. E.: 1972, *J. Geophys. Res.* **77**, 5407.
Holzer, T. E. and Leer, E.: 1973, Preprint Aeronomie Lab. NOAA, Boulder, Colo.
Hundhausen, A. J.: 1973a, *J. Geophys. Res.* **78**, 1528.
Hundhausen, A. J.: 1973b, *J. Geophys. Res.* **78**, 2035.
Hundhausen, A. J.: 1973c, *J. Geophys. Res.* **78**, 7996.
Parker, E. N.: 1960, *Astrophys. J.* **132**, 175.
Sturrok, P. A. and Hartle, R. E.: 1966, *Phys. Rev. Letters* **16**, 628.
Wallis, M. K.: 1973, *Astrophys. Space Sci.* **20**, 3.
Wu, C. S. and Davidson, R. C.: 1972, *J. Geophys. Res.* **77**, 5399.
Wu, C. S., Hartle, R. E., and Ogilvie, K. W.: 1973, *J. Geophys. Res.* **78**, 306.

THE PIONEER 10 PLASMA ANALYZER
RESULTS AT JUPITER

JOHN H. WOLFE

NASA Ames Research Center, Moffett Field, Calif., U.S.A.

1. The Pioneer 10 Mission

The Pioneer 10 spacecraft was successfully launched from Cape Kennedy on March 3, 1972 aboard an Atlas-Centaur launch vehicle which incorporated a TE-364-4 solid propellant third stage. At the time, Pioneer 10 attained the highest injection energy ever achieved as attested to by the fact that the spacecraft only required eleven hours to cross the lunar orbit. After a 21 month flight, the Pioneer 10 spacecraft arrived at its radius of closest approach (RCA) at Jupiter at a distance of approximately 2.8 R_J (Jovicentric Jupiter radii) on December 4, 1973.

The principal scientific objectives of Pioneer 10 are to investigate the nature of the interplanetary medium beyond the orbit of Mars, including the asteroid belt, and to make direct, *in situ*, observations of the planet Jupiter and its environment. The successful flyby of Jupiter by Pioneer 10 achieved the latter objectives; however, the interplanetary objectives are still being pursued in the present post-encounter mission beyond Jupiter. Present estimates are that the Pioneer 10 spacecraft can be utilized for interplanetary observations to a solar radial distance of at least 20 AU.

The Pioneer 10 spacecraft and trajectory details have been reported by Hall (1974) and are only briefly summarized here. The Pioneer 10 spacecraft weighs 258 kg including 33 kg for the 11 on-board experiments. Two additional experiments are performed using the spacecraft S-band communications system. The spacecraft is spin stabilized with a spin rate of 4.8 rpm. The spacecraft spin axis is parallel to the axis of the 2.74 m diameter high gain antenna reflector and is kept pointed toward the earth in order to maximize the communication bit rate. The maximum bit rate used during the Jupiter encounter was 1024 bits per second. Spacecraft spin axis precession maneuvers are required periodically in order to maintain earth pointing and were performed approximately six days prior to and two days after the Jupiter flyby. During the encounter, the earth pointing spacecraft spin axis was oriented at an angle of approximately 9.2° with respect to the spacecraft-Sun line in a direction away from the west limb of the Sun. Electrical power for the experiments and spacecraft subsystems is supplied by four radioisotope thermoelectric generators (RTG) since a conventional solar cell array is not practical for the large solar distance involved in the Pioneer 10 mission. The RTGs are located approximately 2.4 m from the center of the spacecraft at the end of two long booms. Inspection of in-flight data indicates that these RTGs have produced negligible interference with any of the Pioneer 10 experiments. During the encounter the spacecraft approached Jupiter

V. Formisano (ed.), The Magnetospheres of the Earth and Jupiter, 279–296. All Rights Reserved
Copyright © 1975 by D. Reidel Publishing Company, Dordrecht-Holland

in the mid-morning sector of the sunlit hemisphere and exited near Jupiter's dawn meridian. For details of the flyby trajectory see Hall, 1974.

2. The Plasma Analyzer Instrumentation

The Ames Research Center Plasma Analyzer experiment on Pioneer 10 consists of dual, 90°, quadrispherical electrostatic analyzers, multiple charged particle detectors and attendant electronics. This analyzer system is capable of determining the incident plasma distribution parameters over the energy range for protons from 100 to 18 000 eV and from approximately 1 to 500 eV for electrons. A central, cross sec-- tional drawing of the analyzer and detector portions of the experiment is shown in Figure 1. The 'A' detector or high resolution quadrispherical analyzer (the inner

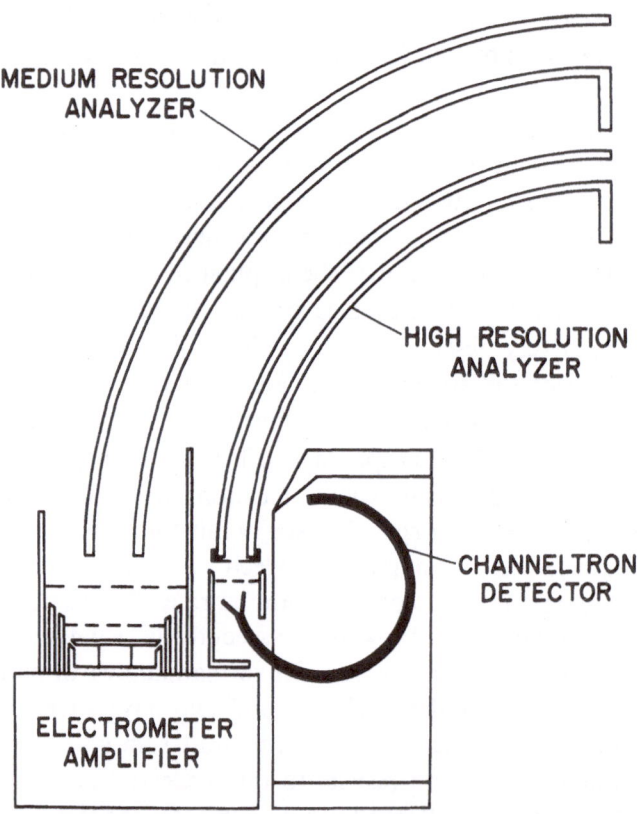

MEDIUM RESOLUTION ANALYZER

HIGH RESOLUTION ANALYZER

CHANNELTRON DETECTOR

ELECTROMETER AMPLIFIER

Fig. 1.   Central, cross sectional schematic of the analyzer and detector portions of the Pioneer 10 Ames Research Center Plasma Analyzer experiment.

analyzer system shown in Figure 1) has an analyzer constant of 9 (charged particle acceptance energy per unit charge divided by the analyzer plate potential) with an analyzer plate mean radius of 9 cm and 0.5 cm separation. The high resolution analyzer is used for ion analysis only and utilizes 26 Bendix type CEM 4012 Channeltrons, operated in the pulse counting mode, for ion detection. The Channeltron detectors are arranged in a semicircle at the base of the analyzer plates and cover the angular range of $\pm 51°$ with respect to the entrance aperture normal. The Channeltrons have an angular separation of approximately 3° near the central portion of the analyzer and approximately 8° separation at the extremes of the analyzer. The Channeltron bias voltage can be changed in two sections (left and right halves) by ground command in eight discrete steps over the range from 2600 to 4400 V. Analysis of flight data has shown that all 26 Channeltrons have operated flawlessly since launch and no appreciable degradation has been observed prior to, during or subsequent to the Jupiter encounter.

The 'B' detector or medium resolution analyzer (the outer analyzer system in Figure 1) has a 12 cm mean radius and 1 cm plate separation giving an analyzer constant of 6. The medium resolution analyzer is used for both ion and electron detection and utilizes five, flat surface, current collectors and electrometer amplifiers. The central three current collectors each have a 15° view width and cover an angular view range of $\pm 22.5°$ with respect to the entrance aperture normal. The two outside collectors each have an angular width of 47.5° and are located at $\pm 46.25°$ with respect to the center of the analyzer.

The Plasma Analyzer experiment is situated on the Pioneer 10 spacecraft such that the entrance apertures view back toward the Earth (and therefore, the Sun) through a wide slit in the back of the spacecraft high gain antenna reflector. The entrance aperture normals are oriented parallel to the spacecraft spin axis thus allowing a complete angular scan of the earthward hemisphere every half spacecraft revolution. The edges of the antenna reflector limit instrument viewing to $\pm 73°$ with respect to the spacecraft spin axis. Although there are a variety of possible operating modes for the experiment, the principal mode utilized during the encounter phase of the Pioneer 10 mission is one in which the energy per unit charge acceptance analyzer potential is stepped every one-half revolution of the spacecraft and all current collectors and Channeltrons are read out at the peak flux roll angle of the spacecraft. Since the medium and high resolution analyzers operate independently, a complete cross check between the two analyzers is possible. The combined analyzer system covers the dynamic range for charged particle fluxes from approximately 1×10^2 to 3×10^9 cm^{-2} s^{-1} and is capable of resolving proton temperatures down to at least 2×10^3 K. Both analyzers were calibrated prior to launch in the Ames Research Center Plasma Ion Calibration Facility. These pre-launch calibrations are utilized in a least-squares fit to the flight data for a variety of possible distribution models in order to determine the plasma ion distribution parameters. Whereas the preliminary report of the Ames Research Center Plasma Analyzer observations for the Pioneer 10 Jupiter encounter (Wolfe *et al.*, 1974) was based on real time data, the results presented here are based on the analysis of the off-line flight data tapes. An isotropic Maxwellian distribution model has been assumed in the fit to the flight data reported here.

3. The Solar Wind-Jupiter Interaction

The first unambiguous indication of the interaction of the solar wind with the Jovian magnetic field occurred on November 26, 1973, at approximately 1946 UT spacecraft time. The telemetry signals were actually received at about 2031 UT on Earth (ground received time or GRT) corresponding to a one-way radio propagation time of approximately 45 min. Note that unless specifically indicated as GRT, spacecraft time will be used throughout this report. At this time the Pioneer 10 spacecraft was inbound toward Jupiter at a Jovicentric radial distance of 108.9 R_J ($R_J = 71\,372$ km). The two solar wind ion spectra shown in Figure 2 were taken in the interplanetary

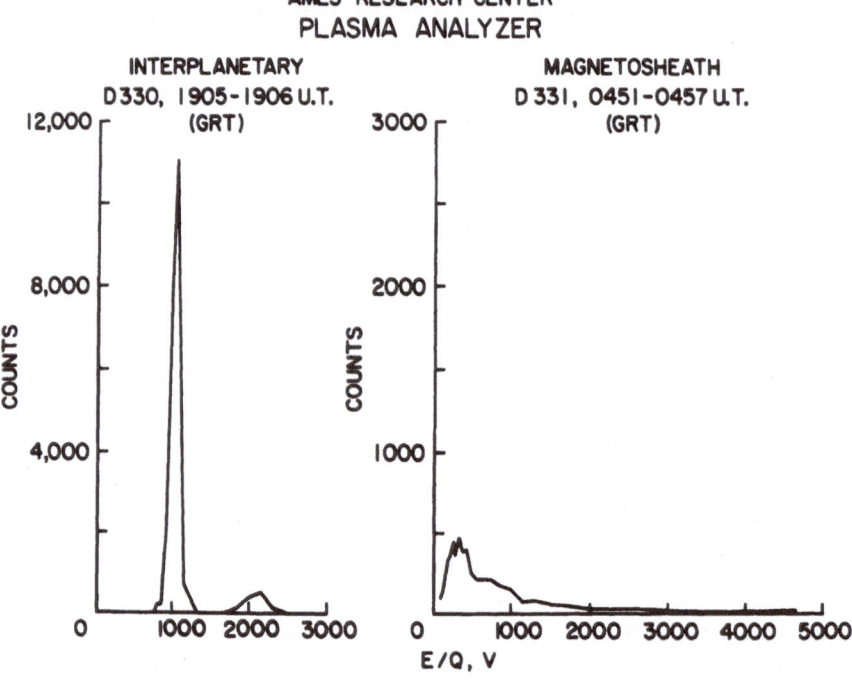

Fig. 2.   Comparison of solar wind ion spectra taken upstream and downstream from Jupiter's bow shock for the inbound portion of the Pioneer 10 Jupiter flyby.

medium (spectrum on the left) at 1905 UT, GRT, on November 26, 1973 (Day 330) about 1 hour and twenty-five minutes before the Jovian bow shock crossing and in the Jovian magnetosheath (spectrum on the right) at 0451 UT, GRT, on November 27, 1973 (Day 331) about 8 hours and twenty minutes after the shock crossing. Although the ion characteristics in the magnetosheath were quite variable, the spectrum shown in Figure 2 is considered to be typical. The ragged appearance of this spectrum is most likely due to fluctuations in the magnetosheath ion characteristics during the period required to obtain the spectrum and are therefore considered to be an artifact in the data caused by sample aliasing. The observation of this drastic change in the ion spectral characteristics, illustrated in Figure 2, is interpreted as the encounter of the Pioneer 10 spacecraft with a detached bow shock wave standing off from Jupiter's magnetosphere and in many respects is quite similar to the case at Earth.

For the interplanetary ion spectrum shown in Figure 2, the proton peak is seen near 1 keV and the doubly charged helium peak near 2 keV. This interplanetary spectrum corresponds to a solar wind convective speed of approximately 440 km s^{-1}, a proton number density of 0.12 cm^{-3} and an isotropic proton temperature of 6.1×10^4 K. It should be noted that this solar wind speed and number density correspond to an anomalously low solar wind dynamic pressure (by about a factor of 4) compared to that normally observed by this experiment in the interplanetary medium near 5 AU. The ion distribution parameters for this first magnetosheath traversal were mostly obtained from the high resolution analyzer. The large flow angle ($\sim 40°$) in the magnetosheath plasma flow direction with respect to the spacecraft spin axis and the high plasma temperature and attendant low density precluded obtaining reliable measurements from the medium resolution analyzer. This large deflection in flow direction from approximately antisunward to a large angle with respect to the spin axis was observed as the spacecraft crossed the bow shock and, with the exception of a 220-min period commencing at approximately 0500 UT on November 27, 1973, persisted throughout the entire magnetosheath traversal. During the above period between 0800 and 0900 UT, the flow directions were both toward the center of the Plasma Analyzer acceptance angle, and in addition the plasma currents were enhanced. Average magnetosheath plasma distribution parameters of 273 km s^{-1} bulk speed, a proton number density of 0.62 cm^{-3} and a proton temperature of 3.5×10^5 K were calculated for this time. The distribution parameters for the magnetosheath spectrum of Figure 2 are similar and here the bulk velocity is approximately 191 km s^{-1} and the isotropic temperature is approximately 2×10^5 K. The magnetosheath flow field characteristics are discussed in more detail in the next section.

At 1953 UT on November 27, 1973, the incident plasma ion flux abruptly dropped below the sensitivity threshold for both the high and medium resolution analyzers. At this time the Pioneer 10 spacecraft was located at a Jovicentric radial distance of 96.4 R_J. This termination of the magnetosheath plasma flow is interpreted as the crossing of the magnetopause boundary and penetration into Jupiter's magnetosphere by Pioneer 10 and is presumed to be due to the exclusion and deflection of the magnetosheath plasma by the equal and opposite pressure exerted by Jupiter's outer magnetic field and its internal gas. As was the case for the bow shock, Jupiter's magnetopause also seems in many ways similar to Earth's.

As the spacecraft proceeded inbound, magnetosheath plasma, flowing at large angles with respect to the spacecraft spin axis, was again observed at 0233 UT on December 1, 1973, corresponding to a radial distance of 54.3 R_J. The observation of magnetosheath plasma persisted for approximately eleven hours and was again abruptly terminated at 1336 UT on December 1, 1973, at 46.5 R_J. At present there are two apparent explanations for the second magnetopause traversal observed during the inbound portion of the Pioneer 10 trajectory. The first would be that the interplanetary solar wind dynamic pressure increased to such an extent that the entire Jovian magnetosphere contracted down to a size such that the spacecraft was again located in the magnetosheath. An alternative possibility is that the topology of Jupiter's magnetosphere is such that a simple change in the interplanetary solar wind flow direction (with little or no change in dynamic pressure) deflected Jupiter's magnetosphere so that the spacecraft was located within the magnetosheath. The present evidence seems to strongly favor the former explanation and is discussed further in the summary section.

During the remainder of the Pioneer 10 traversal of the Jovian magnetosphere, sporadic plasma ion fluxes were observed but their analysis has been complicated by high background rates due to penetrating energetic electrons and protons. Magnetospheric plasma ion observations are very preliminary at this time and are not reported here. Other than these high background rates observed in Jupiter's inner magnetosphere, the Plasma Analyzer experiment successfully withstood Jupiter's intense radiation zones and recrossed Jupiter's magnetopause on the outbound leg of the trajectory at 1153 UT on December 10, 1973 at a distance of 97.9 R_J. In contrast to the inbound portion of the flyby trajectory, where the bow shock was observed once and the magnetopause was crossed three times, during the outbound leg the magnetopause was crossed five times and there were seventeen positively identifiable shock crossings. All of the shock and magnetopause observations during the Pioneer 10 Jupiter flyby are listed in Table I for both the inbound and outbound passes. In this table under spacecraft location, IP refers to interplanetary medium, MSH the Jovian magnetosheath and MS to Jupiter's magnetosphere. For each boundary observed (S for shock and M for magnetopause) the date and spacecraft time in UT and Jovicentric distance in R_J are listed. From Table I it is seen that the last shock crossing occurred at 1928 UT on December 22, 1973 at a distance of 242.6 R_J. Thus the Jupiter encounter for the Plasma Analyzer experiment lasted nearly a month!

Two further plasma observations associated with Jupiter's bow shock may be noted. A period of approximately 14 minutes duration, that began 13 h and 43 min after the last bow shock crossing listed on Table I, exhibits greatly reduced plasma flux and some flow deflection that probably indicates a movement of Jupiter's bow shock near the spacecraft. However, a crossing of the bow shock can not be positively identified here. In addition, 18 min before the first bow shock crossing the solar wind plasma flux was apparently temporarily greatly reduced which could be an interplanetary effect rather than an approach of the bow shock near the spacecraft since a large flow deflection was not observed.

The magnetosheath boundary traversals given in Table I are illustrated in Figure 3, which shows the Pioneer 10 Jupiter encounter trajectory projected onto Jupiter's orbital plane. Each shock (S) and magnetopause (M) location is identified along the spacecraft trajectory at the position where it was observed. Note that in the outbound leg, the point identifying the second magnetopause location actually represents two closely spaced magnetopause crossings and a burst of plasma which could represent two further crossings, and the point identifying the last shock observation represents

TABLE I

Jupiter Magnetosheath Boundary locations observed during the Pioneer 10 flyby. IP refers to the inter-planetary medium, MSH is the magnetosheath, MS is the magnetosphere, S denotes a shock crossing and M a magnetopause crossing

Spacecraft location	Boundary observation	Spacecraft time		Distance R_J
		'73 date	UT	
– Inbound – – – – – – – – – – – – – –				
IP				
MSH	S	26 Nov	1946 ± 2	108.90
MS	M	27 Nov	1953 ± 2	96.36
MSH	M	1 Dec	0233 ± 6	54.32
MS	M	1 Dec	1335.7 ± 2.2	46.50
– Outbound – – – – – – – – – – – – –				
MS				
MSH	M	10 Dec	1153.4 ± 0.5	97.92
MS	M	12 Dec	0943.2 ± 0.5	121.52
MSH	M	12 Dec	0958.2 ± 0.5	121.66
IP	S	12 Dec	1453.2 ± 1.5	124.14
MSH	S	12 Dec	1950.7 ± 6	126.64
MS	M	13 Dec	0158.1 ± 0.5	129.73
MSH	M	14 Dec	1850 ± 1	150.08
IP	S	18 Dec	0328 ± 1	188.87
MSH	S	20 Dec	2145.1 ± 0.5	220.54
IP	S	20 Dec	2233.9 ± 0.5	221.41
MSH	S	21 Dec	0212 ± 2	223.13
IP	S	21 Dec	0643 ± 2	225.27
MSH	S	21 Dec	1027 ± 2.8	227.04
IP	S	21 Dec	1158 ± 2	227.76
MSH	S	21 Dec	1848.5 ± 9.5	230.99
IP	S	21 Dec	1929.2 ± 2.9	231.31
MSH	S	22 Dec	0605 ± 10	236.31
IP	S	22 Dec	1757.6 ± 1.8	241.44
MSH	S	22 Dec	1805 ± 1	241.97
IP	S	22 Dec	1811.8 ± 1.5	242.02
MSH	S	22 Dec	1815.7 ± 0.1	242.05
IP	S	22 Dec	1928.0 ± 2.7	242.62

Note: MS plasma bursts at 0225 and 1345, December 1, and 0947.7, December 12. Greatly reduced MSH plasma flux at 1324.8 ± 0.5, December 1.

five separate crossings (see Table I). The dashed lines in Figure 3 are for illustrative purposes only and are meant to show the extremes in magnetopause and shock locations during the Pioneer 10 flyby. The boundary shapes and shock standoff distances have been determined from the gas dynamic analog (Spreiter et al., 1966). The shock and magnetopause boundaries have been arbitrarily made symmetrical with respect to the Jupiter-solar wind line. The outermost shock and magnetopause boundaries have been scaled to the last shock crossing for the outbound leg. Similarly the innermost shock and magnetopause boundaries have been scaled to the last magnetopause crossing for the inbound leg.

It is interesting to consider the large scale size of Jupiter's magnetosphere and

286   JOHN H. WOLFE

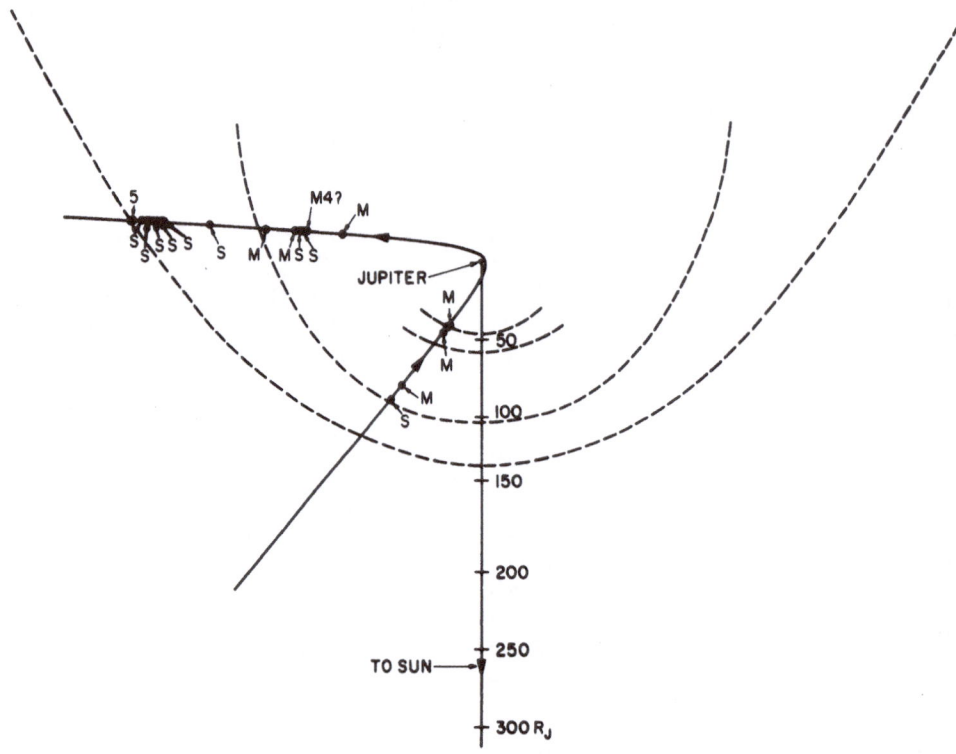

Fig. 3. Locations of the shock (S) and magnetopause (M) crossings on the Pioneer 10 Jupiter flyby trajectory which has been projected onto Jupiter's orbital plane. The inner and outer pair of dahed lines illustrate the observed extremes of position of the magnetopause and standing bow shock. The shape of the boundaries and the shock stand-off distances are based on the gas dynamic analog and scaled to the actual boundary observations.

shock front. For example, the width of the shock front for its largest extent (based on the last shock crossing on the outbound leg and assumed symmetry and shape illustrated in Figure 3) would correspond to a distance of approximately 485 R_J as measured across the dawn-dusk meridian. This is equivalent to a width of 0.23 AU and is more than 2 orders of magnitude larger than the nominal width of the Earth's bow shock. The large extent over which Jupiter's magnetosphere can evidently move indicates that it is extremely responsive to changes in the incident solar wind conditions.

4. The Magnetosheath Flow Field

Figure 4 gives half-hour averages of proton bulk velocities, number densities and isotropic temperatures observed during November 26 and 27, 1973 as Pioneer 10 first crossed Jupiter's bow shock and magnetosheath. The velocities presented do not have the spacecraft velocity subtracted. A correction for this may be estimated by use of the spacecraft velocity components during this two day period, which are

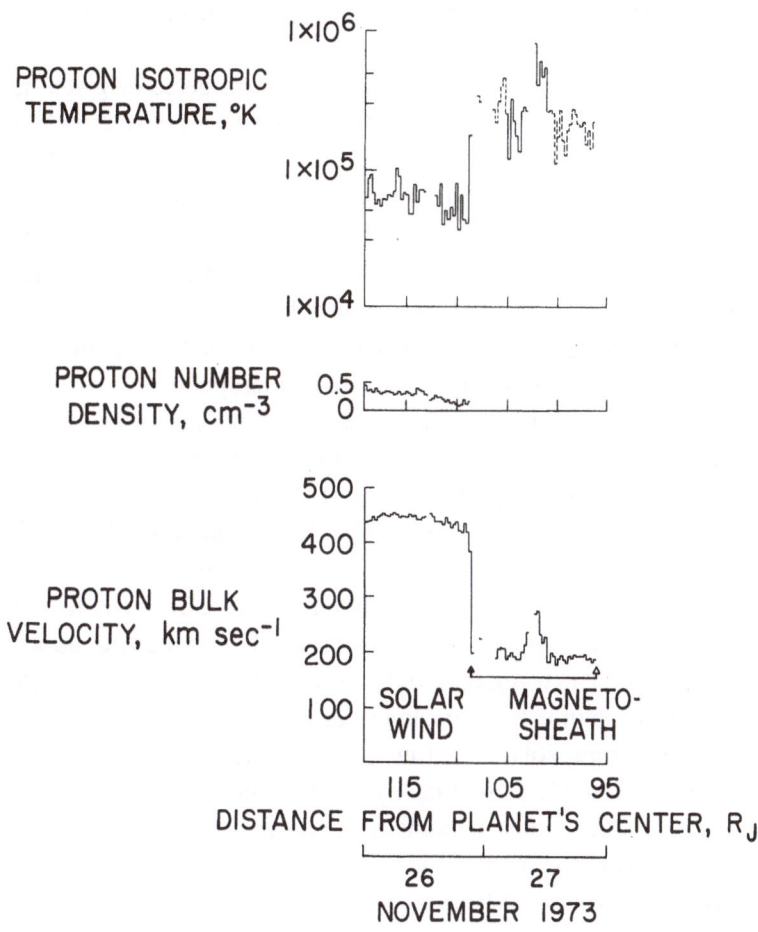

Fig. 4.   Half hour averages of the proton bulk velocities, number densities and isotropic temperatures observed for November 26 and 27, 1973 corresponding to the first crossing of Jupiter's bow shock and magnetosheath. Temperatures indicated with dashes are approximate values derived from scans through the velocity distribution that do not include the peak.

7.2 to 7.5 km s^{-1} in the antisolar direction, and 6.7 ± 0.1 km s^{-1} in the direction of planetary motion, parallel to the ecliptic plane.

The bow shock and magnetopause locations as observed by the Plasma Analyzer are indicated with a solid and an open arrow, respectively, at the bottom of Figure 4. As stated previously, within the inbound magnetosheath, plasma parameters could not be determined by the medium resolution detector, except between 0800 and 0900 UT on November 27, when the plasma flux was enhanced while the flow direction was toward the center of the instrument angular acceptance range. For this time period proton velocities and temperatures from the medium resolution detector were determined by a fit of a Maxwellian distribution to the data. The average proton number density obtained is 0.62 cm^{-3}. These data have not been included in Figure 4. For all magnetosheath times on the figure, half-hour averages of proton bulk velocities and isotropic temperatures from the high resolution detector are given.

Due to the incomplete analysis of the high resolution detector data, values for the ion densities are not yet available.

The plasma parameters from the medium resolution detector given here are calculated by a linear least-squares fit of the flight data to a convecting isotropic temperature Maxwell-Boltzmann distribution, using a representation of the detector response condensed from detailed laboratory calibration data.

The velocities and temperatures from the high resolution detector data are obtained following the formalism of a calculation of the moments of the plasma velocity distribution. From 0244 to 1127 UT on November 27, data from the inner Channeltron that scans through the peak of the proton velocity distribution were used to obtain the proton bulk velocities and isotropic temperatures for Figure 4. Similarly, the data from one of the outermost two Channeltrons were used for the remainder of the magnetosheath times on the figure. The proton counts of the inner Channeltron are integrated for $\frac{1}{512}$, $\frac{1}{128}$, $\frac{1}{64}$ or $\frac{1}{32}$ spacecraft revolution, subject to ground command. Similarly for the outermost Channeltron used here, these counts are integrated, except during special instrument modes, for $\frac{31}{64}$ of a spacecraft revolution. When data from the outermost Channeltron are used here, it is because the statistics provided by the other Channeltrons are felt to be too poor to provide reliable plasma parameters using their data alone. In these cases the velocities derived from the outermost Channeltron are lower limits since this Channeltron does not scan through the maximum of the proton velocity distribution. These derived velocity values are then corrected by dividing by $\cos\delta$, where δ is the angle between the view direction of the outermost Channeltron and the peak of the velocity distribution as determined from the data of the remaining Channeltrons.

The bulk velocity is obtained from

$$N\langle \mathbf{p}\rangle = mN(\langle \mathbf{V}\rangle - \mathbf{u}) = 0$$

and this condition is approximated by

$$u \sim \frac{\sum (n_i/v_i^3)}{\sum (n_i/v_i^4)}.$$

The first equation is written in terms of a velocity value u such that the momentum density of the protons is zero in a reference frame moving with this velocity. In these equations n_i is the count value for the ith velocity (energy) analyzer acceptance value v_i, m is the proton mass, N the proton volume density, $\langle \mathbf{p}\rangle$ the vector average proton momentum and $\langle \mathbf{V}\rangle - \mathbf{u}$ the vector average proton velocity (thermal velocity) referred to the proton bulk velocity u. The isotropic temperature value is obtained from

$$T = \frac{m}{k} \left[\sum (n_i/v_i^4)(u - v_i)^2\right]/\sum (n_i/v_i^4),$$

where k is Boltzmann's constant. When these calculations are performed, an attempt is made to eliminate detector responses due to He^{++} by not using the portion of the spectrum for E/q (energy per unit charge) values two times or greater than that for the peak counts. This is one reason why the magnetosheath temperatures from the high resolution detector are lower limits, since the interplanetary spectra just upstream from the shock indicate a negligibly low He^{++} solar wind abundance at that time (see Figure 2). Thus the high energy portion of a non-thermal distribution would be ignored. Also, the roll-integration of the outermost Channeltron count rates introduces an effect that could broaden the velocity distributions used in the temperature calculation. The medium resolution detector magnetosheath temperatures for 0800 to 0900 UT on November 27 agree with the high resolution detector temperatures given on Figure 4, presumably because the medium resolution detector data have been fit to an isotropic Maxwellian distribution so that the non-Maxwellian portions of the spectrum also tend to be ignored. Consequently, the values for the magnetosheath temperature calculated from the medium resolution detector data also tend to be lower limits.

The medium resolution detector magnetosheath proton bulk velocities determined for the 60-min period beginning at 0800 UT on November 27, 1973, were ~ 25 km s^{-1} lower than the high resolution detector values. This small velocity difference is presumably due to the difference in sensitivity to the non-thermal part of the velocity distribution in the two methods used for velocity calculation.

Gaps in the data of Figure 4 are sometimes due to ground-commanded changes of instrument status into special modes for which the results are not included here. Some gaps are also caused by brief data losses in the ground data network.

The plasma conditions observed during the second magnetosheath traversal (54.3 to 46.5 R_J) were somewhat similar to those observed during the first traversal with one important difference. With the exception of the enhanced speed, temperature and density values observed between approximately 0800 to 1000 UT on November 27 for the first magnetosheath traversal, the speed and temperature values observed during the second traversal were comparable. The density, however, was observed to be almost an order of magnitude higher for the second traversal.

The second magnetosheath traversal proton bulk velocities were in the 165 to 210 km s^{-1} range and the temperature was observed to vary from about 6×10^4 K to 5×10^5 K. Both the higher velocities and temperatures were observed near the end of the traversal. The unaberrated polar and azimuthal flow directions at the beginning of the traversal were $\sim 40°$ southward and $\sim 19°$ in the direction of planetary motion, respectively. Near ~ 0715 UT, there are several samples with lower limit southward flow directions near zero degrees and azimuthal flow directions $\gtrsim 35°$ in the direction of planetary motion. Near the end of the traversal the flow direction is ~ 15–$25°$ southward and ~ 30–$35°$ in the direction of planetary motion. Thus, not only were the densities greater during the second traversal but the flow deflections were also greater than observed for the first traversal.

Figure 5 gives half-hour averages of proton bulk velocities, number densities, isotropic temperatures and hourly averages of unaberrated azimuthal and polar angles for the outbound traversal of Jupiter's magnetosheath by Pioneer 10, during De-

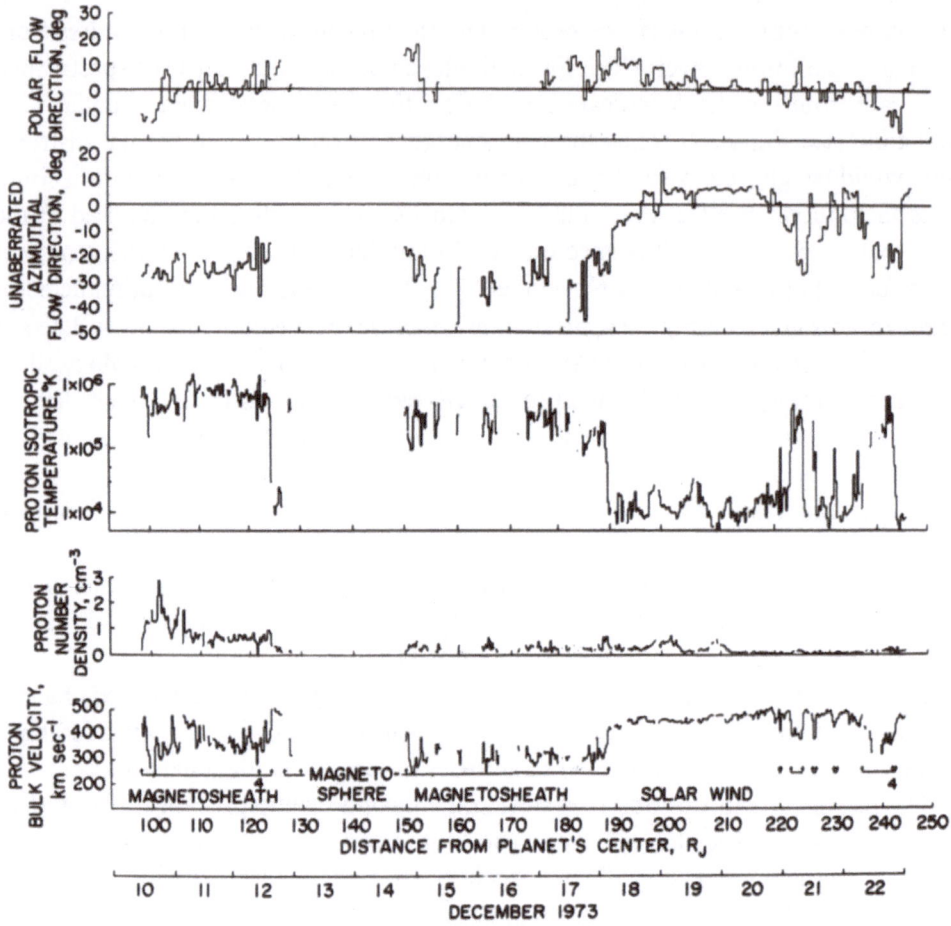

Fig. 5. Half hour averages of the proton bulk velocities, number densities, isotropic temperatures and hourly averages of unaberrated azimuthal and polar angles for the outbound traversal of Jupiter's magnetosheath by Pioneer 10 during December 10 through 22, 1973.

cember 10 through 22, 1973. The flow direction average polar, $\bar{\theta}$, and azimuthal, $\bar{\varphi}$, angles are composed from individual samples θ_i and φ_i using the expressions

$$\sin \bar{\theta} = \frac{\sum \sin \theta_i}{[(\sum \cos \theta_i \cos \varphi_i)^2 + (\sum \cos \theta_i \sin \varphi_i)^2 + (\sum \sin \theta_i)^2]^{1/2}}$$

and $\tan \bar{\varphi} = (\sum \cos \theta_i \sin \varphi_i)/(\sum \cos \theta_i \cos \varphi_i)$. In a spacecraft centered solar ecliptic coordinate system, the polar angles, θ, are positive for southward flow, while azimuthal angles, φ, are positive for solar wind flow deviated in the direction opposite planetary motion. The velocity averages given on this figure have not had the spacecraft velocity subtracted. The correction for this may be estimated using the spacecraft velocity components which are 0.4 and 23.7 km s^{-1} in the antisolar direction and the direction of planetary motion (but parallel to the ecliptic), respectively, at 1200

UT on December 10. These velocity components are 0.7 and 22.6 km s^{-1}, respectively, at 0000 UT on December 22. The times of seventeen bow shock and five magnetosheath crossings, listed in Table I, are shown on Figure 5 and indicated by arrows at the bottom of the figure.

Gaps in the plots of parameters on Figure 5 are sometimes due to ground-commanded changes into experiment modes for which plasma parameter calculations are not available for inclusion in this paper. In addition, at some times, in the magnetosheath, the medium resolution detector currents are reduced to the instrument noise levels, while the high resolution detector data have not been analyzed for this time period. During the times on this figure within the magnetosphere, proton fluxes are not detectable in the data with the standard techniques used for the other portions of the figure.

Inspection of Figure 5 shows large bulk velocity excursions for December 10 and early on December 11. This may be an artifact in the results caused by the response of the computer routine which calculates the plasma parameters, to an apparent non-Maxwellian plasma spectrum with a very broad proton maximum. At various times the computer program weights the higher velocity portion of this broad maximum more or less heavily, and thus calculates higher or lower proton bulk velocities.

The magnetopause crossings from 0943 to 0958 UT on December 12 have the characteristic of relatively gradual disappearance or reappearance of all observable plasma flux as the magnetosphere is entered or left behind.

Perhaps the most striking features in Figure 5 are the much less dramatic changes in velocity and density for the shock crossings further away from the planet as compared to the inbound shock crossing or closer crossings for the outbound leg. Note, however, that relatively large changes in the proton temperature are always observed regardless of shock location. As is the case at Earth, this probably indicates that Jupiter's bow shock becomes weaker for greater and greater angles and distances from the subsolar point. For this reason, the determinations of the shock locations reported here have relied more heavily on the temperature changes rather than on any other parameter, although the flow direction changes are usually very prominent in the high resolution detector data.

In addition, as was the case for the inbound magnetosheath traversals, the flow directions in the magnetosheath seen in Figure 5 are greatly deviated, in general agreement with those expected for plasma flow around a relatively blunt magnetosphere. A large southward component in the magnetosheath flow was observed for the inbound leg when the spacecraft was below Jupiter's orbital plane. Here a large northward component in the flow is seen where for the outbound magnetosheath traversal the spacecraft is above Jupiter's orbital plane.

5. Magnetospheric Plasma

Table II gives estimates of magnetospheric plasma properties using two different methods. The first method assumes pressure balance across the magnetopause, expressed by

$$n_1 k (T_{e1} + T_{i1}) + \frac{B_1^2}{2\mu_0} = n_2 k (T_{e2} + T_{i2}) + \frac{B_2^2}{2\mu_0}.$$

The second method uses the aerodynamic analogy (cf. Spreiter *et al.*, 1966) and is

TABLE II

The estimated Jupiter magnetospheric plasma properties assuming pressure
balance across the Jovian magnetopause

Time (UT)	Magnetosheath			Magnetosphere		
	Thermal pressure (dyn cm^{-2})	Magnetic energy density (erg cm^{-3})	Calculated thermal pressure (dyn cm^{-2})	Calculated beta	Calculated ion number density (cm^{-3})	Magnetic energy density (erg cm^{-3})
Dec 10–1153	12×10^{-11}	44×10^{-11}	12×10^{-11}	0.28	8.4	44×10^{-11}
Dec 13–0158	0.4×10^{-11} [a]	2.9×10^{-11}	0.4×10^{-11}	0.13 ~ 0.2 [b]	0.27 0.43 [b]	2.9×10^{-11}
Dec 14–1850	7.1×10^{-11}	4.1×10^{-11}	7.6×10^{-11}	2.1 ~ 2.8 [b]	5.5 7.3 [b]	3.6×10^{-11}

[a] Measured value not too reliable.
[b] Estimated using aerodynamic analogy.

described below. In the above pressure balance equation, the subscripts 1 and 2 refer to magnetosheath and magnetosphere parameters respectively, n is the plasma ion number density, T_e and T_i are the electron and ion temperatures, respectively, B is the magnetic field magnitude, k is Boltzmann's constant and μ_0, the magnetic permeability of free space, is $4\pi \times 10^{-7}$ H m^{-1}. $T_e \sim T_i$ is assumed, and $T_{i2} \sim 5 \times 10^4$ K (~ 4 eV electrons) is assumed based on the measured magnetospheric electron energy spectra (Intriligator and Wolfe, 1974).

The magnetosheath values for the December 13 crossing are less reliable than the others due to divergence of the plasma flow direction near the outer limit of the medium resolution detector angular acceptance. The December 10 crossing appeared to occur at a time of extreme conditions and in addition the observed magnetic field profile across the magnetopause (E. J. Smith, private communication) appears as if a wide layer was crossed, but fields outside this 'layer' are ignored when the results of the Table are calculated.

The results on Table II obtained with the aerodynamic analogy use Pioneer 10 free-stream plasma parameters obtained closest in time to the indicated magnetopause crossings. Because of the time delay between the magnetospheric and free-stream measurements by Pioneer 10 this method is less reliable due to neglect of possible time variation in the external free-stream conditions. The assumed condition is

$$Kmn_1^* v^{*2} \cos^2 \theta + n_1^* k (T_{e1}^* + T_{i1}^*) + \frac{B_1^{*2}}{2\mu_0} = n_2 k (T_{e2} + T_{i2}) + \frac{B_2^2}{2\mu_0},$$

where m is the proton mass, v the bulk velocity, the asterisks denote free-stream quantities, K is taken as unity and θ is the angle between the magnetopause normal and the free-stream plasma flow direction. θ is obtained from calculations made for the case at Earth given by Spreiter *et al.* (1966). The values of θ were tested against calculated values obtained using the magnetic field measured across the magneto-pause (E. J. Smith, private communication) and assuming the magnetopause is a tangential discontinuity. The Earth analogy values were much larger than the cal-culated values. This result implies a magnetopause body shape comparatively more blunt than the case of Earth.

6. Summary and Conclusions

Of particular interest is the understanding of the topology and dynamics of Jupiter's magnetosphere and its interaction with the solar wind. This, of course, is difficult to do in a single flyby. There are, however, several clues in the Pioneer 10 data which shed some light on this problem. It is clear that in many respects Jupiter's standing bow shock, magnetosheath flow field and magnetopause are similar to the case at Earth. Both the shock and magnetopause at Jupiter are observed to be well defined boundaries and like at Earth, Jupiter's bow shock is a strong shock (high Alfvén Mach number). Jupiter's magnetopause, also like Earth's, is a relatively sharp bound-ary between the planetary field and the magnetosheath flow field wherein it deflects the magnetosheath plasma and excludes it from direct entry into the magnetosphere. Finally, as is the case at Earth, the observed shock normals and magnetosheath plasma flow directions observed for Jupiter are consistent with Jupiter's magneto-sphere presenting a relatively blunt body to the solar wind for the sunward hemi-sphere.

It is cautioned, that Earth analogies may be confused due to the vastly different scale sizes involved. For example, the extent of Jupiter's bow shock, as inferred from the furthest out observed shock crossing, is almost a quarter of an AU wide as mea-sured across the dawn-dusk meridian. This is over 2 orders of magnitude larger than the Earth's bow shock measured in the same fashion. It is suspected that if Jupiter's magnetosphere were scaled down to the size of the Earth's magnetosphere corre-sponding to the geocentric distance to the subsolar point, Jupiter's magnetosphere would be considerably flattened in shape as compared to the Earth's. This is strongly suggested by the manner in which Jupiter's magnetic field lines in the outer mag-netosphere are greatly elongated and stretched out from the planet (Smith *et al.*, 1974a). The degree to which Jupiter's magnetosphere is flattened is impossible to estimate from the data of this single flyby, but the plasma observations can at least place lower limits.

First of all, it is exceedingly unlikely that Jupiter's outer magnetosphere rotates rigidly and wobbles up and down coincident with the rotational period of Jupiter's tilted magnetic dipole (15° tilt reported by Smith *et al.*, 1974a) as suggested by Van Allen *et al.* (1974). The inferred outer magnetosphere plasma densities are sufficiently high such that the measured magnetic field (Smith *et al.*, 1974a) would not be able to contain the plasma much beyond about 20 R_J. The complicating factor seems to be the very narrow latitude extent over which energetic charged particles seem to be confined in Jupiter's outer magnetosphere (Fillius and McIlwain, 1974; Simpson *et al.*, 1974; Trainor *et al.*, 1974; Van Allen *et al.*, 1974). A much more plausible model seems to be that suggested by Smith *et al.* (1974b), where a disturbance field associated with a current sheet is present in Jupiter's outer magnetosphere. This current sheet lies parallel to Jupiter's equatorial plane, contains the observed quasi-trapped energetic particle population, is the plane of symmetry for the flattened magnetosphere and moves up and down in latitude coincident with Jupiter's rotational period.

The question of the degree of flattening for Jupiter's outer magnetosphere still remains. During the inbound portion of the Pioneer 10 trajectory, the magnetopause was first observed at approximately 96 R_J and the spacecraft remained inside the magnetosphere for several days and, of course, many Jupiter rotational periods. Since the magnetosheath was *not* observed during this period, and since the Pioneer 10 spacecraft was 7 R_J below Jupiter's equatorial plane, then it follows that the thickness of Jupiter's magnetosphere must be at least four times this distance or 28 R_J. Likewise on the dawn side of Jupiter for the outbound pass, the spacecraft was within the magnetosphere for many Jupiter rotational periods prior to the last magnetopause crossing at approximately 150 R_J. In this region, Pioneer 10 was 24 R_J above Jupiter's equatorial plane, thus suggesting that here Jupiter's magnetosphere must be at least 96 R_J thick. These are probably conservative lower limits for the magnetospheric thickness, at least since fluctuations in the polar flow direction of the interplanetary solar wind would require the magnetosphere be thicker than the above values in order to avoid detection of the magnetosheath for time periods greater than one Jupiter rotation. Perhaps further detailed analysis and correlation of plasma and magnetic field data could be used to increase these lower limits of the magnetospheric thickness.

One further argument against a disk-like magnetosphere and in favor of a magnetosphere with reasonable thickness, is the second inbound magnetosheath observation made near 50 R_J. If Jupiter's magnetosphere were a disk with the subsolar point near 100 R_J, then it can be argued that a simple shift in the solar wind polar flow direction could deflect the 'magnetodisk' such that the Pioneer 10 spacecraft entered the magnetosheath. If this were true the spacecraft would find itself some 50 R_J downstream from the subsolar point and one would expect the magnetosheath plasma flow to be nearly solar radial. This in fact was not observed, but rather, the flow directions observed during the second magnetosheath traversal were incident from large angles with respect to the solar direction and quite similar to those observed during the first magnetosheath crossing. In addition, Jupiter's outer magnetosphere is apparently a high beta region with inferred thermal plasma densities on the order of a few cm^{-3}. This is supported by the outer magnetosphere magnetic field observations (Smith *et al.*, 1974a) where hourly averaged field strengths are only slowly increasing from about 5–6 γ at 100 R_J to slightly over 10 gamma at 30 R_J. This indicates that the entire outer portion of Jupiter's magnetosphere is highly inflated and therefore highly responsive to changes in the dynamic pressure of the solar wind.

A crude calculation shows that for the estimated internal pressure of Jupiter's outer magnetosphere, an increase in the solar wind dynamic pressure of only a factor of 3 is all that would be required to contract the magnetosphere from 100 R_J down to less than 50 R_J. At the time of the Pioneer 10 encounter, Pioneer 11 was 2.2 AU upstream from Jupiter and almost aligned along the same solar radial (0.8 deg angular difference in solar longitude between Pioneer 11 and Jupiter). Approximately seven days seventeen hours prior to the second inbound magnetosheath traversal by Pioneer 10, a solar wind dynamic pressure increase of approximately 4 was observed by Pioneer 11. The delay time expected for this dynamic pressure increase to reach Jupiter is in excellent agreement with the entry of Pioneer 10 into Jupiter's magnetosheath for the second time during the inbound pass. Inspection of the hourly averaged magnetic field values for this second magnetosheath traversal (Smith *et al.*, 1974a), shows a much higher field strength here as compared to the first traversal, indicating the magnetosheath field had been compressed. It is postulated, therefore, that Jupiter's magnetosphere contracted by at least a factor of 2 in response to an increase in the solar wind dynamic pressure, such that the Pioneer 10 spacecraft became imbedded in Jupiter's magnetosheath for the second time during the inbound leg.

The large number of magnetopause and shock crossings observed during the outbound pass further argues for the great responsiveness that Jupiter's outer magnetosphere must have to changing conditions in the solar wind. For this reason and the arguments in favor of a reasonably thick magnetosphere, the anomalously short distance observed across the first magnetosheath traversal is considered to be best accounted for by an outward expansion of Jupiter's magnetosphere at that time.

A fundamental question remaining concerns the point that if Jupiter's magnetosphere has a reasonable thickness, then why is the energetic particle population constrained to such a narrow disk in the outer magnetosphere? Could the current sheet suggested by Smith *et al.* (1974b) form a sort of magnetic bottle or is there perhaps local acceleration (Simpson *et al.*, 1974)? It is clear that deeper analysis will be required to shed further light on this question as well as observations on future Jupiter flybys (such as Pioneer 11) and orbiter missions.

References

Fillius, R. W. and McIlwain, C. E.: 1974, *Science* **183**, 314.
Hall, C. F.: 1974, *Science* **183**, 301.
Intriligator, D. S. and Wolfe, J. H.: 1974, *Geophys. Res. Letters* **1**, 281.

Simpson, J. A., Hamilton, D., Lentz, G., McKibben, R. B., Mogro-Campero, A., Perkins, M., Pyle, K. R., Tuzzolino, A. J., and O'Gallagher, J. J.: 1974, *Science* **183**, 306.

Smith, E. J., Davis, L., Jr., Jones, D. E., Colburn, D. S., Coleman, Jr., P. J., Dyal, P., and Sonett, C. P.: 1974a, *Science* **183**, 305.

Smith, E. J., Davis, L., Jr., Jones, D. E., Coleman, P. J., Jr., Colburn, D. S., Dyal, P., Sonett, C. P., and Frandsen, A. M. A.: 1974b, *J. Geophys. Res.* **79**, 3501.

Spreiter, J. R., Summers, A. L., and Alksne, A. Y.: 1966, *Planetary Space Sci.* **14**, 223.

Trainor, J. H., Teegarden, B. J., Stilwell, D. E., McDonald, F. B., Roelof, E. C., and Webber, W. R.: 1974, *Science* **183**, 311.

Van Allen, J. A., Baker, D. N., Randall, B. A., Thomsen, M. F., Sentman, D. D., and Flindt, H. R.: 1974, *Science* **183**, 309.

Wolfe, J. H., Collard, H. R., Mihalov, J. D., and Intriligator, D. S.: 1974, *Science* **183**, 303.

PIONEER 10 OBSERVATIONS OF THE SOLAR WIND
AND ITS INTERACTION WITH JUPITER:
PLASMA ELECTRON RESULTS

DEVRIE S. INTRILIGATOR

Physics Dept., University of Southern California, Los Angeles, Calif. 90007, U.S.A.

1. Introduction

Previously we reported (Wolfe *et al.*, 1974a) that during the inbound portion of the Jovian flyby (see Hall, 1974 for a description of the Pioneer 10 flyby geometry) the Jovian magnetosheath was encountered for two extended time intervals: from 2031 UT (Ground Received Time) on November 26, 1973 to 2038 UT on November 27, 1973, corresponding in radial distance from Jupiter of 109 R_J and 96 R_J, respectively; and then from 0318 UT on December 1, 1973 to 1415 UT on December 1, 1973, corresponding in radial distance from Jupiter of 54 R_J and 46.5 R_J. In Wolfe *et al.* (1974b) we summarize the general features of the solar wind interaction with Jupiter. In Intriligator (1975) we summarize the plasma electron observations obtained in the Jovian magnetosphere. In this paper we summarize the plasma electron data obtained in the Jovian magnetosheath.

The inbound electron data from the two extended magnetosheath traversals are of interest since they are in general agreement with the description of the Jovian magnetosheath obtained from the ion data (Wolfe *et al.*, 1974a, b). As in the case of the Earth's bow shock there is heating of the solar wind electrons across the Jovian bow shock. The even higher electron temperatures observed during the second magnetosheath traversal are consistent with the general contraction of the Jovian magnetosphere due to the increase in the external solar wind dynamic pressure.

2. Magnetosheath Observations

The Ames Research Center Plasma Analyzer experiment (Wolfe *et al.*, 1974a, b) on Pioneer 10 measures plasma electrons in the energy range from 2 eV to 500 eV in fifteen electron energy steps. These measurements are obtained with Detector B (Wolfe *et al.*, 1974b). This quadrispherical electrostatic analyzer system has five plate collectors, each having an associated electrometer amplifier. The entrance aperture of the Plasma Analyzer experiment is located on the Pioneer 10 spacecraft such that it views back toward the Earth (and Sun) through a wide slit in the back of the spacecraft high gain antenna reflector. The entrance aperture normals are oriented parallel to the spacecraft spin axis, thereby enabling the detector to make a complete angular scan of the earthward hemisphere every half revolution of the spacecraft. The edges of the antenna reflector limit the field of view of the instrument to $\pm 73°$ with respect to the spacecraft spin axis.

V. Formisano (ed.), The Magnetospheres of the Earth and Jupiter, 297–300. All Rights Reserved

Figure 1 shows an example of a plasma electron spectrum obtained in the Jovian magnetosheath. Upstream of the Jovian bow shock electrons were rarely observed above the instrument noise level. Although the plasma flow conditions in the Jovian magnetosheath were quite variable, the magnetosheath spectrum shown in Figure 1 is considered to be typical. Note that in this and subsequent spectra the electron

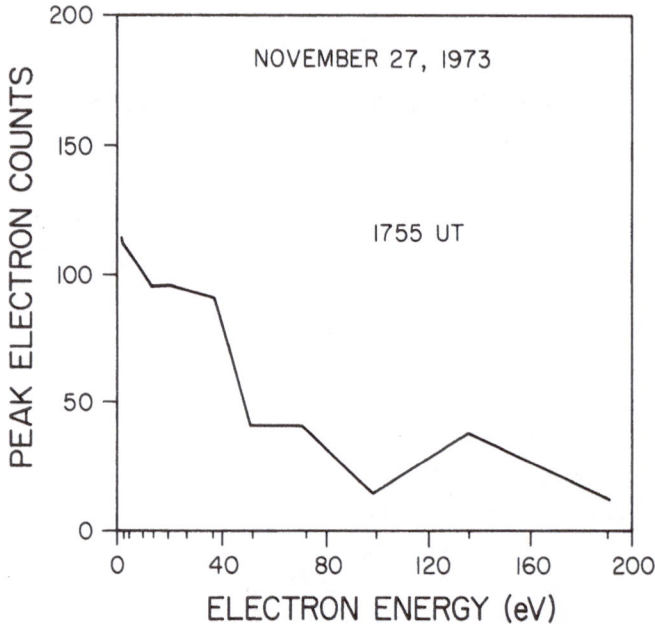

Fig. 1.   Electron spectrum taken in the Jovian magnetosheath during the first extended traversal by Pioneer 10. The vertical axis indicates the electron counts, the digitized output from the Plasma Analyzer per energy channel for the collector recording the peak. Electron fluxes are digitized to nine bit accuracy (0–512) covering the dynamic range from approximately 10^{-14} to 10^{-9} A cm^{-2}. The horizontal axis indicates the energy of the electrons in electron volts. The small vertical lines on the horizontal axis indicate the locations of the individual electron energy channels.

counts are the digitized output from the Plasma Analyzer per energy channel for the collector recording the peak. For the electron measurements the electron fluxes are digitized to nine bit accuracy (0–512) covering the dynamic range from approximately 10^{-14} to 10^{-9} A cm^{-2}. The important feature in the magnetosheath spectrum is the presence of the enhanced high energy tail of the spectrum between ~ 50 eV and 200 eV. The existence of this high energy tail in the magnetosheath electron spectrum indicates that there is heating of the solar wind electrons across the Jovian bow shock. This heating is similar to that observed for the ions across the Jovian bow shock (Wolfe *et al.*, 1974b) and, of course, has been observed for the case of the Earth's bow shock. As Pioneer 10 proceeded inbound the plasma ion data indicated that the spacecraft crossed the Jovian magnetopause at 96 R_J (Wolfe *et al.*, 1974a, b) and entered the outer Jovian magnetosphere. The plasma electron observations in the outer magnetosphere are summarized in Intriligator (1975).

As previously reported (Wolfe *et al.*, 1974a), the Ames Research Center Plasma Analyzer ion observations indicated that on December 1, 1973 the Pioneer 10 spacecraft reentered the Jovian magnetosheath at a radial distance of 54 R_J. The magnetosheath ion flow field persisted for approximately eleven hours and was abruptly terminated at a radial distance of 46.5 R_J. The December 1 magnetosheath plasma ion data and the magnetic field data (Smith *et al.*, 1974) are consistent with the second magnetosheath traversal being associated with a general contraction of the Jovian magnetosphere due to an increase in the external solar wind dynamic pressure (Wolfe *et al.*, 1974b). Figure 2 shows a typical plasma electron spectrum obtained during

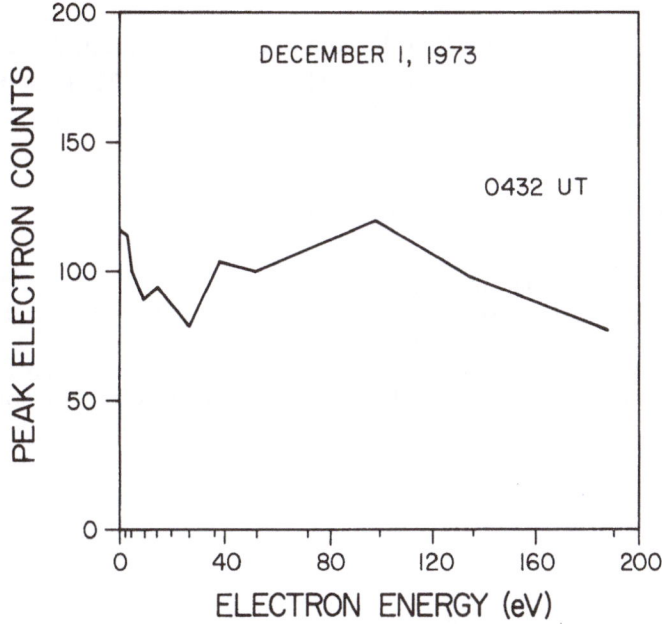

Fig. 2.   Electron spectrum taken during the second extended traversal of the Jovian magnetosheath.

this second inbound magnetosheath traversal. The greatly enhanced high energy tail in this spectrum indicates that the electron temperatures were significantly higher during the magnetosheath traversal on December 1, 1973 than during the earlier traversal. This increase in electron temperature during the December 1, 1973 magnetosheath traversal is consistent with what one would expect for a contraction of the Jovian magnetosphere due to an increase in the solar wind dynamic pressure as has been observed for the case of the Earth's bow shock.

3. Summary

The electrons observed in the Jovian magnetosheath are completely consistent with the ion observations (Wolfe *et al.*, 1974b) of a high Alfvén Mach number bow shock and magnetosheath flow field. These observations are strikingly similar to observations of the solar wind interaction with the Earth and only differ in terms of the scale size of the interaction.

Acknowledgements

This study was carried out at the University of Southern California and was supported by NASA Contract NAS2-7969 with the Ames Research Center and also by the University of Southern California.

The author thanks Dr John H. Wolfe, principal investigator of the Ames Research Center plasma spectrometer on Pioneer 10, for access to the electron data used in this paper.

S. E. Lambert (USC) did much of the computer programming. A number of other USC students N. Cheung, J. Cutter, W. Ho, R. Jourdan, M. Lim, W. D. Miller, W. Montier, and S. Winchester also assisted with the data handling.

References

Hall, Charles, F.: 1974, *Science* **183**, 301.
Intriligator, Devrie, S.: 1975, this volume, p. 313.
Smith, E. J., Davis, L., Jr., Jones, D. E., Colburn, D. S., Coleman, P. J., Jr., Dyal, P., Sonett, C. P., and Fransden, A. M. A.: 1974, *J. Geophys. Res.* **79**, 3501.
Wolfe, John H., Collard, H. R., Mihalov, J. D., and Intriligator, D. S.: 1974a, *Science* **183**, 303.
Wolfe, J. H., Mihalov, J. D., Collard, H. R., McKibben, D. D. Frank, L. A., and Intriligator, D. S.: 1974b, *J. Geophys. Res.* **79**, 3489.

OBSERVATIONS IN INTERPLANETARY SPACE OF RELATIVISTIC ELECTRONS FROM JUPITER

D. L. CHENETTE*, T. F. CONLON, and J. A. SIMPSON*

Enrico Fermi Institute, The University of Chicago, Chicago, Ill. 60637, U.S.A.

The existence of energetic electron bursts in the interplanetary medium outside the bow shock of Earth (Fan *et al.*, 1964; Frank and Van Allen, 1964; Anderson, 1968) led us to search for similar interplanetary electron bursts of Jovian origin as Pioneer 10 approached Jupiter. In this paper, we summarize evidence for the presence of Jovian electrons in interplanetary space as much as 1 AU inside the orbit of Jupiter. The observations and analysis have been discussed in much greater detail by Chenette *et al.* (1974).

In Figure 1, we present a comparison between counting rates responding primarily to 6–30 MeV electrons from identical instruments on Pioneer 10 and Pioneer 11. During the time interval presented, Pioneer 11 was approximately 1–3 AU from

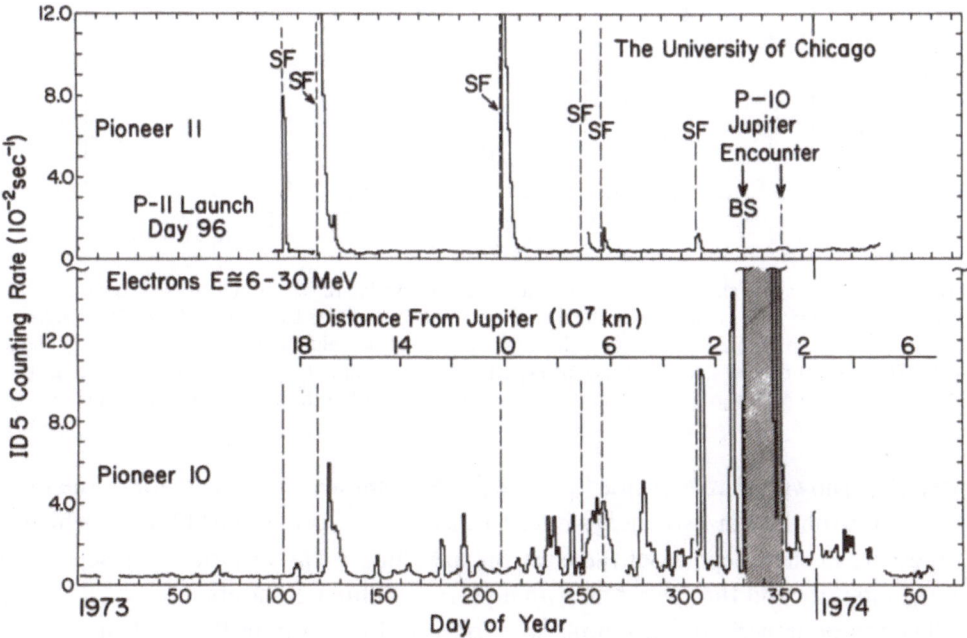

Fig. 1. Comparison of the Pioneer 10 and Pioneer 11 ID5 counting rates which respond primarily to 6–30 MeV electrons during the period shown. The dashed vertical lines labeled SF indicate the times of onset of solar flare particle events. The dashed line labeled BS indicates the time at which Pioneer 10 crossed the bow shock of Jupiter.

* And the College, University of Chicago.

V. Formisano (ed.), The Magnetospheres of the Earth and Jupiter, 301–306. All Rights Reserved
Copyright © 1975 by D. Reidel Publishing Company, Dordrecht-Holland

the Sun and lay approximately on the Sun-Jupiter line, whereas Pioneer 10 was approaching Jupiter. It is clear from the figure that the electron flux at Pioneer 10 showed many more increases than that at Pioneer 11. The electron increases at Pioneer 10 could be correlated with known solar activity only in a few cases, which are marked on Figure 1, whereas essentially all increases in the electron flux observed at Pioneer 11 were correlated with solar activity.

In Figure 2, the behavior of the electron flux during one of the increases at Pio-

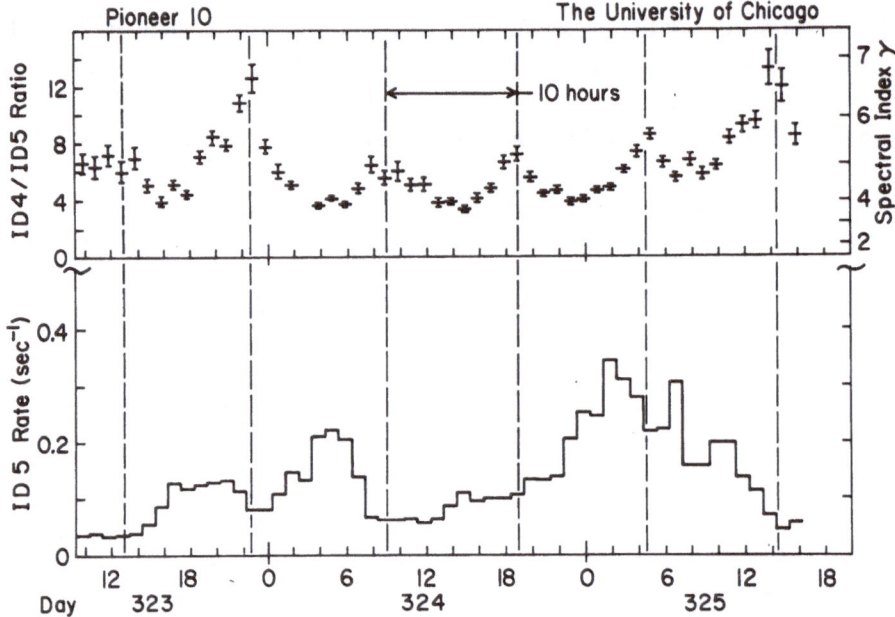

Fig. 2.   The behavior of the electron intensity and spectrum during the electron increase of days 244–245, 1973. The ID4/ID5 ratio is the ratio of the flux of ~3–6 MeV electrons to that of ~6–30 MeV electrons. If a spectrum of the form $E^{-\gamma}$ is assumed, the ratio corresponds to values of γ as shown on the right hand scale. The vertical dashed lines are times of expected flux minima and spectral maxima resulting from extrapolation backward in time from variations observed within the magnetosphere of Jupiter.

neer 10 is shown on an expanded time scale. Also shown is the behavior of the ratio of the electron flux in two energy intervals, from 3–6 MeV and 6–30 MeV, which is a measure of the steepness of the electron spectrum. Clear variations are seen both in the intensity and the spectrum, with a period of about 10 hours.

The ten hour periodic variations were observed in many of the electron flux increases. In Figure 3 we show the results of a power spectrum analysis of the intensity variations of 6–30 MeV electrons for 4 periods at different distances from Jupiter. A peak at Jupiter's rotation frequency is clearly visible even for the period from day 217 to day 246, 1973, when Pioneer 10 was $8-10 \times 10^7$ km from Jupiter.

The ten hour variations observed in the electron spectrum in interplanetary space joined smoothly onto similar variations observed inside the magnetosphere of Jupiter, which have been reported by Simpson *et al.* (1974). In Figure 4, the electron spectral

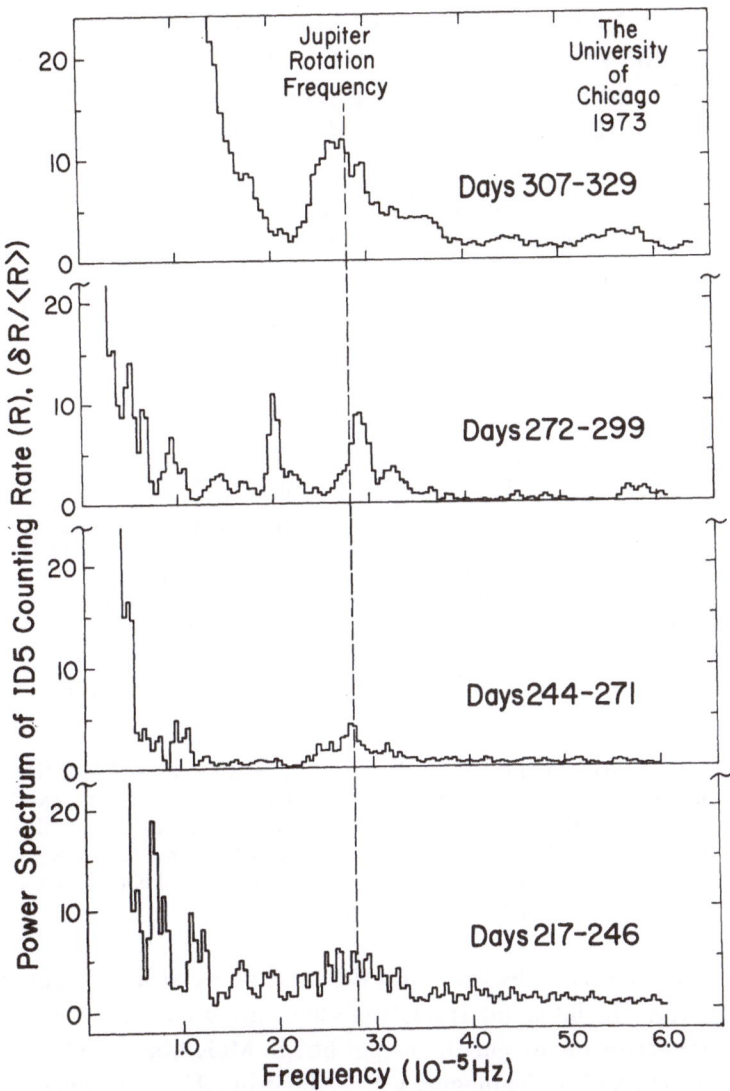

Fig. 3. Power spectrum analysis of the time structure of the ID5 counting rate using the method of Blackman and Tukey (1958). The indicated frequency of the rotation of Jupiter is based on the magnetospheric rotation period, $9^h55^m30^s$ (Carr, 1971).

variations observed in the week preceding the entry into Jupiter's magnetosphere are shown together with those observed during the first two days Pioneer 10 spent within the magnetosphere. The variations continued across the bow shock and magnetopause with no change of phase. This observation seems to cast doubt on the prevailing interpretation of the variations observed inside the magnetosphere as the result of confinement of electrons to the magnetic equatorial plane which alternately approached and receded from Pioneer 10 as a result of the rotation with Jupiter of Jupiter's inclined magnetic dipole (Simpson *et al.*, 1974; McKibben and Simpson, 1974; Trainor *et al.*, 1974; Van Allen *et al.*, 1974; Fillius and McIlwain, 1974). The observation of phase continuity across the magnetopause seems to require either that

Pioneer 10 entered the magnetosphere precisely in the region from which electrons escaped, or that the variation over a large portion of the magnetopause had the same phase so that the variation observed just inside the magnetopause by Pioneer 10 did not depend on the position of Pioneer 10 with respect to the magnetic equator,

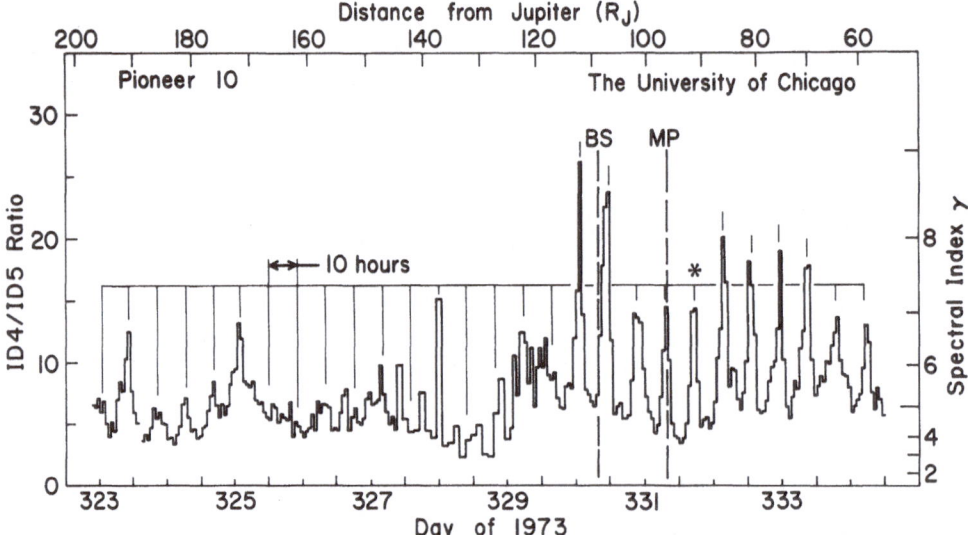

Fig. 4. The ID4/ID5 ratio (spectral index) for electrons from $\sim 200\ R_J$ to $\sim 50\ R_J$ during the inbound pass. From day 326 through day 329 the ID5 counting rate shows no evidence of increased activity due to Jovian electrons and the variations in the spectral index are primarily due to statistical fluctuations. The asterisk near hour 5 of day 332 indicates the time from which the '10-h' tic marks have been calculated for this figure and for Figure 2. 'BS' and 'MP' indicate the times of bow shock and magnetopause crossings, respectively.

but rather could have been observed with the same phase at almost any point near the magnetopause. The latter interpretation seems more natural for explaining the phase coherence across the magnetopause, although McKibben and Simpson (1974) have found evidence within the magnetosphere favoring the confinement of the electron flux to a highly distorted magnetic equatorial plane even near the magnetopause. More study is clearly required to understand the relation between the 10-h variations observed inside the magnetosphere and in interplanetary space.

The coherence between the phase of variations observed in interplanetary space and those observed within Jupiter's magnetosphere extends to great distances from the planet. The dashed lines in Figure 2 indicate the times expected for maximum ID4/ID5 spectral ratio and minimum flux from simply subtracting successively $9^h55^m30^s$ from the times of intensity minima observed just inside the magnetopause on day 332. Even $\sim 8 \times 10^7$ km from Jupiter, the agreement between the observed and extrapolated phase is remarkably good. This observation implies that the electrons must have propagated from Jupiter to the point of observation in much less than 10 hours.

Further evidence for rapid propagation of the particles comes from the study of

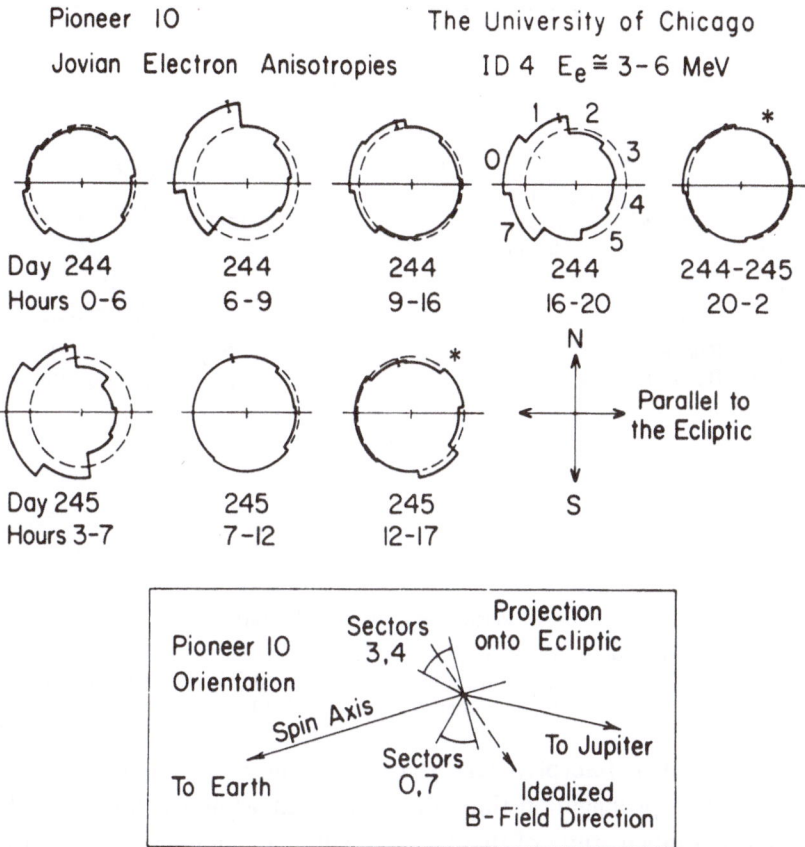

Fig. 5.   The time dependence of the anisotropy of 3–6 MeV electrons (ID4) during the event of days 244–245. The inset shows a schematic representation of the orientation of Pioneer 10 relative to the Earth, Jupiter, and the average interplanetary magnetic field direction (the arrow pointing away from the Sun), and the acceptance cone of our instrument parallel to the ecliptic plane (as viewed from the north). Asterisks indicate times when the amplitude of the anisotropy may be distorted due to saturation effects.

the angular distribution of directions of arrival of the electrons in the event of days 244–245, shown in Figure 5. The anisotropy showed variations with an approximately 10-h period. Comparison with Figure 2 shows that during the rising phase of the 10-h intensity variations, the anisotropy was largest, and in a direction consistent with the interpretation of electrons travelling toward the Sun (or away from Jupiter) along interplanetary spiral field lines.

The relative positions of Pioneer 10 and Jupiter with respect to idealized spiral interplanetary magnetic field lines are shown in Figure 6 for day 245, 1973. It is clear that, in order for the particles escaping from Jupiter to be observed at Pioneer 10 they must have propagated approximately 50 million kilometers perpendicular to the average magnetic field. For reasonable values of interplanetary diffusion coefficients and the derivation as given by Chenette et al. (1974), propagation to such distances across the magnetic field would take ~10 hours. (Unfortunately, due to a numerical error, this value was erroneously reported in Chenette et al. as about 14 days.) The observed variations in the particle anisotropy and the coherency of the phase of

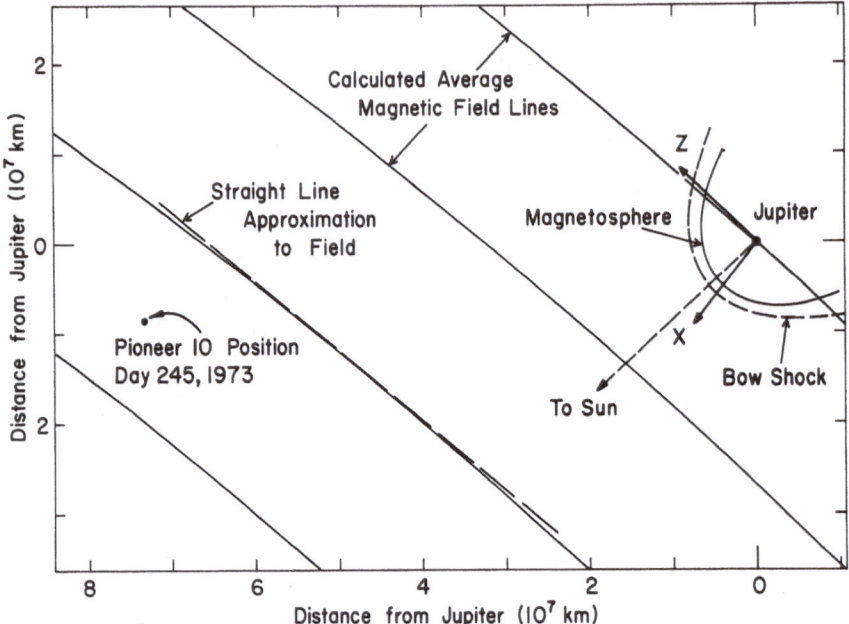

Fig. 6. The position of Pioneer 10 on day 245, 1973 shown relative to Jupiter and the average interplanetary magnetic field. (The size and positions of the bow shock and magnetosphere are drawn to scale after Wolfe *et al.*, 1974.)

variations observed in interplanetary space with those observed inside the magnetosphere imply a propagation time much less than 10 hours, however. Thus it is possible that continued study of these events will yield significant new information concerning the physics of energetic particle propagation in interplanetary space.

Acknowledgements

This research was supported in part by NASA contracts NAS 2-5601 and NAS 2-6551 with the Ames Research Center, NASA Grant NGL 14-001-006 and NSF Grant GA-41692X.

References

Anderson, K. A.: 1968, *J. Geophys. Res.* **73**, 2387.
Blackman, R. B. and Tukey, J. W.: 1958, *The Measurement of Power Spectra*, Dover, New York.
Carr, Thomas, D.: 1971, *Astrophys. Letters* **7**, 157.
Chenette, D. L., Conlon, T. F., and Simpson, J. A.: 1974, *J. Geophys. Res.* **79**, 3551.
Fan, C. Y., Gloeckler, G., and Simpson, J. A.: 1964, *Phys. Rev. Letters* **13**, 149.
Fillius, R. W. and McIlwain, C. E.: 1974, *Science* **183**, 314.
Frank, L. A. and Van Allen, J. A.: 1964, *J. Geophys. Res.* **69**, 4923.
McKibben, R. B. and Simpson, J. A.: 1974. *J. Geophys. Res.* **79**, 3545.
Simpson, J. A., Hamilton, D., Lentz, G., McKibben, R. B., Mogro-Campero, A., Perkins, M., Pyle, K. R., Tuzzolino, A. J., and O'Gallagher, J. J.: 1974, *Science* **183**, 306.
Trainor, J. H., Teegarden, B. J., Stilwell, D. E., McDonald, F. B., Roelof, E. C., and Webber, W. R.: 1974, *Science* **183**, 311.
Van Allen, J. A., Baker, D. N., Randall, B. A., Thomsen, M. F., Sentman, D. D., and Flint, H. R.: 1974, *Science* **183**, 308.
Wolfe, J. H., Collard, H. R., Mihalov, J. D., and Intriligator, D. S.: 1974, *Science* **183**, 303.

ON THE DISTORTION OF THE JOVIAN MAGNETIC FIELD
$R \gtrsim 40\,R_J$ AS DEDUCED FROM CHARGED PARTICLE STUDIES

R. B. McKIBBEN and J. A. SIMPSON*

Enrico Fermi Institute, The University of Chicago, Chicago, Ill. 60637, U.S.A.

In this paper we consider the relationship between the rotation of Jupiter's magnetic field and time variations in the intensity of $\sim$6–30 MeV electrons observed by the University of Chicago experiment on Pioneer 10 in the outer regions of Jupiter's magnetosphere ($20 \gtrsim R \gtrsim 100\ R_J$). This paper summarizes work reported elsewhere in more detail by McKibben and Simpson (1974).

In the outer region of Jupiter's magnetosphere, the 6–30 MeV electron flux showed regular intensity variations with a period of approximately 10 hours, as shown in Figure 1. These variations have been interpreted Fillius and McIlwain (1974),

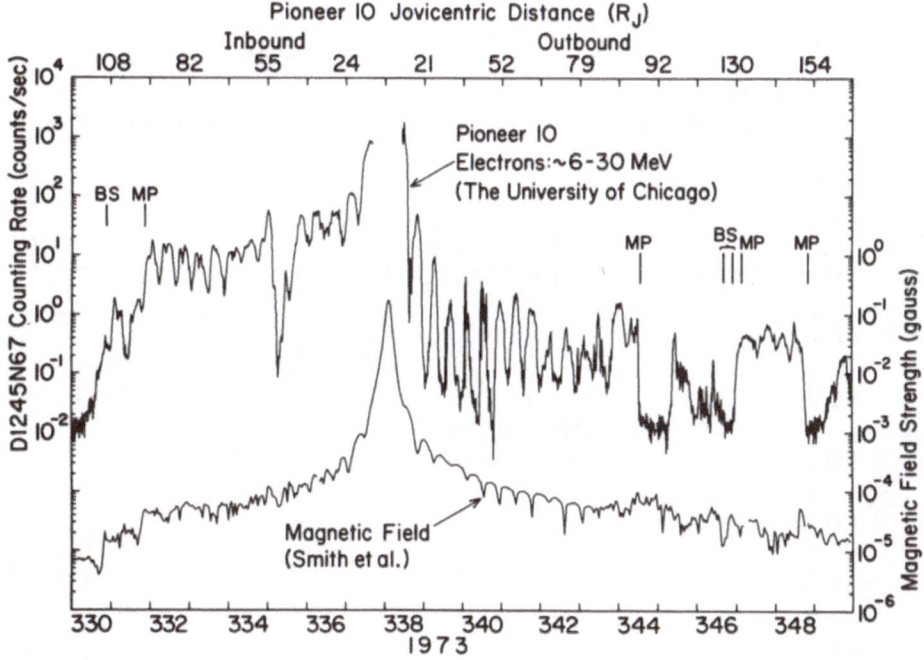

Fig. 1.   The counting rate of $\sim$6–30 MeV electrons measured by the University of Chicago main-telescope, and the magnetic field strength as measured by the helium vector magnetometer on Pioneer 10 (E. J. Smith, private communication). The times of bow shock and magnetopause crossings are indicated as BS and MP respectively. Near periapsis on day 338, the main telescope logic was saturated by the high event rate, and no useful data were returned. The large decrease in counting rate on day 335 is a possible re-entry into the magnetosheath.

* And Department of Physics.

V. Formisano (ed.), The Magnetospheres of the Earth and Jupiter, 307–311. *All Rights Reserved*
Copyright © 1975 by D. Reidel Publishing Company, Dordrecht-Holland

Simpson *et al.* (1974a), Trainor *et al.* (1974), and Van Allen *et al.* (1974) as evidence that the energetic electron flux is confined primarily to the magnetic equatorial plane. Because of the nature of the Pioneer 10 trajectory (Mead, 1974) and the inclination of Jupiter's magnetic dipole (Smith *et al.*, 1974), Pioneer 10 alternately approached the magnetic equator and receded from it to magnetic latitudes of $\sim 20°$ every 10 hours as a result of Jupiter's rotation. If the electron flux were confined primarily to the magnetic equator the maximum flux would be expected when Pioneer 10 was nearest the magnetic equator, and the minimum flux when Pioneer 10 was furthest away from the magnetic equator.

In Figure 2, we show the difference between the observed position of the flux minimum in system III longitude and its expected position, based on confinement of particles to the magnetic equatorial plane for rigid corotation of the magnetic field with Jupiter. Only for $R \gtrsim 40 \ R_J$ is the agreement between expected and observed positions good, so that only for $R \gtrsim 40 \ R_J$ is rigid corotation of the magnetic

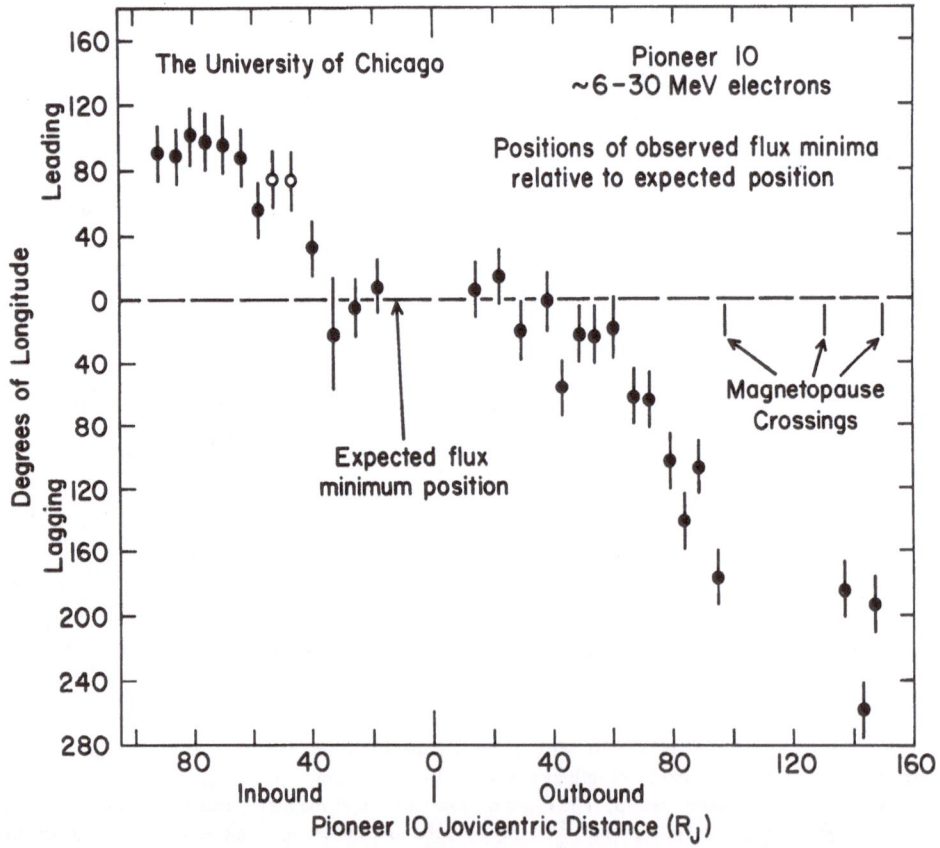

Fig. 2. Positions in longitude of observed flux minima relative to the expected positions during the periods when Pioneer 10 was within the magnetosphere of Jupiter. The open circles correspond to the period of possible re-entry into the magnetosheath, and the times assigned to these minima correspond to the times of maximum spectral index for the electrons, which have been found to be correlated with flux minima (McKibben and Simpson, 1974; Van Allen *et al.*, 1974).

field a good assumption. The large difference observed for $R \gtrsim 40\ R_J$ could be the result of deformations of the magnetosphere due to its interaction with the solar wind or to lack of corotation of the magnetospheric plasma at large radial distances.

An alternative possibility is that the intensity variations are not the result of close confinement of particles to the magnetic equatorial plane, but are a time dependent phenomenon, with the intensity varying with the same phase at all points in the outer magnetosphere. Time dependence is suggested by the following observations. First, from $R \cong 60\text{--}90\ R_J$ inbound, the difference between the observed and expected minimum positions was independent of R. Second, based upon an assumed 9^h55^m-period for the intensity variations independent of spacecraft position the observed minima outbound at $R \cong 90\ R_J$ were approximately in phase with those observed inbound at $R \cong 90\ R_J$, despite the fact that the spacecraft was in different regions of the magnetosphere at $R \cong 90\ R_J$ inbound and outbound. Third, observations of Jovian electron intensity variations in interplanetary space reported by Chenette *et al.* (1974) are most easily interpreted if time dependence rather than spatial dependence for the magnetospheric electron intensity near the magnetopause of Jupiter is assumed.

In Figure 3, we show the difference between observed and expected times for minima assuming a rigid 9^h55^m-period for the variations, referred to the minimum observed at 1530 on day 332. If the intensity were a function only of R and time, then the time difference between expected and observed minima should be the same at

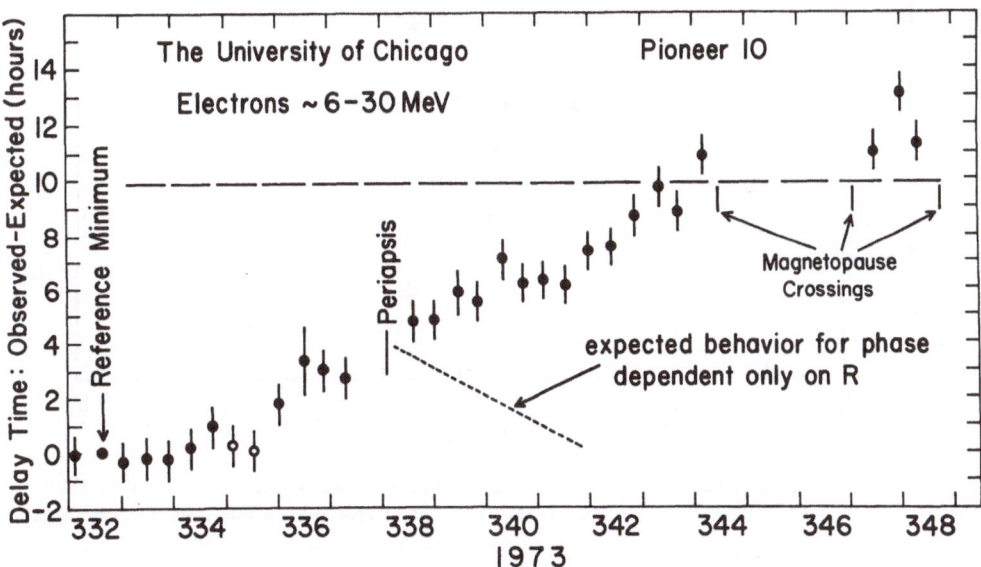

Fig. 3. Times of observed flux minima relative to the expected time for an assumed 9^h55^m periodicity, independent of spacecraft position. The expected times for successive minima are derived by adding successively 9^h55^m to the observed time of the reference minimum (~ 1530 day 332). The heavy dashed line indicates a delay of one full period. The dotted line is the behavior expected for a flux intensity dependent only on radial distance from Jupiter and time, and independent of magnetic latitude or system III longitude. Times of the minima indicated by the open circles were determined in the same manner as for the corresponding points in Figure 2.

the same R inbound and outbound, leading to behavior like that suggested by the dotted line in Figure 3. Instead the time delay continued to increase on the outbound pass, and reached 10 hours on day 343 at $R \cong 90~R_J$. Thus, the apparent 'in phase' behavior of electron intensity variations at $R \cong 90~R_J$ inbound and outbound is in reality the result of a phase difference of 2π rad between the intensity variations at the two points.

Further evidence against a purely time dependent model for the intensity variations is provided by the association between maxima in the electron intensity and minima in the magnetic field strength. This association is clearly evident on the outbound pass, as shown in Figure 1. On the inbound pass at large R the field was disordered except for long time scale averages. Nevertheless, even in this region the counting rate of energetic electrons was anti-correlated with average magnetic field strength, as shown in Figure 4. Such an anticorrelation between particle intensity

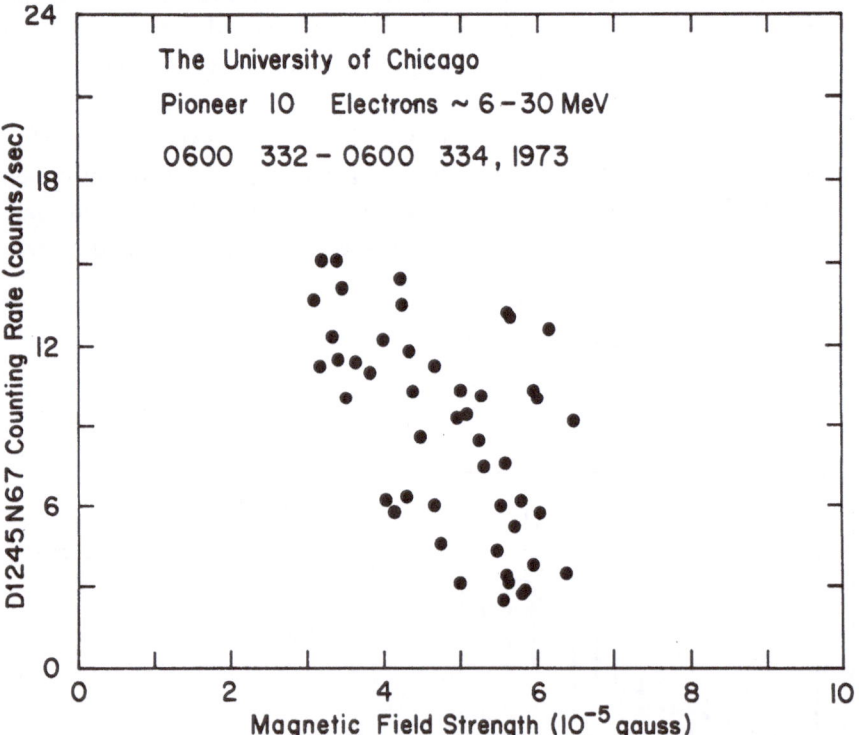

Fig. 4.   The correlation between magnetic field strength and the counting rate of ~6–30 MeV electrons for the radial range 65–92 R_J on the inbound pass. Each point represents a one hour average of data.

and magnetic field strength is consistent with confinement of particles to an extended, if distorted, magnetic equatorial plane.

Thus, considering only observations made within the magnetosphere, we find the behavior of the energetic (6–30 MeV) electron flux to be most consistent with confinement of the particles to a magnetic equatorial plane which is considerably distorted from the shape expected for rigid corotation of the magnetic field with Jupiter. The distortion most probably arises from stresses resulting from the interaction of the magnetosphere with the solar wind and from inertial effects of the magnetospheric plasma. Considerably more study is required in order to determine the precise nature of the distortion and to understand the relation between the intensity variations observed for electrons in interplanetary space and within the magnetosphere of Jupiter.

Acknowledgements

This research was supported in part by NASA contracts NAS 2-5601 and NAS 2-6551 with the Ames Research Center, NASA Grant NGL 14-001-006 and NSF Grant GA 41692X.

References

Chenette, D. L., Conlon, T. F., and Simpson, J. A.: 1974, *J. Geophys. Res.* **79**, 3551.

Fillius, R. W. and McIlwain, C. E.: 1974, *Science* **183**, 314.

McKibben, R. B. and Simpson, J. A.: 1974, *J. Geophys. Res.* **79**, 3545.

Mead, G. D.: 1974, *J. Geophys. Res.* **79**, 3514.

Simpson, J. A., Hamilton, D., Lentz, G., McKibben, R. B., Mogro-Campero, A., O'Gallagher, J. J., Perkins, M., Pyle, K. R., and Tuzzolino, A. J.: 1974, *Science* **183**, 306.

Smith, E. J., Davis, L., Jr., Jones, D. E., Colburn, D. S., Coleman, P. J., Jr., Dyal, P., Sonett, C. P., and Frandsen, A. M. A.: 1974, *J. Geophys. Res.* **79**, 3501.

Trainor, J. H., Teegarden, B. J., Stilwell, D. E., McDonald, F. B., Roelof, E. C., and Webber, W. R.: 1974, *Science* **183**, 311.

Van Allen, J. A., Baker, D. N., Randall, B. A., Thomsen, M. F., Sentman, D. D., and Flindt, H. F.: 1974, *Science* **183**, 309.

PIONEER 10 OBSERVATIONS OF THE JOVIAN MAGNETOSPHERE: PLASMA ELECTRON RESULTS

DEVRIE S. INTRILIGATOR

Physics Dept., University of Southern California, Los Angeles, Calif. 90007, U.S.A.

1. Introduction

In Intriligator (1975) there is a discussion of some of the plasma electron observations obtained by the Ames Research Center Plasma Analyzer experiment on Pioneer 10. The plasma electron observations discussed in Intriligator (1975) are those obtained during the two extended magnetosheath traversals during the inbound passage of the spacecraft. In Intriligator (1975), Intriligator and Wolfe (1974), and Wolfe *et al.* (1974b) there is a brief description of the Ames Research Center Plasma Analyzer experiment. In this paper the plasma electron observations obtained in the Jovian magnetosphere during the inbound portion of the Pioneer 10 flyby trajectory will be discussed. These observations imply the existence of a thermal plasma in the outer Jovian magnetosphere. The thermal plasma is particularly important in the outer Jovian magnetosphere where it is evidently responsible for the large scale inflation of Jupiter's outer magnetic field as indicated by the large deviation from a dipole-like character as observed by Smith *et al.* (1974). Based on these electron measurements and the magnetic field measurements (Smith *et al.*, 1974), we inferred (Wolfe *et al.*, 1974a) the existence of a high beta ($\beta \sim 1$) plasma in the outer magnetosphere by assuming thermal equilibrium between ions and electrons and a pressure balance across the magnetopause.

2. Magnetospheric Observations

At 2038 UT on November 27, 1973 Pioneer 10 crossed the Jovian magnetopause at a radial distance of 96 R_J and entered the outer Jovian magnetosphere (Intriligator and Wolfe, 1974). In the outer Jovian magnetosphere the electron component of a thermal plasma was measured at an energy of a few eV. Figure 1 shows an example of an electron spectrum measured in the outer magnetosphere. This spectrum was obtained at ~ 96 R_J on November 27, 1973 after the first extended magnetosheath traversal. The peak evident here near 4 eV was consistently observed throughout the entire outer Jovian magnetosphere. We cannot conclude the existence of a thermal plasma in the inner magnetosphere where background due to energetic charged particles preclude its observation.

If one assumes that the magnetopause boundary is a tangential discontinuity then the pressure balance across it can be calculated. For the inbound case the thermal component of the magnetospheric electrons was consistently observed near 4 eV

V. Formisano (ed.), The Magnetospheres of the Earth and Jupiter, 313–316. All Rights Reserved

corresponding to a temperature of $\sim 5 \times 10^4$ K. Assuming $T_e - T_i$, the magnetic field values reported by Smith *et al.* (1974) and the magnetosheath ion parameters (where they could be determined) reported by Wolfe *et al.* (1974b) imply that the dayside magnetosphere has a plasma beta near unity corresponding to a number density of a few particles cm^{-3}. It is cautioned that the above value of beta and the correspond-

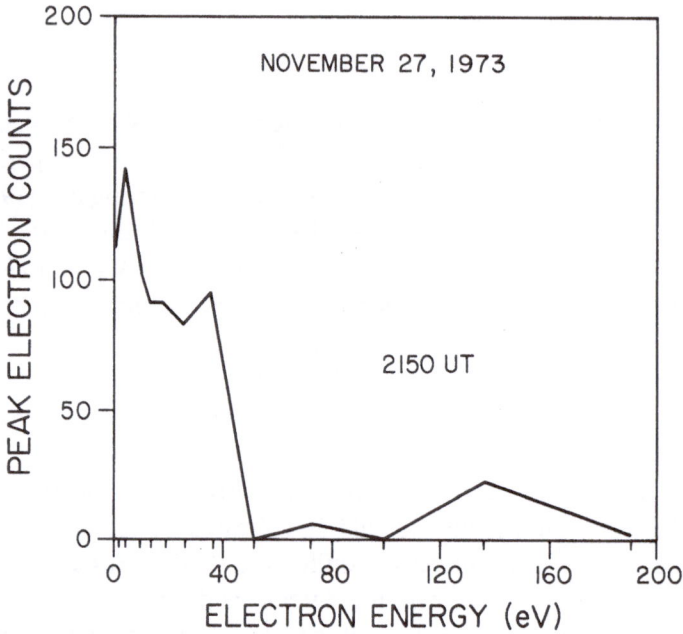

Fig. 1.   Pioneer 10 electron spectrum taken in the outer Jovian magnetosphere. The vertical axis indicates the electron counts, the digitized output from the Plasma Analyzer per energy channel for the collector recording the peak. Electron fluxes are digitized to nine bit accuracy (0–512) covering the dynamic range from approximately 10^{-14} to 10^{-9} A cm^{-2}. The horizontal axis indicates the energy of the electrons in electron volts. The small vertical lines on the horizontal axis indicate the locations of the individual electron energy channels.

ing number density is considered to be an upper limit since the possible magnetospheric pressure contributions from the observed nonthermal plasma electrons and unobservable energetic electrons between 500 eV and ~ 50 keV have not been accounted for.

3. Summary

It is important to discuss the validity of the magnetospheric electron observations obtained during the Pioneer 10 Jupiter flyby. It is recognized that the measurement of low energy charged particles (of a few eV) is exceedingly difficult and subject to error primarily due to spacecraft potential effects. This is particularly true for the case of Jupiter's magnetosphere where the spacecraft is subjected to high intensities of energetic charged particles (Fillius and McIlwain, 1974; Simpson *et al.*, 1974; Trainor *et al.*, 1974; Van Allen *et al.*, 1974). In general, as is the case at Earth, spacecraft charge build-up produces an unknown perturbing effect on these low energy electron measurements; however, at least in the outer portion of the Jovian magnetosphere the consistency of the 4 eV electron peak argues in favor of the dominance of the spacecraft potential by the ambient thermal plasma and the photoelectrons. Note that at 5 AU the flux of photoelectrons is below the instrument threshold. In the inner Jovian magnetosphere, however, where thermal electron measurements are obscured by high background, the possibility of spacecraft charge build-up cannot be excluded. It is argued, however, that large charge build-up probably did not occur since the effects to various spacecraft systems caused by arcing were not observed when the spacecraft passed into the Jovian shadow. This implies the existence of a thermal plasma in the inner Jovian magnetosphere. It is also tantalizing to speculate that the origin of the thermal plasma observed in the outer Jovian magnetosphere is likely to be the inner magnetosphere and perhaps the Jovian ionosphere.

The observation of the electron component of the thermal plasma and the inferred ion component is consistent with this thermal plasma being the primary controlling factor causing the inflation of Jupiter's outer magnetosphere. It is interesting to note that the observations made in the daylight hemisphere for Jupiter's magnetosphere are more reminiscent of the case in the Earth's magnetotail. It is also important to recognize that these thermal electrons were observed everywhere in the outer Jovian magnetosphere and not simply confined to the disc-like configuration observed for the energetic electrons (Fillius and McIlwain, 1974; Simpson et al., 1974; Trainor et al., 1974; Van Allen et al., 1974). This supports arguments in favor of a thick magnetosphere as opposed to a 'magnetodisk' implied by observations of energetic electrons alone.

The thick magnetosphere, the large inflation of the magnetosphere, and the increase in the electron temperatures observed throughout the second magnetosheath traversal all imply the importance of compressional effects and the ease with which Jupiter's magnetosphere responds to relatively minor changes in the solar wind dynamic pressure.

The origin and dynamics of the more energetic plasma electrons are not understood. It is not clear whether these observations are consistent with local acceleration, radial diffusion, temporal effects of dynamic changes in the entire magnetosphere (e.g., its moving in and out), or possibly loss from the inner magnetosphere. Further analysis of the Pioneer 10 data and comparisons with the data obtained during the Pioneer 11 flyby of Jupiter in December 1974 may shed some light on these questions. It is likely, however, that many of the detailed aspects of the plasma of Jupiter's magnetosphere must await completely instrumented eccentric orbiting spacecraft.

Acknowledgements

This study was carried out at the University of Southern California and was supported by NASA Contract NAS2-7969 with the Ames Research Center and also by the University of Southern California.

The author thanks Dr John H. Wolfe, principal investigator of the Ames Research

Center plasma spectrometer on Pioneer 10, for access to the electron data used in this paper.

S. E. Lambert (USC) did much of the computer programming. A number of other USC students N. Cheung, J. Cutter, W. Ho, R. Jourdan, M. Lim, W. D. Miller, W. Montier, and S. Winchester also assisted with the data handling.

References

Fillius, R. W. and McIlwain, C. E.: 1974, *Science* **183**, 314.

Hall, Charles F.: 1974, *Science* **183**, 301.

Intriligator, Devrie S.: 1975, this volume, p. 297.

Intriligator, Devrie S. and Wolfe, John H.: 1974. *Geophys. Res. Letters* **1**, 281.

Simpson, J. A., Hamilton, D., Lentz, G., McKibben, R. B., Mogro-Campero, A., Perkins, M., Pyle, K. R., Tuzzolino, A. J., and O'Gallagher, J. J.: 1974, *Science* **183**, 306.

Smith, E. J., Davis, L., Jr., Jones, D. E., Colburn, D. S., Coleman, P. J., Jr., Dyal, P., and Sonett, C. P.: 1974, *Science* **183**, 305.

Trainor, J. H., Teegarden, B. J., Stilwell, D. E., McDonald, F. B., Roelof, E. C., and Webber, W. R.: 1974, *Science* **183**, 311.

Van Allen, J. A., Baker, D. N., Randall, B. A., Thomsen, M. F., Sentman, D. D., and Flindt, H. R.: 1974, *Science* **183**, 309.

Wolfe, John H., Collard, H. R., Mihalov, J. D., and Intriligator, D. S.: 1974a, *Science* **183**, 303.

Wolfe, J. H., Mihalov, J. D., Collard, H. R., McKibbin, D. D., Frank, L. A., and Intriligator, D. S.: 1974b, *J. Geophys. Res.* **79**, 3489.

CHARACTERISTICS OF JOVIAN TRAPPED ELECTRONS AND PROTONS FOR $R \gtrsim 20\,R_J$ AND THEIR INTERACTION WITH IO

J. A. SIMPSON*, D. C. HAMILTON*, R. B. McKIBBEN, A. MOGRO-CAMPERO**,
K. R. PYLE, and A. J. TUZZOLINO

Enrico Fermi Institute, The University of Chicago, Chicago, Ill. 60637, U.S.A.

Discussion of the energetic particles associated with Jupiter divides naturally into discussion of three regions. These regions are (a) the inner core of Jupiter's magnetosphere ($R \gtrsim 20\,R_J$) where the highest intensity trapped radiation is to be found, and where trapping is almost certainly stable and long term; (b) the extended outer region of the magnetosphere ($2\,R_J \gtrsim R \gtrsim 100\,R_J$) where the energetic particle fluxes are confined primarily to the magnetic equatorial plane, and where the particles are probably not stably trapped, and (c) interplanetary space, where electrons from Jupiter are observed as much as 1 AU upstream from Jupiter in the solar wind.

In this paper, we give a brief summary of observations of the trapped particles in the inner core region of the magnetosphere made with the University of Chicago experiment on Pioneer 10. Companion papers summarize observations of particles found in the outer magnetosphere and in interplanetary space. A detailed treatment of the observations reported here is published elsewhere (Simpson *et al.*, 1974; McKibben and Simpson, 1974; Chenette *et al.*, 1974).

The University of Chicago experiment on board the Pioneer 10 spacecraft contains four sensor systems, three of which were used to study trapped radiation in the region $R \gtrsim 20\,R_J$. Two of the systems are of novel design. One is a silicon detector operated in the current mode surrounded by sufficient beryllium to stop protons with energies $\gtrsim 30$ MeV. Electrons with energies $\gtrsim 3$ MeV were measured by this detector, which had a dynamic range of $> 10^7$, and was capable of measuring fluxes of up to 10^{11} electrons cm^{-2} s^{-1}. Because of the great difficulty in identifying high energy protons in the presence of an intense high energy electron component, $\gtrsim 35$ MeV protons were measured by detection of proton induced nuclear fission of Th^{232}. The third detector system was a low energy particle telescope which measured the intensities, spectra, and angular distributions of 0.5–1.8 MeV/nucleon protons and helium nuclei. Detailed descriptions of these instruments and their behavior in the high intensity radiation zone are given by Simpson *et al.* (1974).

In Figures 1–4 we present the intensity profiles of the $\gtrsim 3$ MeV electrons, $\gtrsim 35$ MeV protons, and 0.5–1.8 MeV protons plotted as a function of magnetic shell parameter L for $L \gtrsim 24$. (The fission cell counting rate is plotted only for $L \gtrsim 10$, since

---

* And Department of Physics.
** Present address: Department of Physics, University of California at San Diego.

V. Formisano (ed.), The Magnetospheres of the Earth and Jupiter, 317–324. All Rights Reserved.
Copyright © 1975 by D. Reidel Publishing Company, Dordrecht-Holland

it did not increase above background outside $L \cong 10$.) We have used the D2 magnetic field model of Smith *et al.* (1974b) in preparing these plots. We have also analyzed our data using the D1 model. In Table I, we give a description of the dependences of the various particle fluxes on L and magnetic latitude, λ, deduced from these data using both the D1 and the D2 magnetic field models. The maximum fluxes

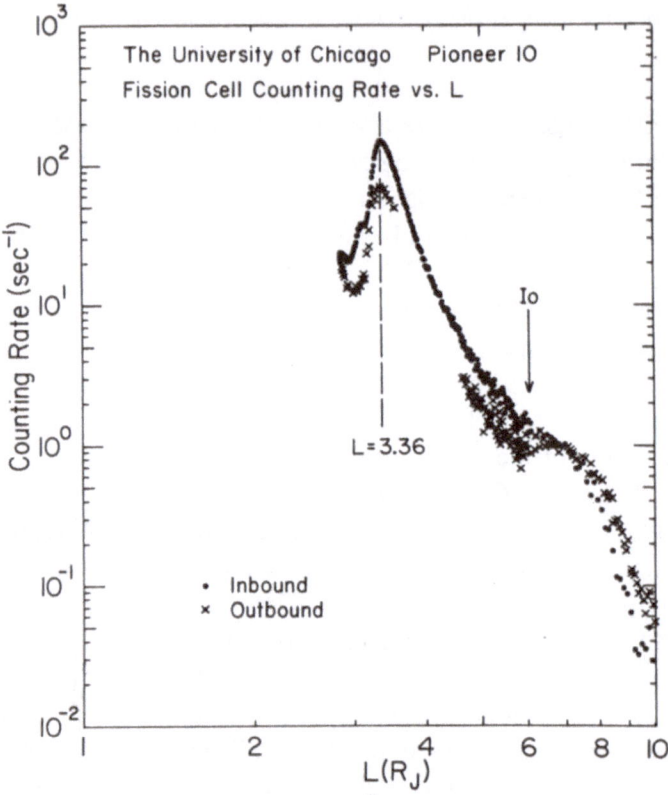

Fig. 1.   The intensity profile of the $\gtrsim 3$ MeV electron flux measured by the electron current detector (ECD) for $L \lesssim 24$, inbound and outbound, as a function of L. Each point represents a one-minute average of the data. For fluxes $\gtrsim 3 \times 10^4$ cm^{-2} s^{-1} on the outbound pass, the leakage current from the detector dominated the current produced by incident electrons.

observed were $\sim 6 \times 10^6$ protons cm^{-2} s^{-1} for $\gtrsim 35$ MeV protons at $L \cong 3.4$ and $\sim 2.5 \times 10^8$ electrons cm^{-2} s^{-1} for $\gtrsim 3$ MeV electrons at $L \cong 3.1$.

From Figures 1, 3 and 4, it is apparent that inside $L \cong 12$, the flux measured inbound and outbound at the same L was the same to within a factor of about 2, whereas outside $L \cong 12$, much larger differences are found. We attributed the inbound-outbound asymmetry for $L \gtrsim 12$ to an azimuthal asymmetry in the particle flux as a result of the distortion of the Jovian magnetic field by the corotating plasma in the magnetosphere and by the solar wind. The smaller intensity differences observed at the same L inbound and outbound for $L \lesssim 12$ are more likely a result of the magnetic latitude dependence of the particle intensity.

TABLE I

Summary of the spatial dependence of energetic particle fluxes for magnetic field models D1[a] and D2[b]

Region of L		Particle type	D1 Model	D2 Model
$20 \gtrsim L \gtrsim 12$	L dependence[c]	~1 MeV protons $\gtrsim$3 MeV electrons	$L_0 \cong 4$ (inbound only) $L_0 \cong 2$ (inbound only)	$L_0 \cong 2.8$ (inbound only) $L_0 \cong 1.8$ (inbound only)
	λ dependence[d]	$\gtrsim$3 MeV electrons	$m \cong 0$ near $L=13$ inbound	
$12 \gtrsim L \gtrsim 7$	L dependence[c]	$\gtrsim$35 MeV protons $\gtrsim$3 MeV electrons	$n \cong 11$ (or $L_0 \cong 0.84$) $L_0 \cong 2.4$	$n \cong 13$ (or $L_0 \cong 0.65$) $L_0 \cong 1.8$
	λ dependence[d]	$\gtrsim$35 MeV protons $\gtrsim$3 MeV electrons	$m \cong -0.5$ $m \cong 0$	$m \cong -2.5$ $m \cong -0.75$
$7 \lesssim L \lesssim 5.5$ (orbit of Io)	Intensity vs L profile	~1 MeV protons	1. Major intensity decrease observed from $L \cong 6.0$–6.2 both inbound and outbound. 2. Good correlation between location in L of major intensity decrease and region of maximum efficiency of absorption by Io. 3. Good correlation between change in slope of intensity decrease at $L=6.5$ and maximum L reached by Io.	1. Major intensity decrease observed at $L=5.8$–6.0 inbound, 5.95–6.15 outbound. 2. Poor correlation between intensity structure and properties of Io's path in L, λ space.
	Anisotropy change	~1 MeV protons	$L \cong 6.1$ both inbound and outbound	$L=6.2$ inbound, $L=6.0$ outbound
$5.5 \lesssim L \lesssim 2.9$	L dependence[c]	$\gtrsim$35 MeV protons $\gtrsim$3 MeV electrons	$n \cong 11$ (or $L_0 \cong 0.3$) $3.5 \lesssim L \lesssim 4.5$ $L_0 \cong 1.8$ $4.7 \lesssim L \lesssim 5.5$	$n \cong 12$ (or $L_0 \cong 0.3$) $3.4 \lesssim L \lesssim 4.3$ $L_0 \cong 1.7$ $4.6 \lesssim L \lesssim 5.5$
	λ dependence[d]	$\gtrsim$35 MeV protons	$m \cong -0.5$ $m \cong -1.0$	$m \cong -2.4$ $m \cong -1.4$ $m \cong 1.4$ $L=3.36$
	Locations of intensity maxima	$\gtrsim$3 MeV electrons ~1 MeV protons $\gtrsim$35 MeV protons	$L=3.66$ inbound, $L=3.88$ outbound $L=3.36$ inbound, $L=3.36$ outbound	$L=3.68$ inbound, $L=3.76$ outbound $L=3.36$ inbound and outbound

[a] D1 dipole description: strength 4.0 g inclination 15°, offset 0.1 R_J north, 0.2 R_J toward system III longitude 170°, tipped towards system III longitude 230° (Smith et al., 1974a).

[b] D2 dipole description: strength 4.0 g inclination 11°, offset 0.03 R_J north, 0.11 R_J toward system III longitude 176°, tipped towards system III longitude 222° (Smith et al., 1974b).

[c] The dependence upon L is assumed to have the form e^{-L/L_0}, except for the $\gtrsim$35 MeV protons, which are also represented as L^{-n}.

[d] The latitude distribution is assumed to have the form $J(\lambda) \propto [B(\lambda)/B(\lambda=0)]^m$.

The effect of absorption of trapped particles by the Jovian satellite Io is apparent in the intensity profiles of all particle species for which we have observations. For $\gtrsim 3$ MeV electrons, the decrease in flux induced by Io was a factor of about 2–3 below the flux expected from inward extrapolation of the behavior observed for $7 \lesssim L \lesssim 12$.

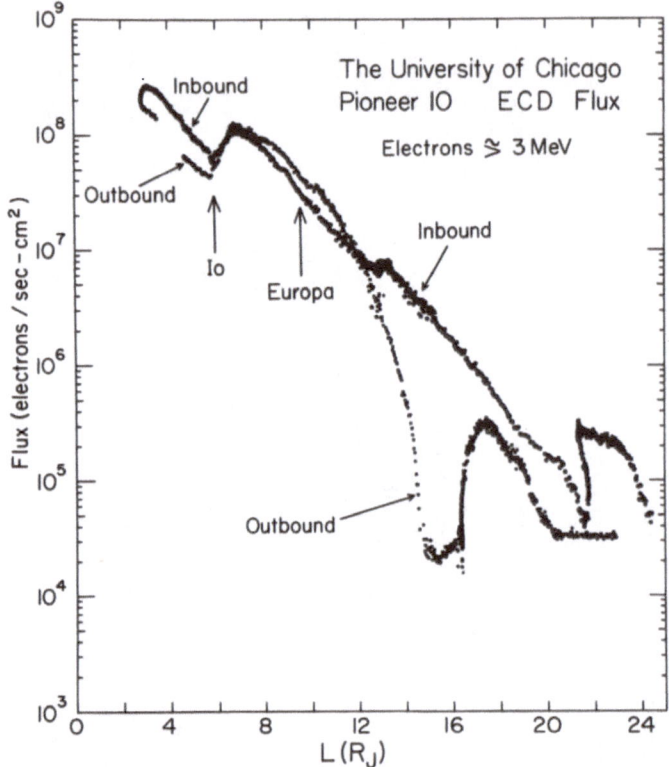

Fig. 2. The counting rate of 0.5–1.8 MeV protons measured by the low energy telescope (LET) for $L \lesssim 24$, inbound and outbound, as a function of L. For $L \lesssim 15$, the response of the counting rate to changes in the incident particle flux was non-linear, and no corrections for this effect have been applied to these data. A full discussion of the non-linearity of the LET counting rate response to very high incident particle fluxes is given in Appendix 1 of Simpson *et al.* (1974).

The angular distribution of the 0.5–1.8 MeV protons changed dramatically at the orbit of Io. Since Pioneer 10 was spinning about an axis approximately parallel to Jupiter's equatorial plane, and the axis of the view cone of the low energy telescope was perpendicular to the spin axis, in the course of each spacecraft rotation, the low energy telescope provided a direct measurement of the character of the pitch angle distribution of the 0.5–1.8 MeV protons, although in this region of L, our measurement of the pitch angle distribution is only a qualitative measurement. As a result of non-linearity in the counting rate response to the very high particle fluxes, we cannot give quantitative results such as the ratio of maximum to minimum flux in an anisotropic pitch angle distribution.

The behavior of the angular distribution of 0.5–1.8 MeV protons as a function of L from $L \cong 7$ to $L \cong 3$ on both inbound and outbound passes is shown in Figure 5. Outside the orbit of Io, $(L \cong 6)$, the low energy proton flux was highest for small pitch angle particles which mirror at high magnetic latitudes. Everywhere inside the orbit of Io, however, the flux was highest for large pitch angle particles which mirror

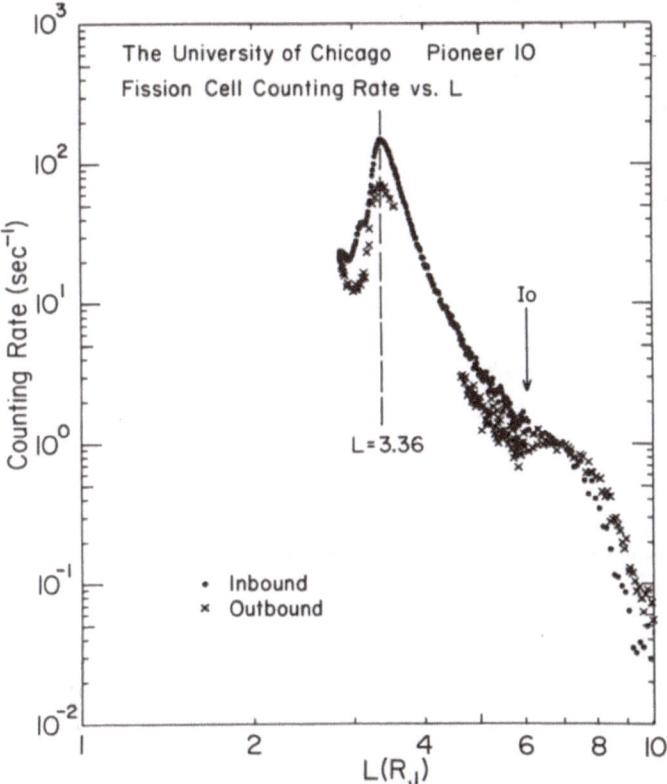

Fig. 3. The fission cell counting rate plotted as a function of L. In the absence of significant background, this counting rate may be converted to a flux of protons $\gtrsim 35$ MeV cm^{-2} s^{-1} by multiplying the counting rate by 4×10^4. Background from direct detection of high Z trapped particles in the detectors may have been significant, however. Simpson *et al.* (1974) give this question full discussion.

near the magnetic equator, and was smallest for small pitch angle particles. We interpret this as the result of preferential absorption of small pitch angle particles by Io. Such preferential absorption is expected, as pointed out by Mead and Hess (1973) as a result of Io's motion in the inclined dipole magnetic field of Jupiter. Simpson *et al.* (1974) have given a thorough discussion of the nature of this motion and of the absorption process. The persistence of the depleted cone of small pitch angle particles inside the orbit of Io may be interpreted to mean that the time required for repopulating the depleted cone by pitch angle scattering is long compared to the time required for particles to diffuse inwards from $L \cong 6$ to $L \cong 3$.

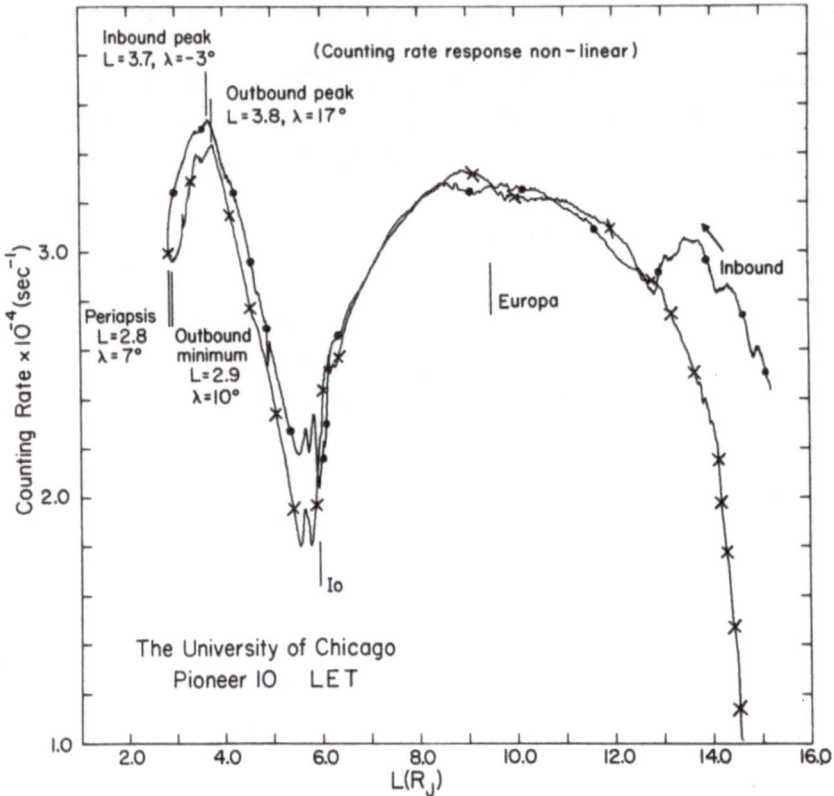

Fig. 4. The counting rate of 0.5–1.8 MeV protons from the LET plotted as a function of L for the region of non-linear response of the LET to intensity changes. Dots and crosses indicate the inbound and outbound measurements, respectively.

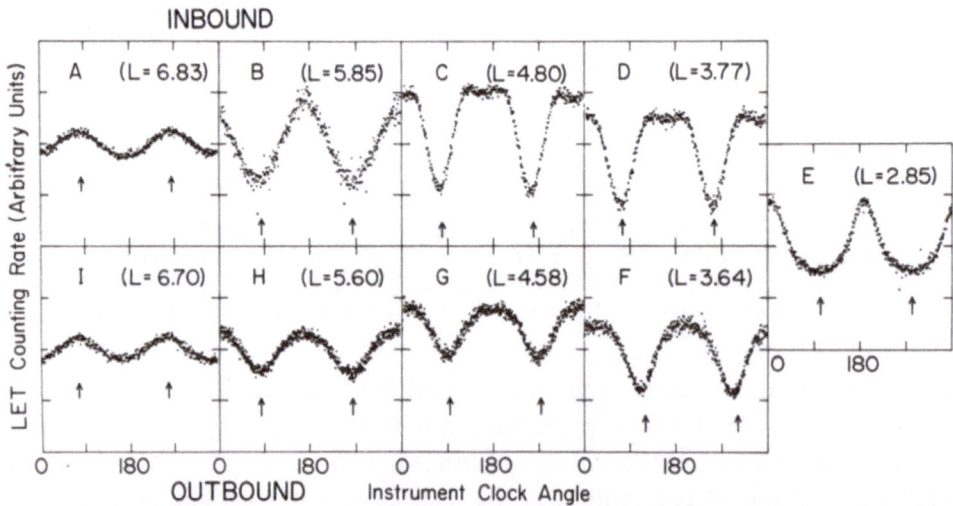

Fig. 5. An overview of the development of the 0.5–1.8 MeV proton anisotropy near and inside the orbit of Io. The series A through I shows the anisotropy at approximately evenly spaced L shells. Instrument clock angle increases with time and is defined to be 0° when the telescope axis rises through the ecliptic plane in the course of the spacecraft's rotation. Arrows show the positions at which the angle between the telescope axis and the magnetic field direction was smallest.

The persistence of the depleted cones gives conclusive evidence that Jupiter's trapped radiation is maintained by inward diffusion of particles across L shells. From the observations of low energy protons at the orbit of Io, and consideration of the nature of the absorption process, Simpson *et al.* (1974) derived a probable value of $\sim 6 \times 10^{-8} \ R_J^2 \ s^{-1}$ for the radial diffusion coefficient for 1 MeV protons at $L = 6$. A diffusion coefficient derived for electrons was of the same order of magnitude. Such a diffusion coefficient is sufficiently large to allow survival of a significant fraction of particles diffusing inward across the orbit of Io, so that no new particle sources inside the orbit of Io are required to explain the high intensity trapped electrons inferred to exist at $L \cong 2$ from decimetric radio observations.

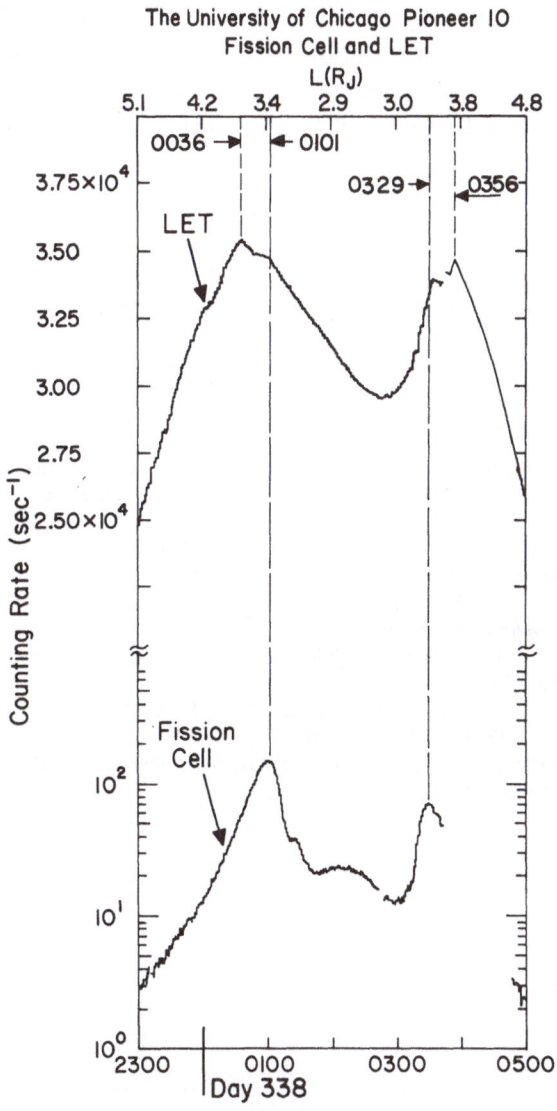

Fig. 6. Counting rates from the fission cell and the LET vs time and L in the vicinity of the proton maxima.

We further note that the radial dependence observed for high energy protons for $L \gtrsim 10$ (see Figure 3 and Table I) is much stronger than that predicted for one possible source of high energy protons, namely CRAND (Thomas and Doherty, 1972). We conclude, therefore, that our observations are not consistent with a CRAND source for these particles, and that the high energy protons observed were accelerated within the magnetosphere of Jupiter, most probably by inward diffusion with violation of the third adiabatic invariant.

An unexpected result of our observations is the discovery of a decrease towards low L in the intensity of $\gtrsim 35$ MeV and ~ 1 MeV protons inside $L \cong 3.5$, so that the maximum flux of protons observed both inbound and outbound occurred at $L \cong 3.5$. Simpson *et al.* (1974) have discussed possible causes for this effect and conclude that neither flux limiting by means of the ion-cyclotron instability (Coroniti *et al.*, 1974; Kennel, 1972) nor absorption by the tiny satellite Amalthea seem likely to explain the decrease. Measurements made by Pioneer 11, which will provide data in to $L \cong 1.8$ offer our best hope of gaining further understanding of this effect.

Acknowledgements

This research was supported in part by NASA contracts NAS 2-5601 and NAS 2-6551 with the Ames Research Center, NASA Grant NGL 14-001-006 and NSF Grant GA-41692X.

References

Chenette, D. L., Conlon, T. F., and Simpson, J. A.: 1974, *J. Geophys. Res.* **79**, 3551.
Coroniti, F. V., Kennel, C. F., and Thorne, R. M.: 1974, *Astrophys. J.* **189**, 383.
Kennel, C.: 1972, *Proc. of the Jupiter Radiation Belt Workshop*, Tech. Mem. 33-543, p. 347, Jet Propulsion Lab., Pasadena, Calif., July.
McKibben, R. B. and Simpson, J. A.: 1974, *J. Geophys. Res.* **79**, 3545.
Mead, G. D. and Hess, W. N.: 1973, *J. Geophys. Res.* **78**, 2793.
Simpson, J. A., Hamilton, D. C., McKibben, R. B., Mogro-Campero, A., Pyle, K. R., and Tuzzolino, A. J.: 1974, *J. Geophys. Res.* **79**, 3522.
Smith, E. J., Davis, L., Jr., Jones, D. E., Colburn, D. S., Coleman, P. J., Jr., Dyal, P., and Sonett, C. P.: 1974a, *Science* **183**, 305.
Smith, E. J., Davis, L., Jr., Jones, D. E., Coleman, P. J., Jr., Colburn, D. S., Dyal, P., Sonett, C. P., and Frandsen, A. M. A.: 1974b, *J. Geophys. Res.* **79**, 3501.
Thomas, J. and Doherty, W. R.: 1972, *Proc. of the Jupiter Radiation Belt Workshop*, Tech. Mem. 33-543, p. 315, Jet Propulsion Lab., Pasadena, Calif., July.

OBSERVATIONS OF JOVIAN ACCELERATED PARTICLES BOTH INSIDE AND OUTSIDE THE JOVIAN MAGNETOSPHERE: RESULTS FROM THE GODDARD-U. OF NEW HAMPSHIRE EXPERIMENT ON PIONEER 10

J. H. TRAINOR, F. B. McDONALD, and B. J. TEEGARDEN

Goddard Space Flight Center, Greenbelt, Md., U.S.A.

and

W. R. WEBBER and E. C. ROELOF

University of New Hampshire, Durham, N.H., U.S.A.

1. Introduction

The Pioneer 10 encounter with Jupiter in December 1973 marked the first time that it has been possible to make *in situ* measurements of a region where particle acceleration and trapping is naturally occurring other than in the Earth's magnetosphere. Preliminary results on the particle and field environment near Jupiter have already indicated that the Jovian magnetosphere is an extremely interesting region which differs in a number of important ways from the Earth's magnetosphere (Trainor *et al.*, 1974; Fillius· and McIlwain, 1974; Simpson *et al.*, 1974; Van Allen *et al.*, 1974; Smith *et al.*, 1974; Wolfe *et al.*, 1974).

Pioneer 10 first encountered the Jovian bow shock on 26 November 1973, and the final bow shock crossing on the outbound pass occurred on 22 December. If the encounter is defined as the time between the first and last bow shock crossing then its duration was 26 days. During this rather long period the spacecraft mapped a range of magnetic latitudes between $\sim 0°$ and $-20°$ on the inbound pass and ~ 0 and $+20°$ on the outbound pass. This variation is due to the tilt of Jupiter's dipole with respect to its spin axis.

The results to be presented here are principally from the Goddard-U. of New Hampshire cosmic ray experiment on Pioneer 10. The telescopes are high resolution dE/dx vs E type instruments capable of providing detailed composition measurements and precise energy spectra. As a cosmic-ray instrument the geometry factors are relatively large yet the charged particle telescopes provided data into $\sim 15\ R_J$. The LET-II provided data on low energy protons over the complete encounter trajectory.

The first part of this paper will deal with measurements of the particle population inside Jupiter's magnetosphere. In the second part we will discuss observations of energetic particles in interplanetary space that have escaped from Jupiter's magnetosphere. This part will be divided into two sections. The first section will present Pioneer data when the spacecraft was outside the bow shock but within ~ 1 AU

V. Formisano (ed.), The Magnetospheres of the Earth and Jupiter, 325–353. All Rights Reserved

of the planet. The data show clearly that Jupiter is a copious source of MeV electrons in interplanetary space. In the second section electron data taken at 1 AU on various IMP spacecraft over the past 9 years will be examined.

The entire low energy spectrum at 1 AU from 0.2–30 MeV is consistent with an $E^{-1.75}$ power law and is remarkably different from the relatively flat differential spectra measured from ~40 MeV to ~1 GeV. Above 1 GeV the measurements approach a power law of ~E^{-3}.

A number of puzzling features exist in the behavior of this low energy component. One of the most significant of these is the frequent occurrence (a total of 18–19 over a four year period) of positive increases of these electrons that could not be associated with discrete solar events. These 'quiet-time' increases represented a factor of 3–5 increase in intensity and lasted from 5–12 days. They displayed a remarkable anti-correlation with low-energy proton events. Qualitatively their amplitude was observed to diminish toward solar maximum.

We believe that a least part of the puzzling features of the low-energy electrons observed at Earth can be explained if some of these electrons are of Jovian origin. We will present evidence in this paper that most of the electron quiet-time increases are of Jovian origin. Furthermore, the source strength appears adequate to supply all the low energy 0.2–40 MeV electrons observed at Earth. However, Jupiter is essentially a point source at 5 AU, and it is not clear if the required flux can be distributed around the Earth's orbit at 1 AU.

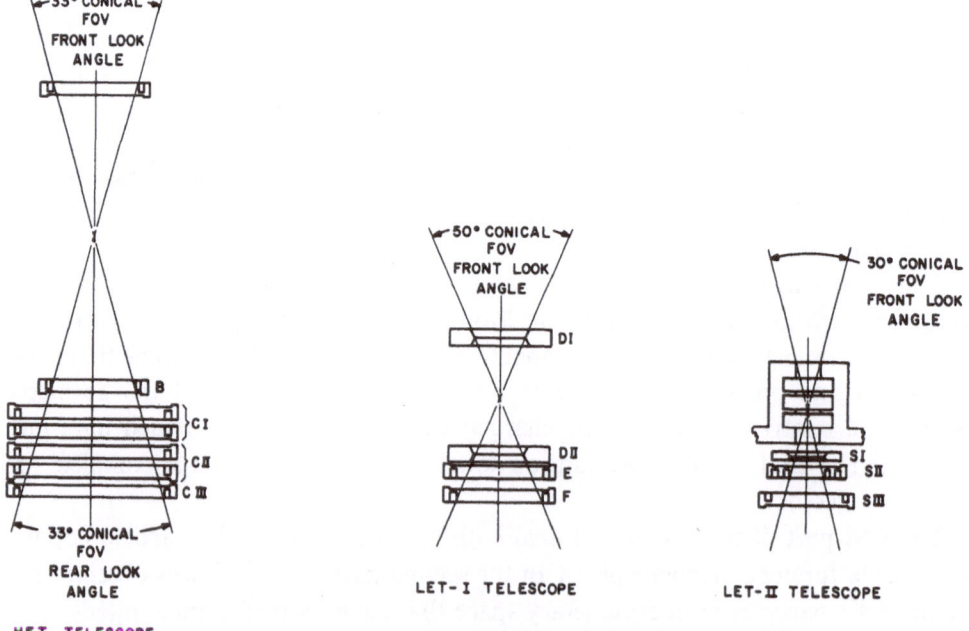

Fig. 1.   Schematic drawing of the solid-state detector telescopes.

2. Experimental Method

This experiment consists of a set of three solid-state telescopes, each designed to complement the others and to cover a broad range in energy, intensity and charge spectra. The three telescopes are shown schematically in Figure 1; they are: (1) the High-Energy Telescope (HET), (2) Low-Energy Telescope-I (LET-I), and (3) Low-Energy Telescope-II (LET-II).

The LET-II telescope was designed to measure low-energy solar-flare particles in interplanetary space and trapped particles in the Jovian magnetosphere. It has a relatively small geometry factor (1.5×10^{-2} cm^2 sr), and can measure fluxes up to $\sim 4 \times 10^6$ cm^{-2} s^{-1} sr^{-1}. It is surrounded by an aluminum and lead shield which will stop electrons up to ~ 25 MeV and protons to ~ 140 MeV. The telescope employs a two-parameter analysis technique to separate electrons and protons. The SI element is a thin (50 μ) silicon surface-barrier detector chosen for its low electron efficiency. The SII element is a 2.5 mm thick lithium-drifted device with an annular guard ring which is part of the same silicon wafer. The purpose of the guard ring is to provide a complete anti-coincidence to insure (1) that particles that stop in SI are uniquely identified and (2) that no particles are allowed to enter SII from the side of the telescope. The SIII element serves always as an anti-coincidence.

The various modes of the LET-II telescope are summarized as follows:

Logic Condition			Energy/Particle
$\overline{\text{SI}}$	$\overline{\text{SII}}$	$\overline{\text{SIII}}$	0.1–2 MeV protons
$\overline{\text{SI}}$	SII	$\overline{\text{SIII}}$	0.1–2 MeV electrons
SI	SII	$\overline{\text{SIII}}$	2.1–21 MeV protons

The electronic threshold of the SI element is chosen such that any electron which comes to rest in SII will be below threshold and any proton which comes to rest in SII will be above threshold. By this technique electron-proton separation is thus obtained. The LET-II telescope has the smallest geometry factor of the three telescopes and is the only one to return useful information during the near-periapsis phase ($\leqslant 15$ R_J) of the encounter.

The LET-I telescope is a double dE/dx vs E instrument capable of measuring detailed charge, mass, and energy spectra. The DI and DII elements are thin (100 μ) surface barrier detectors. For low-energy particles that stop in DII these two detectors perform a dE/dx by E measurement. For higher energy particles that come to rest in E, DI, and DII provide redundant dE/dx measurements, and the total energy is given by the E detector (2.5 mm lithium drift silicon). The energy and particle ranges are summarized as follows:

Logic condition				Energy/Particle
DI	DII	$\overline{E}$		3.2–5.2 MeV nuc^{-1} protons and alphas
DI	DII	E	$\overline{F}$	5.2–21 MeV nuc^{-1} protons and alphas

In addition the telescope is capable of measuring higher charges up to $Z = 16$.

The High Energy Telescope (HET) is composed entirely of lithium-drifted devices all 2.5 mm. thick. As with LET-I the first two elements A and B serve to define the acceptance cone of the telescope and to provide dE/dx measurements except for the lowest energies where the B element gives a total energy measurement. In practice the CI and CII elements are always summed together giving the total energy loss in a stack of four identical detectors. This stack gives the total particle energy in the principal operation mode of the telescope. The CIII element serves the dual purpose of (1) an anticoincidence detector for stopping particles and (2) a dE/dx detector for particles that penetrate the entire telescope. The various modes of operation of the HET are summarized as follows:

Logic Condition	Energy/Particle
A B $\overline{\text{CIII}}$	20–56 MeV nuc^{-1} protons and alphas
	2–8 MeV electrons
A B CIII	>56 MeV nuc^{-1} protons and alphas
	>8 MeV electrons

The HET can also measure higher charges up to $Z = 8$.

The particle anisotropies, or pitch angle distributions are measurements of fundamental importance in understanding the physics of the Jovian magnetosphere. The Pioneer 10 spin axis is always pointed towards the Earth, and our detectors are all pointed perpendicular to the spin axis. Consequently we have the capability of measuring angular distributions in a plane normal to the ecliptic. This is not an optimum situation for interplanetary measurements, but is quite good for pitch-angle distributions in the Jovian magnetosphere. Twelve counting rates encompassing protons and electrons over a fairly broad energy range are each separately divided into eight angular sectors, thus providing a comprehensive set of particle anisotropy measurements.

A major effort has been devoted to understanding the response of all detector systems in the presence of intense particle fluxes. The onset of saturation in the LET-I and HET systems is abrupt and well defined. Negligible corrections are necessary prior to this saturation point. The LET-II response is now sufficiently well understood so that the 1–2 MeV and 14–21 MeV proton fluxes can be determined over the complete trajectory.

There is significant overlap in the response functions of the three telescopes, and it was of great value to observe the consistency between flux measurements made with completely different detector systems. For example, the LET-I and LET-II proton data in the 0.5–2 MeV region are in excellent agreement from 100–16 R_J. This is especially helpful in the outer Jovian Magnetosphere where the nuclear component is small (1–10%) compared to electrons of the same energy. Furthermore, LET-I provides a very sensitive measurement of protons and alphas in the 3–20 MeV range down to an intensity of 10^{-3} protons cm^{-2} s^{-1} sr^{-1} MeV^{-1} up to $\sim 20\ R_J$. A complete description of this instrument will be published in the *IEEE Transactions on Nuclear Science* (Stilwell *et al.*, 1975).

3. Particle Measurements in the Jovian Magnetosphere

3.1. ELECTRON AND PROTON TIME HISTORIES

Figures 2 and 3 give an overview of the Jovian electron and proton fluxes as observed by the experiment, showing particle energies near the upper and lower ranges of the instrument. In discussing these observations, it is convenient to divide the Jovian magnetosphere into 3 regions:

(1) The outer Jovian magnetosphere: This region extends from the bow shock crossing (109 R_J) to $\sim 50\ R_J$. The magnetic field is ~ 8–$20\ \gamma$ and like the Earth's tail is dominated by a neutral plasma sheath which is drawing the field lines outward. It is a region of quasi-trapping and diffusion. Both the electrons and protons show remarkably constant energy spectra, $E^{-\gamma}$ with $\gamma = 1.5$–2.0 and 4 respectively. This suggests that almost no acceleration occurs in this region. There are rapid changes in flux and angular distributions. The high energy electrons (i.e. >6 MeV) show a

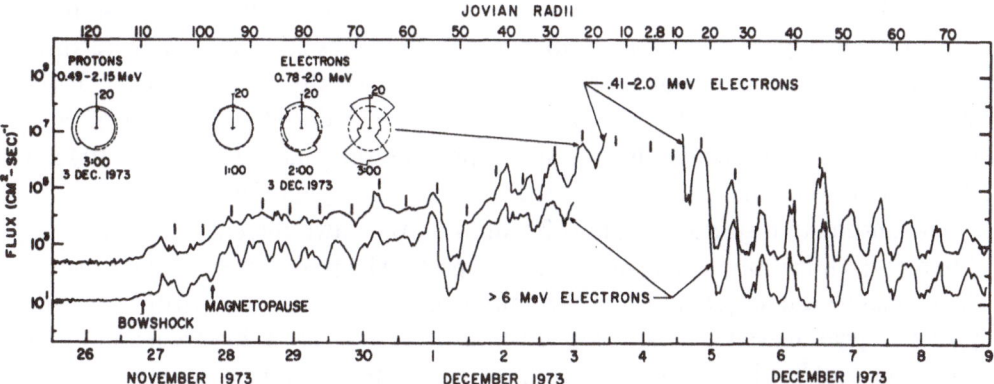

Fig. 2.   Electron time histories during the Pioneer 10 Jovian encounter. Angular distributions are given for selected time periods. Tick marks show points at which Pioneer 10 was closest to the magnetic equator.

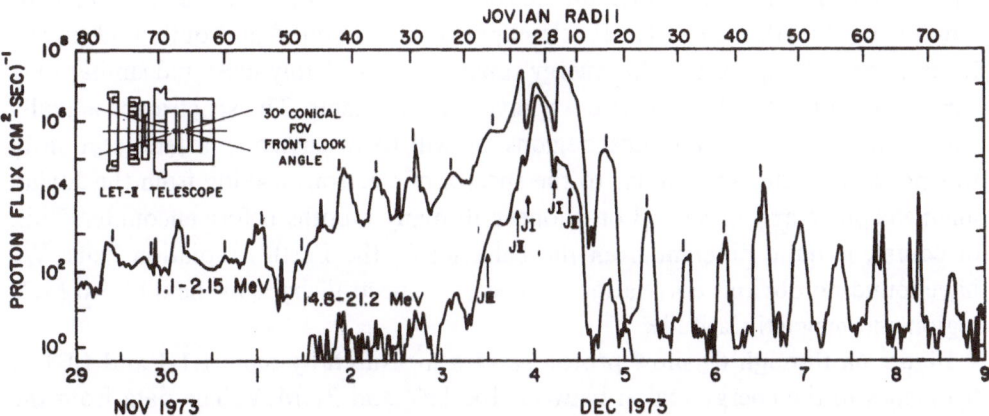

Fig. 3.   Proton time histories in two different energy intervals. Times of crossing of the orbits of the innermost Galilean satellites are shown respectively as JI, JII and JIII.

reasonable 10-h periodicity as expected since the nominal magnetic latitude of the spacecraft should vary with the rotation period of the planet (~ 10 h). This is not nearly as significant for the protons or for the lower energy electrons. The changes in the proton and electron flux are frequently uncorrelated.

(2) The region from ~ 50 to 25 R_J is one of transition between the outer diffusion zone and the point where the field rigidly rotates with the planet. The proton energy spectra begin to change from power law to exponential energy forms on a gradual scale suggesting that some acceleration is occurring. The angular distribution of the protons display large (up to $\sim 70\%$) anisotropies and the hinging effect produced by the transition is strongly evident. The magnetic field is still changing rather slowly, and it is not clear if the particles are stably trapped.

(3) Inside 25 R_J: The particle angular distributions indicate that the field lines are rigidly rotating with the planet within ~ 25 R_J, and this may mark the outer boundary for durable trapping. The rapid decrease in the co-rotation anisotropy provides further evidence that the proton spectra are becoming increasingly flat below 1 MeV. At the outer edge of this region there is a significant increase in the 1.2–2.1 MeV (Figure 3) proton component which climbs steadily until 6 R_J. Inside 6 R_J the proton component is strongly attenuated by the presence of the Jovian moon, Io. For example the 1–2 MeV component is reduced by a factor of 50 by Io absorption. The flux of all proton components then increases until ~ 0115 on December 4 when the spacecraft crossed the magnetic equator at 3.5 R_J.

The outbound trajectory (Figures 2 and 3) on the dawn meredian was strikingly different from the inbound span in many respects. The peak fluxes are near the predicted magnetic equator, however, the dominant feature is the 10-h periodicity which produces peak-to-valley ratios of as much as 5 decades for a 20° excursion in latitude. This implies both electrons and protons are much more concentrated in the low latitude region on the outbound pass.

3.2. ELECTRON AND PROTON SPECTRA

Figures 4 and 5 show the differential electron spectra measured near the magnetic equator by the HET and LET-II telescopes on the inbound and outbound passes, respectively. The spectra in this energy range are remarkably hard and similar over the region outside ~ 25 R_J where we have measurements. The spectra are actually slightly harder in the outermost regions, as will be discussed in a later section of this paper, and are very similar to the spectra of electrons leaking from the Jovian magnetosphere and measured on Pioneer 10 many months before encounter. This, of course, is quite different from the behavior in the Earth's radiation belts. We have found no obvious correlation between the spectral shape in the 0.12–8.0 MeV region and magnetic latitude.

Figure 6a through 6g show proton spectra measured by our LET-I and LET-II telescopes in the energy region between 100 keV and 21 MeV. The data from the magnetosheath region and from the radiation belts in to ~ 40 R_J seem to be well fitted by a simple power law with an exponent of ~ 4 but varying from 4.2–3.0 for

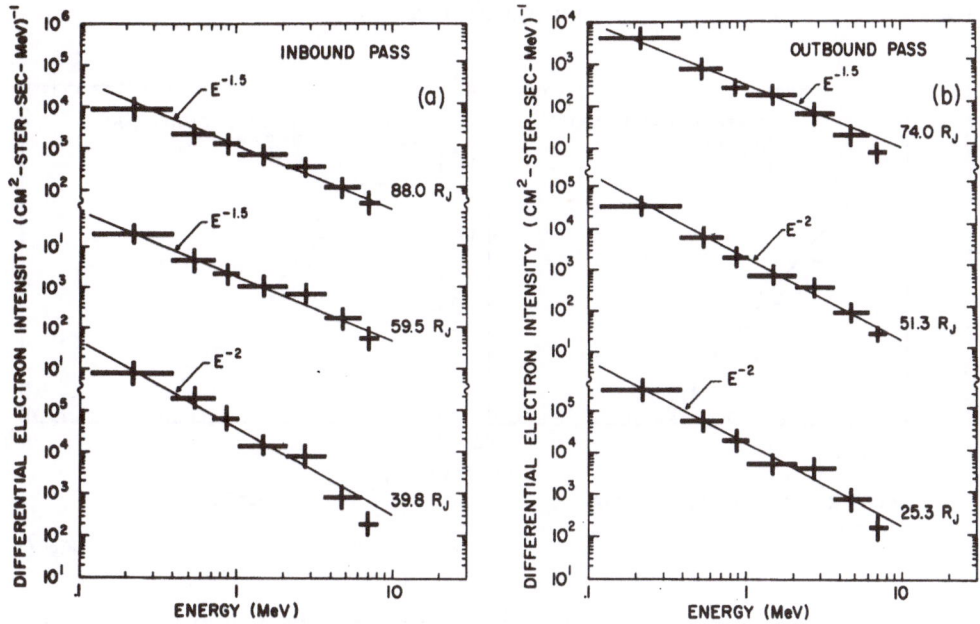

Fig. 4. Electron differential energy spectra on the inbound pass of Pioneer 10 at three different values of R_J.

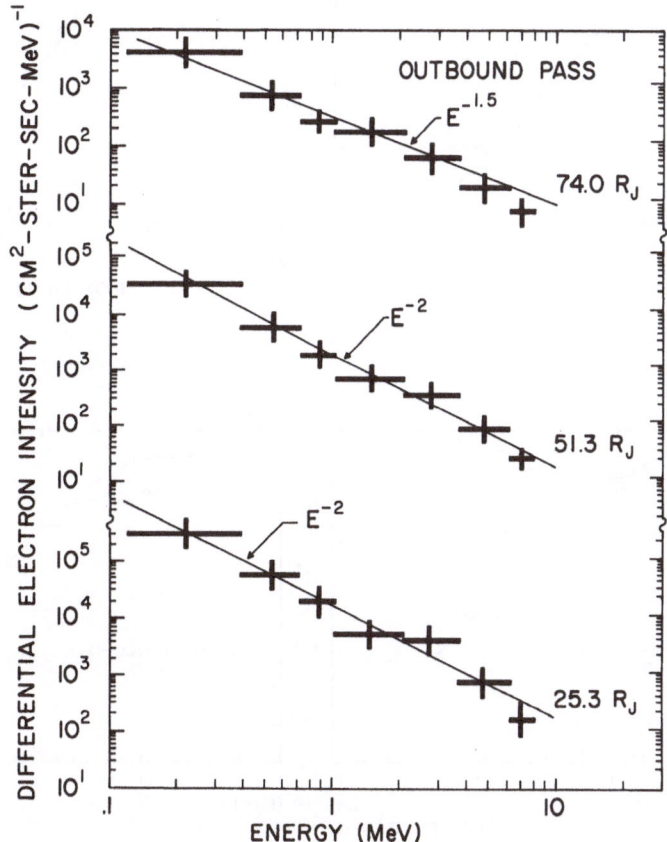

Fig. 5. Electron differential energy spectra on the outbound pass of Pioneer 10 at three different values of R_J.

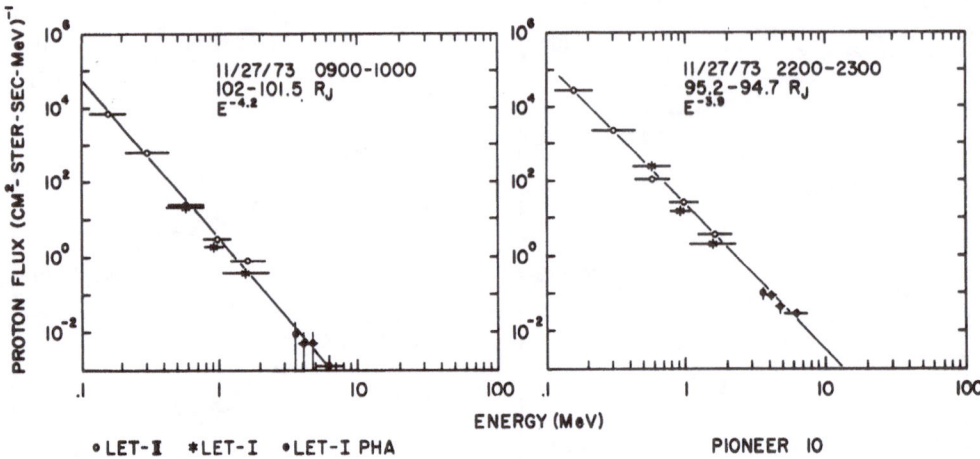

Fig. 6a.   Fig. 6b.

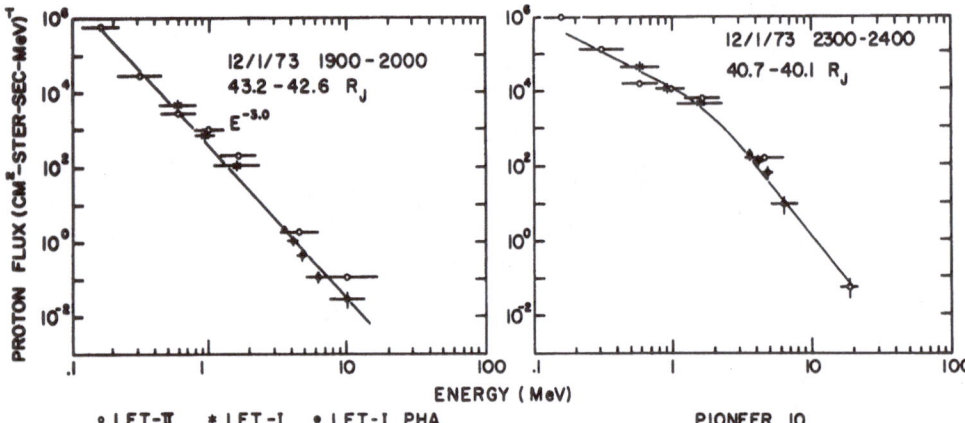

Fig. 6c.   Fig. 6d.

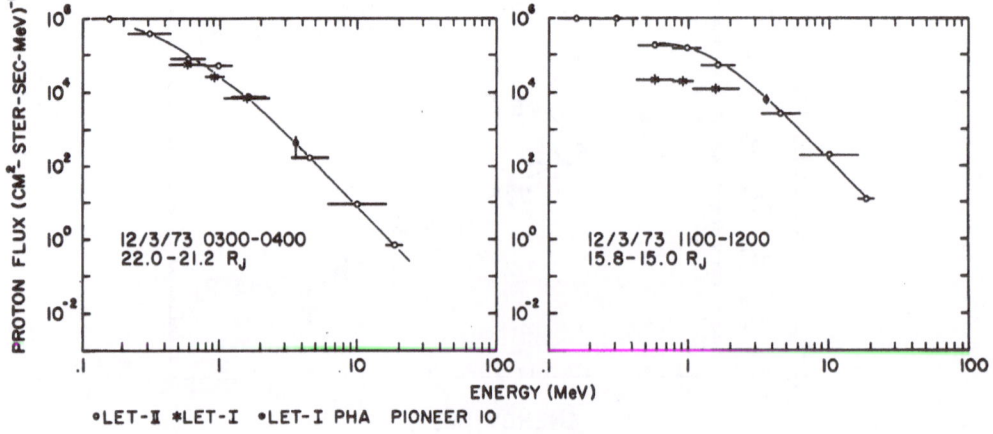

Fig. 6e.   Fig. 6f.

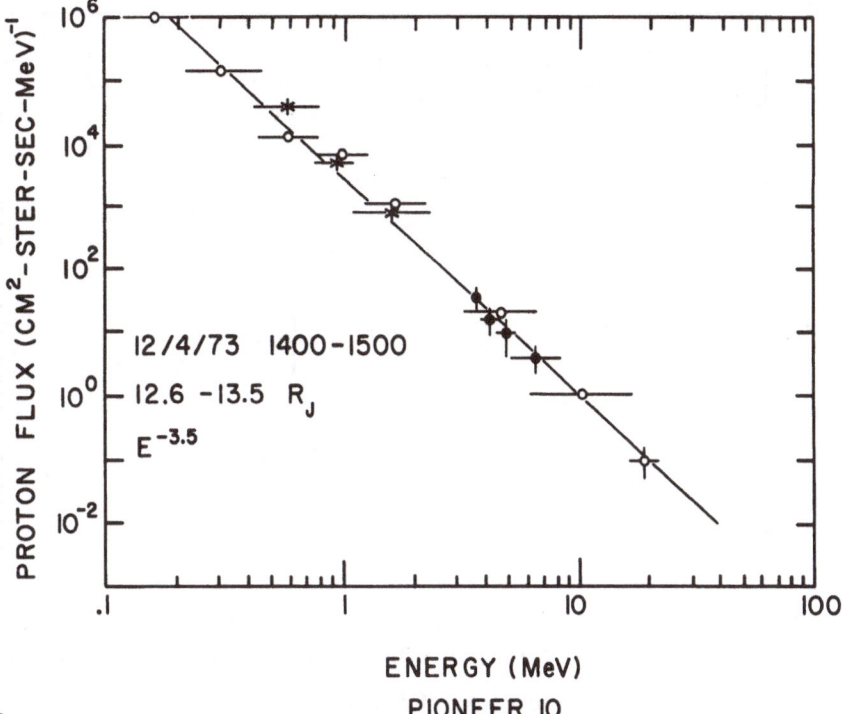

Fig. 6g.

PIONEER 10

Fig. 6a–g.   Proton differential energy spectra of different values of Jovian radius.

brief periods. We interpret this as indicating very little acceleration occurs in the outer magnetosphere.

Beginning with the measurements inside $\sim 41\ R_J$ the spectra are better fitted by an exponential than a power law. Note that the one data point plotted at 10^6 in Figure 6d has been arbitrarily plotted there, since it would have fallen above a flux of $10^6\ \mathrm{cm}^{-2}\ \mathrm{sr}^{-1}\ \mathrm{s}^{-1}\ \mathrm{MeV}^{-1}$, and a correction is required for LET-II data above that figure. None of the LET-II data shown requires a correction of any kind. The spectra near $22\ R_J$ and $15\ R_J$ show a similar exponential form. Note in Figure 6f that two LET-II data points are now in the non-linear region and therefore are not used. Similarly in Figure 6f one can see that the LET-I telescope with its much larger geometrical factor and lack of shielding has now become saturated and the apparent count rates have fallen. Figure 6g for comparison shows the spectra taken in the first flux minimum outbound in the region $12.6–13.5\ R_J$. It is well fitted by a power law of exponent 3.5.

3.3. JOVIAN ALPHA PARTICLES

Unambigous alpha particle identification was obtained in the 3.2–5.6 MeV nuc^{-1} interval using two-parameter analysis in the LET-I telescope. The alpha particle flux in this energy range was quite small in the outer Jovian magnetosphere, making it necessary to average data over fairly long time periods to obtain reasonable statistics. Figure 7 shows the ratio of alpha and proton intensities as a function of Jovian

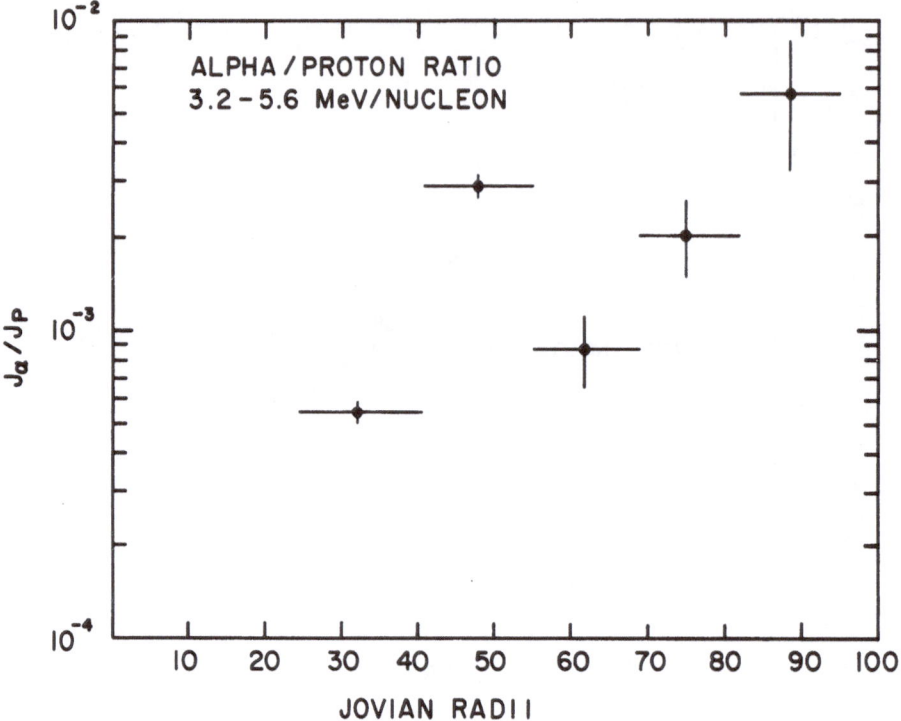

Fig. 7. Alpha-to-proton flux ratio as a function of Jovian radius. Ratio is taken for equal values of energy/nucleon.

radius for the inbound pass of Pioneer 10. A general decreasing trend in the ratio is apparent with one exception, the point between 40 and 55 R_J. This point, however, occurs at the time when Pioneer 10 reentered the magnetosheath (Wolfe *et al.*, 1974). The abnormally high value of the point suggests that conditions near the boundary of the magnetosphere prevailed during this time and is therefore quite consistent with Wolfe *et al.*'s (1974) observations. The ratio varies between $\sim 6 \times 10^{-3}$ and $\sim 6 \times 10^{-4}$ over the region measured. This is to be compared with the solar wind α/p ratio which varies over the range 0.01–0.1 (see, for example Wolfe *et al.*, 1966; Gosling *et al.*, 1967). Thus the α/p ratio in the outer part of the Jovian magnetosphere is closest to the solar wind values.

The α/p ratio in the Earth's magnetosphere shows no L dependence but apparently has a vary strong latitude dependence (D. J. Williams, private communication). The value of the ratio at Earth spans the range 10^{-3}–10^{-4}. With the limited statistical accuracy of our data it is difficult to establish whether a strong latitude dependence exists. The earth observations, however, were made in regions where particles are stably trapped, and such is almost certainly not the case for the measurements presented here.

3.4. PROTON AND ELECTRON ANGULAR DISTRIBUTIONS

Figures 8 through 11 sample the angular distributions measured for two of our logical rates from the LET-II telescope, showing hourly average data for 0.49–2.15 MeV proton and 0.78–1.0 MeV electrons. In the outermost regions of the magnetosphere anisotropies are present and variable in both the electrons and protons. By the time one reaches $\sim 70\ R_J$ (Figure 8), the electrons have already settled down to a nearly isotropic behavior, but the protons are still quite variable as to the magnitude and direction of the anisotropy. The direction of the anisotropy is generally correct for the co-rotation effect which should be from the left in the figures. Figure 9 illustrates

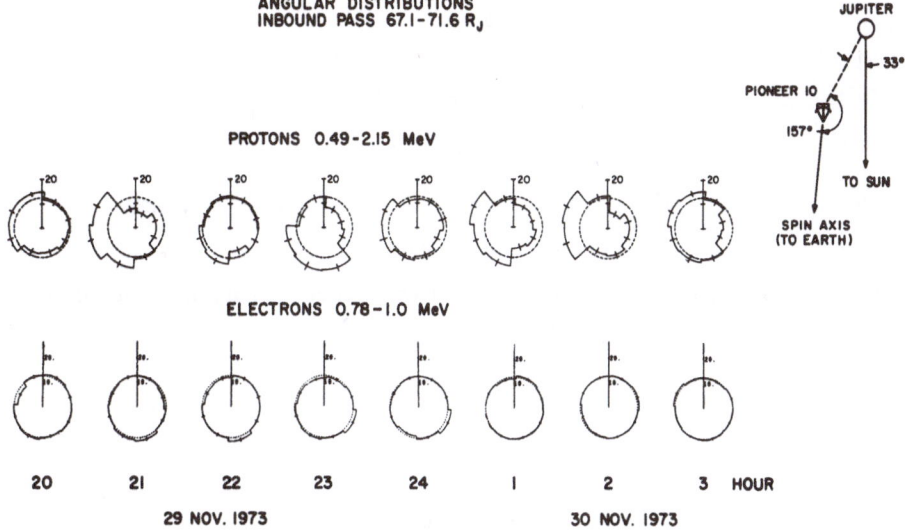

Fig. 8.   Polar plots of angular distributions of proton and electron counting rates. Top of the figure is towards north ecliptic pole. Normal out of page is in direction of spacecraft spin axis.

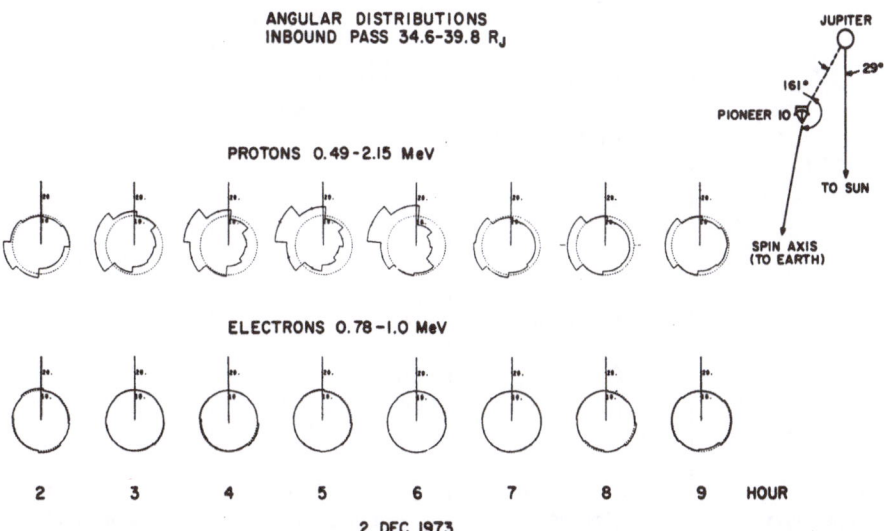

Fig. 9.   Polar plots of angular distributions of proton and electron counting rates. Orientation of plots is the same as in Figure 8.

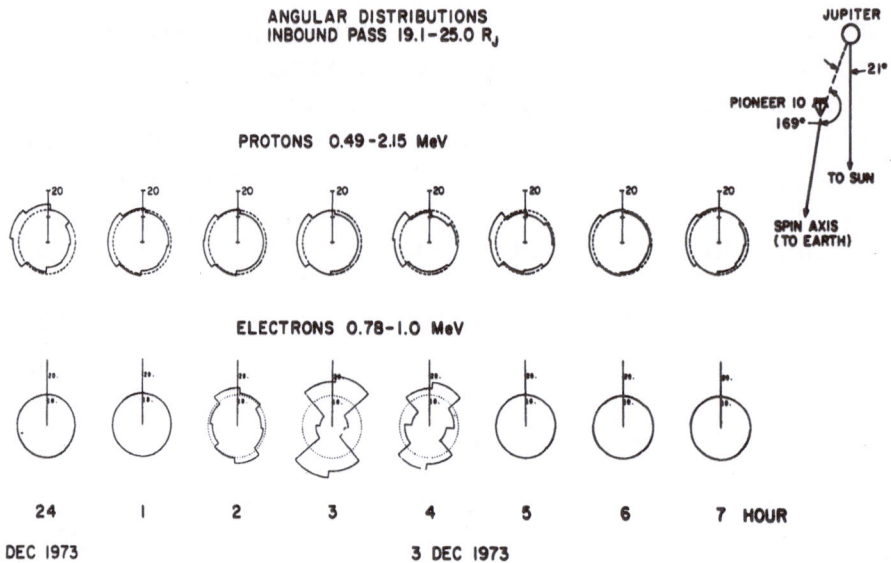

Fig. 10.   Polar plots of angular distributions of proton and electron counting rates. Orientation of plots
is the same as in Figure 8.

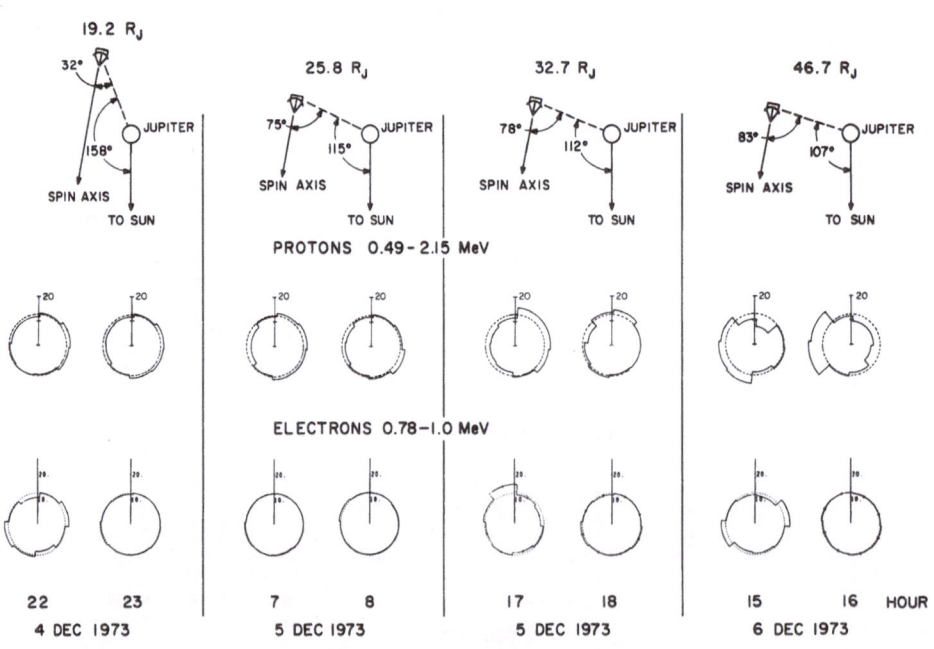

Fig. 11.   Polar plots of angular distributions of proton and electron counting rates. Orientation of plots
is the same as in Figure 8.

data from 40 to 35 R_J showing that the electrons have become even more isotropic, and the proton distributions have settled down to a pattern more regular in amplitude and direction. The rocking of the distribution due to the 10-h periodicity in the magnetic field is quite clear.

Figure 10 illustrates data from 25 R_J to 19 R_J. The proton distributions show a well defined, stable anisotropy from the co-rotation direction. The electrons at the beginning and end of the period show an extremely isotropic behavior; however, there is a huge double-ended anisotropy seen as we approach the expected location of the magnetic equator (See Figure 2). This anisotropy smoothly came and went over a period of almost 4 hours and was aligned along the magnetic lines.

The out-bound data covered in this paper is summarized in Figure 11. These data are all taken near the flux maxima in 10-h periodicity so strongly apparent in the outbound data. Small anisotropies are commonly seen in the electron data, most often a double-ended distribution, and presumably at the location of the magnetic equator. The protons generally show a behavior similar to the inbound pass, except, of course, that the sense of the co-rotational effect is from the right now. In the data shown here outside 30 R_J, more variation in direction is noted outbound than inbound at the same distance, and the proton distribution shown near 47 R_J is most unusual. The co-rotational effect is being completely masked. This behavior began after ~ 1400 UT and persisted for several hours, again while the spacecraft was apparently near the magnetic equator.

3.5. HARMONIC ANALYSIS OF PROTON ANGULAR DISTRIBUTIONS

The great extent of the Jovian magnetosphere combined with the relatively short rotation period means that co-rotation anisotropies should be an important aspect of the particle angular distributions. Furthermore, the weak, non-dipole Jovian magnetic field in the 50–25 R_J region (~ 20 γ) suggests that anisotropies produced by particle intensity gradients may be significant. A number of rates from the three detector systems were sectored into 8 different bins for each spacecraft rotation and then were than summed over 5 rotations. In this section we will be concerned with the LET-II data for the proton energy intervals 1.2–2.15 MeV. In addition there were 4 electron intervals extending between 0.1 and 2 MeV.

The opening angle of the LET-II (30°) is small compared to the 45° sector width so no deconvolution of the data is required.

A harmonic analysis of the form

$$J(\theta) = A_0 + A_1 \sin(\theta - \theta_1) + A_2 \sin 2(\theta - \theta_2)$$

was performed on the 3600 s averages. The resulting values for A_1/A_0, θ_1 and A_2/A_0 for the 1.2–2.2 MeV interval are shown in Figures 12, 13 and 14. Examining the plot of θ_1, there are large systematic fluctuations which steadily decrease and at 23 R_J there is a well defined 'hinge point'. Inside this point there would appear to be rigid rotation of the field with the rotation of the planet. Thus 23 R_J is probably an upper limit on the region of durable trapping. This is also the point where there is a steady

J. H. TRAINOR ET AL.

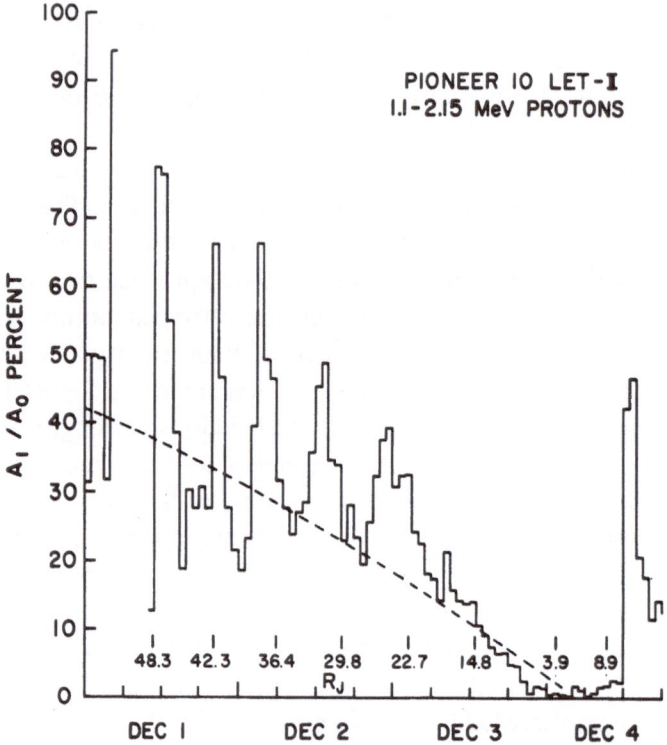

Fig. 12. Magnitude of the first harmonic of the proton 1.2–2.15 MeV angular distribution as a function
of time and Jovian radius, dashed line indicates magnitude of anisotropy predicted
from co-rotation alone.

increase in the 14–21 MeV proton flux. We have not completed the analysis of the
variation of θ_1 in the transition region between 50 and 25 R_J. This region is obviously
'hinged' between the outer diffusion region and that of rigid rotation. However there
are probably other effects present. The changes observed in the low energy (0.4–
1 MeV) proton spectra suggest the particles in this region are being accelerated at
a very slow rate.

A_1/A_0 is a measure of the first order anisotropy. The expected co-rotation anisotropy is given by

$$\xi = \frac{f(v_1) - f(v_2)}{f(v_1) + f(v_2)}$$

$$v_1 = V_r + v$$

$$v_2 = V_r - v,$$

where $f(v) \propto J(E)/E$ is the non-relativistic particle distribution function, v is the
particle velocity and $V_r = \omega r$ is the rotation velocity.

For power-law spectra this is of the form

$$\xi = \frac{E_2^{\gamma+1} - E_1^{\gamma+1}}{E_2^{\gamma+1} + E_1^{\gamma+1}}$$

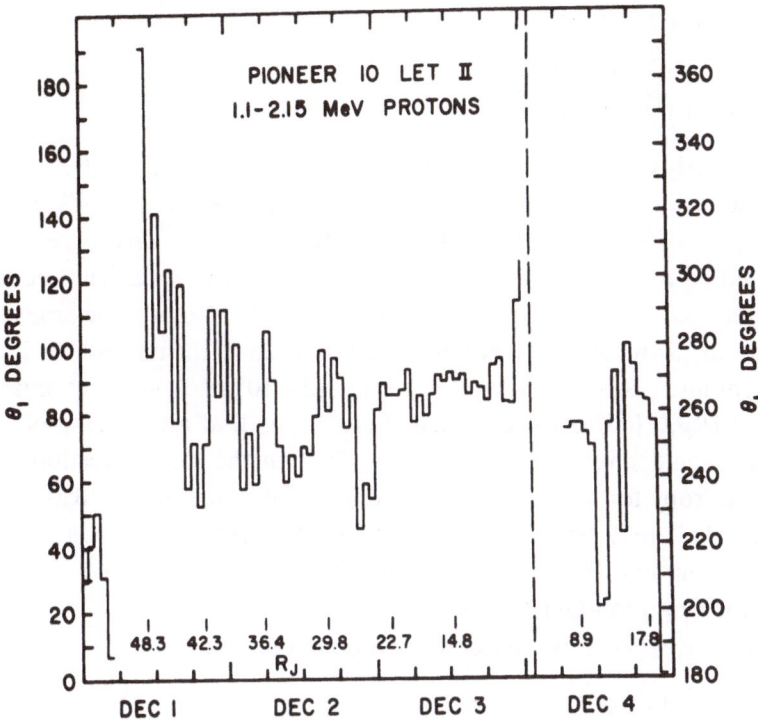

Fig. 13. Direction of the first harmonic as a function time and Jovicentric distance. Note that the scale on the left is different from the scale on the right. The dashed line indicates the regions where the different scales apply.

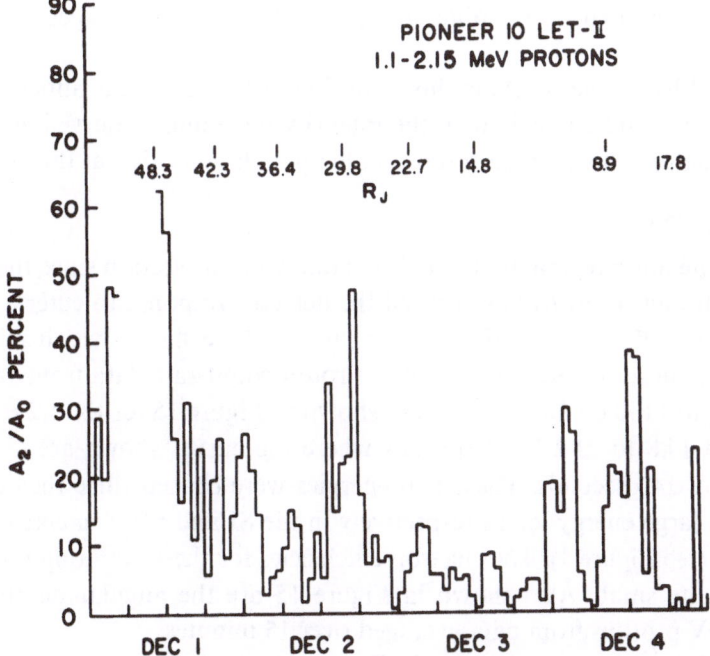

Fig. 14. Amplitude of the second harmonic of the 1.2–2.15 MeV proton counting rate as a function of time and Jovian radius.

and for $\gamma = 4$ this reduces to

$$\frac{(v+V)^{10} - (v-V)^{10}}{(v+V)^{10} + (v-V)^{10}}.$$

For the 1.2–2.2 MeV region $\bar{E} \approx 1.4$ MeV; $v = 1.6 \times 10^7$ m s^{-1}; $V_r = 12.6\, N \times 10^3$, where N is the distance to the observing point measured in units of R_J ($= 7.137 \times 10^7$ m). At 25 R_J, $V_r = 3 \times 10^5$ m s^{-1}; $V_r/v = 0.019$; $\xi = 18\%$; at 50 R_J this increases to 37%. The expected co-rotation anisotropy for $\gamma = 4$ is shown as a dashed line in Figure 12. The agreement with the equatorial values of A_1/A_0 is good. The larger peaks of A_1/A_0 between 25 and 50 R_J are not understood. At 25 R_J the magnetic field is $\sim 20\,\gamma$ and the gyroradius, R_c of a 1.4 MeV proton is 12×10^8 m. A gradient anisotropy of the order of $(R_c/J)\,(\mathrm{d}J/\mathrm{d}R)$ is expected. This means that an increase of 100% per Jovian radii would give an anisotropy of $\sim 20\%$ in the same direction as the co-rotating anisotropy for negative values of $\mathrm{d}J/\mathrm{d}R$ (i.e. value of J increasing with decreasing R_J). This may be the explanation for the large increases in A_1/A_0 which coincide with minima in the particle intensity. For smaller value of v/V_r, A_1/A_0 reduces to the Compton-Getting factor

$\xi = (2 + \alpha\gamma)\, v/V_r$ where α is the following function of energy

$$\alpha = \frac{T + 2m_0 c^2}{T + m_0 c^2}.$$

For electrons the smaller value of α and γ and the smaller value of v/V_r all combine to make the 0.5 MeV electron co-rotation anisotropy ~ 30 times smaller than that of the protons and explain the difference generally observed between the two components.

The second harmonic A_2/A_0 is shown in Figure 14. The peaks anticorrelate with the A_1/A_0 peaks and coincide with the intensity maximum observed as the Jovian magnetic equator. This suggests a more stable population exists at this point.

3.6. PROTONS INSIDE 10 R_J

Protons in the inner region are treated separately in this section since the extremely high flux rates inside 10 R_J have caused the detector response to enter a non-linear regime and significant corrections are required. These flux values inside 6 R_J are considered accurate within a factor of 2. Proton count-rate data from 1830 UT on December 3 to 1130 on December 4 are shown in Figure 15 for the 1.2 to 2.15 MeV protons and 14.8 to 21.2 MeV protons where the curves shown are derived from averages over 144 seconds. These two energies were chosen since they correspond to relatively large energy losses respectively in the SI and SII elements of the LET-II telescope (see Figure 1). The electron efficiencies for these counting rates are consequently quite small. Also shown in Figure 15 are the angular distributions for 1.2–2.15 MeV protons from data averaged over 15 minutes.

The angular distributions shown in Figure 15 for the 1.2 to 2.15 MeV protons are most interesting. The spacecraft was at $\sim 20°$ S magnetic latitude at ~ 1730 and

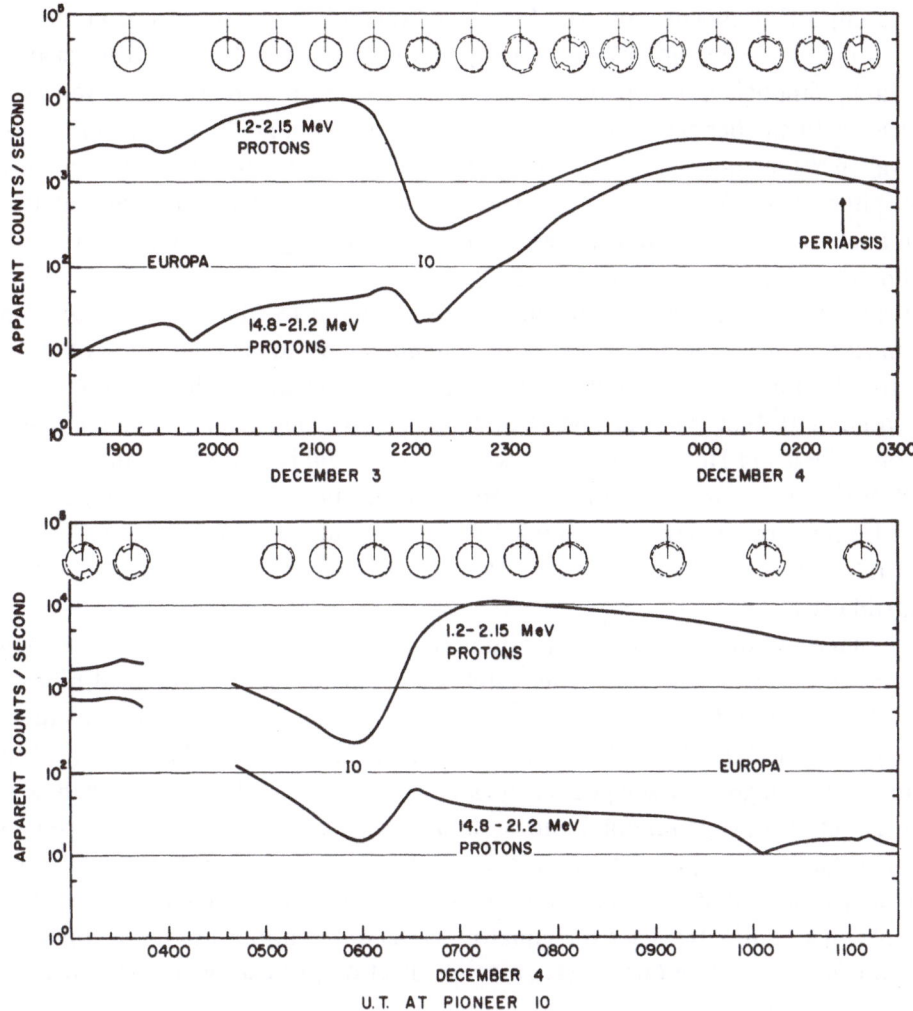

Fig. 15. Time histories of proton count rates at two different energies for the near-periapsis period. 15 min averages of angular distributions are also shown. Locations of orbital crossings of Io and Europa are indicated.

moving inbound towards the magnetic equator which it crossed at ~0100 on December 4. On the outbound leg, we reach a maximum northerly latitude of ~20° at ~0500 and return to near the equator at ~1000 on December 4. A small anisotropy, undoubtedly due to corotation, exists for many hours, up to ~1900 on December 3. Then as the spacecraft moves towards the equator (2000–2130) a small double-ended anisotropy appears with maxima perpendicular to the field lines. (It is difficult to see this effect in Figure 15 due to the photographic reduction.) From our detailed data, this situation has clearly ended by 2230, going over to an X-type distribution from 2300 to 2330, and evolving in a complicated fashion to a bi-lobed distribution near the magnetic equator. The outbound data is the inverse of the above, with the exception that the anisotropies are larger at later times since the spacecraft is much nearer the equator.

Two important conclusions are obvious. As one moves across Io's orbit and observes a flux drop of a factor of 60, the angular distribution of the particles changes very little, although it is apparent that there is a small preference for Io removing particles with smaller pitch angles. Inside the orbit of Io, as the flux in this energy interval increases, it is apparent that particles are at first preferentially added at small pitch angles for a period of 30–45 minutes, and then preferentially at large pitch angles as the spacecraft moves towards the crossing of the magnetic equator. Outbound the inverse sequence is true.

There are several features of the flux curves of Figure 15 which deserve comment. The most obvious is the huge effect of Io in removing protons in the ~1 MeV range. Had the effect not occurred, one could speculate that the flux of these protons could have been $\sim 10^{10}$ cm^{-2} s^{-1} at ~0100 on December 4. Long before encounter several authors (Mead and Hess, 1974; Birmingham *et al.*, 1974) had predicted the sweeping effect of the Jovian moons. While the effects do occur, in detail they are quite different from the predictions, depending upon the moon and the energy range. A later work (Hess *et al.*, 1974) improves on the earlier work. Differences in detail, however, still exist.

The shape and size of the Io effect is quite similar inbound and outbound, although the spacecraft was leading Io in its orbit by less than 90° inbound and by more than 180° outbound. It thus appears that there is little azimuthal dependence of the effect for the 1.2–2.15 MeV protons and only a small one for the 14.8–21.2 MeV protons. These higher energy protons do show a smaller but clear signature in crossing the orbit of Europa inbound and outbound, although the magnitude appears to be somewhat different in the two cases. In the lower energy protons, there is a small effect at Europa inbound when the spacecraft was well off the equator and no observable effect outbound near the equator. These observations are quite consistent, since one expects the moons to have a small effect on particles which are mirroring near the equator.

Simpson *et al.* (1974), have pointed out a small latitude effect for the >35 MeV protons. These protons showed very strong maxima near $L \sim 3.6 \ R_J$. The first peak was largest and occurred essentially at the magnetic equator while the second peak occurred at ~15°N. Our data for the lower energy protons show a maximum in the flux at ~0100 on December 4, near the magnetic equator, but not the striking maximum shown in the >35 MeV data. However, on the outbound trajectory both of our lower energy proton fluxes show slight maxima at the time of the second peak reported by Simpson *et al.* (1974).

The fluxes of protons measured by this instrument inside the crossing of the magnetic equator continue to fall off, but it is obvious from the fluxes at ~0100 and ~0330 when the spacecraft was at $L \sim 3.6$ that the latitude dependence is not large over the latitude range covered by Pioneer 10. The decrease in fluxes inbound to periapsis may well be due to the effects of Amalthea, which orbits at 2.55 R_J. Apparently due to diffusion, the effects of particle removal by Io were seen at least 1 R_J outside the orbit of Io. A similar effect may indeed be present due to Amalthea.

The Pioneer 11 spacecraft will encounter Jupiter in early December, 1974. Two advantages of its trajectory are the high latitude region covered and the very close approach to the planet. One should learn a great deal more about this most interesting inner region if the spacecraft survives.

4. Jovian Electrons Outside the Bowshock

4.1. PIONEER 10 AND 11 MEASUREMENTS

We turn now to a discussion of observations of Jovian accelerated electrons in interplanetary space on Pioneer's 10 and 11. Let us first examine the relative locations of Pioneers 10 and 11 and Jupiter during the period of the measurements. The Pioneer 10 and 11 trajectories are shown in Figure 16. Superimposed on the trajectories are nominal interplanetary magnetic field lines, assuming a constant solar wind speed of 400 km s^{-1}. The location of Jupiter six months prior to encounter is also shown. This figure will be referred to in the following discussion of the data.

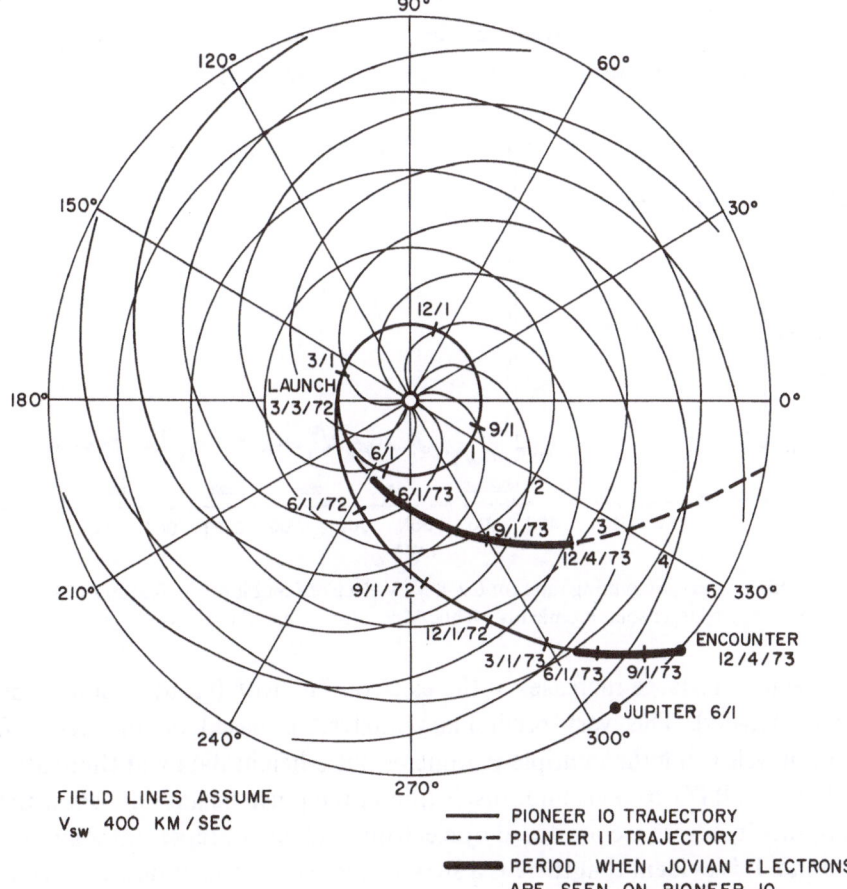

Fig. 16. Pioneer 10 and 11 trajectories are shown in a non-rotating heliocentric coordinate system. Superimposed are idealized spiral magnetic field lines assuming a constant solar wind speed of 400 km s^{-1}. The marks on the 1 AU circle indicate the position of the earth once each quarter.

The Pioneer 10 and 11 daily average electron counting rates in the energy range
6.2–8 MeV are shown in Figure 17(a) and (b) for 1973. Due to the presence of radioisotope power generators (RTG's) on the spacecraft, the background in the electron
counting rates on both Pioneer 10 and 11 is roughly a factor of 40 larger than that
due to true cosmic-ray electrons. Increases over background on these plots there-

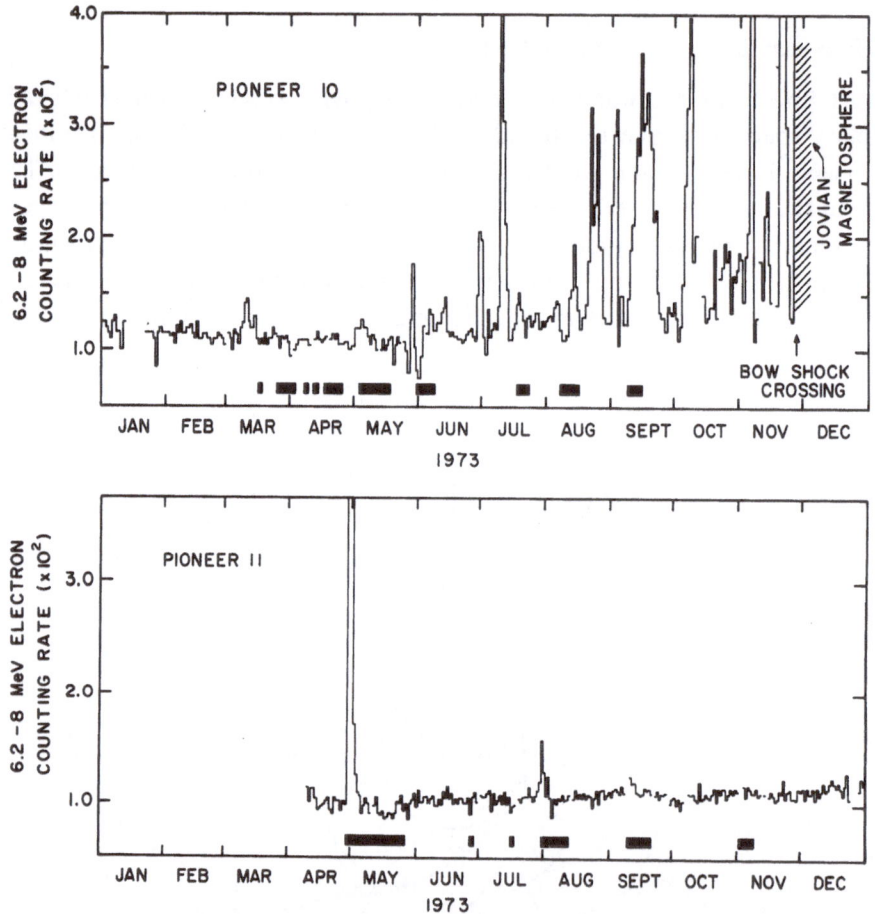

Fig. 17.   The daily average counting rates for 6.2–8 MeV electrons on Pioneers 10(a) and 11(b). The black
rectangles indicate the larger solar cosmic ray events. Most of these are of the low-energy co-rotating type.

fore represent very large increases in the electron intensity (i.e. at least 40 times the
expected intensity). This very preliminary analysis is based on the 'Rates Data'.
Further analysis using the multiple-parameter pulse-height data will eliminate most,
if not all, of the RTG background. Inspection of the pulse-height data show that the
counting-rate increases are produced by electrons and do not represent some spurious
background. It is evident that there is a striking increase in the occurrence of electron
events as Pioneer 10 approaches Jupiter. The black boxes indicate times when solar
protons were present at Pioneers 10 and 11. The largest solar proton event seen on
Pioneer 10 in 1973 was that on 3 May, yet we see that the accompanying electron

increase barely exceeded background. It is unlikely than that the other electron increases on Pioneer 10 in 1973 are of solar origin. Furthermore, in marked contrast to Pioneer 10, there is very little activity seen on Pioneer 11. The only two exceptions (28 April and 30 July) are clearly flare associated. The level of activity steadily increases as Pioneer 10 approaches Jupiter. The small size of the flare associated events on Pioneer 10 and 11 and the lack of any significant Pioneer 11 electron increases show that these electrons are not of solar origin. The steady increase as Pioneer 10 approaches Jupiter as well as the thirteen-month periodicity observed in quiet-time electron increases (as discussed in a later section) indicate these electron bursts are of Jovian origin. Jovian electrons were seen possibly as early as 8 March 1973 and were certainly present by the end of May. During the March–May period Pioneer 10 was between 1.2 and 0.8 AU away from Jupiter. Electrons escaping from Jupiter are therefore seen at quite large distances from the planet.

Let us now refer back to Figure 16 and examine the relative locations of Pioneers 10 and 11 during the times that Jovian electrons are present. Both Pioneers 10 and 11 are moving roughly perpendicular to the interplanetary magnetic field lines during this time. Pioneer 10 does not cross the nominal field line connecting to Jupiter until the actual time of encounter. In fact, during the March–June period Pioneer 10 and Jupiter nearly line up along the same radius to the Sun. Since the electrons are expected to propagate mainly along field lines, some distortion of the field relative to the nominal spiral is necessary for Jovian electrons to be easily seen at Pioneer 10.

During the latter half of 1973 Pioneer 11 is, on the other hand, much more distant from Jupiter's field line. To permit Jovian electrons to propagate to Pioneer 11 would require a very large deformation of the magnetic field line. It is therefore quite reasonable that no large Jovian electron increases are seen on Pioneer 11 during this time.

To show in more detail the time structure and energy dependence of these events we have plotted three electron counting rates at different energies on an expanded scale (6 h av.) in Figure 18. Solar proton events, which are principally low-energy co-rotating types, are again indicated by black boxes. Details of the time structure of these electron events are more evident in this figure. They appear to have roughly equal rise and decay times – typically of the order 1–2 days. The duration of an event is generally of the order of 2–4 days. The longer events, particularly in August and September 1973, appear to be superpositions of shorter events. At this point it is not possible to determine whether this time structure is a feature of the escape process or is a result of varying connection conditions due to changes in the interplanetary magnetic field. The general increase in electron activity in the vicinity of Jupiter is even more strikingly apparent in this figure than in Figure 17. Keeping in mind that the counting rate background level is ~ 40 times the nominal electron intensity, it is evident that Jovian electrons are strongly dominating a region around the planet at least 0.5 AU in extent.

The electron spectrum during the increase of 5 November is shown in Figure 19.

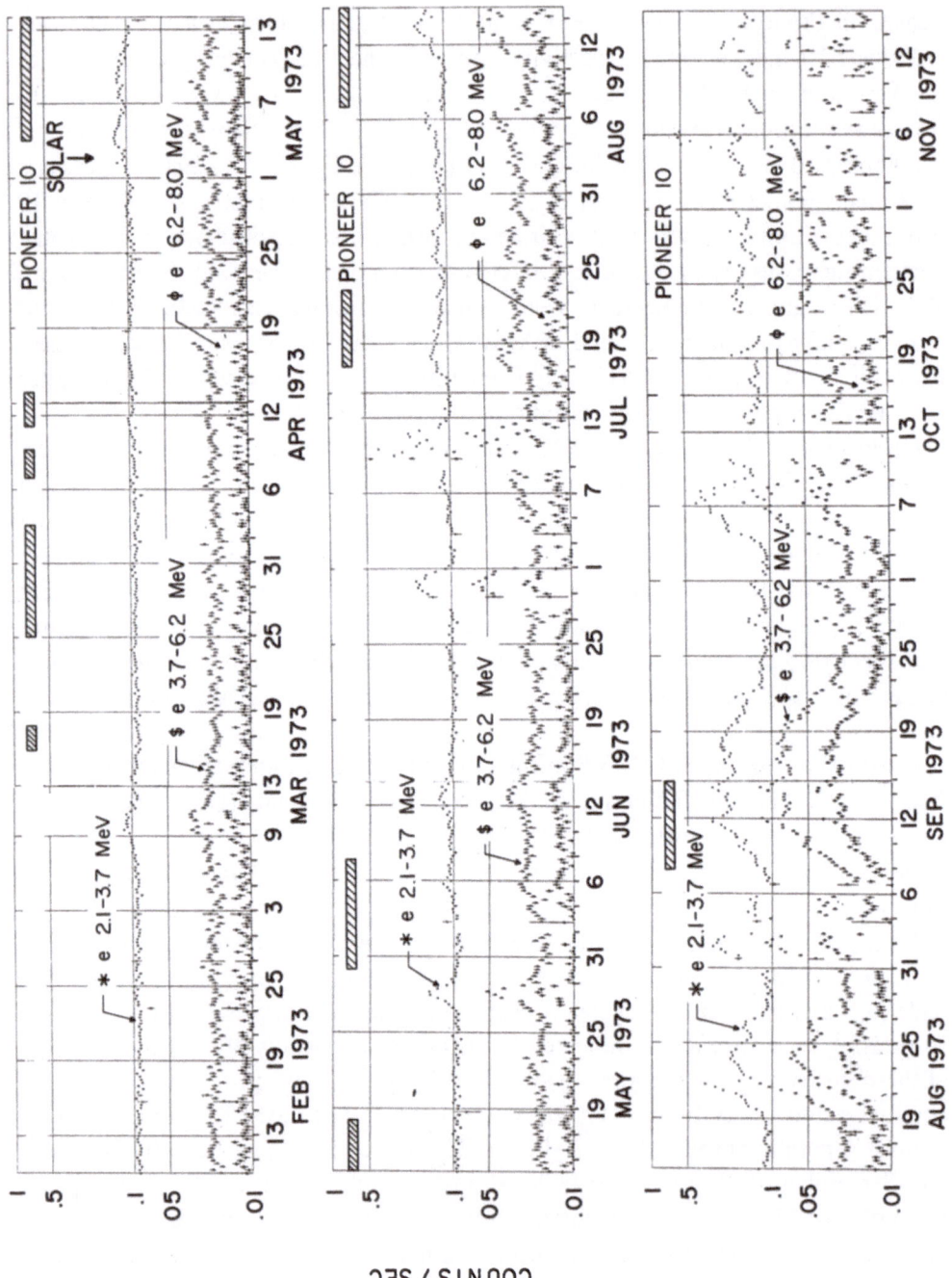

Fig. 18. Six hour counting rate averages are plotted for three electron energy intervals (2.1–3.7 MeV, 3.7–6.2 MeV and 6.2–8 MeV). The cross-hatched rectangles are the same solar proton events of Figure 17. Note that in the November 5 event data points from the two highest energy channels are intermixed with their nearest neighbors.

It was necessary to choose one of the largest events to derive a spectrum in order to insure that the lowest energy points significantly exceeded the radioisotope power supply background. We therefore cannot be certain that this spectrum represents the behavior of all events, particularly the smaller ones at large distances from the planet. The data are consistent with an $E^{-1.5}$ power law over the 0.12–8.0 MeV range.

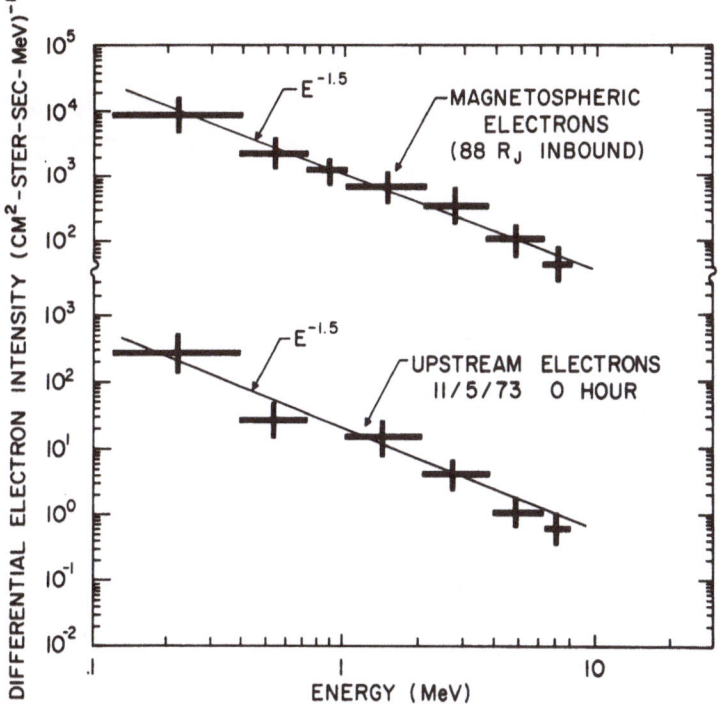

Fig. 19. Electron energy spectra for a large interplanetary electron increase and in the outer Jovian magnetosphere. The data between 0.1 and 2 MeV is from the LET-11 telescope (Trainor *et al.*, 1974). This has a geometric factor of 0.015 cm^2 s^{-1} so meaningful measurements are possible only during the largest electron increases.

This is a preliminary analysis and present uncertainties in the calibration of the detector are responsible for an error of $\sim \pm 0.3$ in the power law exponent. Also shown in Figure 19 is a representative spectrum of electrons in the outer part of the Jovian magnetosphere taken from Figure 4. It was shown earlier that the index of the differential electron spectrum in the outer part of the Jovian magnetosphere ($> 30\ R_J$) is always in the 1.5–2.0 range in the energy interval 0.12–8.0 MeV. The relative similarity of the two spectra in Figure 19 further confirms the Jovian origin and shows that the escape of these electrons from the Jovian magnetosphere is an energy-independent process. As discussed earlier the electrons in the outer Jovian magnetosphere do not appear to be stably trapped with most of them diffusing into interplanetary space.

4.2. OBSERVATION OF JOVIAN ELECTRONS AT THE EARTH

We have seen in the preceding section that Jupiter is a copious source of MeV elec-

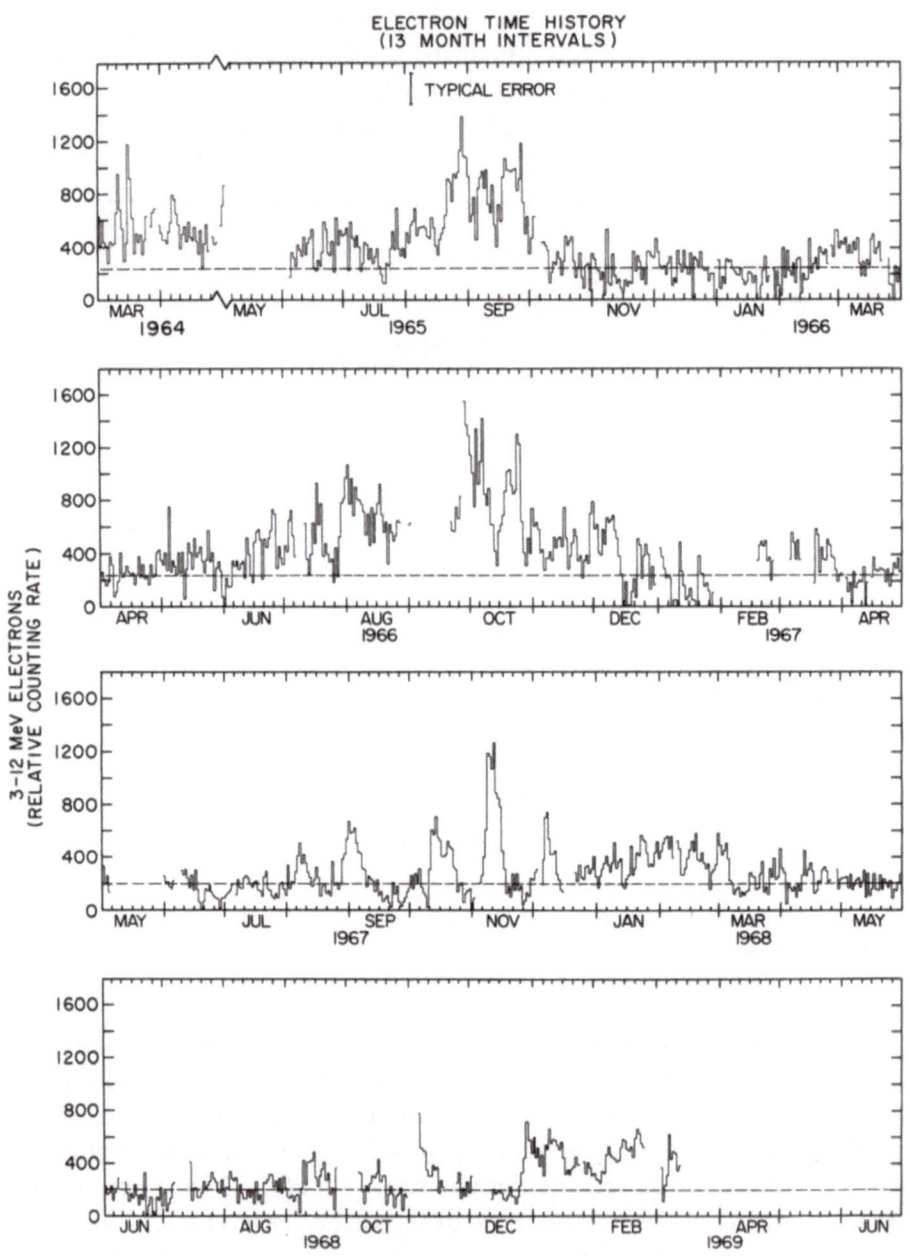

Fig. 20a.

trons. Furthermore, we have established that these electrons are present at least as far as 1 AU away from the planet. The question logically arises, could Jovian electrons be reaching the earth, and, in particular, could they be responsible for the quiet-time increases at 1 AU? Let us first assume that electrons from Jupiter are

ELECTRON TIME HISTORY
(13 MONTH INTERVALS)

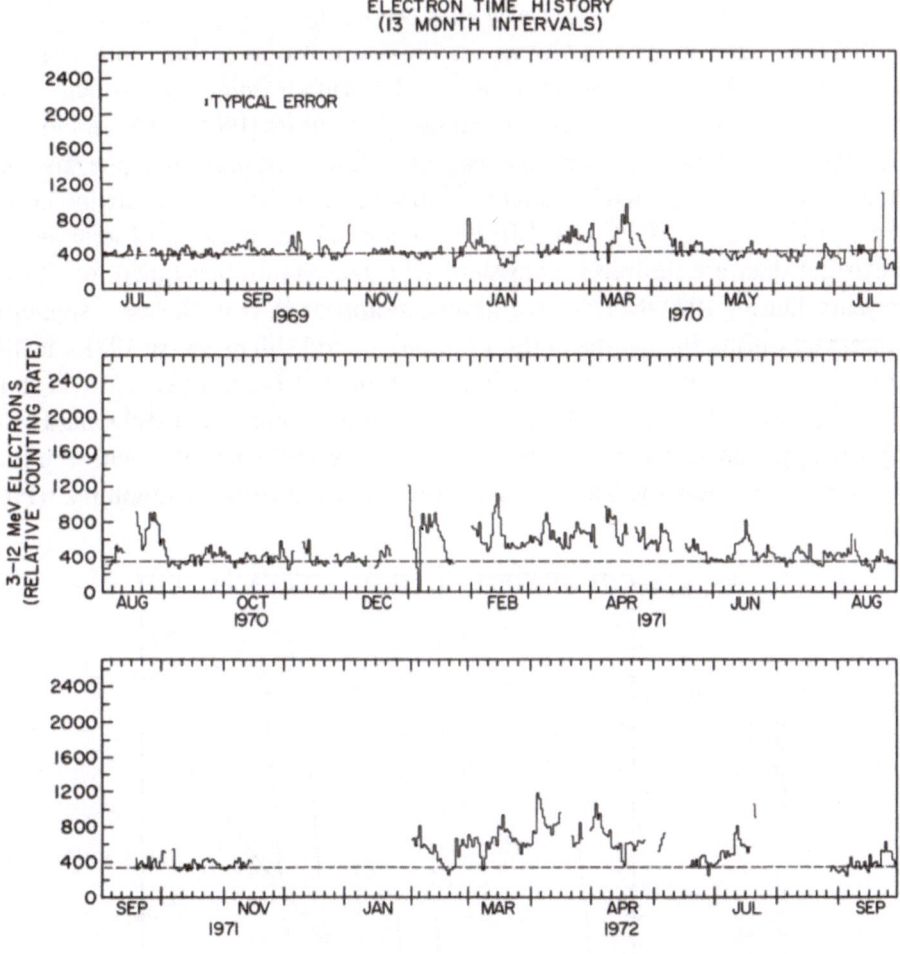

Fig. 20b.

Fig. 20a–b.   3–12 MeV data from IMPs III and IV (McDonald *et al.*, 1972) and IMP V (Van Hollebeke, private communication) plotted in 13 month epochs. The dashed line is a convenience for identifying the electron increases. There still may be major solar contributions in some periods such as 28 July–10 August, 1966 and 20 September–10 October, 1966. The 13 month periodicity is clearly defined and the amplitude decreases over solar maximum (~1969).

strongly tied to field lines intersecting the Jovian magnetosphere. Such electrons would be seen at Earth mainly at times when the earth crossed field lines connecting with Jupiter. If the field were a perfect Archimedean spiral, this would occur every 13 months (since Jupiter's period is 11.86 yr the Earth's synodic period with respect to Jupiter is 13 months). The question then arises, do 13 month periodicities exist in the quiet-time increases at the Earth? In Figure 20 we have plotted the 3–12 MeV electron counting rate from the Earth orbiting satellites IMPs III, IV, (McDonald *et al.*, 1972) and IMP V (Van Hollebeke, private communication) on 13 month epochs over the period 1965–1972. Using >20 MeV protons as an indicator, periods when solar activity was present have been removed. However, there appear to be significant solar MeV electrons present from large corotating events in mid-1966. The large increase in November 1967 may also have some solar contribution. The 13 month periodicity in the electron intensity is immediately obvious. The increases are most pronounced during the first three years (1965–1967) and are generally a factor of 2–5 over the normal background level. The maxima in the envelopes of these increases is separated by roughly 13 months. The data in 1969 are incomplete due to a gap between IMP IV and IMP V coverage and the presence of solar activity. The existing data are, however, consistent with the picture developed for the first three years. During 1970 the effect has almost disappeared. Nonetheless, a significant enhancement during the middle of the 13 month period still exists. In 1971 and 1972 the 13 month variation has increased in magnitude, but has not yet returned to the level of the 1965–1967 period. As has been previously suggested (McDonald et al., 1972) it is apparent that the quiet time increases are anti-correlated with the long-term level of solar activity. The increases were largest near solar minimum, (1965–

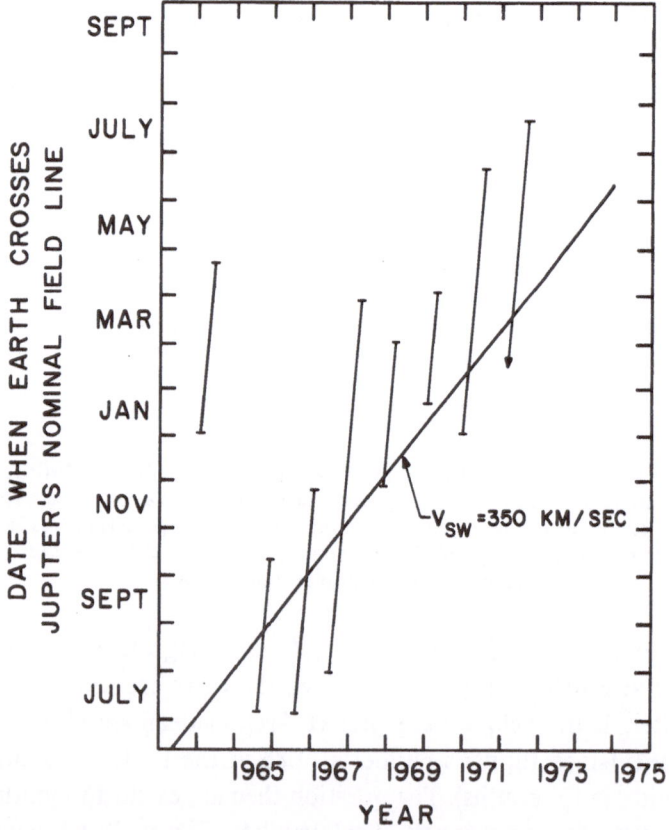

Fig. 21. The vertical bars give the duration of the periods when quiet-time increases were present during the various epochs shown in Figure 20. Arrows indicate when the length of the bar is uncertain due to a data gap. The diagonal line represents the time of the year that an idealized spiral interplanetary magnetic field line would connect the Earth and Jupiter assuming a constant plasma velocity of 350 km s^{-1}.

66), nearly vanished near solar maximum (1969–70), and began to increase as the next solar minimum was approached (1971–72).

In Figure 21 the time of the year that the Earth crosses Jupiter's field line is plotted for each year from 1964–1974. The vertical bars in the plot give the duration of the periods when quiet-time increases were present. With the exception of the 1964 bar all of the periods fall close to or contain the predicted time of crossing Jupiter's field line. There does, however, appear to be a tendency for the field line crossing to occur near the beginning of these periods. Field line distortion will occur due to changes in solar wind speed. Speeds greater than 350 km s^{-1} will produce straighter field lines which in turn will mean that the Earth-crossing of Jupiter's field line will take place earlier in time. Such a mechanism is therefore incapable of explaining the tendency of increases to occur after the field line crossing. Field line co-rotation would tend to sweep particles around azimuthally in the co-rotation direction which in turn could permit electrons to be seen at the Earth after the time that Jupiter's field line is crossed. This mechanism would, however, require long electron containment times (~ 7 days). In view of the short decay times of solar flare electrons ($\lesssim 1$ day) such a mechanism is probably not possible.

The specific details of the quiet-time increases are also of interest. For example, the average duration of the individual increases at Earth is generally on the order of 10 days, which is somewhat longer than the duration of the increases seen on Pioneer 10 near Jupiter.

The magnitude of the increases differs greatly between Pioneer 10 and the Earth. At Earth the increases are typically a factor of 2–5 over the background whereas on Pioneer 10 they are as much as a factor of 300 over background. Thus the increases near Jupiter are ~ 50–150 times as large as the increases near Earth. At this time we have not examined possible models for the propagation of these pulses; however the anti-correlation between quiet-time electron increases and low-energy solar protons pointed out by McDonald et al. (1972), argues that turbulence or scattering effects are most important in this process.

Preliminary analysis of the upstream electron anisotropy in the 2–8 MeV interval on a daily basis shows that anisotropies, if they exist, are $\lesssim 10\%$. Further work can be expected to reduce this limit by at least a factor of 5. The existing limit does, however, show that a significant randomization of the particle motion has occurred over the $\lesssim 0.5$ AU distance between Pioneer 10 and Jupiter at the time of the measurements.

4.3. Interpretation of Results: Can Jupiter be the Source of Quiet-
Time Low-Energy Cosmic Ray Electrons as Well?

The data presented in the preceding actions strongly support the conclusion that most, if not all, the quiet-time increases at 1 AU are of Jovian origin. We now address the question of whether some or all of the low-energy ambient electrons previously thought to be of galactic origin could possibly be coming from Jupiter. It was pointed out in the introduction that there were several puzzling features in the behavior of these particles, including their relatively small long-term modulation relative to higher energy protons and electrons. The spectra of only a few quiet-time increases during 1967 and 1969 have been reported (McDonald *et al.*, 1972). The spectral form of E^{-2} during these events was consistent with the $E^{-1.75}$ spectrum reported by Simnett *et al.* (1969) for the ambient electrons over a broader energy range (3–20 MeV). Webber *et al.* (1972) showed that many of the 1968–69 increases maintained the same spectral form down to ~ 0.2 MeV. McDonald *et al.* used this as an argument that the increases, as well as the nominal electron flux, were most probably of the same origin. The spectral index of the Pioneer 10 increases near Jupiter was ~ 1.5 with a probable uncertainty of a few tenths of a power. The Pioneer 10 spectrum is slightly flatter than the 1 AU spectra. Propagation effects, however, could produce a slight steepening of the electron spectrum as the particles travel inward from 5 to 1 AU.

Referring back to Figure 20, it is apparent that near solar minimum (1965) the integrated flux during each 13-month interval of quiet-time increase electrons at 1 AU is of the same order as the integrated nominal flux of ambient electrons. During 1967 and 1968 the quiet time increases were of the order of 25% of the nominal flux. From this point of view one can infer that Jupiter possesses the necessary source strength to supply these electrons.

Finally, we turn to the question of the propagation of these electrons from Jupiter to the Earth. If Jovian electrons are the source of the low-energy ambient cosmic ray flux, then they must somehow have been transported azimuthally over at least 180° in heliocentric longitude. Diffusion across field lines alone is probably insufficient to do this. Solar wind speed variations can probably only account for spreading the electrons out over 90° (3-month duration). Another possible mechanism is transport due to corotation of the interplanetary magnetic field lines that Jupiter encounters as it moves about the Sun. For this transport mechanism to work at all, containment lifetimes comparable to a solar rotation (i.e., ~ 15 days) are needed. The decay times of solar flare electron events are, however, generally $\lesssim 1$ day.

The 1967–1968 period is of particular interest, since the duration of the period when increases were present was the longest (~ 8 months) and the azimuthal propagation apparently most effective. This period corresponds to rather unusual interplanetary conditions where the dominant sector structure consisted of only two sectors. If in fact sector boundaries are obstacles to particle propagation, as is indicated by the confinement of quiet time increases within sectors (McDonald *et al.*, 1972), then the presence of only two sectors in 1967–1968 could indicate that propagation conditions were particularly favorable.

So far then we have been unable to come up with a satisfactory transport mechanism to fill the inner solar system with Jovian electrons. On the other hand, there are also difficulties with the assumption that this low-energy electron spectrum is due mainly to galactic knock-on electrons. This galactic knock-on contribution can be calculated accurately (Abraham *et al.*, 1966) and has essentially the same spectral shape and is a factor ~ 5 larger than the ambient electron spectrum. Generally, it is expected that the modulation of the low-energy electrons is larger than this, and the large gradients of Jovian electrons between 1 and 5 AU also support the idea of a small diffusion coefficient κ and consequently a large modulation in the equatorial plane. If this is correct, then interstellar knock-on electrons, if they are the dominant component, most probably will have had to enter over the solar poles where κ is presumably larger.

References

Abraham, P. B., Brunstein, K. A., and Cline, T. L.: 1966, *Phys. Rev.* **150**, 1088.

Birmingham, T., Hess, W., Northrup, T., Baxter, R., and Lojko, M.: 1974, *J. Geophys. Res.* **79**, 87.

Cline, T. L., Ludwig, G. H., and McDonald, F. B.: 1964, *Phys. Rev. Letters* **13**, 786.

Fillius, R. W. and McIlwain, C. E.: 1974, *Science* **183**, 314.

Gosling, J. T., Asbridge, J. R., Bame, S. J., Hundhausen, A. J., and Strong, I. B.: 1967, *J. Geophys. Res.* **72**, 1813.

Hess, W. N., Birmingham, T. J., and Mead, G. D.: 1974, preprint, March.

Hurford, G. J., Mewaldt, R. A., Stone, E. C., and Vogt, R. E.: 1973, *13th International Cosmic Ray Conference Papers*, University of Denver, Vol. 1, p. 324.

L'Heureux, J., Fan, C. Y., and Meyer, P.: 1972, *Astrophys. J.* **171**, 363.

McDonald, F. B., Cline, T. L., and Simnett, G. M.: 1972, *J. Geophys. Res.* **77**, 2213.

Mead, G. D.: 1973, private communication to all Pioneer 10 experimenters, October.

Mead, G. D. and Hess, W. N.: 1973, *J. Geophys. Res.* **78**, 2793.

Simnett, G. M. and McDonald, F. B.: 1969, *Astrophys. J.* **157**, 1435.

Simpson, J. A., Hamilton, D., Lentz, G., McKibben, R. B., Mogro-Campero, A., Perkins, M., Pyle, K. R., and Tuzzolino, A. J.: 1974, *Science* **183**, 306.

Smith, E. J., Davis, L., Jr., Jones, D. E., Colburn, D. S., Coleman, P. J., Jr., Dyal, P., and Sonett, G. P.: 1974a, *Science* **183**, 305.

Smith, E. J., Davis, L., Jr., Dyal, D. E., and Sonett, C. P.: 1974b, *Pioneer 10-Jupiter Symposium, Fifty-fifth Annual Meeting*, American Geophysical Union, April 10.

Stilwell, D. E., Joyce, R. M., Trainor, J. H., White, H. P., Jr., Streeter, G., and Bernstein, J.: 1975, to be published in *IEEE Trans. Nucl. Sci.*, NS-22, February.

Teegarden, B. J., McDonald, F. B., Trainor, J. H., Webber, W. R., and Roelof, E. C.: 1975, to be published in *J. Geophys. Res.*

Trainor, J. H., Teegarden, B. J., Stilwell, D. E., McDonald, F. B., Roelof, E. C., and Webber, W. R.: 1974, *Science* **183**, 311.

Van Allen, J. A., Baker, D. N., Randall, B. A., Thomsen, M. F., Sentman, D. D., and Flindt, H. R.: 1974, *Science* **183**, 309.

Webber, W. R., Lezniak, J. A., and Damle, S. V.: 1972, *J. Geophys. Res.* **77**, 2213.

Wolfe, J. H., Silva, R. W., McKibbin, D. D., and Mason, R. H.: 1966, *J. Geophys. Res.* **71**, 3329.

Wolfe, J. H., Collard, H. R., and Mihalov, J. D.: 1974, *Science* **183**, 303.

PIONEER 10: OBSERVATIONS OF ENERGETIC ELECTRONS IN THE JOVIAN MAGNETOSPHERE

B. A. RANDALL

Dept. of Physics and Astronomy, The University of Iowa, Iowa City, Iowa 52242, U.S.A.

1. Introduction

The University of Iowa experiment on Pioneer 10 is a simplified version of the originally proposed instrument. This simplification was due to the constraints specified by the National Aeronautics and Space Administration at the time of acceptance of the experiment. The reduced observational objectives of the experiment were:
  (a) To make an exploratory survey of the intensities, energy spectra, and angular distribution of energetic particles in the magnetosphere of Jupiter;
  (b) To measure the heliocentric radial gradient of the intensity of galactic cosmic rays $E_p \gtrsim 80$ MeV (Van Allen, 1972, 1973; Thomsen, 1974); and
  (c) To study the occurrence, intensity, and angular distribution of solar flare particles and their propagation through the interplanetary medium at large heliocentric distances.
A preliminary report of our Jovian encounter measurements has been published (Van Allen *et al.*, 1974a). The present paper represents a continuing analysis of those data.

2. Description of Instrument

The design of the experiment was guided by the theoretical intepretation by Chang and Davis (1962), Thorne (1963, 1965), and Ortwein *et al.* (1966) of the observed decimetric radio emission from Jupiter. The observations are reviewed by Carr and Gulkis (1969). The theoretical predictions indicated that Jupiter is encircled by a very high intensity of relativistic, trapped electrons.

Previous observational experience with trapped relativistic electrons produced by Starfish in 1962 (Van Allen *et al.*, 1963) also served as a guide in the design. The design objectives were that the instrument should be able to make significant determinations of the intensity of energetic electrons and yet have a very large dynamic range in intensity and be simple, rugged, and very reliable for many years. The instrument should also be insensitive to radiation damage and temperature.

The basic detectors were seven miniature Geiger-Mueller tubes. Four were the EON 6213 type, many of which have been flown on many missions since Pioneer III (1958), and the other three were EON 5107's. The latter type had never been flown before but had proved, with stringent selection and testing, to equal the EON 6213 in reliability and to have a much larger dynamic range.

The seven tubes were in three different physical arrangements. Three EON 6213's (A, B, and C) were mounted in a single block as shown in Figure 1. Detector C is

V. Formisano (ed.), The Magnetospheres of the Earth and Jupiter, 355–373. All Rights Reserved
Copyright © 1975 by D. Reidel Publishing Company, Dordrecht-Holland

in the center and is omnidirectionally shielded. Detectors A, B, and C are similarly shielded on the sides but A and B have thin unidirectional collimators in the $+X$ direction. Each detector is sampled separately; also double (AB) and triple (ABC) coincidences with a resolving time of 1 μs are sampled.

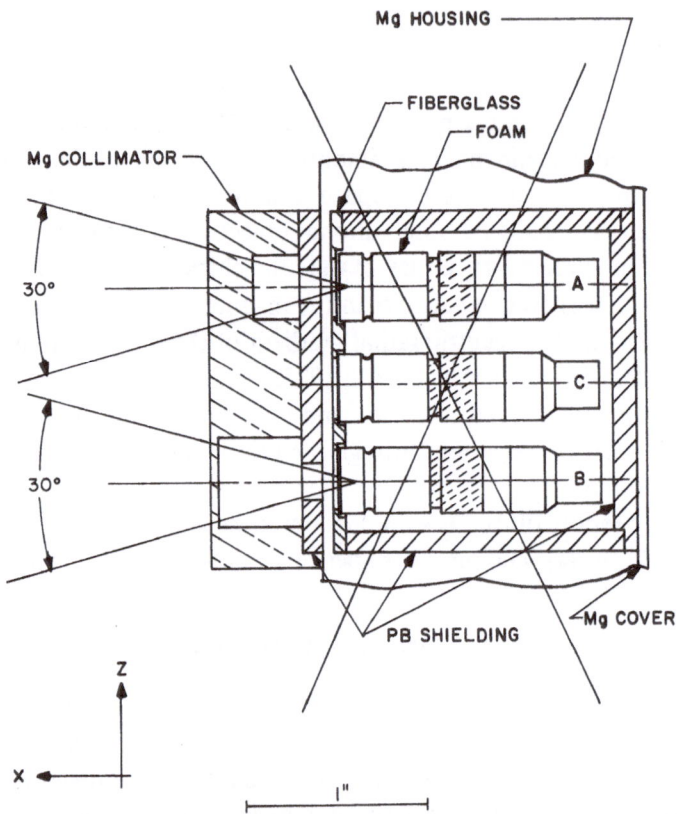

Fig. 1.   Cut-away view of one system of detectors. A, C, and B are miniature, end-window EON 6213 Geiger-Mueller tubes. The Z-axis is parallel to the rotational axis of the spacecraft.

The second assembly comprises an omnidirectionally shielded, triangular array of three EON 5107 tubes (D, E, F) as shown in Figure 2. The rate of D and the triple coincidence rate of DEF are sampled.

The third assembly, detector G, used an EON 6213 with a gold-plated scattering aperture. The purpose of the scatter arrangement was to combine the energy dependence of the scattering aperture and the higher proton threshold of the EON 6213, to obtain a hundredfold reduction in efficiency for non-penetrating protons compared with the efficiency for electrons. The physical arrangement for this detector is shown in Figure 3. The arrangement of the three within the University of Iowa experiment is shown in Figure 4.

The University of Iowa experiment uses 12 bits of each of the 192-bit main science frame. The signal processor, doubly redundant and switchable on command, uses two 24-bit accumulators which are alternately being read out or accumulating. The

24-bits are compressed into 12 bits using a quasi-logarithmic scheme of counting the leading zero order bits, then transmitting this number and the leading significant bits. A complete cycle of University of Iowa data comprises eleven main science frames. This includes a sync word and two samples from G and ABC.

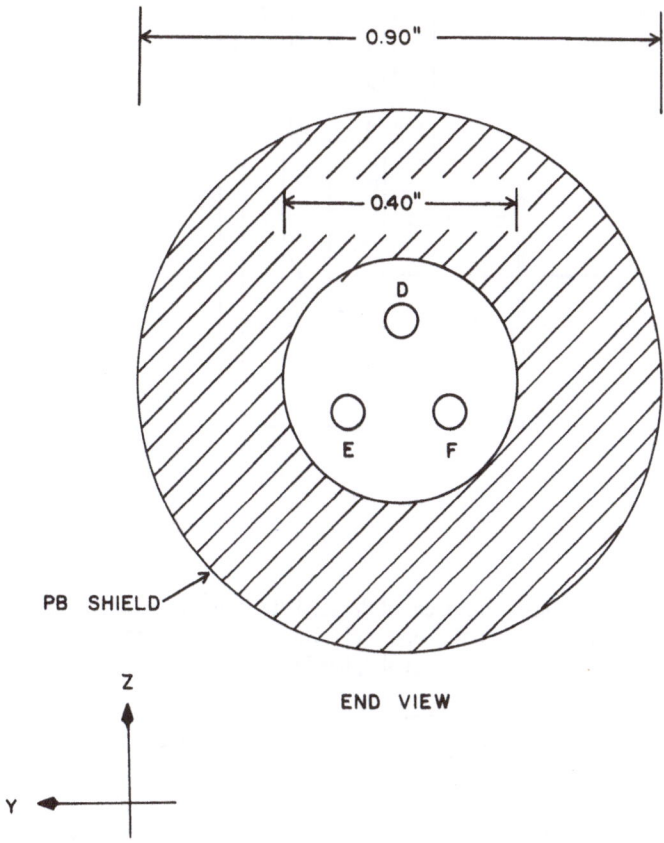

Fig. 2.   Cross-sectional view of the heavily shielded triangular array of miniature, cylindrical EON 5107 Geiger-Mueller tubes D, E, and F.

The accumulating period for each sample depends on the bit rate in the following manner: Period equals 192/Bit Rate. The bit rate was selected by ground command and was 2^N where N was between 4 and 11. During the Jovian encounter the accumulation time was either 0.1875 or 0.375 s from all real time data transmission. When the spacecraft was occulted by the planet, the data were stored at 16 bit per second on the spacecraft and subsequently played back.

The spacecraft was oriented such that the spin axis $(+Z)$ was pointed continuously at the earth with an error of less than 1°. The University of Iowa experiment looks out along the $+X$ axis. In our analysis the roll angle is measured from the ascending node of the spacecraft's equator on the ecliptic to the $+X$ axis. Because of the non-integral relationship of the spacecraft's rotational period of 12.64 s and the sampling period, successive samples are always made at different angles. Ap-

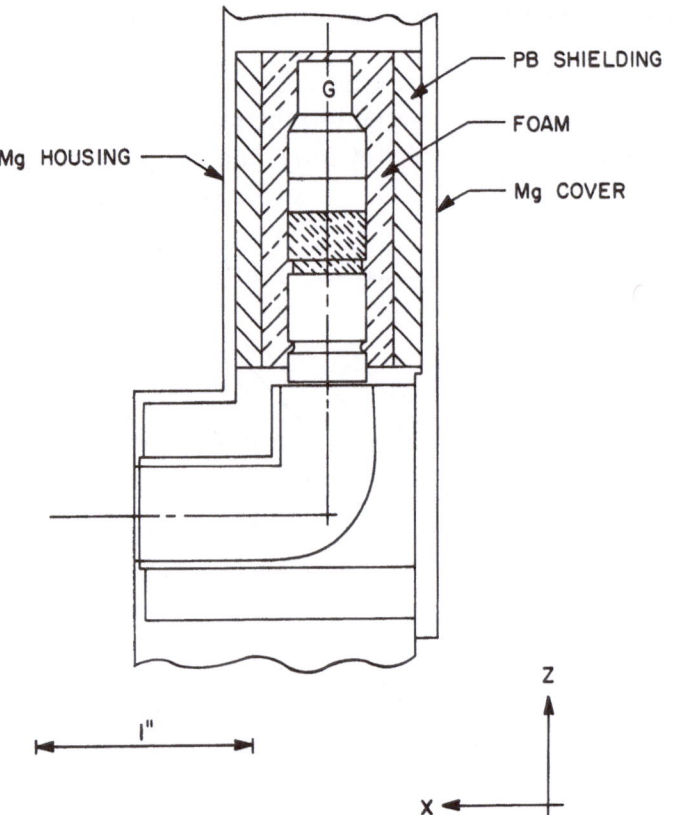

Fig. 3. Cut-away view of the arrangement of the single EON 6213 Geiger-Mueller tube G. Low-energy particles enter the end-window of the tube only after being scattered from the inner walls of the gold-plated elbow.

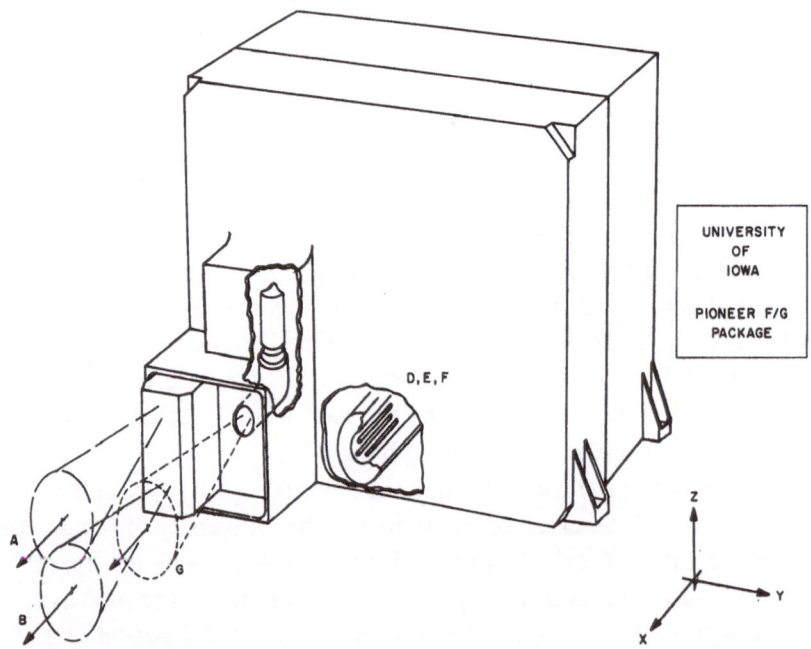

Fig. 4. Sketch of the overall configuration of the University of Iowa instrument.

proximately 52 successive samples produce a complete roll angle distribution with about 7° between each sample.

Our instrument which has a total mass of 1.64 kg and power consumption of 0.78 W has operated continuously for some 26 months with no abnormalities or malfunctions of any kind. This includes the Jovian encounter and postencounter period to date of writing.

3. The Magnetosphere

The Pioneer 10 flyby showed that the radiation belts surrounding Jupiter can be separated into two distinct regions. The region outside of 20 R_J, called the magnetodisc, is characterized by quasi-trapped particles in a magnetic field which does not rotate rigidly with the planet and deviates substantially from a dipolar field. Inside of 20 R_J, the magnetic field becomes more dipolar and particles seem to be durably trapped.

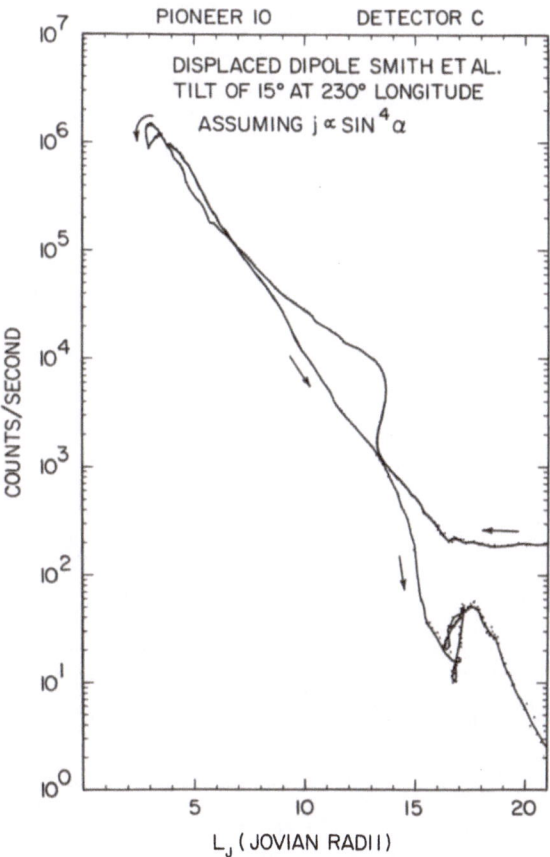

Fig. 5. Inbound and outbound counting rates of detector C ($E_e > 21$ MeV) as a function of L for the Smith *et al.* (1974) model of the Jovian magnetic field. All rates are corrected to magnetic equatorial values assuming a pitch angle distribution $j \propto \sin^4 \alpha$.

To analyze the distribution of particle intensities it was necessary to find a co-ordinate system to organize the data. The best possibility under the circumstances seemed to be a comparison of the data using an L parameter (McIlwain, 1961). To find the L value it is necessary to know the position and orientation of the dipole. The obvious choice was the model inferred by Smith *et al.* (1974) from the direct magnetometer measurements. Preliminary work had indicated that the intensities decreased approximately exponentially with radius and had a Gaussian dependence on magnetic latitude (Van Allen *et al.*, 1974a). The latter indicated that the equatorial pitch angle distribution could be represented by $\sin^m \alpha$. Figure 5 shows an example of the organization of the electron data $E > 21$ MeV by the model proposed by Smith *et al.* (1974). There is a marked lack of the closure between inbound and outbound data beyond $L > 7$ while the particle angular distributions indicate that the intensities should still be ordered beyond this region. The model inferred from decimetric and decametric radio observations (Mead, 1973) is a centered dipole tilted at $10°$ toward $224°$ System III longitude. This model worked much better than the previous one;

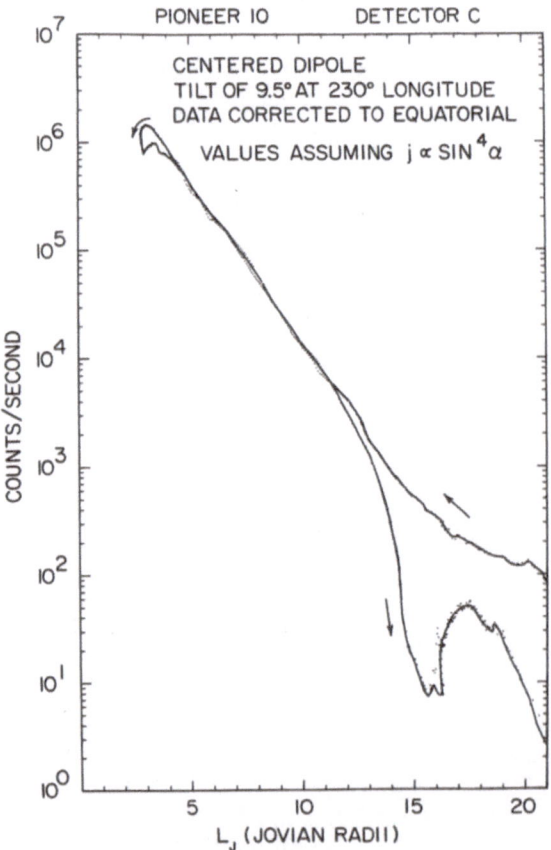

Fig. 6.   A plot similar to Figure 5 but for a centered dipole, tilted at $9.5°$ towards $230°$ System III longitude.

but using a trial and error procedure, it was found that a better fit could be made for a centered dipole with a tilt of 9.5° towards 230° System III longitude. Figure 6 shows the particle intensity $E > 21$ connected to the equatorial intensity using an assumed angular distribution of $\sin^4 \alpha$ for this model. Figure 7 shows a similar plot using a more recent (13 May 1974) model proposed by Smith *et al*. This model does not organize the energetic particle data as well as the centered dipole with a tilt of 9.5° towards 230° longitude.

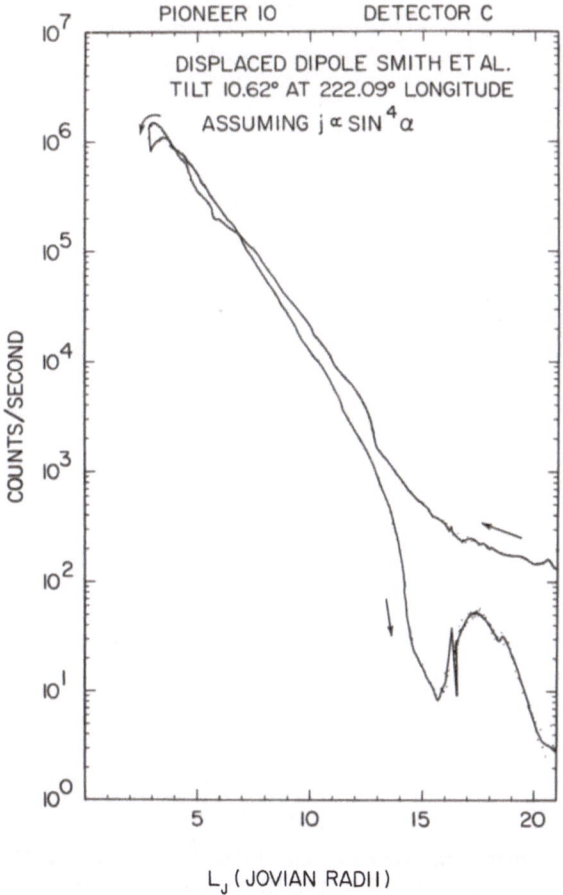

Fig. 7.   A plot similar to Figure 5 but with a newer magnetic model by Smith *et al*. (1974).

For the rest of this paper the model being used is a centered dipole with a tilt of 9.5° towards 230° System III longitude. The trial and error determination of these values indicated that a change of ±0.5° in tilt or ±3° longitude degraded the fit noticeably. The inbound and outbound data for detector C are shown on Figure 8 with no corrections. The exponential decrease of the equatorial data with L is clearly evident.

Examination of families of diagrams such as Figure 6 suggested that the loop inside L of 4 might be due to a monotonic increase in m with decreasing L. Detailed comparison of conjugate inbound and outbound data leads to the following expressions for m as a function of L for detectors C and D. The L dependence of these coefficients is negligible for $L > 5$.

For C: $m = 3.5 + (3.86/L)^8$
For D: $m = 4.0 + (3.567/L)^8$.

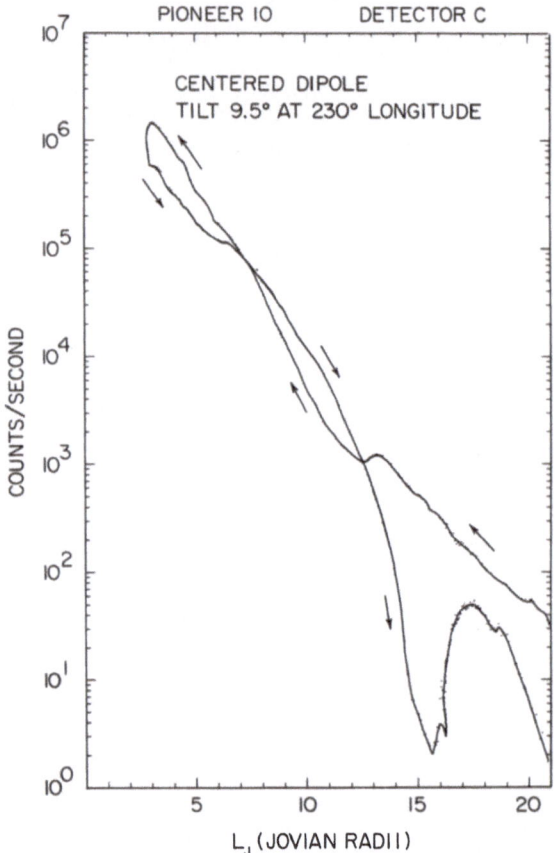

Fig. 8. Inbound and outbound raw counting rate of detector C ($E_e > 21$ MeV) as a function of L for centered dipole, tilted at 9.5° toward 230° System III longitude, show the exponential behavior of the equatorial particle intensities.

Both of these formulae are strictly empirical and there is no assurance that they are valid for lesser values of L. The results of this endeavor are clearly shown in Figures 9 and 10. The closure is very good for both curves and there is a definite tendency for both curves to roll over at closest approach. It is tempting to attribute this effect to a sweeping of particles by Amalthea. A theoretical analysis of phenomena would be useful in assessing validity of this suggestion. The Pioneer 11 spacecraft, which is now targeted for a periapsis of 1.6 R_J, should also contribute to knowledge of this matter.

By least squares fitting to the data of Figures 9 and 10 in the range of $3.5 \leqslant L \leqslant 12\ R_J$, we find

$$J(E_e > 21\ \text{MeV}) = 3.0 \times 10^8 \exp(-L/1.45)\left(\frac{\cos^6 \Lambda}{\sqrt{4 - 3\cos^2 \Lambda}}\right)^{m/2}$$

and

$$J(E_e > 31\ \text{MeV}) = 9.9 \times 10^7 \exp(-L/1.51)\left(\frac{\cos^6 \Lambda}{\sqrt{4 - 3\cos^2 \Lambda}}\right)^{m/2}$$

with J the omnidirectional intensity in electrons $\text{cm}^{-2}\ \text{s}^{-1}$, L in units of R_J, and Λ the magnetic latitude in the dipolar model previously described. The values of m are given in the previous paragraph for C and D, respectively.

A simple power law of the form $dJ/dE \propto E_e^{-\gamma}$ was determined for detectors C and D. The values of γ are plotted as a function of L in Figure 11. The data track very well inbound and outbound from $L = 3.4$ to $L = 14$. Beyond L of 14, statistical fluc-

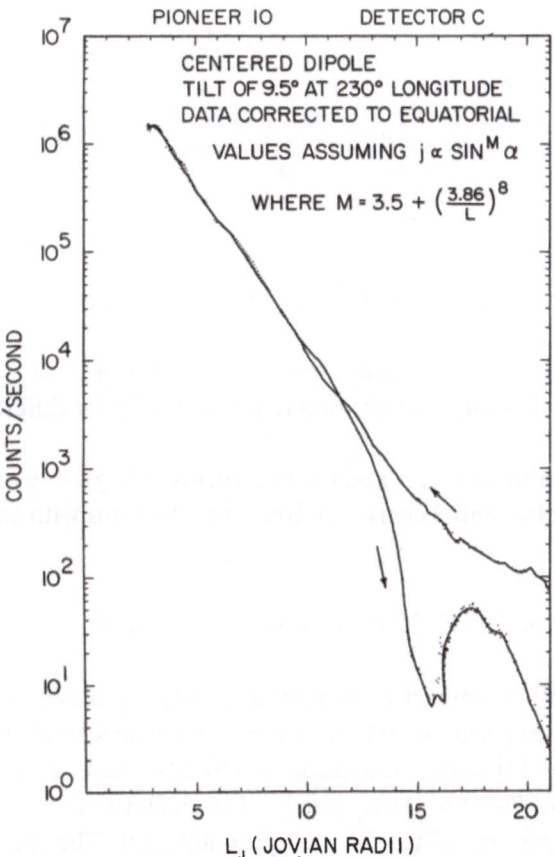

Fig. 9.   An improved version of Figure 6, assuming a simple L-dependence of m in the angular distribution $j \propto \sin^m \alpha$.

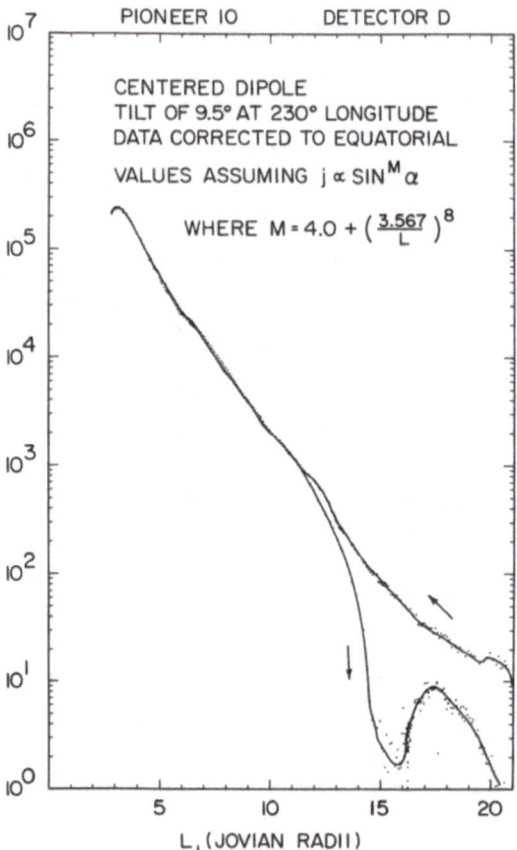

Fig. 10. A plot similar to Figure 9 for detector D ($E_e > 31$ MeV).

tuations in the counting rate produce much scatter. Inside L of 3.4, the slightly different angular distributions produce different values of γ for different magnetic latitudes.

Taking a mean value of $\gamma = 3.5$ and the two curves of Figures 9 and 10 one obtains a mean equatorial differential energy spectrum for electrons with energy greater than 21 MeV of

$$\frac{dJ}{dE} \approx 1.4 \times 10^{12} \, E^{-3.5} \exp(-r/1.45) \text{ cm}^2 \text{ s MeV}.$$

The outbound dip in intensity centered at $L = 15 \, R_J$ shown in all figures is attributed to Ganymede. On the inbound passage the spacecraft was very near the equator (see Figure 14) while Ganymede would have had an immediate effect on particles on the same field line if they mirrored at greater than 8.9° magnetic latitude. On the outbound pass the situation was quite different. The spacecraft crossed its orbit at about 22° magnetic latitude while Ganymede would have affected particles mirroring at greater than 9.2° magnetic latitude on the same field line. The time

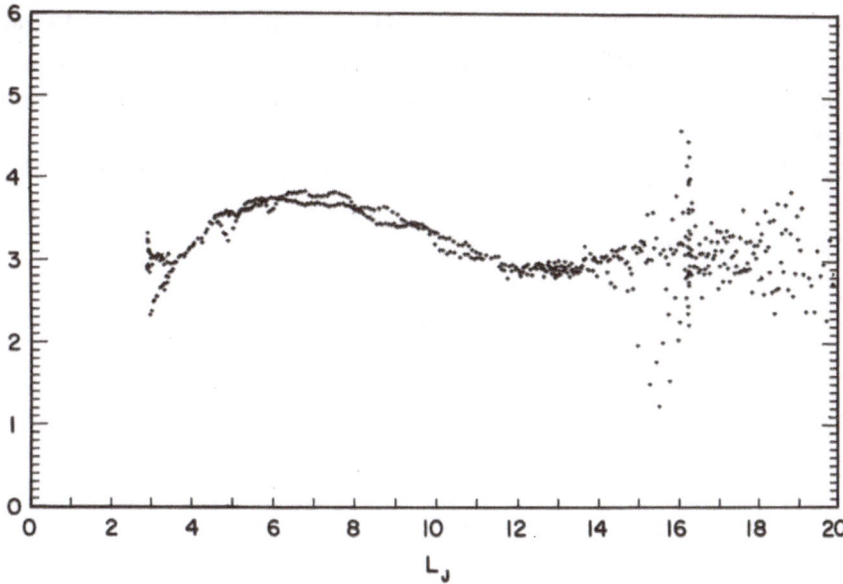

Fig. 11.   The *L*-dependence of the differential spectral index γ in a power law spectrum as derived from the counting rate ratio C/D. The abscissa is based on a 9.5° tilt of the dipole toward System III longitude = 230°.

of this crossing would have been six hours before. The $\bar{E}=0$ drift periods for 21 MeV and 31 MeV electrons are 7.3 and 5 h, respectively, at this radial distance, but Ganymede crossed the magnetic equator during the interim thereby eliminating even the drifting particles. Other satellite crossings show no discernible effect on detectors C and D as is evidenced by Figures 9 and 10.

The following table gives the magnetic latitude of the Pioneer spacecraft at the satellite orbit crossing as well as the magnetic latitude of the satellite and time delay when the satellite crossed the same field line that the spacecraft should have observed. The $\bar{E}=0$ drift period of 21 MeV electrons at each satellite orbit is also indicated. Table I and Figures 13 and 14 give a good view of the geometry of the crossings.

The table makes it evident that the most noticeable effects should occur at the Europa crossing inbound and at the Io and Ganymede crossings outbound.

Figure 12 shows the counting rates observed by detectors G, B, A, C, and D of the University of Iowa Instrument on board Pioneer 10 when the spacecraft was

TABLE I

	Io		Europa		Ganymede	
	in	out	in	out	in	out
Magnetic latitude of Pioneer 10	−5.9°	14.4°	−17.6°	5.0°	−1.7°	22.0°
Magnetic latitude of Satellite	7.6	1.2°	5.0°	−8.3°	8.9	9.2°
Time delay in hours	1.8	5.6	10.5	4.0	0.7	6.2
Drift period for 21 MeV electrons in hours	18.6		11.7		7.3	

within 20 R_J of the planet. All detectors show similar behavior at the outbound Ganymede crossing. Detectors G and B show definite dips at the inbound crossing of Europa orbit. A slight dip is noticeable at the outbound Io orbit in the rates of B, G, and A.

The inbound Ganymede and outbound Europa crossings show up in Figure 12

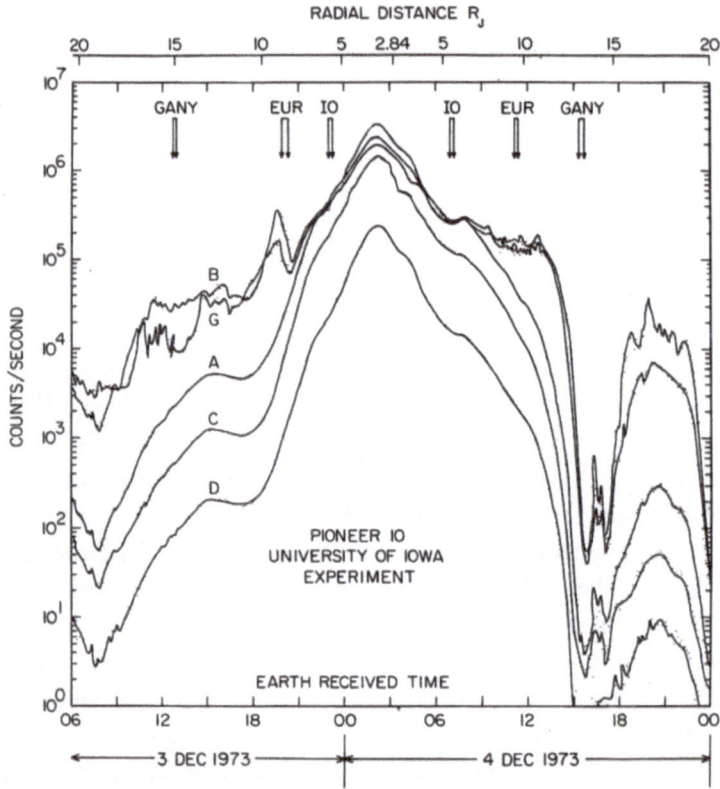

Fig. 12.   The counting rates of particles observed by detectors G, B, A, C, and D in the inner magneto-sphere of Jupiter. The radial distance in Jovian radii and approximate satellite location are indicated.

as possible turbulent wakes in the low-energy particles caused by the passage of the satellite. A thorough analysis of the satellite crossings has not been done, but it would have to be based on a theoretical formulation of combined radial and pitch angle diffusion. The theoretical framework for this has been developed (Roederer, 1970) but to our knowledge no comprehensive calculations have been completed even for the Earth's magnetosphere.

The energetic electron ($E > 21$ MeV) observations indicate that they must radially diffuse very rapidly. That is, they are able to pass the orbit of the satellite in less than the rotation period of Jupiter. The pitch angle diffusion must also take place on the same time scale, since the sweeping effect of the satellites and the drift periods

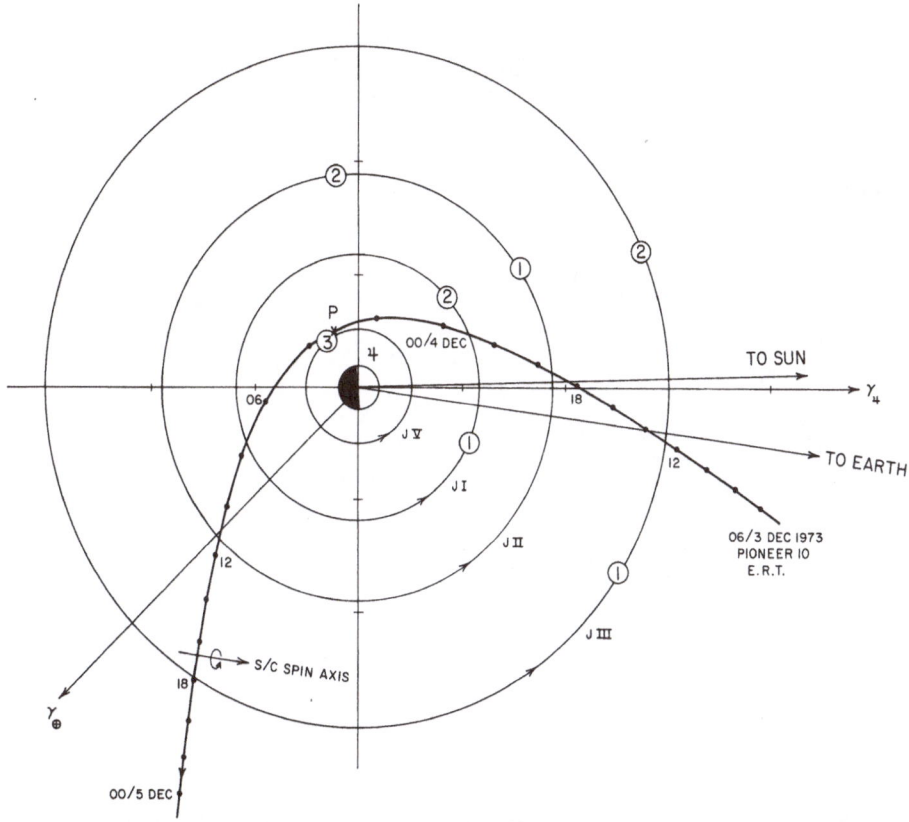

Fig. 13.   Projection on the ecliptic plane of the hyperbolic encounter trajectory of Pioneer 10 and the orbits of the four inner Jovian satellites. The numbers 1 and 2 on the orbits of JI, JII, and JIII show the positions of each of these satellites at the time that the spacecraft crossed the L-shell of that satellite, inbound and outbound, respectively. The number 3 shows the position of JV at the time that the spacecraft was at periapsis, marked P. γ_{4} designates Jupiter's vernal equinox and $\gamma_{\oplus}$ designates Earth's vernal equinox. The ephemerides are courtesy of M. Helton of the Jet Propulsion Laboratory. Note that the spacecraft spin axis is parallel to the planet-Earth line throughout the encounter.

of the electrons limit the replacement time, yet the equatorial angular distribution appears to be proportional to $\sin^4 \alpha$.

The detailed analysis such as Figures 9 and 10 for detectors G, B, and A to find the equatorial intensities, angular distributions, and satellite effects is not yet completed. Since these detectors are directional it is necessary to de-convolve the look angle distribution and with the orientation of the spacecraft and the magnetic field model, to determine the local pitch angle distribution.

4. Magnetodisc

The region outside 20 R_J, referred to as the magnetodisc, was previously characterized as a region of quasi-trapped particles. The initial observations shown in Figure 15 lead one to speculate that the magnetospheres of the Earth and Jupiter are similar with a bow shock, magnetosheath, magnetopause, and so on, but only on a much

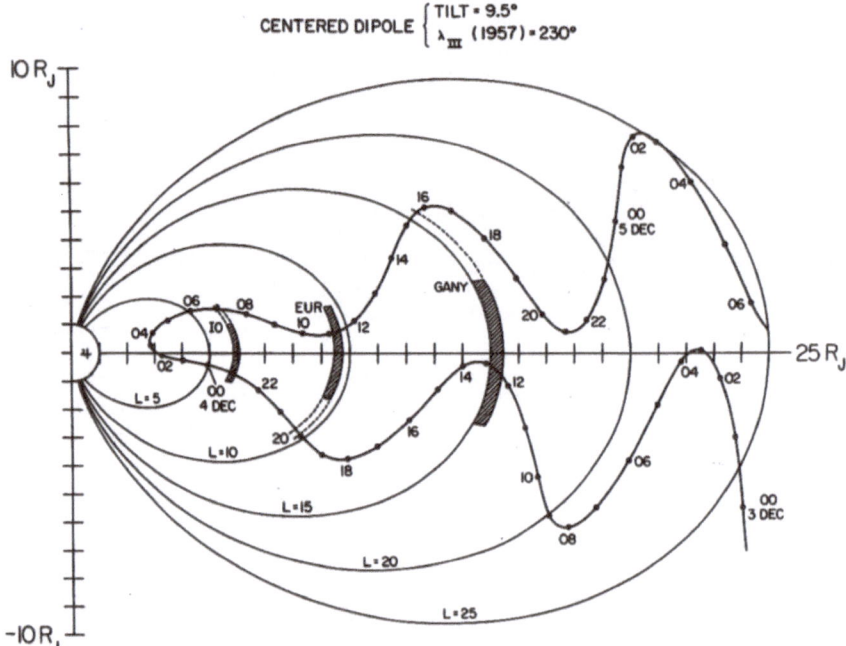

Fig. 14.   The time-labeled trace of Pioneer 10 in magnetic polar coordinates (magnetic meridian plane projection) for the planetary dipolar model as specified. The cross-hatching shows the regions that bound the orbits of Io (JI), Europa (JII), and Ganymede (JIII), respectively, in such a coordinate system.

larger scale. Figure 16 shows that this is not the case; specifically, the particle intensities show a 10-h periodicity in amplitude superimposed upon a gradual increase in magnitude.

The drop-out of particle intensity on December 1 is the most unusual feature of the inbound data. The plasma analyzer experimenters (Wolfe *et al.*, 1974) suggest that an impulsive increase in solar wind pressure caused an inward movement of the sunward magnetopause. If this were the case, the particle detectors should have shown evidence of previously observed intensities. Instead the particle intensity became very low, while the scalar value of the magnetic field remained the same as before. For more discussion of this phenomena it is desirable to investigate the post-encounter data.

Figure 17 shows very sharp modulations in intensity with a 10-h periodicity out to about 80 R_J. Beyond this distance the periodicity becomes more erratic but intensity variations continue (Figure 18) out to almost 175 R_J.

A determination of the System III longitude of the intensity maxima for the outbound data shows a spiraling of the apparent location of the longitude of the dipole tilt. A good fit of the peak intensities for the outbound data was obtained using a centered dipole with a tilt of 9.5° towards 230° longitude out to 30 R_J. Beyond this point, the data could be fit if the apparent longitude of the dipole were decreased from 230° by the following function of radius:

$$3.67 \exp(R/24.6)$$

This organizes the data out to about 80 R_J into a thin ($< 5 \ R_J$) disc-like structure.

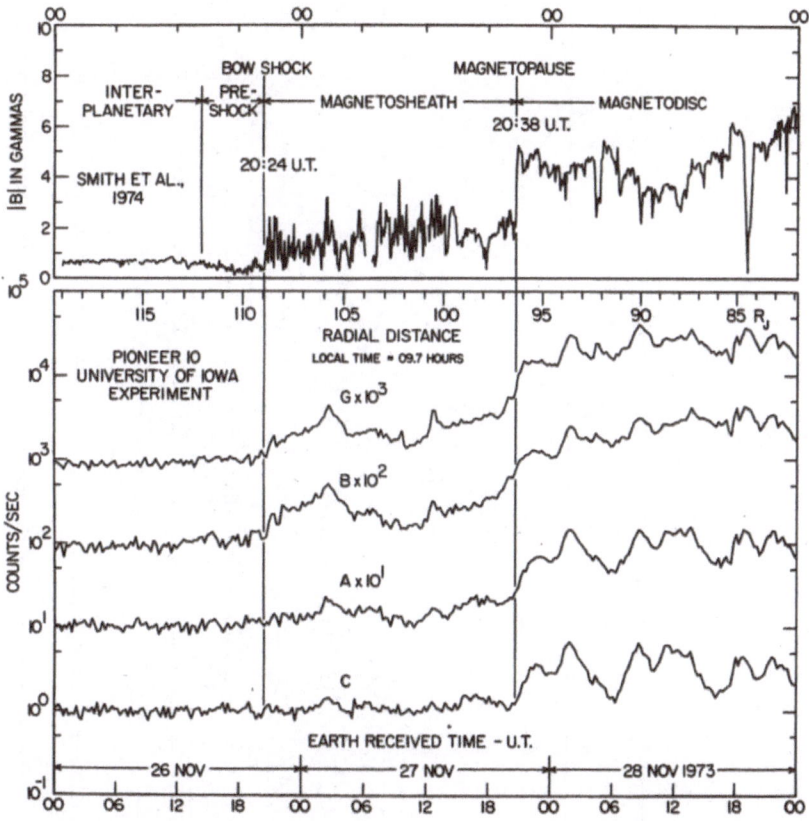

Fig. 15.   Detailed magnetic field strength and electron intensity data associated with the crossing of the bow shock and magnetopause.

The inbound maxima show a completely different story: The apparent longitude of the dipole tilt seems to be constant at value slightly greater than 230° out to 70 R_J. The exception to this occurs for the two maxima after the data drop-out on December 1. Here the apparent longitude of the dipole tilt is shifted to 260°.

This suggests another possible interpretation for the data drop-out, that the Pioneer 10 spacecraft observed a breaking and reconnection of field lines. The breaking resulted in the loss of particles and the subsequent reconnection explains the momentary forward shift of the apparent dipole longitude.

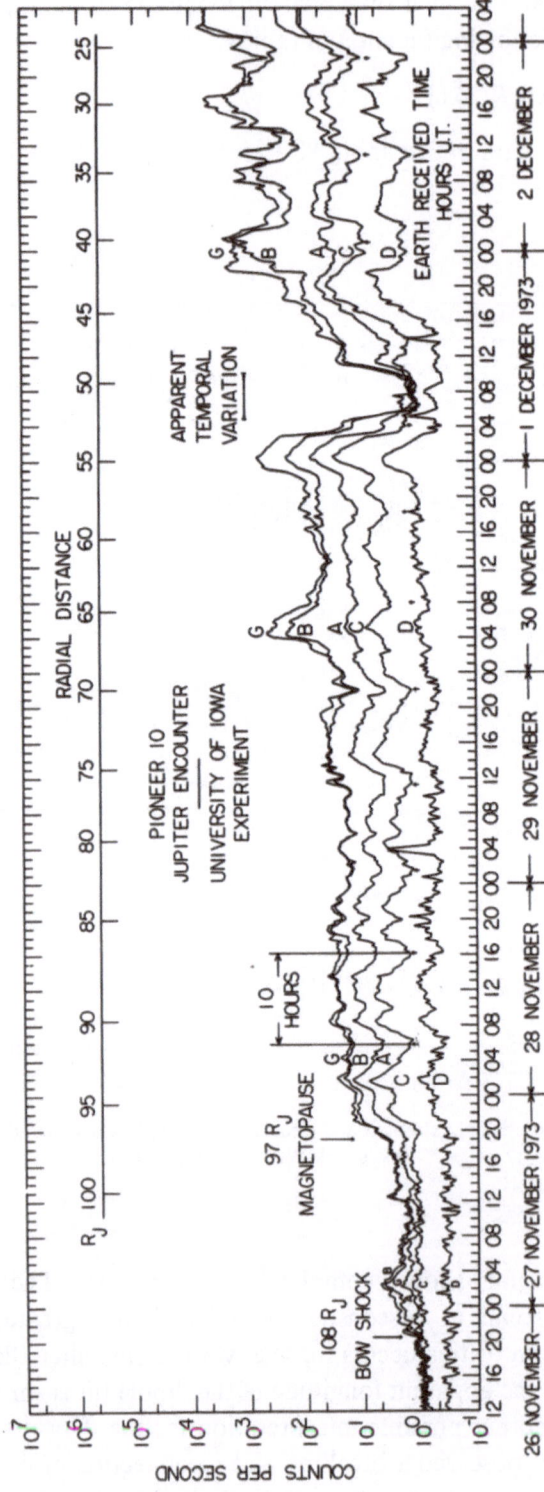

Fig. 16.   Seven days of observations on the inbound leg of the encounter trajectory through the sunward portion of the magnetodisc.

Fig. 16.   Seven days of observations on the inbound leg of the encounter trajectory through the sunward portion of the magnetodisc.

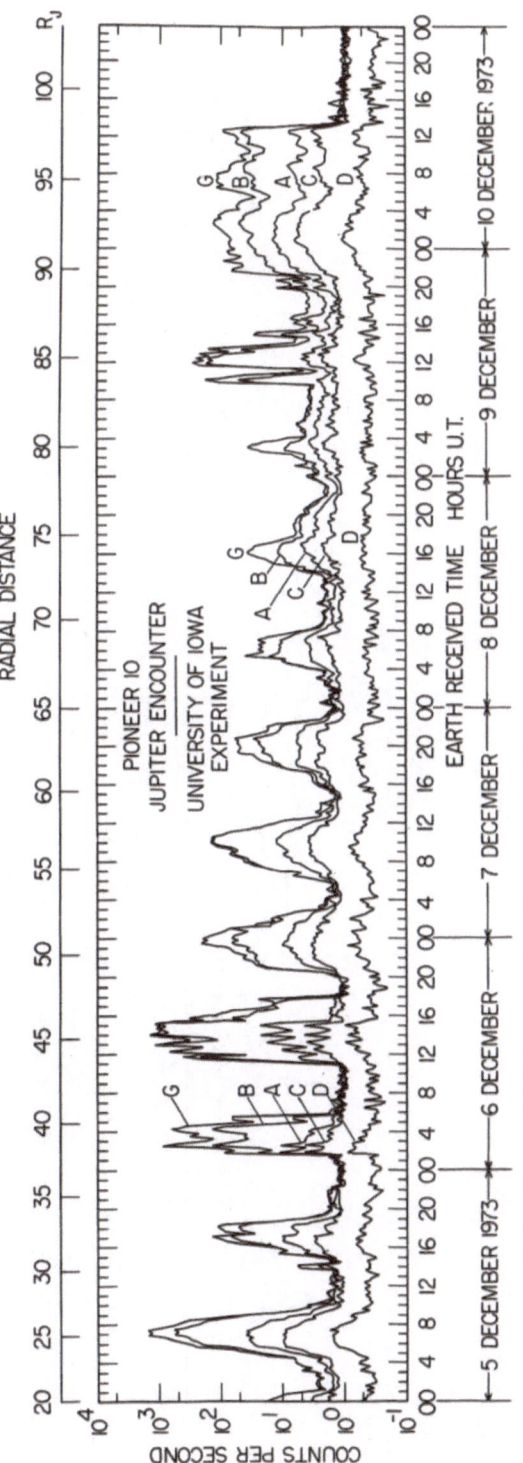

Fig. 17.   Six days of observations on the outbound leg of the encounter trajectory through the pre-dawn portion of the magnetodisc.

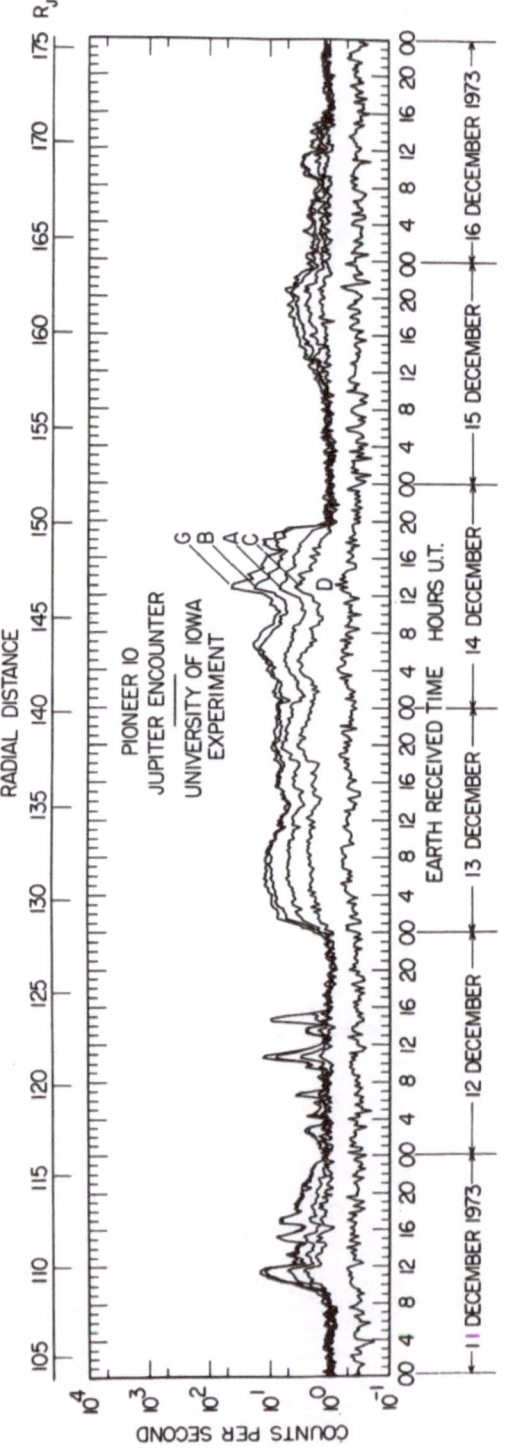

Fig. 18.   Six further days of observations on the outbound leg of the encounter trajectory through the pre-dawn portion of the magnetodisc.

Acknowledgements

I would like to thank Dr J. A. Van Allen and Mr D. N. Baker and Mr D. D. Sentman for allowing this work to be presented prior to more complete joint publication (Van Allen *et al.*, 1974b).

The handling of the Pioneer 10/11 program by the Ames Research Center is gratefully acknowledged.

The project manager for development of the University of Iowa instrument is Roger F. Randall, who designed and developed all the electronics and supervised all engineering aspects. Others at the University of Iowa to whom I am especially indebted are D. E. Cramer, H. D. Owens, R. B. Brechwald, R. J. France, M. Thomsen, and H. R. Flindt.

This work has been supported by Contracts NAS2-5603 and NAS2-6553 with the Ames Research Center of the National Aeronautics and Space Administration and by Contract N00014-68-A-0196-0009 with the Office of Naval Research.

References

Carr, T. D. and Gulkis, S.: 1969, *Ann. Rev. Astron. Astrophys.* **7**, 577.
Chang, D. B. and Davis, L.: 1962, *Astrophys. J.* **136**, 567.
Mead, G. D.: 1973, 'Magnetic Coordinates for the Pioneer 10 Jupiter Flyby', unpublished memorandum, Goddard Space Flight Center, April.
McIlwain, C. E.: 1961, *J. Geophys. Res.* **66**, 3681.
Ortwein, N. R., Chang, D. B., and Davis, L., Jr.: 1966, *Astrophys. J. Suppl. Ser.* **12**, 323.
Roederer, J. G.: 1970, *Dynamics of Geomagnetically Trapped Radiation*, Springer-Verlag, New York, Heidelberg, Berlin.
Smith, E. J.: 1974, Memorandum to Pioneer Particle Investigators, May 13.
Smith, E. J., Davis, L., Jr., Jones, D. E., Colburn, D. S., Coleman, P. J., Jr., Dyal, P., and Sonett, C. P.: 1974, *Science* **183**, 305.
Thomsen, M. F.: 1974, 'The Heliocentric Radial Cosmic Ray Gradient', M.S. Thesis, University of Iowa, May.
Thorne, K. S.: 1963, *Astrophys. J. Suppl.* **8**, 1.
Thorne, K. S.: 1965, *Radio Sci.* **69D**, 1557.
Van Allen, J. A.: 1972, University of Iowa Research Report 72–5, March 27.
Van Allen, J. A.: 1972, *Astrophys. J.* **177**, L49.
Van Allen, J. A., Baker, D. N., Randall, B. A., Thomsen, M. F., Sentman, D. D., and Flindt, H. R.: 1974a, *Science* **183**, 309.
Van Allen, J. A., Baker, D. N., Randall, B. A., and Sentman, D. D.: 1974b, *J. Geophys. Res.* **79**, 3559.
Van Allen, J. A., Frank, L. A., and O'Brien, B. J.: 1963, *J. Geophys. Res.* **68**, 619.
Wolfe, J. H., Collard, H. R., Mihalov, J. D., and Intriligator, D. S.: 1974, *Science* **183**, 303.

IO-ACCELERATED ELECTRONS AND IONS

STANLEY D. SHAWHAN, CHRISTOPH K. GOERTZ, RICHARD F. HUBBARD,
DONALD A. GURNETT, and GLENN JOYCE

Dept. of Physics and Astronomy, The University of Iowa, Iowa City, Iowa 52242, U.S.A.

1. Introduction

Earth-based measurements of the Io-modulation effect on Jovian decametric radio noise emission (Bigg, 1964) and the recently announced observation of intense Sodium-D optical emissions from the vicinity of Io (Brown, 1974) as well as various results from the Pioneer 10 Jupiter flyby (see *Science*, 1974) indicate that the Jovian moon Io (and possibly Europa and Ganymede) play an active role in the Jovian magnetosphere.

We have been developing a model in which Io interacts with the Jovian magnetosphere through plasma sheaths in the vicinity of Io across which particles are accelerated to energies approaching the motional potential across Io in the corotating Jovian magnetosphere (several hundred kilovolts). This model was first suggested by Gurnett (1972) and has been developed in more detail by Shawhan *et al.* (1973a), Hubbard (1973), Shawhan *et al.* (1973b), and Hubbard *et al.* (1974).

In this paper several recent results available from Pioneer 10 are used to revise this Io sheath model. The revised model is then used to suggest an explanation for a number of Earth-based and Pioneer 10 observations and to suggest other phenomena which might be detectable with future experiments.

2. Revised Model Based on Pioneer 10 Results

Two experiments operating during the Pioneer 10 flyby of Jupiter suggest a revision of the Io sheath model and allow several parameters to be better determined. The magnetometer experiment (Smith *et al.*, 1972) has lead to the value of 4 G R_J^3 for the dipole magnetic moment of Jupiter. At Io (at 6 R_J) the magnetic field is approximately 0.02 G. Previously (see Hubbard *et al.*, 1974) the field was assumed in the model to be 0.035 G. This reduced field yields a maximum value of about 400 kV for the potential across Io due to its motion through the Jovian magnetic field. Occultation of the S-band signal by Io has shown a well defined Io ionosphere extending out beyond 750 km in altitude and having a peak density of 6×10^4 electrons cm^{-3} at about 100 km altitude (Kliore *et al.*, 1974a). From the presence of this ionosphere, the experimenters also infer the presence of an atmosphere with a surface density of 10^{10}–10^{12} cm^{-3}. The existence of the Io atmosphere-ionosphere system makes the Io sheath model more plausible because it provides a path of sufficiently high conductivity to close the current system in the vicinity of Io as required by the model. Without the presence of an Io ionosphere it was questionable whether the Io surface

V. Formisano (ed.), The Magnetospheres of the Earth and Jupiter, 375–389. All Rights Reserved

conductivity could be sufficient (see Hubbard *et al.*, 1974). Also preliminary results about the Jovian ionosphere and atmosphere (Kliore and Fjeldbo, 1974) indicate that the Jovian ionosphere has about the predicted electron number density and therefore probably has sufficient conductivity to close the current system in that region (see Hubbard *et al.*, 1974).

Figure 1 depicts the scheme of the revised Io sheath model (see Gurnett, 1972

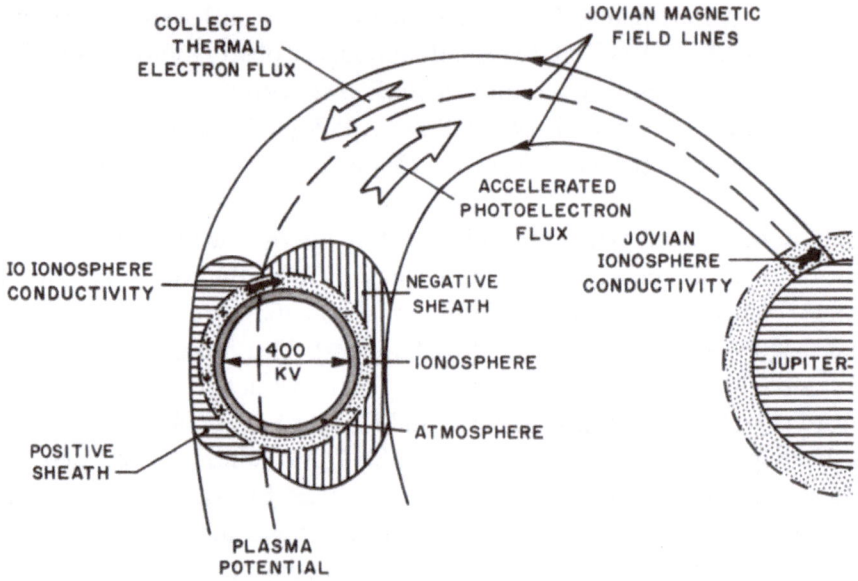

Fig. 1.   Basic Io sheath configuration and electron flux path.

and Hubbard *et al.*, 1974 for original model). Io moving at a velocity of 56 km s^{-1} relative to the Jovian magnetic field of 0.02 G develops a $\mathbf{v} \times \mathbf{B}$ electric field directed toward Jupiter. This electric field produces a potential difference of about 400 kV across Io. Plasma sheaths are assumed to form at the top of the Io ionosphere to make the transition from the plasma moving with Io to that corotating with Jupiter. At the top of the ionosphere on the face of Io toward Jupiter, the plasma potential is negative with respect to the Jovian plasma potential. The opposite face has a positive plasma potential. Therefore we call these transition regions negative and positive sheaths, respectively.

The negative sheath accelerates Io ionospheric electrons down the Io magnetic flux tube toward the Jovian ionosphere. Thermal plasma electrons are accelerated through the positive sheath into the Io ionosphere. Because of the dynamic resistance of these sheaths, the motional potential is dropped across these two regions. Accelerated particles can then attain energies of several hundred keV. These two electron fluxes constitute a current system which can be closed at the ends of the Io flux tube within the Jovian and the Io ionosphere. The relative surface area on Io covered by the negative and the positive sheaths is determined by the requirement of continuity in the total current; the electron flux times the sheath area must be equal on the two faces of Io. The relative areas then determine the faction of the motional potential available for inward and outward acceleration of particles (see Hubbard, 1973; Hubbard *et al.*, 1974).

Several model parameters can be determined from assumptions based on the Pioneer 10 observations of the Io ionosphere. The dusk dayside ionosphere shows a peak density of 6×10^4 electrons cm^{-3} in the range of 50–150 km altitude and an extent of about 750 km with a scale height of $\sim$220 km (Kliore *et al.*, 1974a). Preliminary results for the dawn nightside ionosphere show a peak of 9×10^3 electrons cm^{-3} at $\sim$50 km and an extent of $\sim$200 km (Kliore *et al.*, 1974b). These profiles are presented in Figure 2. The discontinuity in scale height of the dayside ionosphere

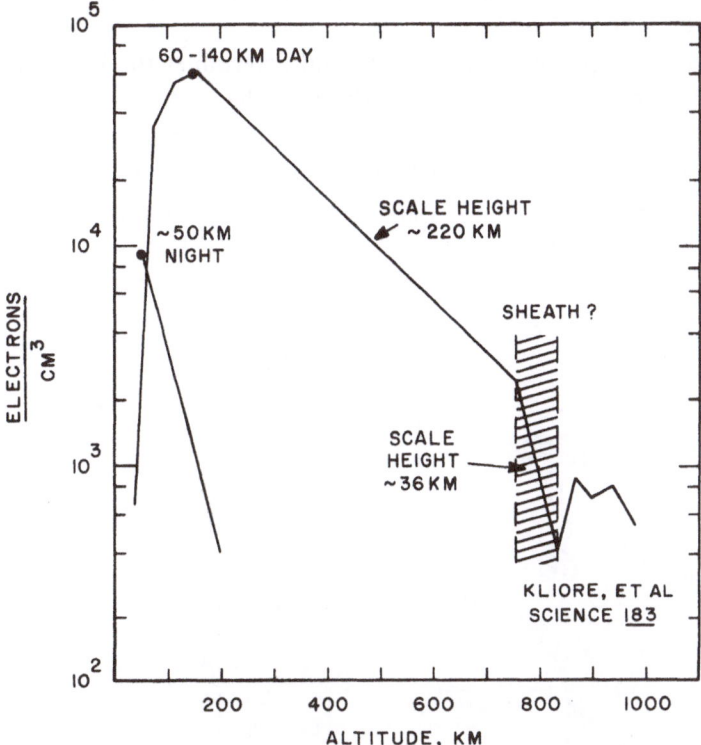

Fig. 2.   Io dayside and nightside ionosphere (Kliore *et al.*, 1974a, b).

at 750 km is consistent with a sheath at this altitude. Taking the density of 2.5×10^3 electrons cm^{-3} and a temperature for sodium ions of 400 K (Kliore *et al.*, 1974b) at the lower side of the sheath with $\frac{1}{4}$ of the flux directed upward, the maximum topside electron current density would be 1×10^{-5} A m^{-2} emitted from Io at this point. This current density corresponds to an electron flux of 6×10^9 electrons cm^{-2} s^{-1}.

Measurements of the thermal plasma density and temperature were not made by Pioneer 10. An upper limit to the plasma density is available in the vicinity of Io from Figure 2. If it is assumed that the density above 850 km could be thermal magnetospheric plasma then an upper limit density would be ~ 500 cm^{-3}. The electron flux that can be accelerated into the Io ionosphere above the outward face is related to this density n_e and the electron temperature T_e by

$$F_e \sim n_e T_e^{1/2}.$$

Assuming $n_e = 200$ cm^{-3} and $T_e = 10^5$ K the flux would be 10^{10} electrons cm^{-2} s^{-1} which is comparable to the outward accelerated flux.

Combining the measured Io ionosphere profile with the inferred atmosphere profile (10^{12} cm^{-3} surface density with 100 km scale height), the Io ionospheric conductivity can be calculated. Following the procedure of Webster *et al.* (1972) a value of 260 mhos is obtained for an equatorial belt region. This conductance is more than an order of magnitude larger than the conductance necessary to cause a significant potential drop at Io.

Since the Io ionosphere moves with Io and determines the Io 'surface' electrical properties, it seems reasonable to consider that the motional potential associated with Io is that developed across the Io ionosphere. For an ionosphere extending 750 km in altitude, the potential might be increased to ~ 580 kV.

A close-up view of the Io system is shown in Figure 3 roughly to scale. A significant atmosphere is inferred from the presence of an ionosphere with a maxima in the

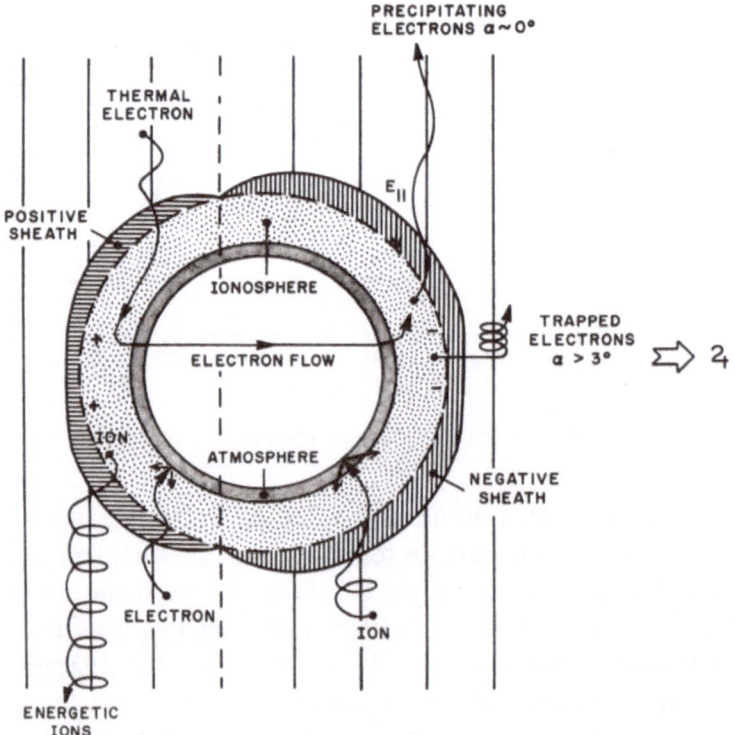

Fig. 3.   Particle acceleration at Io.

density above the surface. We assume that the sheaths could be located at the top
of the ionosphere. Characteristic sheath thicknesses have been estimated by Hubbard
et al. (1974) to be in the range of 10 to 50 km. Since the sheaths separate two plasma
regions with a strong current flowing between them, these sheaths are probably
'double layers' as described by Block (1972). From the work of Knorr and Goertz
(1974) and of Goertz and Joyce (1975) a typical thickness for the double layer can
be calculated. The relationship between the potential across a double layer and its
thickness is given by

$$\frac{\phi}{L^2} \lesssim 0.14,$$

where ϕ is the potential in kT_e/e and L is the length in Debye lengths. For a number
density of 2.5×10^3 cm^{-3}, a temperature of 400 K and a potential of 300 kV, the
sheath thickness must be greater than $\frac{1}{4}$ km. For both thickness estimates the sheath
size is larger than or comparable to a 300 keV electron gyroradius (~ 1 km) but
comparable to or smaller than a 300 keV ion gyroradius (~ 50 km).

A qualitative picture of particle acceleration in the vicinity of Io is included in
Figure 3. As mentioned earlier, the electrons carry the current in this dc circuit which
is somewhat similar to the model of Goldreich and Lynden-Bell (1969). Although
the ion currents are insignificant, the fluxes of accelerated ions may be significant
and important to the Io related chemistry. The major features of particle acceleration
are as follows:

2.1. FACE TOWARD JUPITER (NEGATIVE SHEATH)

2.1.1. *Electrons*

In the mid and high latitude regions of the Io face toward Jupiter a significant electric
field component should exist in the magnetic field direction. Ionospheric electrons
from this region can be accelerated to energies up to several-hundred keV. Because
of the small random thermal energies, these electrons have pitch angles well within
the Jovian atmospheric loss cone ($\alpha \lesssim 3°$) and are therefore beamed down the Io
flux tube. This beam of $\sim 10^9$ electrons cm^{-2} s^{-1} from $\sim \frac{1}{2}$ of Io's area could constitute a current of 10^7 A which carries up to 10^{13} W of power.

In the equatorial latitude range the electric field has a significant component directed transverse to the magnetic field lines and if the sheath region is thin enough,
some electrons could gain a substantial amount of perpendicular energy which would
trap them on field lines just inside Io's orbit. These electrons could diffuse inward
toward Jupiter and contribute to the synchrotron emitting electrons in the hard
trapping region (see Shawhan *et al.*, 1973b).

2.1.2. *Ions*

This same negative sheath can accelerate ions to several hundred keV energies into
the Io atmosphere toward the Io surface. An estimate of the proton flux can be

made by assuming the same parameters as for the magnetospheric electrons ($T_i = 10^5$ K, $n_i = 200$ cm^{-3}). The flux is then $\sim 2 \times 10^8$ protons cm^{-2} s^{-1}. These fluxes exceed energetic proton fluxes measured by Pioneer 10 (*Science*, 1974) and may be significant for ion-sputtering of the Io surface.

2.2. FACE AWAY FROM JUPITER (POSITIVE SHEATH)

2.2.1. *Electrons*

Thermal plasma electrons from the magnetosphere can be accelerated through the positive sheath region to several hundred keV energies. As indicated in Figure 3 these electrons are directed into the ionosphere and atmosphere of Io. The fluxes may be in the range of 10^9 to 10^{10} electrons cm^{-2} s^{-1} which is sufficient to cause optical emissions, impact ionization and heating. Primary and secondary electrons are conducted into the negative sheath region to complete the electron current circuit.

2.2.2. *Ions*

Ionospheric ions are accelerated through the sheath away from Io. Because of the small sheath size compared to a gyroradius most of these ions should gain perpendicular energy and be trapped on field lines just outside Io's orbit. If Sodium ions are assumed (Kliore and Fjeldbo, 1974) the flux could be as high as 3×10^7 ions cm^{-2} s^{-1} for energies up to several hundred keV.

TABLE I

Table of physical parameters

Parameter	Value
1. Maximum sheath potential (Maximum particle energies)	400 kV across Io 580 kV across Io ionosphere
2. Characteristic sheath thickness at ~ 750 km altitude above Io ionosphere	$\frac{1}{4}$ to 50 km
3. Io ionospheric conductance (height integrated)	260 mhos
4. Maximum current density in Io flux tube near sheath; Maximum electron flux in Io flux tube near sheath < 580 keV	1×10^{-5} A m^{-2} 6×10^9 electrons cm^{-2} s^{-1}
5. Maximum proton flux available for sputtering < 580 keV	2×10^8 ions cm^{-2} s^{-1}
6. Maximum electron flux precipitating into Io atmosphere for thermal plasma around Io of ~ 200 cm^{-3} at 10^5 K	$\sim 10^{10}$ electrons cm^{-2} s^{-1}
7. Maximum outward flux of energetic Na$^+$-ions	3×10^7 ions cm^{-2} s^{-1}
8. Maximum power carried down Io flux tube	$\sim 10^{13}$ W

Important quantities deduced for this revised Io sheath model based on the Pioneer 10 measurements are summarized in Table I. These quantities are used later in the explanation of observed phenomena related to the Io-Jupiter interaction.

3. Related Pioneer 10 and Earth-Based Observations

A variety of observations from Earth-based and Pioneer 10 measurements seem to be associated with the Io-Jupiter interaction and related to the Io sheath model for this interaction. In Figure 4 the Pioneer 10 trajectory, the position of Io at the

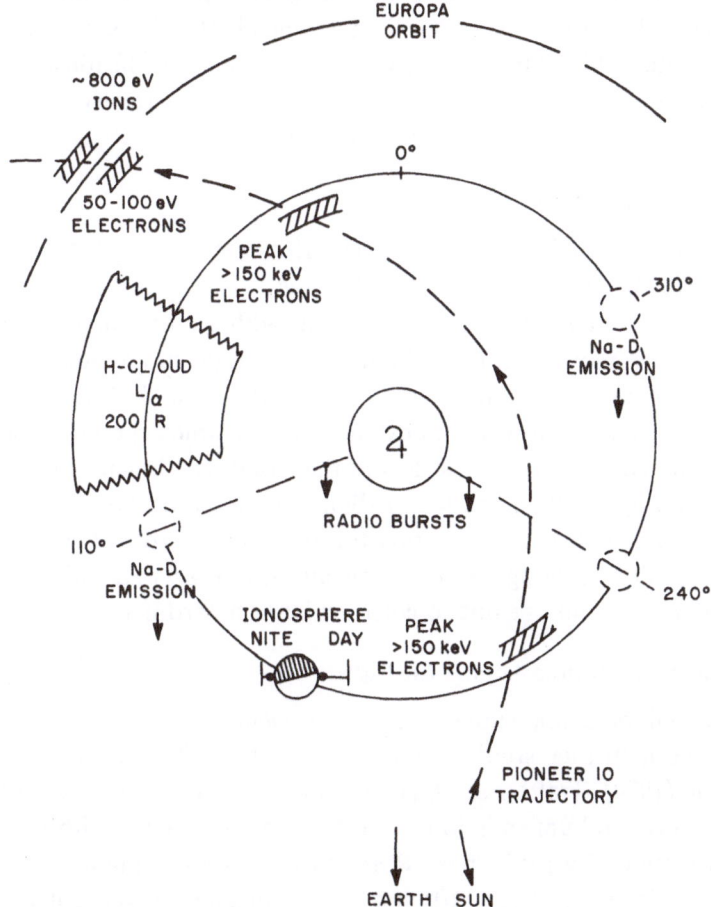

Fig. 4.   Summary of related Pioneer 10 and Earth-based observations.

time Pioneer 10 crossed the Io orbital radius ($\sim 6 \ R_J$), and a summary of observed phenomena are schematically depicted. The important characteristics of these phenomena are as follows:

3.1. PEAK IN > 150 keV ELECTRONS

From their experiment on Pioneer 10 McIlwain and Fillius (1974) observe a peak in the >150 keV electrons at $\sim 5.6\ R_J$ on the inbound and outbound pass which they attribute to injection by Io. The flux of these electrons above the background is about 2×10^7 cm^{-2} s^{-1} and the pitch angle distribution appears to be peaked in the perpendicular direction so that these electrons are trapped (Fillius, 1974).

3.2. NON-THERMAL ELECTRONS AND IONS

In crossing the orbit of Europa on the outbound pass of Pioneer 10, Frank and Ackerson (1974) have detected the presence of nonthermal electrons and ions using data from the plasma analyzer experiment (Wolfe *et al.*, 1974). These data show that electrons with a peak energy of 50 to 100 eV are present just inside the orbit of Europa and that ions with ~ 800 eV energy exist just outside the Europa orbit, A very preliminary estimate yields a number density of ~ 100 cm^{-3} for the ions. The possibility of non-thermal electron and ion populations with the same relative orbital positions at Io and Ganymede is also indicated by the data.

3.3. HYDROGEN Lα EMISSION

From a preliminary analysis of the Pioneer 10 ultraviolet photometer data Judge and Carlson (1974) suggested that Io might have a hydrogen Lα glow of 10 kR and that Jupiter is surrounded by a hydrogen torus with a mean diameter equal to the Io orbit and an uv intensity of several hundred rayleighs in Lα. Subsequent analysis (Judge, private communication, as quoted in McElroy and Yung, 1974) indicates that all the radiation comes from an extended cloud which precedes and follows Io in its orbit. The Lα intensity is 200 R and the cloud has dimensions of 120° in the Io orbital plane and less than one Jovian diameter perpendicular to the orbital plane. McElroy and Yung (1974) conclude that this emission is probably due to resonance scattering of sunlight although atmospheric air glow at Lα and emissions at Lβ and Balmer-α if detected could be due to corpuscular bombardment.

3.4. IO-ASSOCIATED SODIUM-D LINE EMISSION

The intense sodium-D line emission from the vicinity of Io announced by Brown (1974) have been further studied and interpreted by Brown and Chaffee (1974), Matson *et al.* (1974), Trafton *et al.* (1974), and McElroy and Yung (1974). Brown concluded that the sodium emissions were definitely associated with Io and deduced a column density of 2×10^{12} cm^{-2}. Peaks in the time varying emissions were observed to occur for Io at about 110° and 310° from superior geocentric conjunction as shown in Figure 4. Trafton *et al.* found that the sodium cloud extended to more than $10''$ in radius from Io, and that the emission was stronger close to the orbital plane and especially on the face of Io toward Jupiter. Matson *et al.*, Trafton *et al.*, and McElroy and Yung agree that emission is probably due to resonant scattering of sunlight. Matson *et al.* and McElroy and Yung suggest that the sodium exists as an impurity in perhaps ammonia ice on the Io surface which is released by ion sputtering of the surface. The observed cloud requires a flux of $\sim 10^7$ atoms cm^{-2} s^{-1} from Io which could be provided by a flux of $\sim 10^8$ protons cm^{-2} s^{-1} or a lower flux of heavier ions which may include a cascade process (Matson *et al.*, 1974).

3.5. Io IONOSPHERE

A significant Io ionosphere has been reported by Kliore *et al.* (1974a, b) for both the illuminated side (solar zenith angle = 81°) and the dark side as shown in Figure 2. The illuminated side ionosphere has a scale height consistent with the presence of Na^+ ions (Kliore and Fjeldbo, 1974; McElroy and Yung, 1974). However, the dark side ionosphere has a significantly lower scale height and the peak density occurs at a lower altitude. McElroy and Yung consider six different models to explain the Io ionosphere. They conclude that dayside and nightside results can be made consistent if there are finite vertical drifts (upward during the day and downward at night ~ 1 km s^{-1}) which could be due to evaporation or condensation of gases on the surface or due to motional electric fields. Another plausible model requires incident corpuscular radiation such as 10 keV electrons or 10 MeV protons of $\sim 3 \times 10^7$ cm^{-2} s^{-1} on the dayside and about 1% of this flux on the nightside.

3.6. Io-MODULATED DECAMETRIC RADIO EMISSION

The earliest evidence for Io's interaction with the Jovian magnetosphere comes from the Io modulation of the Jovian decametric radio bursts (Bigg, 1964). Bursts of up to 10^8 W and less than 40 MHz in frequency are observed for Io at orbital positions of $\sim 90°$ and 240° from superior geocentric conjunction and for several ranges of Jovian central meridian longitude. These bursts are thought to be due to beamed radiation from a very small source region near the Jovian ionosphere at the foot of the Io magnetic flux tube (see Warwick, 1967; Carr and Gulkis, 1969). Recently Europa-modulation of bursts at ~ 1 MHz has been reported (Carr, 1974).

4. Io Sheath Model Explanation of Observations

The particle acceleration mechanism which we propose for the vicinity of Io seems to explain, at least qualitatively, the experimental results enumerated in the previous section. These explanations are summarized schematically in Figure 5.

4.1. ELECTRONS > 150 keV

Energetic electrons with energies < 580 keV fluxes $< 6 \times 10^9$ cm^{-2} s^{-1} and small pitch angles are predicted to exist within the Io flux tube inside the Io position (see Gurnett, 1972; Shawhan *et al.*, 1973a, b). It is felt that a significant fraction of these electrons could gain energy perpendicular to the magnetic field and be trapped so that they would be observed at Io's orbit. As suggested earlier, the mechanism for perpendicular energy gain could be acceleration through a sheath thinner than a gyroradius perpendicular to the Jovian magnetic field. Also electrons could be backscattered at some point in the Io flux tube due to an instability (Goertz, 1973a) and consequently trapped. Consideration has not been given to distortion of the magnetic field in the vicinity of Io due to the current system (Goertz, 1973b). This distortion may be important for creating a trapped electron population. To calculate the possible equilibrium flux of trapped electron requires an experimental estimate of the energetic electron diffusion coefficient which is not yet available.

4.2. NON-THERMAL ELECTRONS AND IONS

The separation of non-thermal particle populations with ions just outside and electrons just inside the moon's orbit follows from the polarity of the sheath acceleration. For all of the moons the electric field would be directed toward Jupiter. The

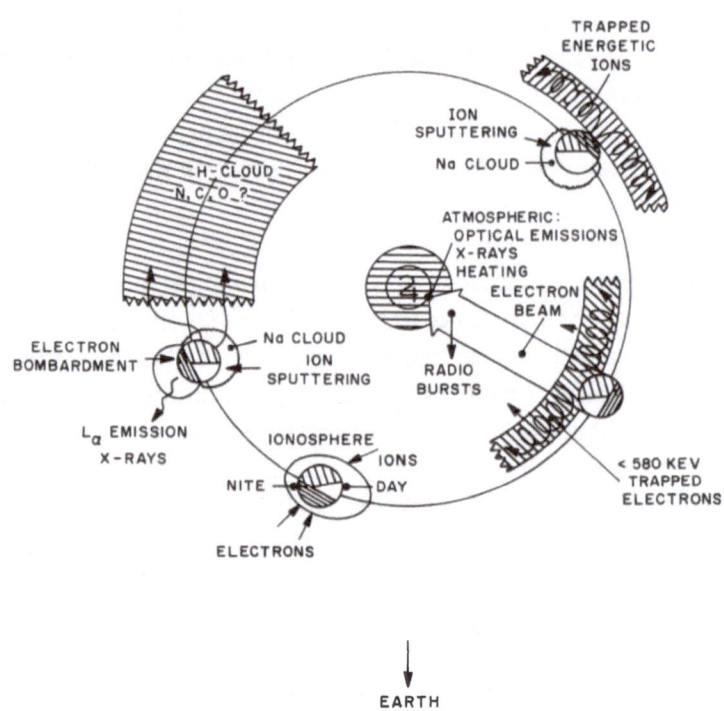

Fig. 5.   Consequences of Io-accelerated particles.

particle energies observed at Europa by Frank and Ackerson (1974) suggest a sheath system with a total potential of several kilovolts. From the motion of Europa in the Jovian magnetic field a maximum potential of ~ 200 kV might be expected ($\sim 40\%$ Io, see Dermott, 1970). The lower observed potential can be explained if Europa has little or no ionosphere and a low surface conductivity. A similar situation may exist at Ganymede.

4.3. HYDROGEN Lα EMISSION

According to McElroy and Yung (1974) the presence of hydrogen in the partial torus associated with Io can be explained by a thermal escape flux of 10^{11} atoms cm^{-2} s^{-1} which could be maintained by photolysis of atmospheric NH_3. We suggest that the inward accelerated ion flux which impacts the Io surface facing Jupiter may be important for the liberation of NH_3 into the atmosphere by sputtering. Clouds of other components such as C, N and O might also be expected. The limited extent of the hydrogen cloud requires a rapid loss process. McElroy and Yung (1974) suggest that hydrogen can be lost by charge exchange with $\sim 10^9$ cm^{-2} s^{-1} fluxes of low energy protons. Io could be producing fluxes of $\sim 10^8$ protons cm^{-2} s^{-1} with energies up to several hundred keV from the face away from Jupiter which may contribute to the loss process.

Io accelerated electrons and ions impinging on the ionosphere and atmosphere should produce other observable optical emissions.

4.4. Io-ASSOCIATED SODIUM-D OPTICAL EMISSIONS

Ion sputtering of the Io surface seems to be the most plausible explanation for the release of sodium into the vicinity of Io (Matson et al., 1974; McElroy and Yung, 1974). Our model accelerates ions into the face of Io toward Jupiter. This is the face for which Trafton et al. (1974) observed the more intense sodium emissions. A flux of sodium atoms of 2×10^7 atoms cm^{-2} s^{-1} is necessary to maintain the observed sodium cloud (McElroy and Yung, 1974) which requires an incident energetic proton flux of 10^8 cm^{-2} s^{-1}, a lower flux of heavier energetic ions or a cascade process (Matson et al., 1974). According to our calculations the proton flux accelerated by Io would be 10^8 cm^{-2} s^{-1} and 3×10^7 cm^{-2} s^{-1} if sodium ions. As pointed out by Matson et al., the sheath thickness is less than an ion gyroradius so that it could be that liberated sodium ions are energized and could reimpact the surface to liberate more sodium atoms and ions in a cascade process. The primary energetic ion fluxes as observed with Pioneer 10 (Science, 1974) are insufficient to provide the necessary sodium flux.

Matson et al. (1974) and Trafton et al. (1974) suggest that the sodium emission is excited by resonant scattering of sunlight. McElroy and Yung (1974) favor a combination of sunlight scattering and of collisional excitation in the atmosphere with light scattering in the extended could. From our model energetic electrons are accelerated into the Io atmosphere on the face away from Jupiter. The sporadic nature of the enhanced sodium emissions (Brown, 1974) may be explained by requiring a significant production of Na by ion sputtering and sufficient excitation by electron impact as viewed from the Earth. These conditions vary with Io's orbital position and with its latitude with respect to the magnetic equatorial plane.

4.5. Io IONOSPHERE

McElroy and Yung (1974) show that the two observations of Io's ionosphere at ~ 0700 and ~ 1900 local Io time cannot be compatible unless strongly time dependent particle influxes or bulk motions of the ionosphere are invoked. It is not obvious that such variations are related to the difference between the illuminated and the dark sides of Io. We propose a different explanation in the context of the Io sheath model for which the ionosphere formation may be related to the particular face of Io – toward or away from Jupiter. On the face toward Jupiter ('illuminated' side when observed by Pioneer 10), ions are injected which tends to inflate the ionosphere. On the face away from Jupiter ions are lost so the ionosphere tends to be deflated. Also it could be assumed that the ionosphere is maintained by photoionization (or impact ionization) of ion sputtered sodium. On the face toward Jupiter the accelerated ions provide a directed flux of 10^8 cm^{-2} s^{-1} for sputtering so that a significant ionosphere would be expected. On the face away from Io primary protons from the radiation environment are decelerated so that only ions with energies greater than several hundred keV could impact the surface. The flux of these ions is $\lesssim 10^7$ cm^{-2} s^{-1} as measured by various experiments on Pioneer 10 (*Science*, 1974). Consequently a less significant ionosphere would be expected.

4.6. Io-MODULATED DECAMETRIC RADIO EMISSION

As yet a detailed model of Io-modulated decametric radio emission has not been developed in the context of the Io sheath model. Energetic electrons with energies up to several hundred keV energies constitute a current of $\sim 10^7$ A in the Io flux tube and transport 10^{13} W toward the Jovian ionosphere. The emission mechanism needs to be only 10^{-4} to 10^{-5} efficient to explain the power in the decametric bursts. One suggestion is that the bursts may be related to instabilities or to the formation of double layers along the Io flux tube which produce coherent emission from a small source region (Shawhan *et al.*, 1973a). Radio emission associated with Europa may be explicable by a similar mechanism.

4.7. JUPITER-ASSOCIATED X-RAYS

Remote measurements have been made to detect Jovian X-rays using the Uhuru (Hurley, 1975) and Copernicus (Vesecky *et al.*, 1975) satellites. Upper limits to the X-ray flux at the Earth in the energy range of 2–6 keV are 5×10^{-4} cm^{-2} s^{-1} and in the range 0.6 to 1.9 keV 8×10^{-3} cm^{-2} s^{-1} respectively. An order of magnitude estimate can be made from the Io sheath model for the X-ray flux from the Jovian atmosphere at the root of the Io flux tube and from the Io atmosphere. For both cases an electron flux $\sim 10^9$ electrons cm^{-2} s^{-1} is associated with an area of $\sim 10^{17}$ cm^2. Assuming a constant energy spectrum to 300 keV (Hubbard *et al.*, 1974) the specific source intensity is approximately 3×10^{23} electrons s^{-1} keV^{-1}. For a photon efficiency of $\sim 10^{-2}$ the photon intensity is approximately 10^{22} photons s^{-1} keV^{-1}. The corresponding flux at the Earth would be 10^{-5} photons cm^{-2} s^{-1} keV^{-1} which is more than an order of magnitude below the detection limit of either experiment. To detect these X-rays it seems that a Jupiter orbiter experiment located outside of the hard trapping region is necessary.

5. Further Consequences of the Sheath Model

If the precipitating electron beam, associated with the Io flux tube, exists and reaches the Jovian atmosphere, then ionization, optical emissions and atmospheric heating should result as the beam is dissipated in the atmosphere. Estimates of these effects are made to see if they are significant and to suggest further experimental observations.

5.1. BEAM DISSIPATION AND ATMOSPHERIC IONIZATION

A model calculation has been carried out to determine the penetration depth of the Io-related precipitating particle beam and the resulting impact ionization. The initial electron spectrum at the top of the Jovian atmosphere is assumed to have a flux of 10^9 electrons cm^{-2} s^{-1} keV^{-1} up to 300 keV where account has been taken for the field line convergence. A model atmosphere due to McElroy (1973) was used. Energy loss was assumed to be by impact ionization only (30 eV for each collision) with a cross section varying as $E^{-1} \ln E$ and having a value of 2×10^{-19} cm^2 at 100 keV.

The beam was completely dissipated at about 0 km altitude (the Jovian cloud tops) so that the associated effects would occur above this altitude. In the ionospheric region (100–400 km), the impact ionization rates are several orders of magnitude below the photoionization rates. The maximum contribution to the electron density is 10^4 electrons cm^{-3} at 200 km (for a radiative recombination coefficient of 6.6×10^{-12} cm^3 s^{-1}) which is 10% of the photoionized density.

5.2. OPTICAL EMISSIONS

Optical emissions associated with the probable atmospheric constituents (H, H_2, He, CH_4 and NH_3) are expected due to energetic electron excitation. These electrons could be those precipitating due to Io or those precipitating from the radiation belts.

For the Io-related electrons a power flux of 10^{13} W spread over 10^5 km^2 (10^{-5} of the disk area) yields a maximum energy flux of 10^5 erg cm^{-2} s^{-1}. Rees (1973) has carried out model calculations that indicate an upper limit Hα (6563 Å) intensity of about 10^3 kR for this input energy flux. Dulk and Eddy (1966) searched for Hα emission from Io-related electrons. With a threshold of 1.2 kR no aurora were observed. Although their entrance slit aperture is not specified it would have to cover only 10^{-3} of the disk area to be able to observe the predicted optical intensity.

Based on Rees' calculations and the 10^5 erg cm^{-2} s^{-1} energy flux, the most intense radiations with 10^4 kR could occur for the Lyman series from H_2, for the Werner series from H_2 and for some singlet line emissions from He especially at 10830 Å.

5.3. ATMOSPHERIC HEATING

Energy not lost from the Io electron beam by radio emissions, by ionization or by optical emissions goes into heating the Jovian atmosphere. Assuming that the other losses are insignificant, the 10^5 erg cm^{-2} s^{-1} energy flux, at least locally, probably exceeds other energy sources. Consequently, a hot spot or warm strip may exist associated with the foot of the Io field line. Such a feature would have a linear dimension of $R_J/500$ so that it would be observable with the Pioneer 10 and 11 IR radiometer (resolution of $R_J/100$, Chase et al., 1974). This localized heat source might be important in understanding the atmosphere dynamics.

6. Summary

By using Pioneer 10 results concerning the Jovian magnetic field and the Io ionosphere we have quantitatively revised the Io sheath model for the interaction of Io with the Jovian magnetosphere by accelerated particles. This model has then been invoked to explain qualitatively and somewhat quantitatively a number of experimental results derived from Earth-base and the Pioneer 10 flyby observations. The agreement between the model and the experimental observations suggests that this Io sheath model is plausible and that it can be used to make quantitative predictions to test against further experimental results.

Acknowledgements

We thank Dr J. A. Van Allen for helpful discussions during the course of this work. This work was supported in part by the Atmospheric Sciences Section, U.S. National Science Foundation under Grant GA-31676 and U.S. National Aeronautics and Space Administration under Grants NGL-16-001-002 and NGL-16-001-043.

References

Bigg, E. K.: 1964, *Nature* **203**, 1008.
Block, L. P.: 1972, *Cosmic Electrodyn.* **3**, 349.
Brown, R. A. (preprint): 1974, in A. Woszczyk and C. Iwaniszewska (eds.), 'Exploration of the Planetary System', *IAU Symp.* **65**, 527.
Brown, R. A. and Chaffee, F. H., Jr.: 1974, *Astrophys. J. Letters* **187**, L125.
Carr, T. D. and Gulkis, S.: 1969, *Ann. Rev. Astron. Astrophys.* **7**, 577.
Carr, T. D.: 1974, private communication.
Chase, S. C., Ruiz, R. D., Munch, G., Neugebauer, G., Schroeder, M., and Trafton, L. M.: 1974, *Science* **183**, 315.
Dermott, S. F.: 1970, *Monthly Notices Roy. Astron. Soc.* **149**, 35.
Dulk, G. A. and Eddy, J. A. (abstract): 1966, *Astron. J.* **71**, 160.
Fillius, R. W.: 1974, paper presented at the Neil Brice Memorial Symposium *The Magnetospheres of Earth and Jupiter*, Frascati, Italy, 28–31 May, not in these proceedings.
Frank, L. A. and Ackerson, K. L.: 1974, Univ. of Iowa, private communication.
Goertz, C. K.: 1973a, *Astrophys. Letters* **13**, 95.
Goertz, C. K.: 1973b, *Planetary Space Sci.* **21**, 1431.
Goertz, C. K. and Joyce, Glenn: 1975. *Astrophys. Space Sci.* **32**, 165.
Goldreich, P. and Lynden-Bell, D.: 1969, *Astrophys. J.* **156**, 59.
Gurnett, D. A.: 1972, *Astrophys. J.* **175**, 525.
Hubbard, R. F.: 1973, University of Iowa Dept. Physics and Astronomy Res. Rpt. 73–27 (M.S. Thesis).
Hubbard, R. F., Shawhan, S. D., and Joyce, Glenn: 1974, *J. Geophys. Res.* **79**, 920.
Hurley, K. C.: 1975, this volume, p. 241.
Judge, D. L. and Carlson, R. W.: 1974, *Science* **183**, 317.
Kliore, A., Cain, D. L., Fjeldbo, G., Seidel, B. L., and Rasool, S. I.: 1974a, *Science* **183**, 323.
Kliore, A. J., Cain, D. L., Fjeldbo, G., Seidel, B. L., and Rasool, S. I. (abstract): 1974b, **EOS 55**, 339.
Kliore, A. and Fjeldbo, G.: 1974, JPL, private communication.
Knorr, G. and Goertz, C. K.: 1974, *Astrophys. Space Sci.* **31**, 209.
Matson, D. L., Johnson, T. V., and Farale, F. P. (preprint): 1974, JPL, Space Sciences Division, submitted to *Astrophys. J. Letters*, 1 February.
McElroy, M. B.: 1973, *Space Sci. Rev.* **14**, 460.
McElroy, M. B. and Yung, Y. L.: 1974, Center for Earth and Planetary Physics, Harvard University,

(preprint), submitted to *Astrophys. J.*

McIlwain, C. E. and Fillius, R. W. (abstract): 1974, *EOS* **55**, 404.

Rees, M. H.: 1973, AIAA/AGU Space Science Conference: *Exploration of the Outer Solar System*, Denver, Colo., 10–13 July, 1973; and private communication.

Science: 1974, Volume **183** (special issue on preliminary results from Pioneer 10), pp. 301–324.

Shawhan, S. D., Hubbard, R. F., Joyce, G., and Gurnett, D. A.: 1973a, in R. Grard (ed.), *Photon and Particle Interactions with Surfaces in Space*, D. Reidel, Dordrecht, p. 405.

Shawhan, S. D., Gurnett, D. A., Hubbard, R. F., and Joyce, Glenn: 1973b, *Science* **182**, 1348.

Smith, E. J., Davis, L., Jones, D. E., Colburn, D. S., Coleman, P. J., Dyal, P., and Sonett, C. P.: 1974, *Science* **183**, 305.

Trafton, L., Parkinson, T., and Macy, W., Jr.: 1974, *Astrophys. J. Letters*, **190**, L85.

Vesecky, J. F., Culhane, J. L., and Hawkins, F. W.: 1975, this volume, p. 245.

Warwick, J. W.: 1967, *Space Sci. Rev.* **6**, 841.

Webster, D. L., Alksne, A. Y., and Whitten, R. C.: 1972, *Astrophys. J.* **174**, 685.

Wolfe, J. H., Collard, H. R., Mihalov, J. D., and Intriligator, D. S.: 1974, *Science* **183**, 303.

DÉNOUEMENT OF JOVIAN RADIATION BELT THEORY

F. V. CORONITI

Depts. of Physics and Planetary and Space Science, University of California,
Los Angeles, Calif. 90024, U.S.A.

1. Introduction

Prior to the Pioneer 10 encounter a concerted theoretical effort was made to construct models of the Jovian Van Allen belt electron fluxes. For the most part, the physics underlying these models is similar to terrestrial radiation belt physics. Thus Jupiter provided an opportunity to test whether the physics of the Earth's radiation belts is applicable to other planetary magnetospheres. Even if only partially successful, we can then contemplate exporting this physics to the unseen magnetospheres of planets orbiting other stars.

Since several groups worked virtually simultaneously on the theoretical models, it seems appropriate to give a quasi-historical review of the evolution of the basic theoretical ideas. Ahead of most of his terrestrial magnetospheric colleagues, Neil Brice recognized the importance of Jupiter, and made two important contributions on the hydromagnetic structure of the Jovian magnetosphere (Brice and Ioannidis, 1970) and on the distribution of cold plasma within the magnetosphere (Ioannidis and Brice, 1971). After the JPL Jupiter Radiation Belt Workshop in July, 1971 (Section 2), theorists focused their attention on constructing models of the Jovian electron fluxes. It was quickly recognized that three physical processes would be dominant at Jupiter: (1) radial diffusion transport (Brice, 1972); (2) the limitation of particle fluxes by plasma wave turbulence (Kennel, 1972; Thorne and Coroniti, 1972); and (3) particle losses from collisional sweep-up by the Galilean satellites (Mead, 1972). The first problem was to find a radial diffusion mechanism which was fast enough at low L-shells to avoid total particle absorption by the satellites and to balance the synchrotron energy loss. Here Brice and McDonough (1973) made the key suggestion that ionospheric dynamo electric fields could drive radial diffusion (Section 3). This idea was tested by constructing radial diffusion models of the synchrotron radiation in order to deduce the radial diffusion coefficient. Finally an attempt was made to construct a self-consistent model of the electron fluxes which included all three of the above physical processes and which estimated the flux levels in the Jovian outer zone (Section 5).

Although the analysis of the Pioneer 10 data is not yet complete, it is instructive to compare the predictions of the theoretical models with the Pioneer 10 measurements assuming, of course, some theoretical license in evaluating consistency. The attempt will be to use the data to comment on the theory. Section 6 discusses the inner zone data which yields the clearest evidence that radial diffusion is the dominant transport process at Jupiter. We also estimate the synchrotron flux density using the

V. Formisano (ed.), The Magnetospheres of the Earth and Jupiter, 391–410. All Rights Reserved
Copyright © 1975 by D. Reidel Publishing Company, Dordrecht-Holland

Pioneer 10 results. In Section 7 the observed outer zone electron fluxes are favorably compared with the qualitative and quantitative predictions of the whistler mode stable trapping model (Coroniti, 1974). In Section 8 the outer zone electron precipitation flux is estimated and the suggestion is advanced that precipitation could affect the structure of the Jovian ionosphere. Section 9 discusses satellite sweep-up in light of the Pioneer 10 data, and Section 10 estimates the electron precipitation energy flux which could be incident on Io and Europa. Section 11 comments on what we did not learn from Pioneer 10.

2. Goal of the Model

The only pre-encounter data on the Jovian Van Allen belts was the spectrum (flat between 200–3000 MHz) and energy flux density (7×10^{-23} erg cm^{-2} s^{-1} Hz^{-1}) of the synchrotron radiation detected in the Jovian inner zone ($1.2 < L < 6$) (Carr and Gulkis, 1969). From an interpretation of the decametric radiation, Warwick (1967) estimated the dipole magnetic field strength as 10–12 G. With this value, the energy of the synchrotron radiating electrons could be estimated as 3–10 MeV with an integral flux at $L = 2$ of $J = 10^7$–10^8 el cm^{-2} s^{-1}. The JPL Jupiter Radiation Belt Work-

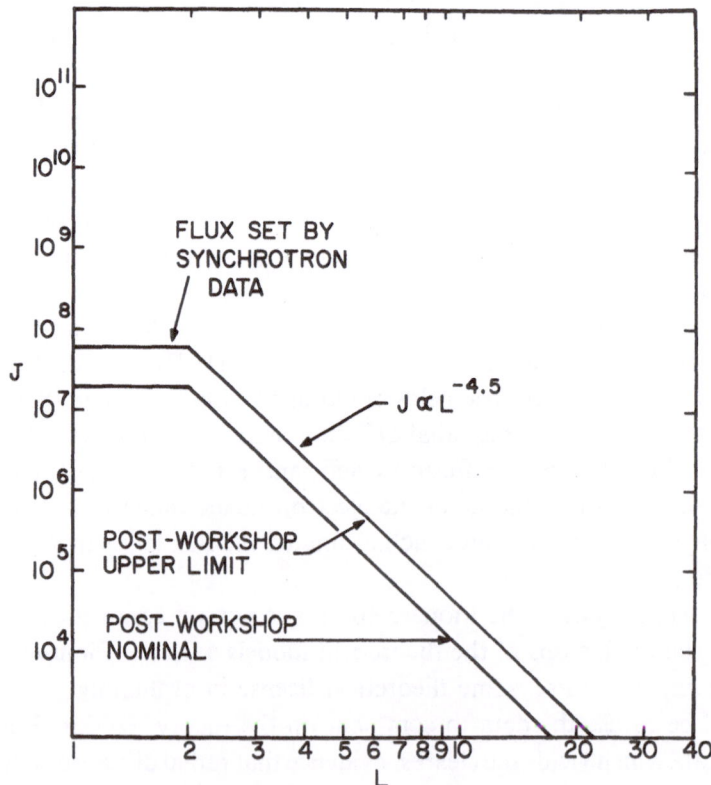

Fig. 1.   *Ad hoc* model of the Jovian relativistic electron flux adopted at the JPL Jupiter Radiation Belt Workshop in July, 1971 (Beck, 1972).

shop in July, 1971 adopted an *ad hoc* electron flux model which had $J = 3 \times 10^7$ cm^{-2} s^{-1} at $L = 2$ and in which J decreases as $L^{-4.5}$ for $L > 2$ (Figure 1, Beck, 1972). This model predicted outer zone $(20 > L > 6)$ fluxes of $J < 10^6$ cm^2 s. The source of the electrons was commonly assumed to be the solar wind since a typical solar wind electron has a first invariant $\mu \sim 10$ MeV G^{-1} which, if conserved, implied an energy of ~ 10 MeV at $L = 2$.

The goals of the post-Workshop theoretical efforts were to understand the transport of electrons from the solar wind to the inner zone and to construct a better outer zone electron flux model which would be consistent with the synchrotron radio data. All models discussed herein adopted the $B \approx 10$ G value, which is a factor of 2.5 larger than deduced by Smith *et al.* (1974) from the Pioneer 10 magnetometer data.

3. Cross-*L* Transport

The fastest radial transport process in the Earth's outer zone is internal hydromagnetic convection from the tail. For Jupiter, however, Brice and Ioannidis (1970) had shown that the plasmapause coincided with or was beyond the expected 50 R_J location of the magnetopause. Hence convection could not transport electrons inside 50 R_J. This left the possibility that the cross-*L* transport would be by inward radial diffusion. Since the dominant electron drift is corotation, radial diffusion requires electromagnetic fluctuations at the Jovian rotation period, 10 hours. The diffusion coefficient D_{LL}, however, must be large enough to balance the approximately one year synchrotron energy radiation lifetime at $L = 2$. Both Nagada-Mead (1965) magnetic diffusion with $D_{LL} \propto L^{10}$ and electrostatic convection diffusion (Birmingham, 1969) with $D_{LL} \propto L^6$ are too slow at the low *L*-values (Jacques and Davis, 1972).

Brice and McDonough (1973) suggested that atmospheric neutral winds at ionospheric heights could set up fluctuating dynamo electric fields which would then map into the magnetosphere. If the electric field is independent of latitude, the diffusion coefficient scales as $D_{LL} \propto L^3$, which is a much weaker *L*-dependence. Taking a modest neutral wind speed of 130 m s^{-1} and assuming that the wind velocity changes every 5 hours, Brice and McDonough (1973) estimated an upper limit of $D_{LL} = 3 \times 10^{-8}$ L^3 s^{-1}, which yields sufficiently rapid diffusion rates at $L = 2$ to balance synchrotron losses.

Coroniti *et al.* (1973) and Coroniti (1974) suggested that a possible source for the ionosphere neutral winds was solar wind-convection coupling to the polar cap. If a convection electric field is suddenly applied to the ionosphere, the ions move in the electric field or Pedersen direction but have a $\mathbf{J} \times \mathbf{B}$ stress which is exerted against the neutrals. The neutral gas is accelerated to the convection velocity in a time $\tau = (n_n M_n / n_i M_i) \nu_{in}^{-1}$ (Fedder and Banks, 1972) where ν_{in} is the ion-neutral collision frequency. In Figure 2 τ is plotted against altitude for the model photon-produced ionosphere of Atreya *et al.* (1974). Since $\tau \ll 10$ h, convection could stimulate a neutral wind throughout the polar cap. Coroniti (1974) noted that the anti-solar polar cap neutral wind matched a natural tidal eigenmode of the atmosphere, and hence could drive a planetary wide neutral wind system. For a simple model of solar wind electric field variations, Coroniti estimated a diffusion coefficient of $D_{LL} = 2 \times 10^{-10} L^3 \text{ s}^{-1}$.

The rapid coupling between convection and the ionospheric neutrals occurs because the Jovian ionosphere is similar to the Earth's F-region where the v_{in} is less than the ion cyclotron frequency and the inertia of the neutral gas is small. Models

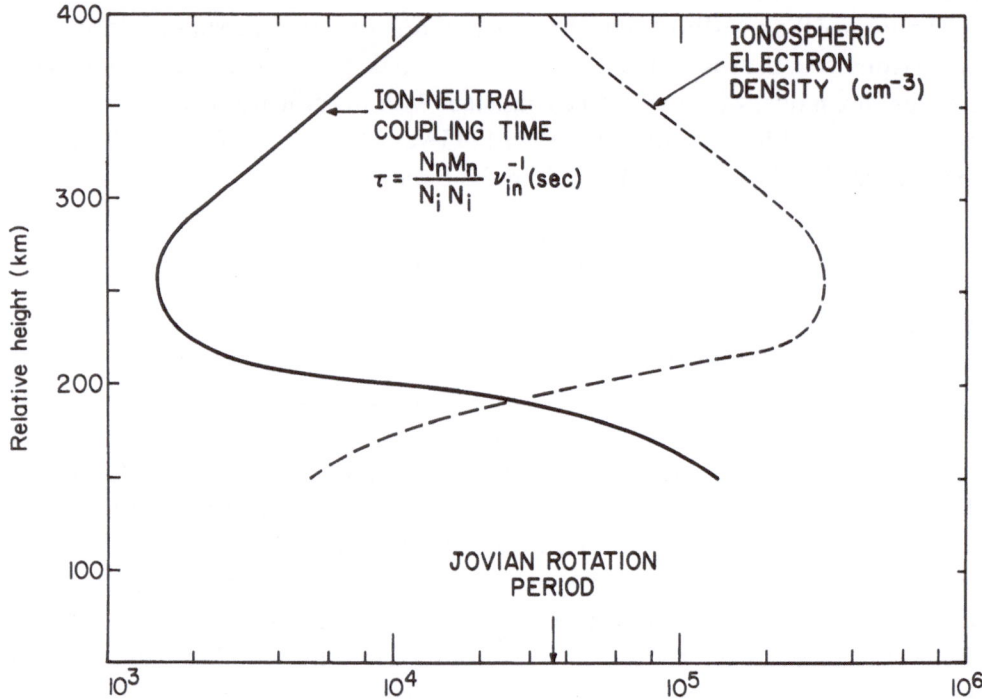

Fig. 2.   Model of Jovian ionospheric electron density assuming solar photo-production (Atreya *et al.*, 1974) and the neutral atmospheric response time in the ionosphere to a change in the convection electric field.

of the Jovian ionosphere suggest that the ionosphere lacks an E-region. This low inertia suggests that the high-latitude Jovian ionosphere may not be able to enforce corotation of magnetospheric field lines at large L; i.e., if the magnetosphere exerts a substantial spin-down torque on the ionosphere, the ionosphere may cease to corotate (Kennel and Coroniti, 1975).

At the inner Galilean satellites, the diffusion rates are sufficiently slow that collisional sweep-up might constitute an important particle loss (Mead, 1972). Calculations by Mead and Hess (1973) and Hess *et al.* (1974) demonstrated severe flux reduction for the Nagada-Mead and Birmingham diffusion rates. Radiometric data, however, suggested that the Jovian dipole was inclined by 10° to the spin axis (Warwick, 1967). Particles which mirror within ±10° of the magnetic equator have a small probability of colliding with the satellites and hence can diffuse past the satellites. Coroniti (1974) and Hess *et al.* (1974) found that Io would reduce the electron fluxes by factors of 10–20 if the conductivity of Io were so small that the field lines were not excluded from intersecting the satellite. The sweep-up loss at Europa was some-what smaller.

4. Radial Diffusion Models of the Synchrotron Data

As a test of the radial diffusion hypothesis several groups attempted to fit the synchrotron spectrum with a radial diffusion-synchrotron emission model. Jacques and Davis (1972) first demonstrated that $D_{LL} \propto L^3$ was sufficient rapid to diffuse electrons into low L-shells on the synchrotron energy lifetime. Birmingham et al. (1974) fit Berge's (1966) 10.4 cm data, as analyzed by Beard and Luthey (1973) (Figure 3). Their

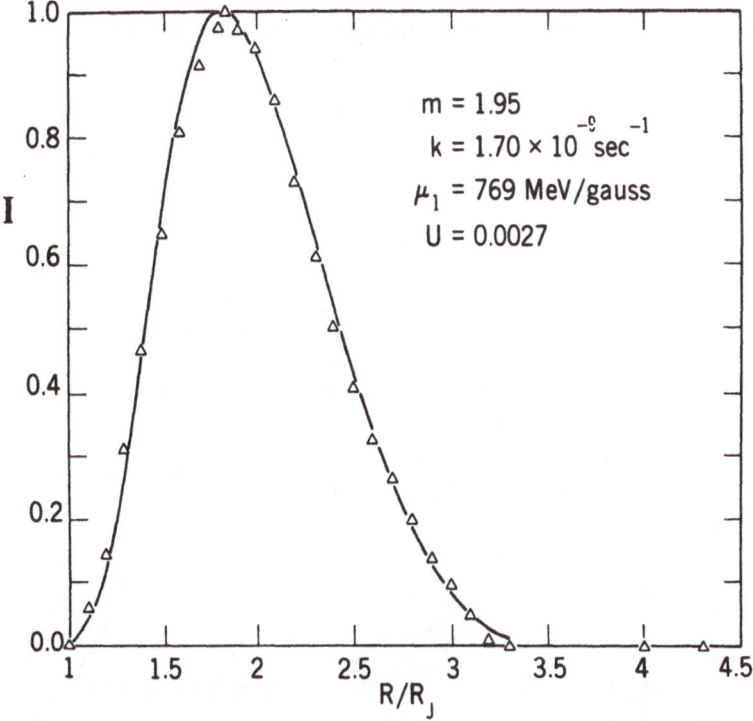

Fig. 3. Radial diffusion modeling (solid curve) of Berge's (1966) 10.4 cm synchrotron data (triangles) assuming a radial diffusion coefficient of $D_{LL} = 1.7 \times 10^{-9} L^{1.95}$ and delta function injection at $\mu = 769$ MeV G^{-1} (Birmingham et al., 1974).

calculations assumed a delta function in the first invariant $\delta(\mu - \mu_1)$ and a constant, but arbitrary, electron flux at a distance greater than 6 R_J. The diffusion coefficient and initial magnetic moment μ_1 were varied until an optimum fit with the volume emissivity I at 10.4 cm was obtained; their results are $D_{LL} = 1.7 \times 10^{-9} L^{1.95 \pm 0.5}$ s^{-1} and $769 < \mu_1 < 1836$ MeV G^{-1}. No attempt was made to fit the synchrotron spectrum at other frequencies, nor was the absolute electron flux determined. The dependence of D_{LL} on L is consistent with the Brice-McDonough ionospheric dynamo origin of the radial diffusion.

Prior to these calculations it was generally accepted that the synchrotron electrons had $\mu \sim 3$–10 MeV G^{-1} (Warwick, 1970; Beck, 1972). However in diffusing into $L = 1.5$–2, synchrotron radiation degrades the electron energy faster than the energy gained by transport into the stronger magnetic field. Thus much higher values of μ, 100–2000 MeV G^{-1} are required the outer zone in order to explain the high frequency synchrotron radiation; Coroniti *et al.* (1973) and Williams and Brice (1973) reached a similar conclusion. The presence of substantial fluxes of > 1 MeV electrons in the outer zone was therefore anticipated theoretically.

A much more ambitious calculation was carried out by Stansberry and White (1974). They attempted to fit not only the strip-scan brightness at a fixed frequency, but the flux density-frequency spectrum and polarization. In order to fit the frequency spectrum, an *ad hoc* loss process was needed at low L-shells. Their optimum fit to all radiation parameters (Figure 4) had a diffusion coefficient of $D_{LL} = 2 \times 10^{-8} L^3$ s^{-1},

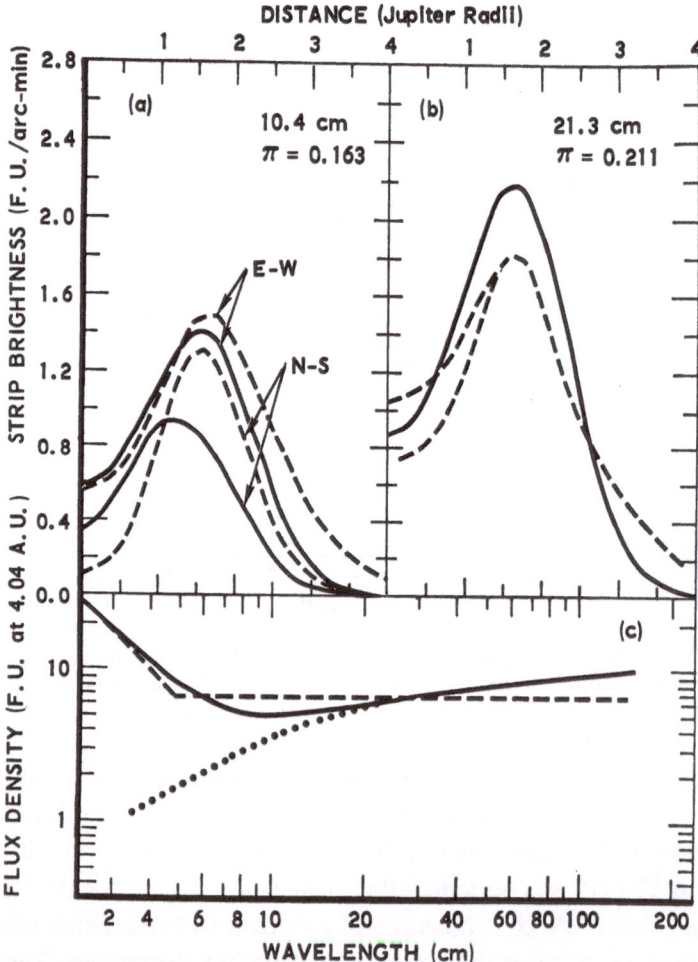

Fig. 4.   Radial diffusion model of equatorial strip brightness at 10.4 cm (a) and 21 cm (b) for east-west and north-south polarization. π is the calculated degree of polarization. Dash lines are Berge's (1966) (a) and Branson's (1968) (b) synchrotron data. Calculated total (solid) and non-thermal (dotted) flux density vs wavelength (c); dash curve is an approximation to radio data (Stansberry and White, 1974).

although coefficients in the range $2.5 \times 10^{-10} L^3 < D_{LL} < 4 \times 10^{-9} L^3$ also yielded reasonable fits. The optimum fit had an initial Gaussian injection spectrum peaked at 100 MeV G^{-1} with a 50 MeV G^{-1} width. The rapid Gaussian fall-off above 100 MeV G^{-1} implies that very few electrons have values in the range $769 < \mu < 1828$ MeV G^{-1} which Birmingham et al. (1974) found gave a good fit to the 10.4 cm data. Stansberry and White (1974) calculated that the maximum electron flux was $J = 1.4 \times 10^9$ cm^{-2} s^{-1} at $L = 2.7$ and was above 3×10^8 cm^{-2} s^{-1} from $1.6 < L < 6.0$. The high diffusion coefficient and large fluxes, compared with Warwick's estimate, might be due to the low energy and steep energy spectrum of the model, although Stansberry and White claim, but do not show, that a high energy model peaked at 600 MeV G^{-1} yielded essentially the same results as the low energy model.

5. Attempt at a Self-Consistent Model

Both of the above calculations demonstrated that the synchrotron radiation could be consistent with a radial diffusion coefficient which scales as a low power of L. However, the various attempts to either unfold (Warwick, 1970; Beard and Luthey, 1973) or model (Birmingham et al., 1974; Jacques and Davis, 1972; Williams and Brice, 1973; Stansberry and White, 1974) the synchrotron radiation lead to quite different quantitative values for the radial diffusion coefficient, the mean electron energy, the energy spectrum, and the electron fluxes. This discrepancy is reasonable since the models used to invert the synchrotron data contained more free parameters than independent measurements; hence the inversions are not unique. Furthermore, these calculations could not determine the electron fluxes beyond about $L = 6$.

At the same time Coroniti (1973, 1974) attempted to construct a self-consistent electron flux model following a program set forth by Kennel (1972) and Thorne and Coroniti (1972) at the JPL Jupiter Workshop. In the outer zone $(L > 6)$ whistler mode turbulent precipitation limits the fluxes to a stably trapped level which is independent of the electron source strength. After including satellite sweep-up losses, these fluxes are used as an initial condition for an inner zone radial diffusion-synchrotron radiation model similar to that of Birmingham et al. (1974) and Stansberry and White (1974).

As in the Earth's magnetosphere, the electron fluxes in the far outer zone $(L > 20)$ are probably highly variable due to fluctuations in the solar wind source and possible local acceleration processes. The very small loss cone at high L suggests that only a low intensity of microscopic plasma turbulence is required to maintain the electron flux isotropic inside the loss cone. Hence it is reasonable to assume that the electrons precipitate at the strong pitch angle diffusion rate (Kennel and Petschek, 1966). For a dipole field the minimum precipitation lifetime on strong diffusion is (Kennel, 1969)

$$T_{\min} = \frac{2L^4 R_J \gamma}{c(\gamma^2 - 1)^{1/2}}, \qquad \gamma = (1 - v^2/c^2)^{-1/2}. \tag{1}$$

At large L, $T_{\min}$ is so long that few electrons are lost as they diffuse inward. However, eventually $T_{\min}$ becomes comparable with the radial diffusion scale time, approximately D_{LL}^{-1}; near this L-shell precipitation reduces the electron flux to the stably trapped level set by whatever mode of plasma turbulence is dominant. For $D_{\mathrm{LL}} \sim$ $\sim 2 \times 10^{-10} L^3 \, \mathrm{s}^{-1}$, $D_{\mathrm{LL}} T_{\min} \approx 1$ occurs at $L \approx 20$ for a wide range of electron energies (γ). Inside $L = 20$, although radial diffusion increases the flux by phase space compression, the electron fluxes will be reduced to the stably trapped level on a time scale less than D_{LL}^{-1}. Hence in the near outer zone ($6 < L < 20$) the electron fluxes should be close to the stably trapped limit wherever pitch-angle scattering instabilities are possible. Furthermore, the fluxes should be relatively steady and not reflect rapid temporal variations in the source.

Since whistler turbulence theory has successfully described the inner zone electron fluxes at Earth (Lyons *et al.*, 1972) a reasonable first attempt for Jupiter was to calculate whistler mode stable trapping levels. To have unstable cyclotron resonant interactions of energetic electrons with whistlers requires the presence of modest quantities of cold plasma to give the whistler a low parallel phase speed (Kennel and Petschek, 1966). Fortunately, Ioannidis and Brice (1971) had previously constructed a model for the cold plasma density at Jupiter based on the escape of ionospheric photoelectrons. Using this model, whistler stably trapped limit fluxes were calculated for two assumed components of the electron energy spectrum. Since high values of μ are required to explain the wide bandwidth of the synchrotron radiation, Coroniti *et al.* (1973) suggested that the most likely electron source would be convection injection from a postulated Jovian magnetic tail neutral sheet. The low neutral sheet magnetic field strength implies that even post-shock thermal solar wind electrons could have $\mu \approx 100$ MeV G^{-1}. In order to be self-consistent a convection source should be included since the inward radial diffusion was driven by convection stimulated ionospheric neutral winds. In addition, a high energy component with a power law spectrum above $\mu = 10^3$ MeV G^{-1} was included; the high μ electrons are required to generate the high frequency synchrotron radiation. The source of the superthermal electrons might be local acceleration within the magnetosphere or the trapping of > 10 keV solar cosmic ray electrons (Lin, 1970) in the neutral sheet.

The stably trapped electron flux levels are shown in Figure 5. For both energy components whistler mode turbulence limits the flux to $J \sim 3$–1×10^8 cm^{-2} s^{-1} in the region $7 < L < 10$; the flux gradually decreases to $J \gtrsim 6 \times 10^7$ cm^{-2} s^{-1} by $L = 20$. Subsequently Brice and Axford and Mendis (1974) concluded that the Ioannidis and Brice (1971) cold plasma density should be reduced by a factor of 5. This reduction raised the stably trapped flux by less than a factor of 2. Inside $L = 2$ the rapid decrease in the cold plasma density in the Ioannidis and Brice model increases the whistler phase speed to the speed of light which stabilizes the cyclotron resonance interactions. With $D_{\mathrm{LL}} = 2 \times 10^{-10} L^3 \, \mathrm{s}^{-1}$, the radial diffusion time to cross the diameter of Io is comparable to Io's synodic time. Io was assumed to present a bare geometrical obstacle to the electrons and to sweep-out all electrons with mirror points above $\pm 10°$ of the magnetic equator. The resulting flux reductions at Io and Europa are included in Figure 5.

Inside $L=6$ the electron distribution function was transported inward with a radial diffusion equation which included synchrotron radiation energy losses. Near $L=2$ the flux of electrons with energies above 1 MeV was $J \sim 6 \times 10^7$ cm^{-2} s^{-1}, the inner zone flux remained below 8×10^7 cm^{-2} s^{-1} at all L (Figure 5). Following the rough

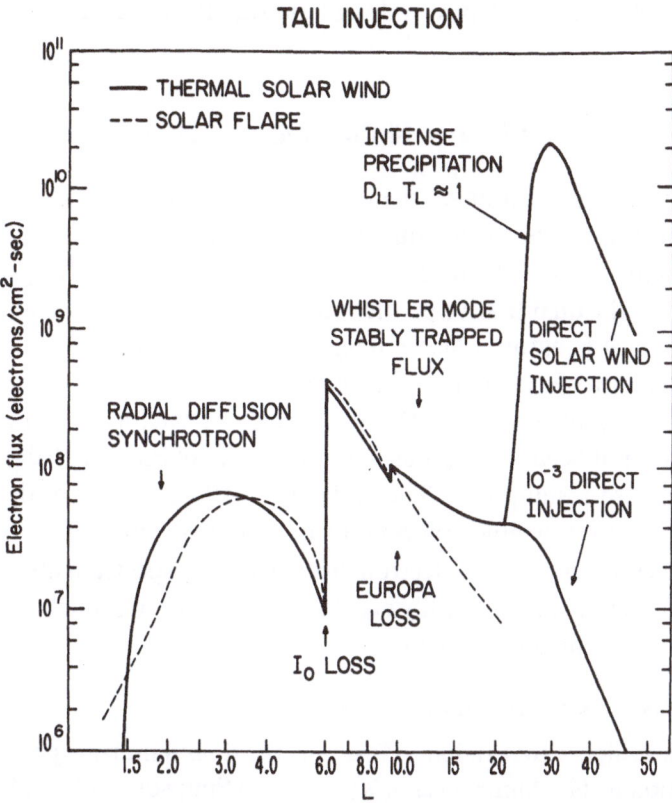

Fig. 5. Summary of the radial diffusion-stably trapped limit model (Coroniti, 1974) assuming the electrons are injected by convection from a tail neutral sheet; thermal solar wind (solar flare) electrons have $\mu \approx 100$ MeV G^{-1} ($\mu > 10^3$ MeV G^{-1}). For $L > 20$ electrons follow temporal variations in the injection source. Between $20 > L > 7$ the radial diffusion time scale is longer than the minimum precipitation lifetime, $D_{LL}T_L < 1$. Whistler mode turbulent precipitation maintains the electron fluxes near the stably trapped limit. The flux is reduced by collisional sweep-up at Io and Europa. Inside $L=6$ the flux is determined from a radial diffusion-synchrotron energy loss kinetic equation.

estimation procedure outlined by Warwick (1970), the synchrotron frequency bandwidth extended up to 400 MHz for the thermal solar-wind deutrons (initial $\mu \sim 100$ MeV G^{-1}), while the initial $\mu \sim 10^3$ MeV G^{-1} energetic electron component was required to account for the 3000 MHz radiation. The synchrotron flux density at Earth (erg cm^{-2} s^{-1} Hz^{-1}) was estimated to be within a factor of 2 of that observed, with the two electron spectral components contributing in roughly equal amounts.

Aside from the rough numerical consistency with the decimetric data, the stably trapped limit-radial diffusion model is qualitatively consistent with the observed long term stability of the synchrotron radiation. Whistler mode turbulence should maintain the electron flux levels in the near outer zone approximately constant since temporal variations in the electron source should be damped by precipitation inside of about $L=20$. The radial diffusion time scale in the inner zone is many years so that several years would be required to change the electron flux profiles. Thus the synchrotron radiation region should behave dynamically like the Earth's inner Van Allen zone which is known to possess long term stability even during magnetic storm conditions.

6. Post-Pioneer 10 Comparisons – Inner Zone

In comparing the pre-encounter theoretical models with the presently available Pioneer 10 measurements, precise quantitation agreement is less significant than determining whether or not the data is consistent with the underlying physics of the models. After all the quantitative predictions are dependent on several basic assumptions: (1) a surface equatorial field strength of 10 G; (2) the radial diffusion coefficient; (3) the energy spectrum and pitch-angle distribution of the injected electrons; (4) the cold plasma density; and (5) the efficiency of satellite sweep-up. At present the models have not been recomputed with the benefit of Pioneer 10 data in removing or modifying these assumptions. Thus a factor 3–10 agreement of the models with data is probably a reasonable criterion for quantitative consistency.

We first review the inner zone electron flux and energy spectrum observations made by Pioneer 10. Here the theoretically expected physics was reasonably simple – radial diffusion and synchrotron energy loss.

6.1. INNER ZONE ($L<6$) ELECTRON FLUXES

The strongest evidence that radial diffusion is the dominant transport process in the Jovian radiation belts is Simpson et al.'s (1974) and Simpson's (1974) observation that the >3.0 MeV integral electron flux scaled as $L^{-4.8}$ inside $L=6$. For loss-free radial diffusion the integral relativistic electron flux should scale as $L^{-4.5}$ (Davis, 1972). Between $3<L<6$ synchrotron energy losses for 3 MeV electrons should be small and precipitation losses should be negligible since whistler waves are stable. Hence the observed $L^{-4.8}$ dependence of the >3 MeV electron flux is consistent with the radial diffusion hypothesis. Furthermore the electrons are confined to a narrow region around the equatorial plane (Van Allen et al., 1974) which is also consistent with cross-L radial diffusion.

TABLE I

Experiment	$J(\text{cm}^{-2}\,\text{s}^{-1})$ $L=3$	$\mu(\text{MeV G}^{-1})$	$\Delta\nu(\text{MHz})$ $L=3$
Simpson et al., 1974	$J(>3\text{ MeV})=5\times10^{8}$	140	60
Van Allen et al., 1974	$J(>8\text{ MeV})=10^{8}$	840	353
Van Allen et al., 1974	$J(>21\text{ MeV})=2\times10^{7}$	5400	2.26×10^{3}
Fillius and McIlwain, 1974	$J(>50\text{ MeV})=1.3\times10^{7}$	2×10^{4}	1.2×10^{4}

Table I contains the integral electron flux above the detector energy threshold measured at $L = 3.2$ by the Pioneer 10 energetic particle experimenters. Also shown is an estimate of the first invariant μ based on the threshold energy. For comparison at $L = 3$ Stansberry and White (1974) predicted a flux of $J \approx 10^9$ cm^{-2} s^{-1} for electrons with $50 < \mu < 150$ MeV G^{-1}. Coroniti's (1974) inner zone model had $J = 8 \times 10^7$ cm^{-2} s^{-1} for both the $\mu = 10^2$ and 10^3 MeV G^{-1} energy populations. Since the synchrotron flux density S scales as $S \propto BJ$, the fluxes in both models should probably be increased by 2.5 to compensate for the reduction of B from 10 to 4 G. In any case the observed fluxes at 3 and 8 MeV are roughly consistent with the theoretical models.

Although the presence of high energy or μ electrons in the inner zone was anticipated by the radial diffusion modeling of the synchrotron radiation, the existence of significant fluxes of extremely high μ electrons was not. The highest value of μ which was discussed in the pre-encounter models was 1836 MeV G^{-1} (Birmingham et al., 1974) which corresponds to 14.5 MeV at $L = 3$. The model energy spectra of both Stansberry and White (1974) and Coroniti (1974) predicted a negligible flux at $\mu = 5400$ MeV G^{-1}, to say nothing about $\mu = 2 \times 10^4$ MeV G^{-1}. If we make a rough fit to the observed flux-energy spectrum by assuming that $J(>\gamma) \propto \gamma^{-n}$, then the fluxes from 3 to 21 MeV are fitted by $n \approx 1.7$; from 21 to 50 MeV, the spectrum appears to harden and a value of n closer to one is obtained. The synchrotron frequency spectrum from a thin magnetic shell of radiating electrons is proportional to $\nu^{-(n-1)}$ (Westfold, 1959). Since the observed synchrotron flux density is reasonably flat between 200 and 3000 MHz (Carr and Gulkis, 1969), the obvious choice for the electron energy spectrum which the theorists should have made is $n = 1$. In this case hindsight is truly better than foresight.

Also shown in Table I is the synchrotron frequency bandwidth $\Delta \nu = 10 B(L) \gamma^2$ MHz (Warwick, 1970) at $L = 3.2$ for γ equaling the threshold energy of the various detectors. Clearly the observed electrons have no difficulty in radiating at all the observed synchrotron frequencies. Recall, however, that the maximum synchrotron power at 10.4 cm occurs at $L = 1,8$,

6.2. ESTIMATE OF SYNCHROTRON FLUX DENSITY

We now attempt to estimate an upper limit to the synchrotron flux density detected at Earth which would be emitted by the Pioneer 10 electrons. The synchrotron intensity emitted by an omni-directional flux of relativistic electrons is (Warwick, 1970)

$$I = \frac{1.3 \times 10^{-23} B(L) J}{c} \text{ erg cm}^{-3} \text{ s}^{-1} \text{ Hz}^{-1} \text{ sr}^{-1}. \tag{2}$$

To obtain the total power/Hz we must integrate I over the emission region which we estimate as $1.6 < L < 3.2$. The differential volume is approximately $4\pi\theta_0 R_J L \, dL$ where we have assumed cylindrical symmetry and where θ_0 is the magnetic latitude below which the electrons mirror. Van Allen et al. (1974) give $\theta_0 \approx 16°$ at $L = 3.0$. To obtain an upper limit to the flux density, we assume that the electron flux scales inward as $L^{-4.5}$, the loss-free radial diffusion scaling. Taking $B(L) = B_0 L^{-3}$ with $B_0 = 4\,G$, and dividing by $R^2 = (4\ \mathrm{AU})^2$, the synchrotron flux density observed at Earth is

$$S_{up} = \frac{1.3 \times 10^{-23} J(L=3.2)\, B_0 4\pi\theta_0 R_J^3}{cR^2} \int_{1.6}^{3.2} \frac{dL}{L^{13/2}} (3.2)^{4.5}$$

$$S_{up} = 1.46 \times 10^{-30} J(L=3)\ \mathrm{erg\ cm^{-2}\ s^{-1}\ Hz^{-1}}. \tag{3}$$

A lower estimate but not a lower limit, to the synchrotron flux density would be to assume that the electron flux remained roughly constant from $1.6 < L < 3.2$. The theoretical radial diffusion-synchrotron radiation calculations suggest that $J \approx$ constant may be a reasonable approximation (see Figure 5). In this case we find

$$S_{10} = 1.8 \times 10^{-31} J(L=3)\ \mathrm{erg\ cm^{-2}\ s^{-1}\ Hz^{-1}}. \tag{4}$$

Table II contains the results for both models of J vs L scalings. Recall that the

TABLE II

	S_{up}	S_{10}	$\Delta\nu$(MHz)	
			$L = 1.8$	$\gamma =$ const
$J(>3\ \mathrm{MeV})$	7.3×10^{-22}	9×10^{-23}	337	
$J(>8\ \mathrm{MeV})$	1.5×10^{-22}	1.8×10^{-23}	1980	(1241)
$J(>21\ \mathrm{MeV})$	2.9×10^{-23}	3.6×10^{-24}	1.3×10^4	(800)
$J(>50\ \mathrm{MeV})$	$1.9 \times 10^{\ 23}$	2.3×10^{-24}	6.7×10^4	(4000)

synchrotron spectrum is $S = 7 \times 10^{-23}\ \mathrm{erg\ cm^{-2}\ s^{-1}\ Hz^{-1}}$ from 200–3000 MHz. Also included is the synchrotron frequency bandwidth at $L = 1.8$ assuming that the electron energy, not μ, remains constant. Radial diffusion solutions indicated that for low energy electrons (<5 MeV) the synchrotron energy loss was roughly compensated by the betatron acceleration resulting from transport to a higher magnetic field strength (Coroniti, 1974). At higher energies, the more rapid synchrotron energy loss reduced the energy by about 4 (Coroniti, 1974); the synchrotron bandwidth assuming a factor 4 reduction in the energy is given in parentheses.

Two conclusions can be drawn from Table II. First, the observed Pioneer 10 electron energies can account for the synchrotron bandwidth. Second, except for the >3 MeV case, the synchrotron flux density calculated by assuming $J \propto L^{-4.5}$ more nearly agrees with observed flux density than does the $J =$ constant assumption. Hence it is likely that inside of Pioneer 10's closest approach the electron fluxes continue to increase at a rate approaching $L^{-4.5}$. Pioneer 11 is intended to cross the dipole equator on the outbound pass at $L = 1.9$. If the $L^{-4.5}$ scaling holds even approximately, Pioneer 11 could encounter electron fluxes a factor of 10 higher than those measured by Pioneer 10.

7. Post Pioneer 10 Comparison – Outer Zone $(6 > L > 20)$

In the outer zone the theoretical models predicted the physics to be more complex. In addition to radial diffusion, whistler mode turbulence was expected to limit the electron flux levels and satellite sweep-up losses were predicted to significantly reduce the electron fluxes. In this section we compare the measured Pioneer 10 electron fluxes and spatial profile with the theoretical expectations. In Section 8 we speculate on the intensity of electron precipitation.

The electron integral flux > 0.1 MeV just outside Europa was 3×10^7 cm^{-2} s^{-1} sr^{-1} or probably $J \sim 2$–3×10^8 cm^{-2} s^{-1} (Fillius and McIlwain, 1974) while the integral flux > 3 MeV was $J \approx 5 \times 10^7$ cm^{-2} s^{-1} (Simpson et al., 1974). The whistler mode stably trapped flux for both the $\mu = 100$ and 10^3 MeV G^{-1} energy models was $J \approx 10^8$ cm^{-2} s^{-1} at Europa (Coroniti, 1974); with $B_0 = 4.0$ G at the surface (Smith et al., 1974) $\mu = 10^2$ MeV G^{-1} (10^3 MeV G^{-1}) corresponds to 0.35 (1.75) MeV. The stably trapped flux is numerically proportional to the magnetic field strength (Kennel and Petschek, 1966) so that the flux of 10^8 cm^{-2} s^{-1} should be scaled down by a factor 2.5. However, a factor 5 reduction in the cold plasma density raised the stably trapped flux by a factor of 2 above 10^8 cm^{-2} s^{-1} (Coroniti, 1974). Probably the stably trapped flux is uncertain by factors of 2 to 4. At 15 R_J, the flux > 0.1 MeV was $\sim 2 \times 10^7$ cm^{-2} s^{-1} (Fillius and McIlwain, 1974) while the stably trapped flux model with $\mu = 100$ MeV G^{-1} ($= 0.1$ MeV) gave 4×10^7 cm^{-2} s^{-1}. Thus in the region outside Europa the Pioneer 10 observations are in fair quantitative agreement with the stably trapped flux calculations.

Aside from the overall quantitative consistency of the Pioneer 10 data with theory, there is an even more striking and significant qualitative consistency. Except for satellite and pitch-angle effects, the radial flux profiles inside 20 R_J are remarkably smooth. In contrast energetic electrons in the Earth's outer zone exhibit complex spatial and temporal structure which arises from substorm convection injection, local acceleration, and drifting plasma clouds. The most fundamental conclusion of the radial diffusion-stable trapping model for the Jovian outer zone electrons is that the fluxes should be temporally steady and that the flux should monotonically increase with decreasing L. For $L < 20$ whistler mode precipitation maintains the fluxes near the stably trapped limit even if the far outer zone source is temporally unsteady. Since the precipitation lifetime is less than the radial diffusion time scale, the electron fluxes should follow the gradual increase of the stably trapped flux with decreasing L and not the more rapid L^{-6} ($L^{-4.5}$) (Davis, 1972) increase expected for loss-free radial diffusion of non-relativistic (relativistic) electrons. The lower energy electrons whose fluxes are near the stably trapped limit have this more gradual behavior; high energy electrons have an insufficient flux to destabilize whistlers and can follow the more rapid L-shell increase.

In summary, after comparing the Pioneer 10 data with pre-encounter theory, the radial diffusion-stable trapping hypothesis remains a viable model for understanding the dynamics of the Jovian outer zone radiation belts.

8. Estimate of Electron Precipitation Fluxes

From about 10 to 16 R_J, Fillius and McIlwain (1974) detected an almost constant integral flux above 0.1 MeV of $J \approx 6 \times 10^7$ cm^{-2} s^{-1}. Radial diffusion compression would increase the non-relativistic integral electron flux as L^{-6} (Davis, 1972). Since the electron fluxes are close to the predicted stably trapped flux, a plausible explanation for the rough constancy of J is that precipitation losses just balanced radial diffusion compression. In steady state the non-relativistic integral flux satisfies the equation

$$L^2 \frac{\partial}{\partial L} \left[\frac{D_{LL}}{L^2} \frac{\partial}{\partial L} \right] (L^6 J) = \frac{JL^6}{T_L},$$

(5)

where T_L is an energy averaged precipitation lifetime. If $D_{LL} = DL^3$ and J is approximately constant, the lifetime is

$$T_L = \tfrac{1}{36} DL.$$

(6)

The precipitation flux J_p is then given by (Coroniti and Kennel, 1970)

$$J_p = J \frac{T_{\min}}{T_L} = \frac{J\gamma}{(\gamma^2 - 1)^{1/2}} \frac{720 \, R_J L^5}{c}.$$

(7)

For $L = 15$, $\gamma = 1.2$ (0.1 MeV), and D in the range 2×10^{-9}–2×10^{-10} s^{-1} (Coroniti, 1974; Stansberry and White, 1974), the precipitation flux is $J \sim 1.5 \times 10^5$–1.5×10^6 cm^{-2} s^{-1}.

The whistler mode magnetic amplitude δB which is required to sustain the above precipitation losses can be estimated from the lifetime T_L. The pitch-angle diffusion coefficient $D_{\alpha\alpha}$ is (Kennel and Petschek, 1966) $D_{\alpha\alpha} = \Omega_- (\delta B/B)^2$ where $\Omega_- = eB/mc$ is the equatorial electron cyclotron frequency. In weak diffusion ($J_p \ll J$), the diffusion coefficient is roughly equal to the inverse of the lifetime

$$D_{\alpha\alpha} = \Omega_- \left(\frac{\delta B}{B} \right)^2 \approx \frac{1}{T_L} = 36 \, DL.$$

(8)

For $L = 15$ and $D = 2$–20×10^{-10} s^{-1}, we find $B = 0.25$–0.8 mγ. The whistler frequencies will probably lie in the range 1–10 kHz between $L = 10$ and $L = 20$. Although modest, whistler amplitudes in 0.1–1.0 mγ range are readily detectable by standard VLF magnetic loop or short electric dipole antennae.

With the above precipitation flux, we can estimate the precipitated energy flux as 0.03–0.3 erg cm^{-2} s^{-1}. Since the fluxes of electrons below 0.1 MeV could be substantially higher than 6×10^7 cm^{-2} s^{-1}, the above estimate may well be a lower limit to the precipitated electron heat flux. In any case a precipitation energy flux of $\gtrsim 0.1$ erg cm^{-2} s^{-1} could make a substantial contribution to the production of the ionospheric electron density, and, since energetic electrons can penetrate to low altitudes, could even modify the ionospheric chemical structure. Hence in order to construct an accurate model of the Jovian ionosphere it will be necessary to measure the elec-

tron precipitation fluxes or, more easily, the intensity of the microscopic plasma wave
turbulence which is responsible for the precipitation.

9. Satellite Losses

A central question which was raised by the pre-encounter theories was the extent
to which the satellites reduced the electron and proton fluxes by collisional sweep-up
(Mead, 1972). The basic issue was whether or not radial diffusion was sufficiently
rapid to transport the particles past the satellite orbits before the probability of a
collision reached unity.

The sweep-up losses of electrons at Io and Europa were apparently less severe than
had been anticipated (Mead and Hess, 1973; Hess *et al.*, 1974). Fillius and McIlwain
(1974) report a decrease of about 5 in their low energy detector (> 0.1 MeV) at Europa
and smaller decreases at higher energies at Europa and Io. Van Allen *et al.*'s (1974)
data appear to have about a factor of 2 decrease in their > 70 keV channel at Europa.
The satellite-associated decreases for energetic electrons (> 1 MeV) appears to be
much smaller than at lower energies (Van Allen *et al.*, 1974; Simpson *et al.*, 1974).
Van Allen *et al.* (1974) found that the > 8 MeV and > 21 MeV electron fluxes de-
creased with magnetic latitude θ as $\exp(-\theta^2/\theta_0^2)$ with θ_0 about $15°$. Since the dipole
tilt angle is inclined by $15°$ with respect to the spin axis (Smith *et al.*, 1974), a large
fraction of the relativistic electrons have mirror points which do not intersect the
plane of the satellites' orbits and hence will not be absorbed. Furthermore, the elec-
tron's magnetic gradient drift velocity $v_D = \frac{3}{2}(mc^3 L^2/eB_0 R_J)\,(\gamma^2 - 1)/\gamma$ is opposed to
the corotation drift. At Europa (Io) a 22 MeV (30 MeV) electron has the same synodic
rotation period as the satellite. Hence electrons with 10's MeV energy may have only
weak interactions with the satellites even if the radial diffusion coefficient is small.

A complication which was not included in the satellite loss calculations is that at
least Io has a dense ionosphere (Kliore *et al.*, 1974a, b). The presumably high ionospheric conductivity might sustain currents which distort or exclude the Jovian dipole
field from the satellite surface. Since the electron gyroradius is small compared to
the satellite diameter, significant magnetic field distortions would prevent electrons
from colliding with the surface, thus preventing satellite sweep-up.

The electron fluxes in the region between Europa and Io are not inconsistent with
the concepts of stable trapping and radial diffusion transport. The low energy electrons (< 1 MeV) probably interact more strongly with the whistler turbulence and
have a less anisotropic pitch-angle distribution. Satellite collisional losses would
therefore be more effective for the low energy electrons than for the more anisotropic
electrons at higher energies.

10. Electron Heat Flux to Io and Europa

Although the dynamics of the satellite collisional losses may be complex, it is still
interesting to estimate the electron energy flux which might be precipitated on the

satellites Io and Europa. Assuming that the lower energy electrons (< 3 MeV) are not too anisotropic, the total number of electrons on a 1 cm^2 flux tube is approximately

$$\delta N/\delta A = JLR_J/c. \tag{9}$$

Let's assume for simplicity that a satellite can sweep-out an entire flux tube; i.e., we ignore the 15° inclination between the satellite orbital plane and the magnetic equator. The total number of electrons hitting the satellite per second is then

$$\dot{N} = v_{rel}d(J_L R_J/c), \tag{10}$$

where v_{rel} is the relative velocity between the satellite and the drifting electrons and d is the satellite diameter. For low energy electrons, v_{rel} is approximately $v_{rel} \approx \Omega_s LR_J$ where Ω_s is the synodic rotation frequency. Measuring J in units of 10^8 cm^{-2} s^{-1} (J_8), we have

$$\dot{N}(\text{Io}) = 2.8 \times 10^{23} \, J_8 \, \text{s}^{-1}$$
$$\dot{N}(\text{Eur}) = 7 \times 10^{23} \, J_8 \, \text{s}^{-1}$$

If we normalize the mean energy to 1 MeV (T_1), the total power $\dot{N}T$ absorbed by the satellite is

$$\dot{W}(\text{Io}) = 4.5 \times 10^{19} \, J_8 \, T_1 \, \text{erg s}^{-1}$$
$$\dot{W}(\text{Eur}) = 1.1 \times 10^{20} \, J_8 \, T_1 \, \text{erg s}^{-1}.$$

The above calculation undoubtedly grossly overestimates the precipitated energy. However, even if geometrical factors reduce W by a factor of 100, the total energy input from electrons to Io and Europa is still comparable to the total energy dissipation rate of a substorm at Earth.

At present we can only speculate on the consequences of such a high energy input. Pioneer 10 measurements indicated that Io has an ionosphere with an electron density of 6×10^4 cm^{-3} on the sunlit side and $< 10^4$ cm^{-3} on the dark side (Kliore *et al.*, 1974b). If distributed over the entire surface of Io, the electron precipitation heat flux would be 1–10 erg cm^{-2} s^{-1}. Clearly precipitation heating at this rate must be an important factor in determining the structure of Io's ionosphere. Although the day-night asymmetry (Kliore *et al.*, 1974b) of the ionospheric electron density has the obvious interpretation associated with photo-production, another possibility exists. If Io sweeps-up electrons and the radial diffusion rate is slow, the electron flux is higher on the sunlit or outer side than on the dark or inner side; i.e., the electron flux has an outward radial gradient across Io. Thus if electron precipitation produces Io's ionosphere, the day-night asymmetry in the ionospheric electron density would be due to the asymmetry in the precipitation flux caused by sweep-up.

Io's atmosphere may not be dense enough to keep the energetic electrons from penetrating to the surface. In this case Io would charge, and the ionosphere, which may extend to the surface, might be part of the Io sheath (Gurnett, 1972).

11. Discussion

At the present stage of the data analysis, the Pioneer 10 measurement can be interpreted as being consistent, in the broadest sense, with the physics of radial diffusion, stable trapping, and satellite collisional losses. However, much remains to be done, both theoretically and experimentally, before the details of these physical processes are adequately understood. This section attempts to summarize the conclusions reached so far and to comment on areas of future research.

The most convincing Pioneer 10 observation that radial diffusion is the dominant transport process is the $L^{-4.8}$ dependence of the electron flux in the inner zone. In the region $3 < L < 6$ all loss processes should be weak and radial diffusion theory predicts that $J \propto L^{-4.5}$ for relativistic electrons. This observation, however, does not establish a value or scaling with L of the radial diffusion coefficient. Hence whether ionospheric dynamo electric fields or some other process drives the radial diffusion is not resolved. Perhaps a careful analysis of the electron flux gradients near the Galilean satellites can provide at least an estimate of the magnitude of the radial diffusion coefficient.

The inner zone electron fluxes are near the theoretically expected level, but of course the synchrotron emission data placed a rather strong constraint on the range of possible fluxes. The electron energy spectrum of $\gamma^{-(1-2)}$ was a surprise, but should not have been. For many years astrophysicists were aware that a γ^{-1} energy spectrum produced a flat synchrotron frequency spectrum. Almost all high energy astrophysical radio sources are consistent with a γ^{-n} electron energy spectrum, with n being a small number; even cosmic ray protons have $n \approx 2.7$. One of the central problems of high energy astrophysics, however, is that the physical processes which produce such a power law spectrum are poorly understood. Hence if the origin of the energetic electron spectrum in the Jovian radiation belts can be resolved, the answer might be of general astrophysical significance.

In the near outer zone the sweeping effect of the Galilean satellites proved to be a relatively minor loss process for energetic electrons. The large dipole tilt angle of 15°, the flat pitch-angle distribution, and the slow total drift speeds probably combine to allow most of the electrons to escape collisions with the satellites. The lower energy electrons apparently suffered a larger collisional loss, perhaps because interactions with plasma wave turbulence maintained their pitch-angle distribution near isotropy. A rough estimate indicated that the satellites might absorb electron energy at a rate of 10^{18-20} erg s^{-1}. If true, the ionosphere's and surface properties of the satellites must certainly be affected.

The question of whether Io and Europa are significant sources of radiation belt electrons (Gurnett, 1972; Hubbard et al., 1974) is still open. The $\mathbf{v} \times \mathbf{B}$ potential drop across Io (Europa) is 350 kV (150 kV). If this entire potential were given to electrons by acceleration through a sheath around the satellites, and if all the predominantly parallel energy were converted into perpendicular energy, the resulting first invariant would be $\mu \approx 40$ MeV G^{-1} (25 MeV G^{-1}). Thus although the satellites may be a source of low energy electrons, they are unlikely to produce the electrons which are important for synchrotron radiation.

In the region $9.4 < L < 20$ where satellite influences are weak, the observed fluxes of > 0.1 MeV and > 3.0 MeV electrons are roughly consistent with the calculated whistler mode stably trapped limit. The important point here is that the fluxes were not $< 10^5$ cm^{-2} s^{-1} as predicted by the JPL Workshop model, nor were they at the 10^{10}–10^{11} cm^{-2} s^{-1} upper limit corresponding to magnetic field saturation. The radial flux profile is smooth, as was expected if radial diffusion and stable trapping are the dominant processes. Furthermore, the > 0.1 MeV and > 3.0 MeV fluxes were relatively flat between 16 and 10 R_J, whereas loss-free radial diffusion predicts a L^{-6} and $L^{-4.8}$ dependence respectively. Hence it is quite likely that the flux levels of these lower energy electrons are determined by stable trapping, which predicted a much gentler increase with decreasing L. The higher energy electrons probably do not interact strongly with the wave turbulence, and can therefore have a much stronger L dependence, as was observed. If precipitation limits the flux increase, the precipitation heat flux into the ionosphere could be as large as 0.1–1.0 erg cm^{-2} s^{-1}. At this intensity precipitation should modify the structure of the ionosphere.

What questions did Pioneer 10 leave unresolved? The payload was not instrumented to measure non-streaming particles in the energy range 0–70 keV, nor was there a VLF magnetic loop or electric dipole antenna to detect the local microscopic plasmawave turbulence. Without knowledge of the cold or thermal plasma distributions, stably trapped limit fluxes cannot be calculated with any more assurance than before encounter. The measurement of a single whistler or other VLF emission such as the electrostatic modes at $\frac{3}{2}$ and $\frac{5}{2}$ of the electron cyclotron frequency would have settled the question of whether or not the concept of stable trapping is correct. In addition, from a measurement of the plasma wave intensity and frequency spectrum, a more confident and accurate calculation of the electron precipitation fluxes could be performed. With the presence of such intense radiation belt fluxes, precipitation seems inevitable. Knowledge of the electron precipitation flux is crucial to the construction of an accurate model of the Jovian ionosphere. As Kennel and Coroniti (1975) argue in a companion paper, the structure of the ionosphere may determine whether corotation can be enforced on distant magnetic field lines and what type of convective flow pattern exists inside the Jovian magnetosphere.

The origin of the high energy or μ electrons is also not clear. At least on the dayside energetic electrons appear to be escaping from the inner radiation belt and travelling outwards along the thin magnetic disk (Simpson et al., 1974; Van Allen et al., 1974). The structure of the magnetic disk resembles a solar wind-type outflow solution (Hill et al., 1974; Michel and Sturrock, 1974). Hence it is unlikely that the source of the radiation belts is inward diffusive transport from the dayside magnetopause. The very high first invariants of the energetic electrons suggests that the electrons originate in regions of low magnetic field strength. An obvious possibility is that solar wind electrons are entrained in the neutral sheet of Jovian magnetic tail, energized by the reconnection process, and injected into the radiation belts by internal magnetospheric convection on the nightside. A Jupiter fly-by or orbiter which passed through the distant nightside magnetosphere is essential to resolving this fundamental Jovian, and possibly astrophysical, question. Of course, stochastic acceleration by plasmawave turbulence might also produce the required high μ values. Here again, plasmawave measurements are essential.

Finally, a comment on the Pioneer 10 proton observations. Coroniti et al. (1974) attempted to set a theoretical upper limit on the proton fluxes with $\mu = 10$–100 MeV G^{-1} by combining the stably trapped limits for the ion cyclotron wave in the outer zone and for a quasi-electrostatic loss-cone mode in the inner zone. The Pioneer 10 proton data is consistent with this upper limit, but probably in a meaningless way: the observed proton fluxes are a factor of 20 to 100 below the stably trapped limit at all L-shells. There is, however, a difficulty in interpreting the Pioneer 10 data as evidence for the absence of stable trapping; the low energy (< 3 MeV) proton detectors apparently have saturation problems. Both Trainer et al.'s (1974) 0.44–2.0 MeV detector and Simpson et al.'s (1974) 0.5–1.8 MeV detector saturated inside 15 R_J. Also Trainer et al.'s (1974) > 1.23 MeV and 3.3–21 MeV integral channels may have been saturated inside 5 R_J. At $L = 3$, $\mu = 10$–100 MeV G^{-1} corresponds to proton energies of 1.5–15.0 MeV, while at $L = 15$ the energy range is 10^{-2}–10^{-3} MeV. Thus the proton energies for which the stably trapped limit calculations were performed may not have been adequately measured. Calculations should probably be made for the high energy protons in order to determine whether the observed fluxes are near the stable trapping limit.

Acknowledgements

Prof. C. F. Kennel made substantive contributions to the sections on electron precipitation and the precipitation heat flux to the Jovian satellites. The comparison of the theoretical models with the Pioneer 10 data benefited from many illuminating discussions with C. F. Kennel and F. L. Scarf. This work was supported by NASA grant NGL 05-007-190.

References

Atreya, S. K., Donahue, T. M., and McElroy, M. B.: 1974, *Science* **184**, 154.
Axford, W. I. and Mendis, D. A.: 1974, *Ann. Rev. Earth Planetary Sci.* **2**, 419.
Beard, D. B. and Luthey, J. L.: 1973, *Astrophys. J.* **183**, 179.
Beck, A. J. (ed.): 1972, *Proc. Jupiter Radiat. Belt Workshop*, NASA TM 33-543.
Berge, G. L.: 1966, *Astrophys. J.* **146**, 767.
Birmingham, T. J.: 1969, *J. Geophys. Res.* **74**, 2169.
Birmingham, T., Hess, W., Northrup, T., Baxter, R., and Lujko, M.: 1974, *J. Geophys. Res.* **78**, 87.
Branson, N. J. B. A.: 1968, *Monthly Notices Roy. Astron. Soc.* **139**, 155.
Brice, N.: 1972, in A. J. Beck (ed.), *Proc. Jupiter Radiat. Belt Workshop*, NASA TM 33-543, p. 283.
Brice, N. M. and Ioannidis, G. A.: 1970, *Icarus* **13**, 173.
Brice, N. and McDonough, T. R.: 1973, *Icarus* **18**, 206.
Carr, T. and Gulkis, S.: 1969, *Ann. Rev. Astron. Astrophys.* **7**, 577.
Coroniti, F. V.: 1973, *AIAA Bull.* **10**, 204.
Coroniti, F. V.: 1974, *Astrophys. J. Suppl.* **27**, 261.
Coroniti, F. V. and Kennel, C. F.: 1970, *J. Geophys. Res.* **75**, 1279.
Coroniti, F. V., Kennel, C. F., and Thorne, R. M.: 1973, *Trans. Am. Geophys. Union* **54**, 466.

Coroniti, F. V., Kennel, C. F., and Thorne, R. M.: 1974, *Astrophys. J.* **189**, 383.

Davis, L., Jr.: 1972, in A. J. Beck (ed.), *Proc. Jupiter Radiat. Belt Workshop*, NASA TM 33-543, p. 517.

Fedder, J. A. and Banks, P. M.: 1972, *J. Geophys. Res.* **77**, 2328.

Fillius, R. W. and McIlwain, C. E.: 1974, *Science* **183**, 315.

Gurnett, D. A.: 1972, *Astrophys. J.* **175**, 525.

Hess, W. N., Birmingham, T. J., and Mead, G. D.: 1974, *J. Geophys. Res.* **79**, 2877

Hill, T. W., Dessler, A. J., and Michel, F. C.: 1974, *Geophys. Res. Letters* **1**, in press.

Hubbard, R. F., Shawhan, S. D., and Joyce, G.: 1974, *J. Geophys. Res.* **79**, 920.

Ioannidis, G. and Brice, N. M.: 1971, *Icarus* **14**, 360.

Jaques, S. A. and Davis, L., Jr.: 1972, CalTech. Rept.

Kennel, C. F.: 1969, *Rev. Geophys.* **7**, 379.

Kennel, C. F.: 1972, *Proc. Workshop on Jupiter's radiat. Environ.*, JPL.

Kennel, C. F. and Coroniti, F. V.: 1975, this volume, p. 451.

Kennel, C. F. and Petschek, H. E.: 1966, *J. Geophys. Res.* **71**, 1.

Kliore, A., Cain, D. L., Fjeldbo, G., Seidel, B. L., and Rasool, S. I.: 1974a, *Science* **183**, 323.

Kliore, A. J., Cain, D. L., Fjeldbo, G., Seidel, B. L., and Rasool, S. I.: 1974b, *Trans. Am. Geophys. Union* **55**, 339.

Lin, R. P.: 1970, *Solar Phys.* **12**, 266.

Lyons, L. R., Thorne, R. M., and Kennel, C. F.: 1972, *J. Geophys. Res.* **77**, 3455.

Mead, G. D.: 1972, in A. J. Beck (ed.), *Proc. Jupiter Radiat. Belt Workshop*, NASA TM 33-543, p. 271.

Mead, G. D. and Hess, W. N.: 1973, *J. Geophys. Res.* **78**, 2793.

Michel, F. C. and Sturrock, P. A.: 1974, *Planetary Space Sci.*, submitted.

Nakada, M. P. and Mead, G. D.: 1965, *J. Geophys. Res.* **70**, 4777.

Simpson, J. A.: 1974, *Trans. Am. Geophys. Union* **55**, 340.

Simpson, J. A., Hamilton, D., Lentz, G., McKibben, R. B., Mogro-Compero, A., Perkins, M., Pyle, E. D., and Tuzzolino, A. J.: 1974, *Science* **183**, 309.

Smith, E. J., Davis, L., Jr., Jones, D. E., Colburn, D. S., Coleman, P. J., Jr., Dyal, P., and Sonett, C. P.: 1974, *Science* **183**, 305.

Stansberry, K. G. and White, R. S.: 1974, *J. Geophys. Res.* **79**, 2331.

Thorne, R. M. and Coroniti, F. V.: 1972, *Proc. Workshop on Jupiter's Radiat. Environ.*, JPL.

Trainor, J. H., Teegarden, B. J., Stilwell, D. E., McDonald, F. V., Roelof, E. R., and Webber, W. R.: 1974, *Science* **183**, 311.

Van Allen, J. A., Baker, D. N., Randall, B. A., Thomsen, M. F., Sentman, D. D., and Flindt, H. R.: 1974, *Science* **183**, 309.

Warwick, J. W.: 1967, *Space Sci. Rev.* **6**, 841.

Warwick, J. W.: 1970, NASA CR 1685.

Westfold, K. C.: 1959, *Astrophys. J.* **130**, 241.

Williams, G. J. and Brice, N. M.: 1973, *Trans. Am. Geophys. Union* **54**, 446.

MAGNETOSPHERES OF EARTH AND JUPITER
AFTER PIONEER 10

A. PRAKASH and N. BRICE

Center for Radiophysics and Space Research and School of Electrical Engineering, Cornell University, Ithaca,
N.Y. 14850, U.S.A.

Long before the Pioneer 10 flyby of Jupiter, it was known from Earth based observations of decimetric and decametric radio emissions from Jupiter that the planet has a large magnetic field (1–10 G). Assuming a radial flow of the solar wind at constant velocity, with density decreasing as the distance squared, and the interplanetary field decreasing linearly with distance, Brice and Ioannides (1970) obtained a model of Jupiter's magnetosphere by scaling the magnetosphere of the Earth. For a surface magnetic field of 10 G, they predicted the magnetopause to be at 53 R_J at the subsolar point and at 80 R_J on the dawn side, on the basis of a balance between the solar wind pressure and the planetary magnetic field pressure. One of the first surprises of the Pioneer 10 encounter with Jupiter in November, 1973, was that a magnetopause was observed way out at 96 R_J on the inbound trajectory. The other major surprise, in so far as the large scale structure of the magnetosphere is concerned, was the observation of a 10 h periodicity in the detection of magnetospheric plasma. In addition, it was found that the outer region of Jupiter's dayside magnetosphere exhibited phenomena which resembled more the phenomena observed in the geomagnetic tail.

In this paper we point out possible reasons for the remarkable differences observed between the magnetospheres of Earth and Jupiter and give a model of Jupiter's magnetosphere which sheds light on the observations of Pioneer 10 mission and gives some idea of the interesting features one should look for during the forthcoming Pioneer 11 flyby of Jupiter. But a crucial test of the model requires a flyby in which plasma detectors look out towards the planet to detect magnetospheric plasma predicted to be flowing out (at Alfvén speed) from a region surrounding the planet, a possibility for the Mariner Jupiter (MJS) mission planners to consider.

In their discussion of Jupiter's magnetosphere, Brice and Ioannidis (1970) derived an important result viz., the plasmasphere in Jupiter's magnetosphere would extend almost to the boundary of the scaled ('classical') magnetosphere. More specifically, they showed that whereas the region of corotating plasma (the plasmasphere) for the case of the Earth extends to less than 5 R_E (Carpenter, 1966), in the case of Jupiter, with a stronger magnetic field, faster rotation and a weaker solar wind, the plasmapause would coincide with the magnetopause boundary at dusk and be about 1 R_J inward of the magnetopause at dawn, the dawn-dusk magnetopause boundaries being at 80 R_J from Jupiter's center in the classical model. (Similar results have been obtained independently by Vasyliunas in the paper presented at this conference.) This result was obtained by using the idea, known to be successful for the case of the Earth (Brice, 1967), that at the plasmapause boundary the magnetospheric ('convective') electric field generated by the interaction of the solar wind with the magnetic field lines of the planet by reconnection (Dungey, 1961) or by viscous interaction (Axford and Hines, 1961), is about equal to the corotation electric field. The corotating plasmasphere so obtained is shown by the closed dashed curve in Figure 1.

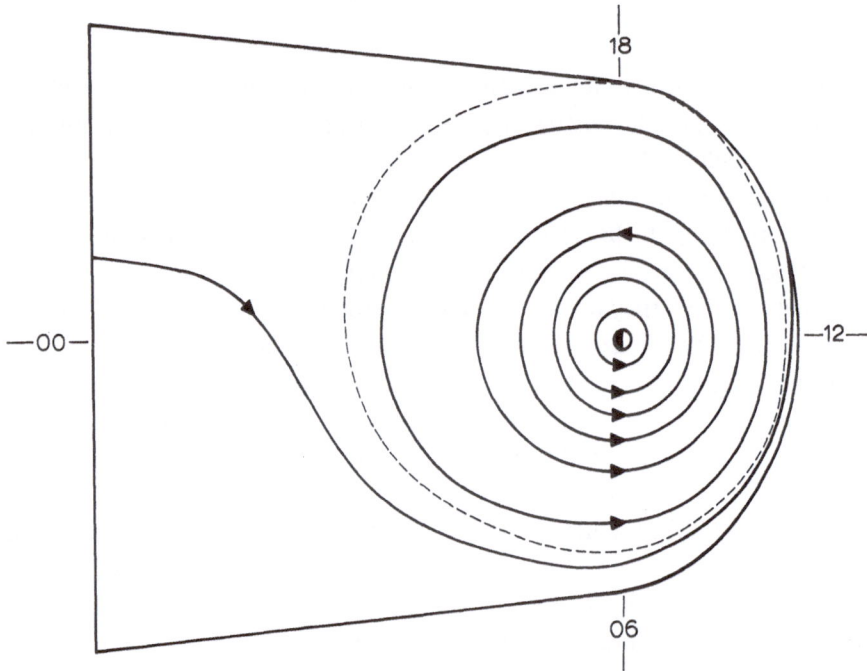

Fig. 1.   Boundary of the corotating plasma for Jupiter's magnetosphere as obtained by Brice *et al.* (1970) is that shown by the dashed curve.

Such a corotating plasma in the plasmasphere will experience gravitational, centrifugal and magnetic forces, in addition to a pressure gradient. For a centered aligned dipole, with a rotation period of 10 hours, the locus of points where components of centrifugal and gravitational forces along the magnetic field lines cancel out (Brice and Ioannidis, 1970) is shown by the dashed lines in Figure 2. The equatorial distance at which these components balance is approximately $2 R_J$, which is slightly less than the synchronous orbit, $2.2 R_J$. (For the case of the Earth this distance is approximately $6.5 R_E$ (Angerami and Thomas, 1964), which is outside the plasmapause and hence of no consequence for the plasma in the plasmasphere.) Now the height of potential barrier, for Jupiter's case, ranges from ~ 6 eV to ~ 13 eV for field lines with equatorial distances $(R = L R_J)$ from $L=2$ to $L=50$. Ionospheric plasma exceeding these energies will be thrown out by the centrifugal force along the field lines. The ionospheric plasma which so reaches the outer magnetosphere is primarily that which originates as non-Maxwellian photoelectrons. Unlike the Earth, the contribution from thermal Maxwellian distribution for Jovian ionospheric temperatures of a few hundred degrees is small.

The inertial forces exerted by the plasma loading the field lines beyond $2 R_J$ will result in the formation of a flattened disc of corotating plasma (Gold, 1962; Melrose, 1967; Piddington, 1967). The latest interpretation of the magnetic field measurements by Pioneer 10 is that of a skew, and off-centered, dipole tilted at 11° to the axis of rotation (Smith *et al.*, 1975). However, it seems that the proposal by Van Allen *et al.* (1974) that Jupiter's magnetosphere consists of a flattened, skew (with the 11° tilt) 'magnetodisc', rotating rigidly out to 96 R_J has some fundamental difficulties. This is

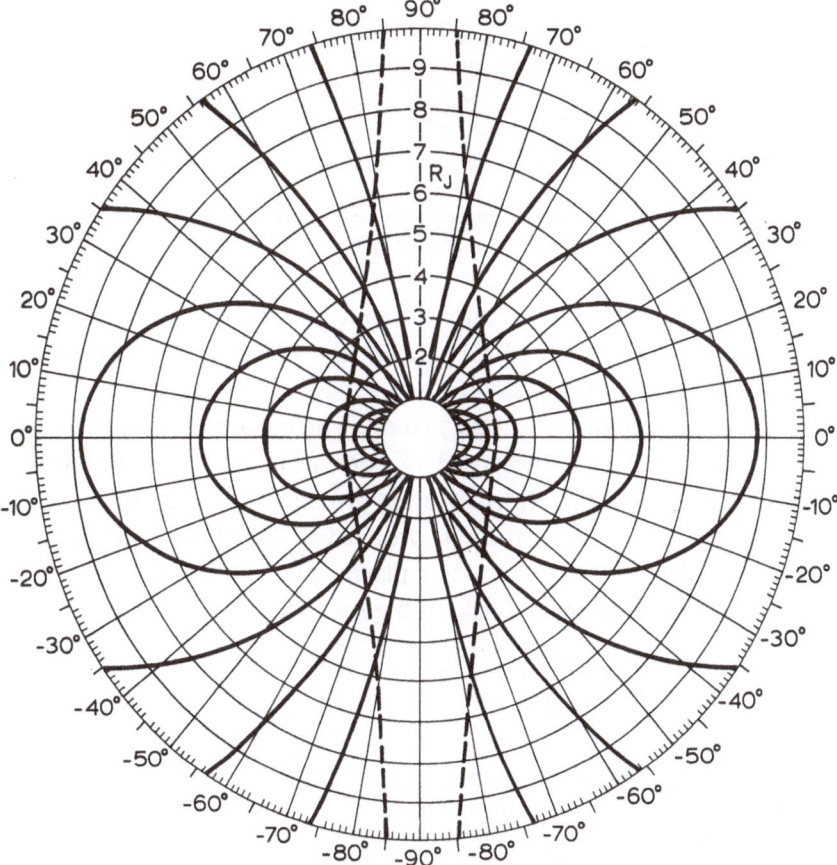

Fig. 2.   Meridional section of Jupiter's dipole field (Brice *et al.*, 1970). At points on the dashed lines, components of gravitational and centrifugal force along the field lines cancel.

shown in Figure 3 in which equilibrium of an element of plasma located in the magnetic equatorial plane at a distance R from the axis of rotation is considered. The centrifugal force is along a line perpendicular to the axis of rotation, but the magnetic force is at another inclination. Consequently, a balancing of the centrifugal force by the parallel component of the magnetic force would leave the vertical component of the magnetic force unbalanced. This unbalanced component points towards the spin equatorial plane. Thus, with increasing R, the skew disc would have a tendency to become parallel to the spin equatorial plane. If this is accomplished around 30 R_J, then beyond this distance the rigidly rotating disc would deviate by less than 5 R_J from the rotational equatorial plane.

We propose a model of Jupiter's magnetosphere in which the *inner part* consists of such a warped magnetodisc. The modification of the dipole field lines would be accomplished by a ring current. (The ring current in Earth's magnetosphere is of a

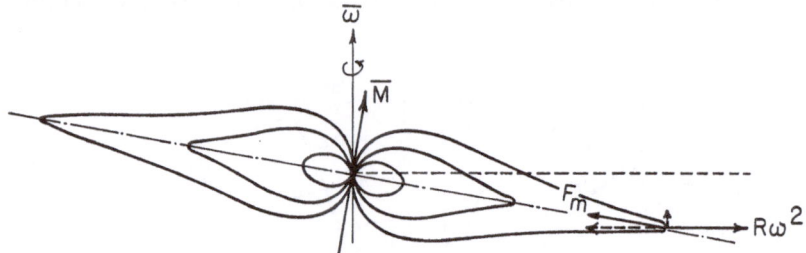

Fig. 3.   Forces acting on plasma at large distances from the spin axis of a skew rotating magnetodisc tend to warp the disc toward the spin equatorial plane.

different origin.) The current sheet model deduced from Pioneer 10 magnetic field measurements by Smith *et al.* (1975) satisfies the above warped magnetodisc re-requirement but its extent is questionable.

A limitation on the extent of the rigidly rotating magnetodisc (warped or otherwise) will arise from the fact that the speed of corotation, R, increases with distance ($\omega R = 12.5L$ km s^{-1} = $0.6338L$ R_J/h). As an approximation, when this speed exceeds the local Alfvén speed, the plasma cannot corotate with the planet. This gives a limit on the maximum number density (N cm^{-3}) of plasma that can corotate at an equatorial distance L R_J, where the magnetic field is B gauss, as

$$N = 2.56 \times 10^{10} \left(\frac{B}{L}\right)^2 \tag{1a},$$

or, for a dipole field,

$$N = 2.56 \times 10^{10} \times \frac{B_0^2}{L^8}, \tag{1b}$$

where B_0 is the equatorial field. This limit for critical plasma density is just about the same as that obtained by equating the plasma energy density of corotation with the magnetic field energy density. The criterion is justified by Michel *et al.* (1974).

For Jupiter, with a surface field of 4 G (Smith *et al.*, 1974), the Alfvénic limit on maximum density, given by Equation (1b), is shown by the dashed line in Figure 4. Also, to give an idea of the order of magnitude of the Alfvén speeds involved, we note that for $N = 1$ cm^{-3} the Alfvén speed at $L = 30$ is 300 km s^{-1} and at $L = 60$ it is 37 km s^{-1}, for a dipole field.

Now, because of the photoelectron input, the equatorial plasma density increases with distance in the region beyond the synchronous orbit radius. An equilibrium distribution in which the loss of plasma form the magnetosphere by diffusion is

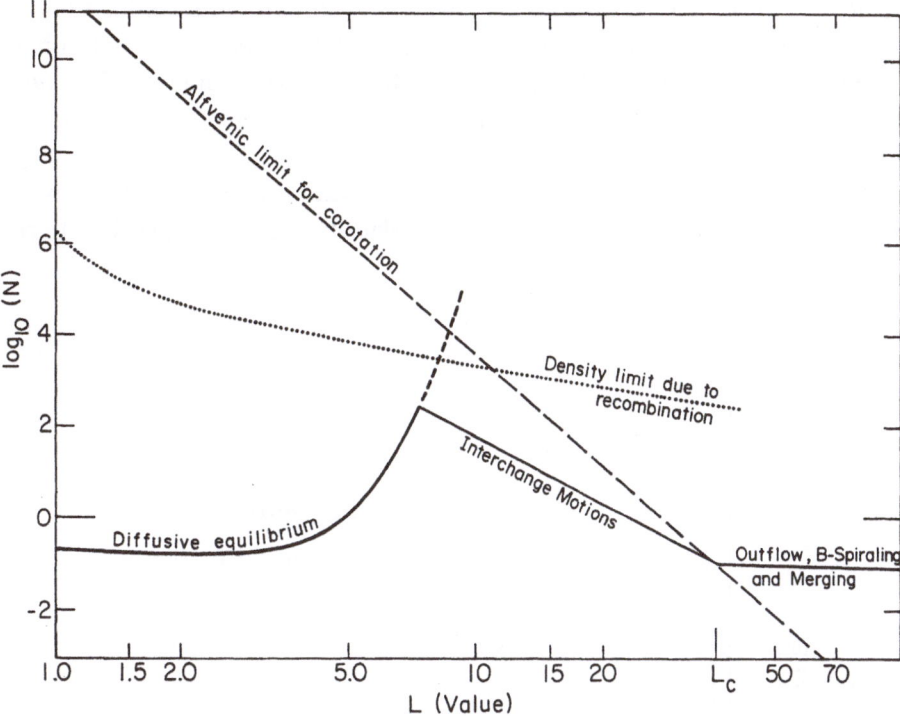

Fig. 4. An estimated plasma density distribution in Jupiter's equatorial plane. Assuming a dipole field of surface strength 4 G gives the dashed line as the density (as a function of L) at which the Alfvén speed equals the corotation speed. Existence of the critical L_c within the plasmasphere is a new feature of the present model of Jupiter's magnetosphere.

balanced by the photoelectron input was obtained by Ioannidis and Brice (1971) by using for the photoelectron flux leaving Jupiter's ionosphere a scaled down version of the photoelectron escape flux for Earth. They found that beyond $5\,R_J$ the diffusive equilibrium density increases rapidly. However, a recent calculation (Swartz, 1974) of the photoelectron escape flux for Jupiter shows that the flux above 10 eV used by Ioannidis and Brice was about 50 times too large, but the flux used for 6–10 eV electrons was right within a factor of two. We are, therefore, doing a fresh calculation of the diffusive equilibrium. For the present purposes we assume that the diffusive equilibrium curve of Ioannides and Brice, drawn in Figure 4, is approximately correct to $\simeq 9\,R_J$. For a later discussion, Jupiter's photoelectron escape flux obtained by Swartz (1974) is shown in Figure 5.

There are reasons to believe that it would be incorrect to assume that the equatorial plasma density would keep increasing with L until it reaches the Alfvénic limit for corotation. Since Jupiter, like the Earth, is surrounded by an insulating atmosphere, interchange motions of magnetic tubes are possible (Piddington, 1969), just as in the case of the Earth (Gold, 1959). Interchange motions would cause a density drop proportional to L^{-4} (or $L^{-20/3}$, at most). It is not easy to calculate the value of L at which interchange motions would begin. But since the Alfvén limit for corotation goes as L^{-8}, the density drop due to interchange motions would be overtaken by the

Alfvénic limit at some $L = L_c$. In Figure 4 we have taken this intersection to occur at $L_c \simeq 40$, but in the present model this has to be regarded as a parameter to be obtained from spacecraft observations. From L_c it is only necessary to draw the L^{-4} curve to obtain an approximate plasma density distribution between $L = L_c$ and the inner region of diffusive equilibrium. In Figure 4 this procedure gives the point of origin of interchange motions at N greater than 200 but L greater than that of the orbit of Io ($\sim 6\ R_J$). At the orbit of Io, one then has plasma density slightly smaller than $10\ cm^{-3}$.

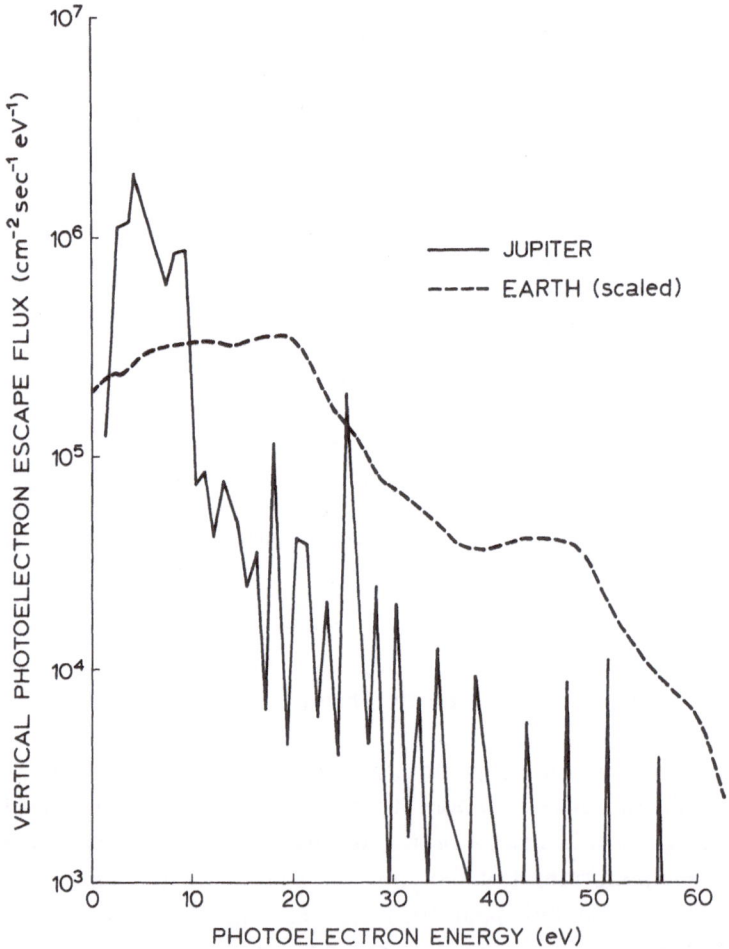

Fig. 5. Continuous line is the recently calculated (Swartz, 1974) photoelectron escape flux from Jupiter's ionosphere and the dashed one is scaled from Earth's case.

This is smaller than a simple estimate of $\sim 100\ cm^{-3}$ (McDonough, 1974) based on the idea that a third of a torus of hydrogen, instead of a complete one at Io's orbit, which has apparantly been detected by Pioneer 10 (Judge and Carlson, 1974), could be due to loss of hydrogen via charge exchange collisions with a corotating plasma in Jupiter's magnetosphere. Finally, the plasma density everywhere in the magnetosphere is below the recombination limit.

Thus, the plasma distribution in the magnetosphere is such that the density rises with increasing L beyond the synchronous orbit, the rise continues to a value of about 200 cm^{-3} until interchange motions become operative and lead to a L^{-4} decrease in density. This decrease in density continues, and the plasma remains corotating, until a critical value $L=L_c$ is reached where the corotation speed becomes equal to the local Alfvén speed and the plasma can no longer corotate rigidly. So, the boundary of rigid corotation is at a distance $R=L_c R_J$ from the center of the planet in the spin equatorial plane. This is the edge of the 'magnetodisc'.

According to this model, interesting things will happen at L_c, for there is a continual supply of plasma from inside. In the first approximation of a steady magnetosphere, the plasma would leave the corotating disc at $L=L_c$ and would flow out at the Alfvén speed, carrying the field lines with it. The magnetic field lines carried by the plasma coming off the rotating magnetodisc would be wound up into a spiral form (which would start out as a hypocycloid from $L=L_c$), as shown in Figure 6, which is a view from above the N-pole, with L_c now taken to be equal to 50 to take into account field distention. It shows the computed spiraling of the magnetic field lines for our model. Pioneer 10 trajectory is also shown. For the outbound passage the model will give the $\sim 35°$ spiraling (or 'lagging') of the field lines which has been deduced in the outer part of the magnetosphere from the magnetic field data (Smith et al., 1975), if L_c of 74 is used. But Fillius has pointed out that in system III plot his minima seem to 'click in' when the spacecraft comes around 50 R_J inbound. Compression by solar wind can give L_c on dayside smaller than L'_c on nightside. (See also independent work of Hill et al. (1974), Kennel and Coroniti (1975), and Scarf (1975).)

In Figure 6, the open arrows pushing the spiral field lines are the predicted directions of Alfvénic flow within the magnetosphere. The plasma detector on Pioneer 10 was, unfortunately, looking away from Jupiter. It seems important to have a plasma detector facing Jupiter on some future Jupiter missions in order to detect this most fascinating possible feature predicted by our model. (A likely variation of the plasma ejection mechanism at L_c will be discussed below.)

The dotted curve in Figure 6 is the 'classical' magnetosphere obtained by scaling the Earth's magnetosphere (Brice and Ioannidis, 1970; Dryer et al., 1971). These classical models overlooked the possibility of Alfvénic flow beyond $L=L_c$ within the Jovian magnetosphere. In our model, because of the pressure of the outflowing magnetospheric plasma and the fact that the magnetic field beyond $L=L_c$ will fall slower than L^{-3} ($B \propto B_0 L_c^{-2} L^{-1}$ now), the actual magnetopause would be much further out than the classical magnetopause, as is indicated by the Pioneer 10 data (Wolfe et al., 1974a). The dashed curves in Figure 6 have been drawn at distances where major magnetopause crossings were observed either on inbound or outbound trajectory. (The second inbound crossing of the magnetopause was near the solid circle in Figure 6 and has not been shown by a dashed curve to avoid overcrowding the diagram.) We will not go into an exact calculation of the new magnetopause here. However, in Figure 7, the 'minimal' effect of the outflowing plasma in the X-direction, such that the solar wind pressure is balanced at 96 R_J on the sunward side, is shown by the sunward distention

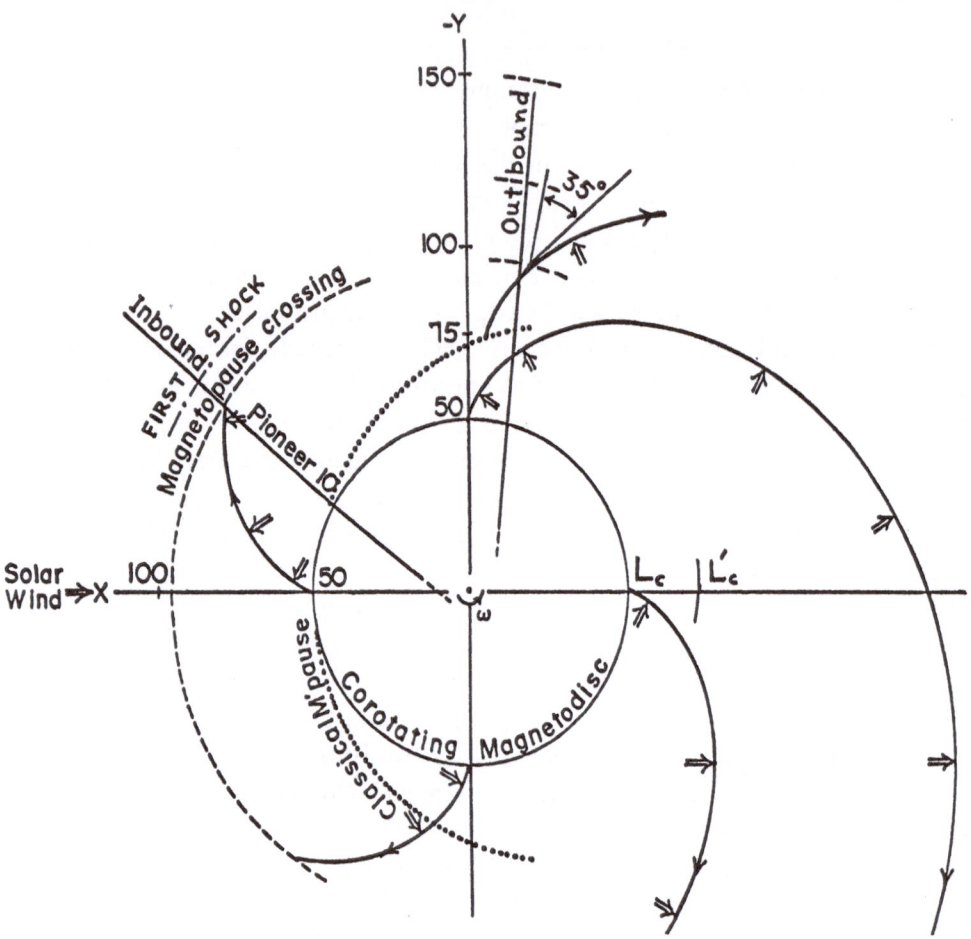

Fig. 6. Spiraling of the magnetic field lines in the outer magnetosphere resulting from flow of plasma (open arrows) at Alfvén speed from the edge of the corotating magnetodisc, according to Prakash's model calculations. The view is from the top. The dotted line is the classical magnetopause obtained by scaling from Earth's case. The dashes are at distances where Pioneer 10 underwent magnetopause crossings, in-bound or outbound. The outflow speed from 50 R_J is 625 km s^{-1}.

of the classical magnetopause. The outer dashed line represents a 'maximal' distention of the magnetopause obtained by taking the dawn side extent of the magnetosphere to be 1.5 times the new subsolar extent. (The final crossing of the magnetopause by Pioneer 10 at 105.08 R_J (Wolfe *et al.*, 1974b) is consistent with this boundary.)

It should be noted that the value of L_c used in Figure 7 is 30 rather than 50 of Figure 6. This has been done so that the magnetic field spiraling for two values of L_c could be compared. Since the location of L_c is dependent on the plasma density distribution in the inner magnetosphere, the degree of spiraling of the field lines in the outer magneto-sphere will be altered whenever the plasma density distribution changes, for example, during substorms. The erratic behavior of the spiral angles which can be deduced from the Pioneer 10 data (Smith *et al.*, 1975; Fillius *et al.*, 1974b) could be due to this effect. The variation of L_c with fluctuations in the density distribution would make the

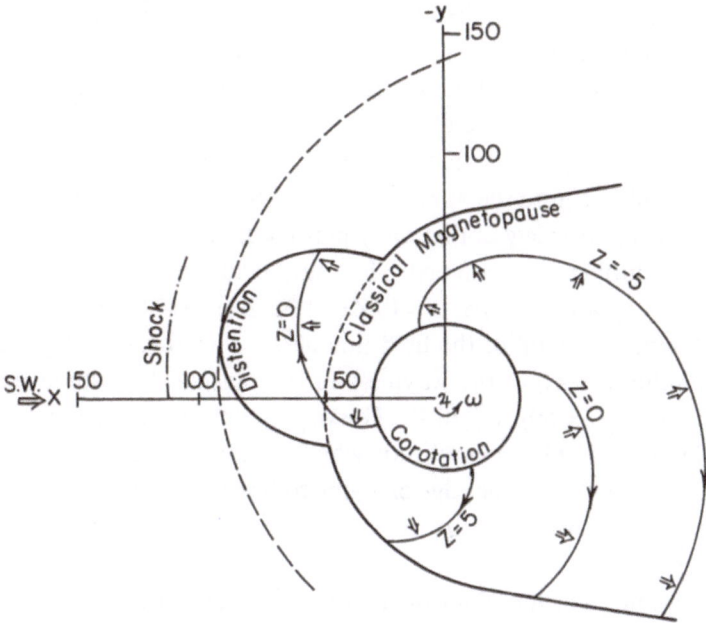

Fig. 7. Spiraling of field lines when the critical value L_c is 30. Release of corotating plasma from L_c would cause a distention of the magnetosphere beyond the classical magnetopause obtained by scaling from Earth's case.

outer magnetosphere subject to frequent changes and could be one reason for the 'spongy' nature (Wolfe *et al.*, 1974a) of the outer magnetosphere.

The tilt in the magnetic dipole of rapidly rotating Jupiter will lead to another interesting effect, as explained in Figure 8. For simplicity we have omitted the stretching of the field lines and warping of the disc and the figure is drawn for a tilted dipole field. The R_0 in this diagram is to be thought of as $L_c R_J$. We see that the locus of the points where the centrifugal force has a given value forms a tilted curve (shown dotted in Figure 8) on a cyclindrical surface of radius R_0. Thus, when the plasma comes off the magnetodisc, at a distance R_0 from the spin axis, it will come out horizontally (parallel to the spin equatorial plane) along such a tilted circular curve. When the dipole tilt is

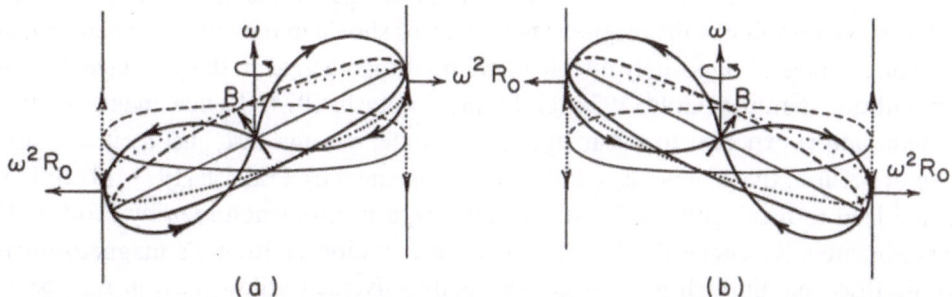

Fig. 8. The dotted curve is the locus of points where the centrifugal force has a given value, for a skew dipole with angular velocity ω. The wobbling of this curve with planetary rotation implies that the plasma ejected at R would look like a wavy plasma disc of amplitude shown by the vertical dotted arrows.

towards the sun (Figure 8 (a)), the plasma will come out below the (spin) equatorial plane on the dayside and above the equatorial plane on the nightside. After half a rotation of Jupiter (i.e., after 5 hours), the opposit would be true (Figure 8 (b)). Thus, the wobbling of the dotted curve of Figure 8 with the planetary rotation would have the result that that the plasma which is flowing out horizontally from L_c at the local Alfvén speed would acquire the appearance of a wavy plasma disc.

Figure 6 is then a top view of this wavy plasma disc, if we do not assume the skew (tilted) dipole of Jupiter to be aligned. The spiral field lines in Figure 6, therefore, do not lie in the $Z=0$ plane. One has $Z=0$, $Z=+ve$, $Z=-ve$, $Z=0$, for each $\pi/2$ difference in the 'point of origin' of the field line at L_c. The absolute upper limit on the maximum amplitude ($|Z|$) of the waviness of the plasma disc would be $L_c \tan 11°$, which for Figure 6 ($L_c=50$) is $<10\ R_J$. The warping of the magnetodisc, as discussed earlier, would in fact imply an amplitude which is much smaller ($<5\ R_J$). In Figure 7, the Z values are the values at the edge of magnetodisc, under the assumption of a larger dipole tile ($\sim 15°$) as was initially deduced from Pioneer magnetic field data by Smith *et al.* (1974).

In the noon-midnight meridian plane, the spiraling of the field lines in the wavy plasma disc, beyond the corotating magnetodisc, would show up like that displayed schematically in Figure 9. In this diagram we have accentuated the new features which can occur in Jupiter's magnetosphere. Figure 6 and 7 were exact, but Figure 9 is only schematic. In the meridian section the wavy nature of the plasma disc can be better appreciated than in Figure 7. Although several crests have been shown on the sunward side in Figure 9, in an exact version of this diagram there would be only a fraction of 'wavelength' by the time the plasma reaches 96 R_J. The warped magnetodisc of the inner magnetosphere is also shown in Figure 9, together with the ring current. The tailward plasma disc has been labelled as plasma sheet. The flow of plasma in the plasma disc region is horizontally outwards. A fixed point in space, lying within the 'amplitude' of the plasma disc in the outer magnetosphere, would be periodically immersed in the outflowing plasma. It is possible that some of the periodicities observed by Pioneer 10 in the outer magnetosphere may be due to such an effect. However, because of the equatorward warping of the inner magnetodisc, discussed earlier, the Pioneer 10 trajectory may not have intersected such a region for a prolonged duration. Possibly, Alfvén waves are generated in higher field lines.

The wavy spiraling of the magnetic field lines, as shown in the outer magnetosphere in Figure 9, presses oppositely directed field lines together, as in the geomagnetic tail but without spiraling (Gold, 1974; Gold and Prakash, 1974). Hence, magnetic field annihilation is expected to occur in Jupiter's outer plasma disc, just as it was first seen to occur in the geomagnetic tail at lunar distances by Prakash (1971, 1972). This would lead to production of bursts of particles, a phenomenon seen by Pioneer 10 experimenters (Chenette *et al.*, 1974). The outer region of Jupiter's magnetosphere would thus contain high β plasma. The steady Alfvénic outflow may, in fact, be an idealisation which may not hold for a significant fraction of the time. Instead of a uniform plasma flow, bulges of plasma could form at L_c and these would move outward carrying entangled field lines which are subsequently cut off by reconnection. This would especially be the case when a rapid and uneven increase of density takes place inside L_c, as may occur during substorms or by heating of ionosphere due to dragging of field lines by Io. In both the steady outflow version of our model and that

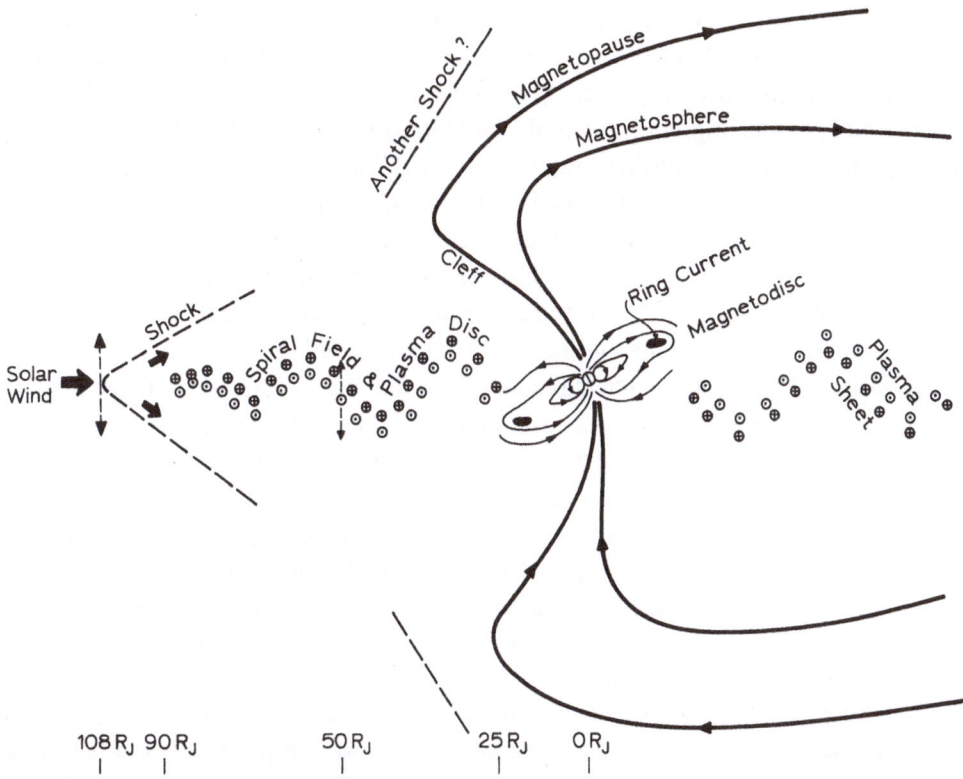

Fig. 9. Schematic representation of the noon-midnight meridional section of Prakash's model magneto-sphere. In the very interior the magnetosphere is just like the Earth's, but at increasing radial distances stretching of field lines begins and a skew but warped magnetodisc forms a corotating *inner* magnetosphere. The outer magnetosphere consists of a wavy plasma disc, with spiraling field lines and horizontally flowing plasma. If the shock generated by this disc in the solar wind intersects the classical magnetopause, another shock surface at high latitudes would be formed.

of the break up of the outer boundary at L_c by ejection of blobs, the energy released in the outer magnetosphere of Jupiter is derived ultimately from the rotation of the planet. (See also Gold (1962) and Scarf (1945).)

Since, in the steady outflow version of our model, the rate at which the plasma is lost from the edge of the magnetodisc must equal the rate at which plasma is being supplied in the interior of the magnetodisc, if the plasma supply rate in the interior is that given by Figure 5 continuous line then the plasma disc beyond L_c would be quite thin (less than 0.01 R_J thick). But Figure 5 could be an underestimate because it does not include contributions arising from auroral particle activity or from the influence of Io. On the other hand, loss due to absorption by satellites is also not included. Simpson *et al.* (1974) found energetic particle population at 50 R_J to be contained within a disc of thickness ± 6 R_J. We can use this to argue that ejection of plasma blobs, rather than a steady outflow of plasma, may be a more likely state of affairs at L_c. For if we use the photoelectron escape flux of Figure 5, it would take 1–6 hours to fill a ring of thickness 12 R_J and radial width 1 cm for $L_c = 50$–30. Thus a steady outflow at Alfvénic speeds from $L = L_c$ circle could not be sustained. It would, therefore, seem more likely that plasma is released every few hours in the form of blobs, or as an intermittent planetary wind. Future Jupiter missions with low energy plasma detectors facing Jupiter should be able to distinguish between a steady (or quasi-periodic) Alfvénic outflow and the ejection of blobs of plasma from L_c.

The plasma in the outer region of Jupiter's magnetosphere being of limited thickness, it is quite possible (Figure 9) that the subsolar shock due to this disc intersects the classical magnetosphere at some large $|Z|$. Another shock surface would then be required at high latitudes. The overall effect to be expected is a weakening of the first shock as it approaches the high latitudes and a joining with a strong bow shock. Since the plasma ejected in the outer magnetosphere has the Z coordinate dependent on the orientation of the Jovian dipole, the shocks could undergo an intermittent movement in the Z direction. These features are indicated in Figure 9 and they could be responsible for some of the 17 multiple shock crossings observed (Wolfe *et al.*, 1974b) in the outbound pass of Pioneer 10. The periodic vertical displacements at the nose (dashed arrows) can also cause periodicities in high latitude magnetosheath flow.

Because of the strong viscous interaction of the solar wind on the dawn side of the magnetosphere, the spiral field lines are expected to get more tightly wrapped up near the dawn side. This has not been shown in our diagrams. Also, in Figures 6 and 7 the field lines have been arbitrarily cut off at one of the observed or expected magnetopause boundaries. Because of the significant east-west component of the field lines in the outer magnetosphere, the merging rate with the solar wind field would depend on the east-west component of the sector field, unlike the case of the Earth where the north-south component of the solar wind field is the controlling factor. The merging of the wobbling spiral field of Jupiter with the sector fields on the dayside could account for periodic upstream flux of energetic particles detected by Pioneer 10 (Simpson *et al.*, 1974). Also, the two colliding streams at nose can form a double shock.

Last but not the least important point to note is that Jupiter's dipole is not only skew (tilted) but-off-centered as well (Smith *et al.*, 1974). This would cause a periodic drainage of trapped or quasi-trapped particles as the dipole center approaches the dayside of the magnetosphere once every 10 hours with each planetary rotation. (The situation is analogous to the South African anomaly on Earth.) It is quite possible that the clear-cut 10 h periodicities in the charged particle flux during the outbound pass of Pioneer 10 arise from such a time dependent effect rather than from the tilt of the dipole. Future observations will clarify this point.

This then is our concept of the magnetosphere of Jupiter before Pioneer 11!

Acknowledgements

This work was supported in part by National Aeronautics and Space Administration under Grant NGR 33-010-161 and National Science Foundation under Grant GA-36916.

References

Angerami, J. J. and Thomas, J. O.: 1964, *J. Geophys. Res.* **69**, 4537.
Axford, W. I. and Hines, C. O.: 1961, *Can. J. Phys.* **39**, 1433.
Brice, N. M.: 1967, *J. Geophys. Res.* **72**, 5193.
Brice, N. M. and Ioannidis, G. A.: 1970, *Icarus* **13**, 173.
Carpenter, D. L.: 1966, *J. Geophys. Res.* **71**, 693.
Chenette, D. L., Conlon, T. F., and Simpson, J. A.: 1974, *J. Geophys. Res.* **79**, 3551.
Dryer, M., Rizzi, A. W., and Shen, W. W.: 1973, *Astrophys. Space Sci.* **22**, 329.
Dungey, J. W.: 1961, *Phys. Rev. Letters* **6**, 47.
Fillius, R. W. and McIlwain, C. E.: 1974a, *Science* **183**, 314.
Fillius, R. W. and McIlwain, C. E.: 1974b, *J. Geophys. Res.* **79**, 3596.
Gold, T.: 1959, *J. Geophys. Res.* **64**, 1219.
Gold, T.: 1962, *Estratto da Rendiconti della Scuola Internazionale di Fisica E. Fermi*, Varenna, p. 181.
Gold, T.: 1974, *Bull. Am. Astron. Soc.* **6**, 378.
Gold, T. and Prakash, A.: 1974, *Annual Rep. C.R.S.R.*, p. 6.
Hill, T. W., Dessler, A. J., and Michel, F. C.: 1974, *Geophys. Res. Letters* **1**, 3.
Ioannidis, G. and Brice, N.: 1971, *Icarus* **14**, 360.
Judge, D. L. and Carlson, R. W.: 1974, *J. Geophys. Res.* **79**, 3623.
Kennel, C. and Coroniti, F. V.: 1974, this volume, p. 451.
McDonough, T. R.: 1974, *Proc. IAU Colloq.* **28**, to be published.
Melrose, D. B.: 1967, *Planetary Space Sci.* **15**, 381.
Michel, F. C. and Sturrock, P. A.: 1974, *Planetary Space Sci.*, in press.
Piddington, J. H.: 1967, Univ. Iowa Res. Rep. 67–63, unpublished.
Piddington, J. H.: 1969, *Cosmic Electrodynamics*, Wiley-Interscience, New York.
Prakash, A.: 1971, *EOS Trans. Am. Geophys. Union* **52**, 904.
Prakash, A.: 1972, *J. Geophys. Res.* **77**, 5633.
Scarf, F. L.: 1975, this volume, p. 433.
Simpson, J. A., Hamilton, D., Lentz, G., McKibben, R. B., Mogro-Campero, A., Perkins, M., Pyle, K. R., and Tuzzolino, A. J.: 1974, *Science* **183**, 309.
Smith, E. J., Davis, Jr., L., Jones, D. E., Coleman, P., Colburn, D., Dyal, P., and Sonett, C. P.: 1974, *Science* **183**, 305.
Smith, E. J., Davis, L., Jones, D., Coleman, P., Colburn, D., Dyal, P., Sonett, C., and Frandsen, A.: 1975, *J. Geophys. Res.* **80**, in press.
Swartz, W. E.: 1974, private communication.
Van Allen, J. A., Baker, D. N., Randall, B. A., Thomsen, M. F., Sentman, D. D., and Flindt, H. R.: 1974, *Science* **183**, 309.
Wolfe, J. H., Collard, H. R., Mihalov, J. D., and Intriligator, D. S.: 1974a, *Science* **183**, 303.
Wolfe, J. H., Mihalov, J. D., Collard, H. R., McKibben, D. D., Frank, L. A., and Intriligator, D. S.: 1974b, *J. Geophys. Res.* **79**, 3489.

IS JUPITER THE COSMIC RAY SOURCE
IN OUR SOLAR SYSTEM?

G. PIZZELLA

Università di Roma, Italy and Laboratorio Plasma nello Spazio del C.N.R. Frascati, Italy

1. Introduction

In debating the problem of the cosmic ray origin one piece of experimental information has been, up to now, somewhat neglected. This is the Earth magnetosphere with
its Van Allen belt which is a good example of a high energy particle source. It is
well known that a complete theory of the Van Allen belt has not been formulated
yet.

Theories have been proposed that the Van Allen particles come from the solar wind
(Nakada *et al.*, 1965) or from the Earth ionosphere (Pizzella, 1972). However, in all
cases it appears certain that an acceleration mechanism operates in the Earth magnetosphere up to at least 50 MeV for protons, perhaps 500 MeV, highest energy in the
belt.

It is therefore conceivable that these accelerated particles leave with an unknown
mechanism (perhaps a mechanism connected to the centrifugal forces due to the planet
rotation) the Earth magnetosphere and go into space to contribute to the cosmic ray
fluxes. A simple calculation (Pizzella, 1970), based on geometrical consideration only,
shows that these particles could have a differential momentum spectrum of the type
$p^{-2.5} \, dp$, which is extremely close to the observed spectrum.

In our solar system there exists at least another object, in this respect, of the Earth
type. This is Jupiter which has a strong magnetic field and a magnetosphere with a
Van Allen belt. Computation of the highest possible proton energy in the Jupiter
magnetosphere (Mihalov, 1972) gives about 10^{12} eV. Furthermore it is well known
that Jupiter emits electromagnetic radiation. This suggests that this planet is some
sort of active source for the radiation and, perhaps, also a source of high energy
particles.

Following these lines of thought we have here investigated (Pizzella and Venditti,
1973) whether Jupiter contributes to the cosmic ray proton fluxes observed at the
Earth surface with neutron monitors which, near the equator, respond, on the average,
to primary radiation of about 30 GeV protons.

2. Data Analysis

The data we use are daily averages supplied by the Rome Cosmic Ray Group (Pizzella
and Venditti, 1973) and obtained by the following stations:

V. Formisano (ed.), The Magnetospheres of the Earth and Jupiter, 425–432. All Rights Reserved

Station	Rigidity threshold (GV)	Period of time
Kodaikanal	17.5	Jan 61–June 64
Ahmedabad	15.9	Jan 61–June 64
Huancayo	13.5	Jan 61–Dec 65
Mina Aguilar	12.5	Jan 61–Dec 65
Rio de Janeiro	11.7	Jan 61–Dec 62
Mt. Norikura	11.4	Jan 61–Dec 65
Buenos Aires	10.6	Jan 61–Feb 65
Roma	6.3	Jan 61–Dec 65

Corrections for pressure, temperature, altitude and instrumental effects have been made. Care has been taken to use only those stations which were found to be free of large instrumental efficiency changes which, in all other cases, were properly considered (Bachelet *et al.*, 1971).

Of all available data we have used only those obtained by high cut-off stations since high energy particles are less scattered by magnetic irregularities. Furthermore we consider here only the period January 1961–December 1965 although we have at our disposal also data for the period January 1957–December 1960. The reason for this limitation is that the years 1957 through 1960 are highly disturbed by the solar activity. On the contrary years 1961 through 1965 are remarkably quiet; in this last period we have only eliminated the data for the epochs: days 190–225 of year 1961 and days 280–310 of year 1963 when we have two Forbush events. In order to normalize for the various stations and to correct for the solar cycle modulation, we have defined the quantity:

$$F(t) = \frac{\text{Intensity}(t) - \text{FIT}(t)}{\text{FIT}(t)},$$

where t is the time, for each day, and FIT (t) is the value given at day t by first or second order degree polinomials which, for each station, least square fit the daily averages. We now correlate these $F(t)$ with the position of Earth and Jupiter in the solar system. To do this we first define the coordinate systems. We have an intertial system centered in the Sun with the x-axis directed towards the γ point. The angle θ_J gives the jovian heliographic longitude, θ_E gives the Earth heliographic longitude (which means that $\theta_E = 180°$ on March 21st). The angle $\theta_{EJ} = \theta_E - \theta_J$ gives the Earth longitude in a reference system with center in the Sun and coorotating with Jupiter.

We consider the data both in the inertial and in the jovian systems, that is a function of θ_E and θ_{EJ} respectively. This is done in Figure 1 obtained as follows. On the horizontal axis we have respectively θ_{EJ} and θ_E in degrees. On the vertical axis we have $F(t)$ averaged for all stations and in given angle intervals as indicated in the figure. The points obtained with FIT (t) by straight lines (one line for each station) can be compared with those obtained with FIT (t) by parabolas; it is evident that there is no effect due to the way at fitting the original data. Comparing the points obtained by averaging at $\theta_i \pm 10°$ with those obtained by averaging at $\theta_i \pm 30°$ (θ_i by steps of $10°$) it is also evident that there is no effect due to the particular angle interval, except some filtering. We note a well defined, although rather broad, peak in the jovian system and a sinusoidal behaviour in the intertial system with an amplitude smaller than the previous peak. The variation of cosmic rays in the inertial system has already been observed by Barker and Hatton (1971) and represents the well known seasonal effect.

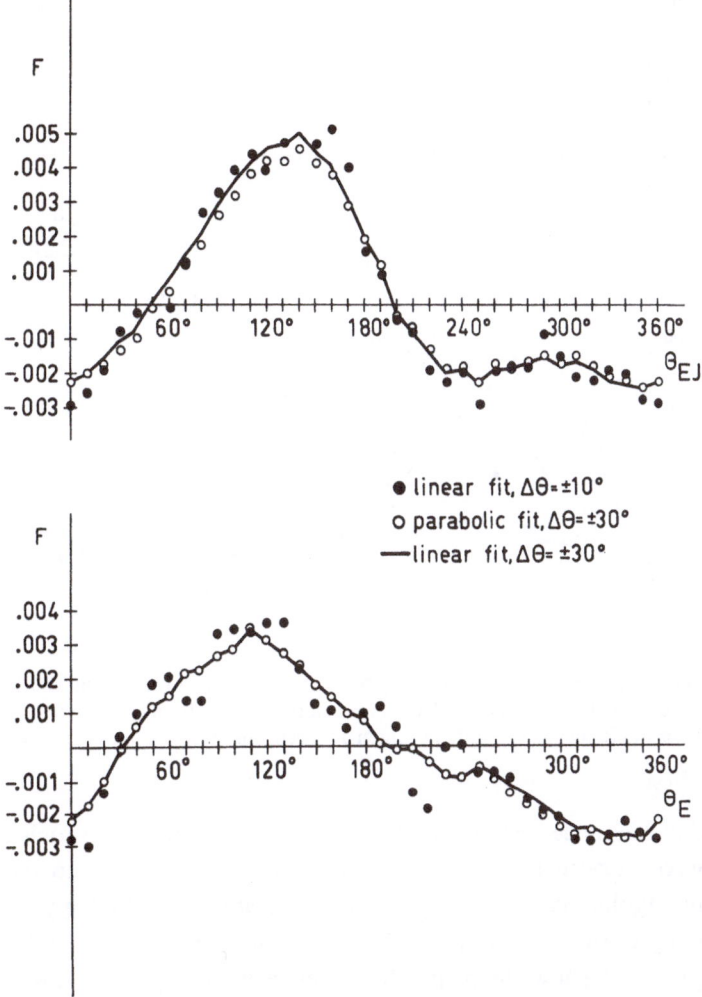

Fig. 1.   Cosmic ray modulation curves in the jovian and in the inertial systems. See text for explanations.

We now suggest that the modulation observed in the inertial system could be due to Jupiter. To give some proof we consider two periods of time (1961–62 and 1963–64–65) having, each period, approximatively the same number of data. We then assume that the cosmic ray intensity is modulated by the function $F_J(\theta_{EJ})$ representing the Jovian contribution and which is just given by the uppercurve of Figure 1. Introducing the coordinate transformation $\theta_{EJ} = f(\theta_E)$ it is then possible for any period of time to obtain the function $F_E(\theta_E) = F_J[\theta_{EJ}(\theta_E)]$ and compare it with the lower curve of

Figure 1 for the entire period 1961–1965 and with two similar curves for the two periods 1961–62 and 1963–65. The comparison is done in Figure 2 and shows that Jupiter could indeed be responsible for the observed cosmic ray variation in the inertial system.

We would like finally to ask another question:

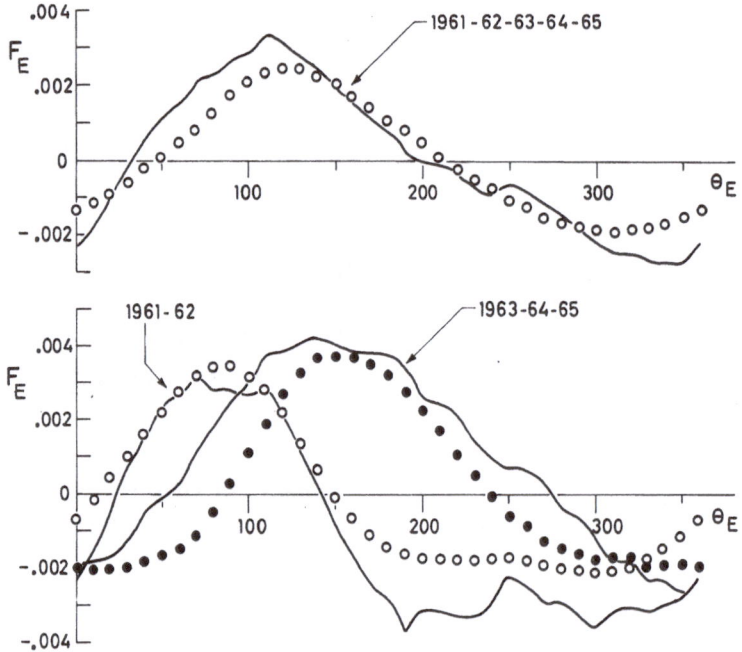

Fig. 2. The continuous lines are the cosmic ray modulation curve shown in Figure 1 (lower curve) and similar curves for periods 1961–62 and 1963–65. The data points have been calculated under the hypothesis that Jupiter modulates the cosmic ray intensity according to the upper curve of Figure 1.

Is the Jovian reference system a preferred system? To answer to such a question we have considered various reference systems rotating with respect to the inertial one with different angular velocities ranging from 10 deg yr^{-1} to 70 deg yr^{-1}, the average angular velocity of the Jovian system being, for the period 1961–1965, $\omega_j = 32.3$ deg yr^{-1}. In Figure 3 we show the results for seven reference systems. Two curves, those labelled by 0 and 30, obviously coincide (except for the phases) with the curves shown in Figure 1. We see now that the curve labelled by 30 is among those with a larger and more defined peak.

To put this on a more quantitative basis, for each curve we have computed the quantity $\sqrt{\langle F^2 \rangle}$, which should be a good indicator for distinguishing among the peaks the peak which is more pronounced. Such a quantity, which is obviously a function of the reference system angular velocity ω, was computed for many other values of ω in addition to those showed in Figure 3. The results have been plotted in Figure 4. A maximum exists at the Jovian angular velocity $\omega_j = 32.3$ deg yr^{-1} which leads us to believe that the chance that the preferred reference system had nothing to do with Jupiter is small.

Finally we would like to recall that a data analysis of some of these data corrected for the coronal green light effect has been done by Cini-Castagnoli (1973) and yields the same results.

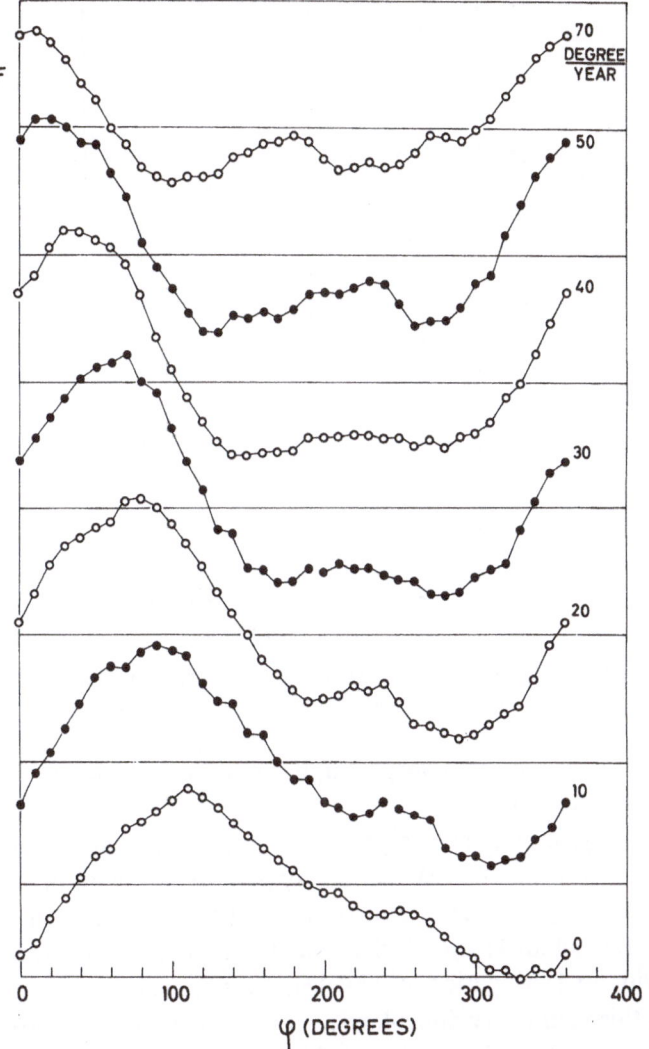

Fig. 3. Cosmic ray modulations curves obtained in various reference systems which rotate with angular velocities 0, 10, 20, 30, 40, 50, 70 deg yr^{-1} with respect to the inertial system, starting with a angle of 0° on day 1st January 1961.

3. Discussion

The criticism can be put forward that the correlation with Jupiter be only apparent, as the solar cycle of about 11 years could modulate the seasonal effect. Although it is not possible at the moment to reject completely this modulation hypothesis it seems

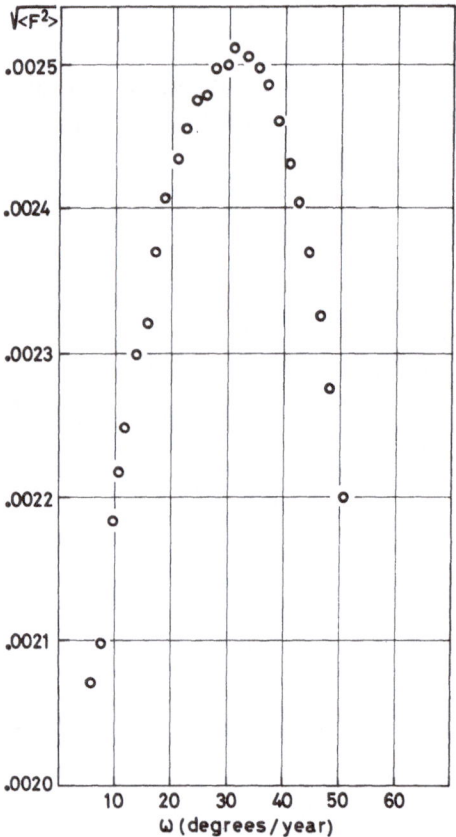

Fig. 4.   The quantity $\sqrt{\langle F^2 \rangle}$ obtained from the data presented, partly, in Figure 3 vs the reference system angular velocity. It is evident that the jovian system (32.3 deg yr^{-1}) is a preferred reference system.

to us that this would require a very complicate model which, at a first attempt, we were not able to figure out.

In fact, if an 11 yr modulation is present then, as suggested by Teegarden (private communication) we would expect cosmic ray periodicities with periods of 13 and 11 months. Now, looking to Figure 2, it seems evident that the 13 month periodicity exists, but no trace of an 11 month periodicity is seen. In other words, more than a modulation with an 11 yr period we see clearly a phase-shift of, roughly, one month each year and this seems very difficult to justify on the basis of the solar cycle.

It would be very interesting to extend this analysis over a larger number of years. However we would like to point to a problem which certainly will arise if the same data analysis technique is employed. Namely, in order to have a jovian contribution which gives a peak at the same phase angle, that is about $\theta_{EJ} = 130°$, we need to have a stationary state of the interplanetary space. Clearly θ_{EJ} depends on the interplanetary magnetic field lines of force. During the years 1961 through 1965, as already indicated, the solar activity was extremely low in terms of cosmic ray observations; therefore it is conceivable that the interplanetary magnetic field did not change much during such a period and so we were able to obtain the consistent set of data shown in Figure 2.

After 1965 the solar activity was larger and we have a rough indication (from a quick look of additional cosmics ray data) that θ_{EJ} changed, which means that the geometry of the magnetic field lines of force was changed. We notice that from the Parker spiral field only, we would expect $\theta_{EJ} \sim 240°$, which means, if our interpretation of the data is correct, that beyond the Earth the interplanetary field is affected by other processes (i.e., shock fronts, interstellar neutral hydrogen ionized by the Sun ultraviolet light (Coleman and Winter, 1971).

Finally we would like to discuss our interpretation of this cosmic ray data analysis in the light of the Pioneer 10 experimental results.

The idea that, if cosmic ray come from Jupiter, their flux would increase by going nearer to Jupiter is not correct because the Liouville theorem states that the directional differential momentum flux is constant. In this sense the fact that the cosmic ray flux is nearly constant (Thomsen, 1974; Simpson, 1974) is not in contradiction with the hypothesis, which we put forward, that Jupiter is a source of cosmic rays trapped by the interplanetary magnetic field (Alfvén, 1972). In this case we might expect an homogeneous cosmic ray density up to a solar distance of the order of one or a few Jupiter distances from the Sun.

From Figure 1 we have, according our interpretation, a Jupiter contribution $F \sim 0.004$ for θ_{EJ} between 60° and 180°, roughly. Thus, if we assume that Jupiter be the only cosmic ray source in our solar system, we infer a trapping lifetime, due to the magnetic interplanetary field, of the order of a few months for 30 GeV protons if these protons need to fill up a sphere with a radious of 10 AU.

That Jupiter could be an active particle source is also suggested by the experiments of Simpson (1974) and Teegarden (1974).

Overall the general idea that Earth-type magnetospheres be sources of particles and fields rather than sinks seems to be more and more experimentally proved.

Acknowledgements

I thank prof. N. Iucci for supplying the cosmic ray data and for useful suggestions.

I thank prof. G. Cini-Castagnoli and dott. V. Formisano for their constant interest in this reasearch and for useful discussions.

References

Alfvén, H.: 1972, *Phys. Today* **5**, 20.
Bachelet, F., Iucci, N., Parisi, M., and Villoresi, G.: 1971, Internal Note Laboratory for Space Plasma, Rome, LPS-71-7.
Barker, M. C. and Hatton, G. J.: 1971, *Planetary Space Sci.* **19**, 549.
Chenette, D. L., Conlon, T. F., and Simpson, J. A.: 1975, this volume, p. 301.
Cini-Castagnoli, G.: 1973, *Denver Cosmic Ray Symposium*, August.
Coleman, P. J., Jr. and Winter, E. M.: 1971, *Proc. Asilomar Conf.*, March, p. 698.
Mihalov, J. D.: 1972, *Planetary Space Sci.* **20**, 1345.
Nakada, M. P., Dungey, J. W., and Hess, W. N.: 1965, *J. Geophys. Res.* **70**, 3529.
Pizzella, G.: 1972, *Nuovo Cimento* **8B**, 35.
Pizzella, G.: 1970, *Nature* **226**, 434.

Pizzella, G. and Venditti, G.: 1973, *Denver Cosmic Ray Symposium*, August.
Simpson, J. A.: 1974, *Asilomar Conference*, March.
Trainor, J. M., McDonald, F. B., Teegarden, B. J., Webber, W. R., and Roelof, E. C.: 1975, this volume, p. 325.
Thomsen, M. F.: 1974, *Asilomar Conference*, March.

THE MAGNETOSPHERES OF JUPITER AND SATURN

FREDERICK L. SCARF

Space Sciences Dept., TRW Systems Group, one Space Park, Redondo Beach, Calif. 90278, U.S.A.

1. Introduction

On the basis of radio observations, it was known many years ago that Jupiter had a large intrinsic magnetic field with a substantial population of energetic trapped electrons. Radio astronomers were able to ascertain the field polarity, and they determined that the field was tilted and offset with respect to the spin axis (Carr and Gulkis, 1969). The full body of radiometric observations indicated that Jupiter had an extensive magnetosphere immersed in a conventional type of solar wind and, to a very large extent, the Pioneer 10 encounter data confirm these earlier expectations for the dipolar component of the planetary field and for the trapped electrons in the inner belt. Pioneer 10 has also provided new information on the energetic protons and on the field distortions and particle spectra farther from the planet, but in fact, Pioneer was not instrumented to shed any direct information on the most basic questions concerning the dynamics of the Jupiter magnetosphere or the origin of the energetic trapped particles and the radio emissions. These fundamental questions involve plasma physics, and Pioneer 10 carried no specific magnetospheric plasma physics instrumentation, so that we know with certainty only a little more than we did before encounter concerning the dynamics of the Jupiter magnetosphere. One important new development is that analysis of the observed field distortions strongly indicates that the significant insights of Neil Brice and his coworkers were fundamentally correct (Brice and Ioannidis, 1970; Ioannidis and Brice, 1971; Brice, 1972; Brice and McDonough, 1973). That is, it is evident that the high rotation rate of Jupiter must fling some plasma population outward to form a high-β outer magnetosphere, and it is also clear that this plasma controls the overall dynamics and configuration of the Jupiter magnetosphere.

Pre-encounter analysis of the Saturn magnetosphere is much more conjectural for two reasons:

1.1. Scaling to Saturn

Since we now have no definitive radio observations to confirm that Saturn does have any intrinsic magnetic field or energetic particle population, we must be prepared for for the possibility that the Saturn plasma environment is grossly different from that of Jupiter. For instance, it is possible that the heliosphere boundary is located within 10 AU, or that charge-exchange interactions strongly reduce the solar wind-planetary interaction effects at the orbit of Saturn. For this case, even if Saturn does have an intrinsic magnetic field, the field configuration could resemble that of a vacuum dipole much more than that of a conventional magnetosphere. At the other extreme, we note

V. Formisano (ed.), The Magnetospheres of the Earth and Jupiter, 433–449. All Rights Reserved

that present information does not preclude a Venus-like interaction in which a strong solar wind flows into the ionosphere of an unmagnetized planet. However, although no real facts or persuasive logic can be used to select one of these extreme models or any other specific model, it seems reasonable on an intuitive basis to assume that the magnetospheres of Jupiter and Saturn are 'similar' in a gross but basic sense. That is, these two planets have similar atmospheres, gravity effects, rotation rates, and solar illumination, and we can expect to apply basic principles for formation of the baseline ionospheric and exospheric plasma that can be scaled from Jupiter to Saturn in a straightforward manner. Moreover, most analyses of the heliosphere boundary location suggest that Saturn is still immersed in a supersonic solar wind having characteristics that are readily extrapolated using the current observations made within 5 AU.

1.2. Understanding of Outer Planet Magnetosphere Dynamics

Even if the scaling problem is solved to everyone's satisfaction, we must ask if we do understand how Jupiter works well enough to be able to extrapolate to Saturn. It is difficult enough to extrapolate for an Earth-like magnetosphere, because we know that all elements (magnetic field and thermal plasma, intermediate (auroral) energy plasma, energetic particles, local and propagating plasma wave fields, convection electric fields, ionospheric currents, neutral atmosphere-ionosphere interactions, etc.) form parts of a closely coupled feedback system in which strong perturbation effects can be exerted over widely-separated regions (e.g., current systems in the ionosphere can regulate convection deep in the magnetosphere, pitch angle scattering caused by plasma waves at many Earth radii near the equator can modify high latitude ionospheric structure, produce a 'slot' in the total energetic particle distribution, lead to SAR red arcs and proton aurora, etc.). For the outer magnetosphere of Jupiter it is difficult enough to be confident of the basic mechanisms without complete plasma physics measurements. The task is even more difficult when we come to analyze the inner magnetosphere where plasma sheath corrections may be needed even for energetic particle measurements. The most formidable problems arise in the vicinity of the Jupiter satellite orbits because the satellites interact strongly with the rotating magnetosphere and they may well be strong sources or sinks of thermal and energetic particles. The reliability of any Saturn generalization is particularly weak in terms of satellite effects because the Jupiter and Saturn satellites have very different orbital distributions; moreover, the anticipated magnetospheric effects should depend in detail on the surface and body characteristics of the satellites and the presence or absence of an atmosphere.

In order to assess models for these two outer planet magnetospheres, we clearly have to be concerned with the question of scaling, and we also have to conduct a critical reevaluation of the earlier Jupiter ideas, based on analysis of Pioneer 10 encounter data. The next section contains a summary of some of the earlier ideas on outer planet magnetospheres and a gross comparison between the predictions of these models and Jupiter observations on Pioneer 10. In the following sections we then turn to a more detailed evaluation of these theoretical models. With the assumption that the (unmeasured) thermal plasma is accurately given by the most recent calculations by Brice or by Axford and Mendis (1974) some models for the Jupiter field distortion, the position of the magnetopause, and spacecraft charging effects are analyzed. It appears that the Brice or Mendis-Axford electron distributions can form a basis for explaining many of the observations beyond about $L=10$, although these 'explanations' do include an assessment that Pioneer 10 did charge to significant negative voltages (hundreds to thousands of volts) in crossing the $L=10$ region. However, it appears that the Brice or Mendis-Axford density profiles cannot readily account for all observations in the neighborhood of the orbit of Io. For instance, within the $L \simeq 6-8$ region, it appears likely that Pioneer 10 was surrounded by an additional significant low energy plasma of unknown origin, or that Pioneer 10 experienced electrical breakdown on the outer insulated surfaces, so that photo-emission from the sunlit insulators served to keep the spacecraft potential to low levels.

The final section contains a discussion of the requirements on the observations that will provide valid and unambiguous comparisons of the two magnetospheres.

2. Background

The most detailed theoretical models of outer planet magnetospheres were developed by Brice (1972), Thorne and Coroniti (1972), and Kennel (1972), for presentation at the 1971 JPL Jupiter Radiation Belt Workshop. These authors assumed that Jupiter has a strong planetary dipole field (of order 1–10 G at the surface), that the thermal plasma distribution is given by the Brice-Ioannidis 1970 distribution of exospheric photoelectrons, that magnetosheath electrons and protons are injected with high values of $\mu = \kappa T_\perp / B$ across a porous high-β outer boundary (resembling a plasma-pause, rather than a magnetopause), and that these high-μ particles convect and diffuse inward (conserving μ) to attain very high energies in the inner belt of Jupiter. It was assumed that the resultant trapped particle fluxes were limited by local plasma instabilities (as on Earth) and by synchrotron radiation losses.

In several more recent theoretical reports, improvements, and extensions of these basic outer planet magnetosphere models were developed. Coroniti (1974) considered high-μ particle injection in the magnetic tail rather than at the subsolar magnetopause; Brice (private communication, 1973) reevaluated the expected thermal plasma density and he proposed a reduction in $N(L)$ by a factor of five from the original Brice-Ioannidis model (we refer to this as the Brice 1973 model); Axford and Mendis (1974) computed a new thermal density profile for Jupiter; and Scarf (1973) applied the Radiation Workshop ideas to Saturn; he showed that for a one-gauss Saturn surface field, the corresponding trapped electrons produce synchrotron radiation that connot be detected by Earth-based radio telescopes. Figure 1 summarizes our post-encounter knowledge of the field and plasma environment of Jupiter. While the field has been well measured in to 2.85 R_E, Smith *et al.* (1974), we have essentially no

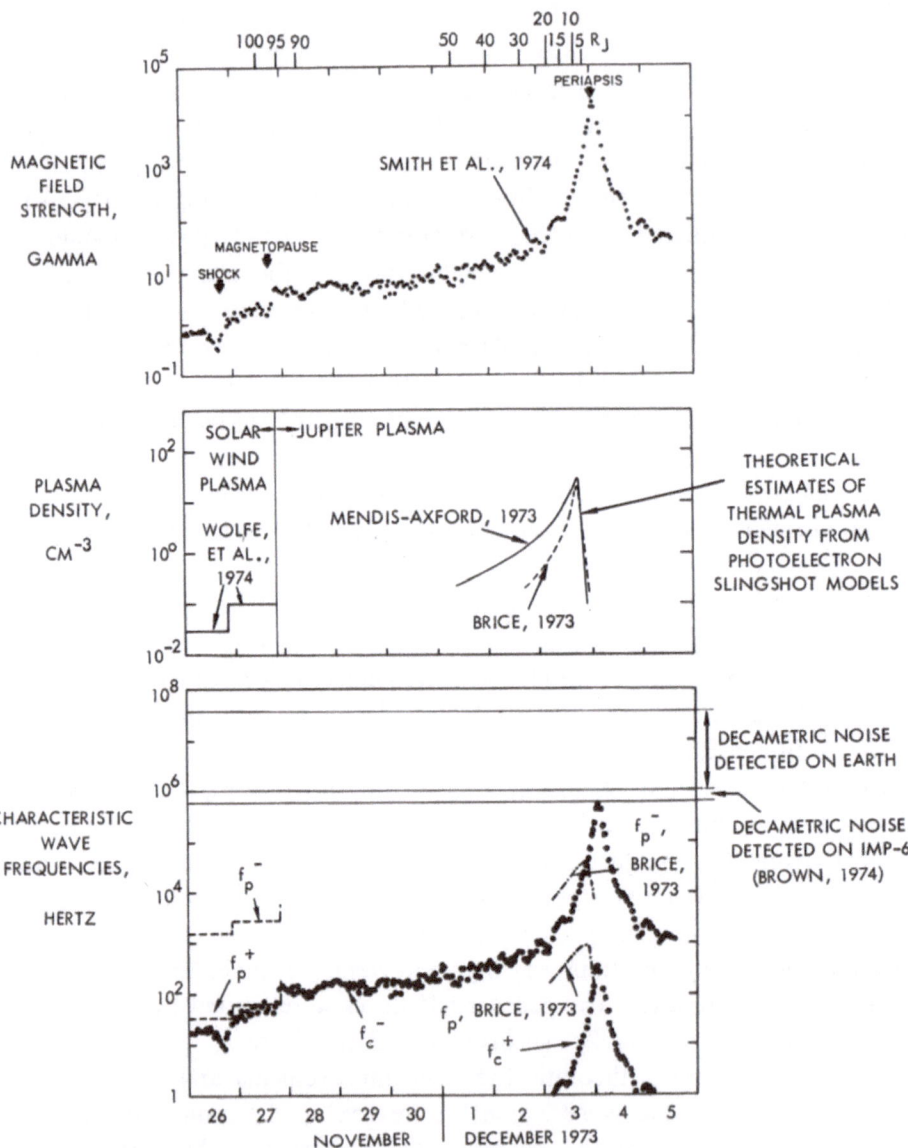

Fig. 1.   A summary of our present knowledge of the plasma environment of Jupiter. The field has been well mapped into 2.85 R_J, but our knowledge of the plasma density within 96 R_J is essentially theoretical. Radio astronomy observations give no direct information on local parameters above the ionosphere.

direct information on the thermal plasma density within 96 R_J (Wolfe *et al.*, 1974). We only know that the plasma was dense enough to distort the field strongly, and our plan involves tentative use of the theoretical Brice (1973) or Mendis-Axford profiles, to see how far one can get with these minimum or baseline plasma distributions. Note that the recent Jupiter measurements from IMP 6 Brown (1974) extending down to 600 kHz (bottom panel) still give no local information; the appropriate plasma wave measurements are all at frequencies below about 20 kHz along the Pioneer 10 trajectory.

Information on the cold plasma density profile can be obtained by analyzing the trapped energetic particle distribution, the field distortions, and the spacecraft charging with respect to the plasma. In his presentation at this conference, Coroniti (1975) noted that the observed fluxes were actually close to the stable trapping flux values that would be expected if the cold plasma profile was given by the Brice (1973) model. However, one cannot really derive a density estimate by comparing trapped fluxes with computed fluxes in the absence of any additional information. That is, since no plasma wave measurements were made here, there is no measurement of the local turbulent diffusion coefficient. Moreover, for the whistler mode instability, an increase in N(thermal) primarily leads to a drop in the resonant energy, so that lower energy particles interact with the waves and are precipitated by being scattered into the loss cone. This effect is hard to sort out even if differential analyzers are used for energetic particle measurements, but with integral detectors, the task is almost impossible. Accordingly, we confine further discussion to analysis of the field distortions and spacecraft charging effects associated with thermal plasma.

3. Implications of the Brice-Mendis-Axford Plasma Models

Following Gleeson and Axford (1974), we set up some simplified equations for the corotating thermal plasma. The radial current gives the necessary acceleration for circular motion, but it also yields a non-dipolar field distortion. We assume given forms for $B(L)$, $N(L)$, and ask where one runs into the large field perturbations.

Near the equator the Jupiter field points southward and in order for plasma to corotate with the magnetic field, a radial electric field (given by $v \times \mathbf{B}$) must produce an inward acceleration equal to mV_{CR}^2/r, where $V_{CR} = \omega r$ is the corotation speed. This means that electrons drift inward across B, protons drift outward, and a net radial current develops. The appropriate expression for the current is

$$j = Ne\, v(\text{Drift}) = Ne(v_D^+ + |v_D^-|) = N(m_+ + m_-)\, V_{CR}^2/r\, Bz. \qquad (1)$$

Since $\mathbf{V} \times \mathbf{B} = \mu_0 \mathbf{j}$, this current gives rise to a field distortion, and the entire problem must be solved in a self-consistent manner.

Gleeson and Axford (1974) assumed reasonable functional forms for $j(r)$ and they attempted to derive the self-consistent $N(r)$, $B(r)$ distributions. Our approach is less ambitious. We assume that we know $N(r)$ [specifically we use $N \simeq N$ (Brice, 1973) for $L < 10$ and $N \simeq N$ (Mendis-Axford) for $L \geqslant 10$, and we also use the Pioneer 10 dipole field model $[B_z(L) \simeq 4$ gauss$/L^3]$ near the equator. We then compute the perturbation field, $\Delta B(L)$, to see if the Brice-Mendis-Axford cold plasma model predicts a significant non-dipolar contribution at the 'correct' L-shell. That is, Hill *et al.* (1974) recently noted that the field measured on Pioneer 10 can be well fitted by the expression

$$B(L) = 4 \text{ gauss} \left[\frac{1}{L^3} + \frac{1}{(30)^2 L} \right],$$
(2)

and we test the adequacy of the Brice-Mendis-Axford model by asking if this density distribution does predict a significant field distortion starting near $L = 30$.

The results of this calculation are summarized in Figure 2. It is assumed that

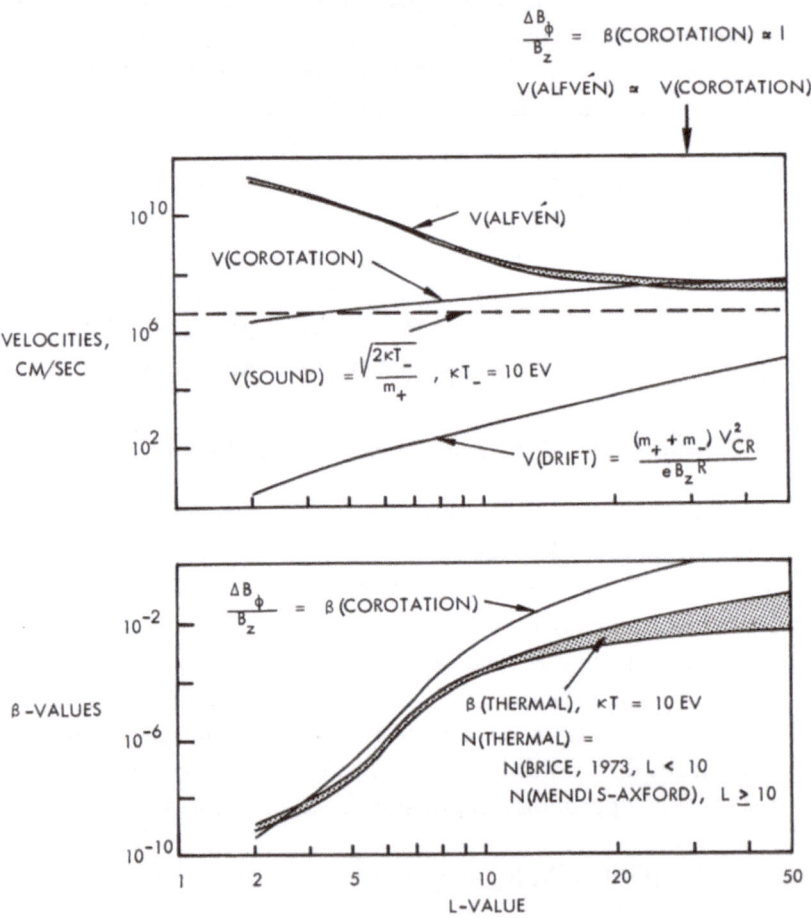

Fig. 2.   Calculations of the field distortion, various velocities and β-values, based on use of Equations (1) and (2), together with the Brice-Mendis-Axford plasma density profile.

$\mathbf{B} \simeq B_z(r)\, \mathbf{i}_z + \Delta B_\phi(r)\, \mathbf{i}_\phi$ near the equator, and Equation (1) then gives $\Delta B_\phi / B_z = \mu_0 N(m_+ + m_-)\, V_{CR}^2 / B_z^2 \simeq \beta$ (corotation). This curve is shown in the bottom panel, along with $\beta(\text{thermal}) = \mu_0 N \kappa T / B_z^2$ (both the dipolar and best-fit fields are used to compute the range of β-thermal values, and the shaded area shows how β varies with the change in field).

It can be seen that corotation of the Brice-Mendis-Axford thermal plasma produces a negligible field distortion in the entire inner magnetosphere. Even at $L=10$, Figure 2 shows that ΔB_ϕ is only about one-percent of the total field. However, the field perturbations rise rapidly with increasing L, and Figure 2 illustrates that $\Delta B_\phi \sim B_z$ near $L=30$. We interpret this to mean that Equation (2) appears to be explainable in terms of the assumed Brice-Mendis-Axford exospheric photoelectron distribution.

The upper panel in Figure 2 contains other significant information on velocity profiles that go with this model. The radial drift velocity associated with the current is clearly small in comparison with all other speeds, whereas the ion sound wave phase velocity, the corotation speed, and the Alfvén velocity are all comparable beyond $L=20$. It is noteworthy that the corotation speed equals the Alfvén velocity at $L=30$. It is reasonable to expect that the super-Alfvénic corotation near and just beyond $L=30$ will generate large scale magnetic disorder in this β (corotation) $\simeq 1$ region, and that this will 'break' the corotation and provide a 'break' in the field profile, as given by Equation (2).

One possible dynamical model involves the onset of subsonic radial flow near $L=30$. If the plasma particles stop rigid corotation when $\Delta B_\phi/B_z=\beta$ (corotation) $\simeq 1$, they will certainly continue outward, and the total corotation energy will partially go into streaming and into heat. Since $V_{CR}=375$ km s$^{-1} \gg (\kappa T_-/m_+)^{1/2}$ (the ion sound wave speed at $L=30$, see Figure 2), currents associated with local magnetic disorder in the high-β region can readily drive two-stream instabilities that lead to anomalous resistivity and efficient plasma heating. We therefore assume

$$\left[\frac{m_+ V_{CR}^2}{2}\right]_{L=30} \rightarrow \left[\frac{mu_r^2}{2}+\kappa T_+\right]_{L=30} \simeq [\kappa T_+]_{L=30}, \tag{3}$$

and for u_r small but finite, $Nu_r r^2 =$ constant so that $N \sim r^{-2}$; then $B \sim r^{-1}$ (see Equation (2)) implies constant β (thermal) beyond $L=30$ if the proton temperature does not change.

This isothermal subsonic or 'solar breeze' type of model can actually do a fair job of explaining why the Pioneer 10 experimenters detected the magnetopause at 96 R_J on November 27, 1973. At $L=30$ we use $N=1.0$ cm^{-3} (from the Brice-Mendis-Axford distribution), $B=20$ γ (observed by Smith et al., 1974), and $\kappa T_+ \simeq m_+ V_{CR}^2/2$ with $V_{CR}=375$ km s^{-1}. This gives $[B^2/8\pi+N\kappa T_+] \simeq 2.8 \times 10^{-9}$ ergs cm^{-3} at $L=30$ and with this model the corresponding outward pressure at $L=96$ R_J would be 2.8×10^{-10} ergs cm^{-3}. In the inner magnetosheath on November 27, the field strength was about 3γ (Smith et al., 1974), and at approximately 0815 UT, November 27, the sheath flow speed was 240 km s^{-1}, the proton number density was 0.56 cm^{-3} and the proton temperature was at least as high as 1.2×10^5 K (Wolfe et al., 1974).*

At 0815 UT on November 27, the measured magnetosheath parameters then give $[B^2/8\pi+N\kappa T_+ +Nmu^2/2]=2.7 \times 10^{-10}$ ergs cm^{-3}, which provides an inward pres-

---

* Although these numbers are cited for a period 12 h before encounter with the magnetopause, they are clearly more appropriate for use in the pressure balance equation than the upstream solar wind parameters measured before 1900 on the *previous day*.

sure balance in striking agreement with the outward pressure balance derived using the isothermal subsonic flow model.

This agreement should not be taken too seriously, and it certainly must not be interpreted as confirmation of any specific feature of the plasma model discussed above. However, we can conclude that there is no feature of the Brice-Mendis-Axford density profile that obviously contradicts the Pioneer 10 measurements of magnetic field distortions or the measurement of a subsolar magnetosphere boundary location at 96 R_J on November 27, 1973.

4. Extrapolation to Saturn

If the Jupiter model presented above has any validity, it is relatively simple to scale it for Saturn. In the Brice-Mendis-Axford calculations, solar ultraviolet produces the magnetospheric plasma by generating photoelectrons that escape the gravitational attraction and populate the region beyond the planetary synchronous orbit. As noted by Scarf (1973), this gives a Saturn thermal plasma density profile that is reduced by approximately a factor of four (with respect to Jupiter) when expressed in dimensionless L-coordinates. The thermal plasma temperature is still about 10 eV, and the corotation speeds at Saturn and Jupiter are approximately the same, at equal L-values. The solar wind density is also decreased by a factor of about four, assuming no significant interstellar charge-exchange effects.

During the past year, much more information has been acquired on the dipole moments of solar system bodies, and it is worthwhile to ask how the assumption of a one-gauss surface field for Saturn fits into the present classification. Dolginov *et al.* (1973) analyzed the Mars 2, 3 data and evaluated the Martian dipole moment. Ness *et al.* (1974) detected magnetic perturbations at Mercury and they estimated a surface field value near 200 γ. Finally, the Pioneer 10 observations (Smith *et al.*, 1974) provided the conclusive determination of the Jupiter dipole moment already cited. These recent evaluations give new data points that can be used with semi-empirical models of planetary magnetism (such as the magnetic 'Bode's Law' that relates magnetic moment to planetary angular momentum with a constant ratio). Figure 3 shows an updated plot of angular momentum (L) and L to μ ratio for the Sun, Mercury, Venus (upper limit to μ), Earth, Mars, and Jupiter, and it can be seen that the range of variation of L/μ for six solar system elements is actually very restricted, in comparison with the 12 order of magnitude range of variation in L-value. Figure 3 also shows the L/μ-value for an assumed one-gauss surface field at Saturn. It can be seen that this point falls near all the others, and we conclude that the recent information doesn't provide any basis for abandoning the $\mu_S = 1$ gauss-R_S^3 assumption.

With these assumptions for the dipole moment and the thermal plasma distribution, it is a straightforward matter to construct a Saturn model along the lines of the Jupiter model discussed above. We use $B_S(L) = 0.25 \, B_J(L)$, $N_S(L) = 0.25 \, N_J(L)$, $V_{CR}^S(L) = V_{CR}^J(L)$ and we find that $\beta_S(L) = 4 \, \beta_J(L)$ (both thermal and corotation β-values) and that at Saturn the Alfvén point is near $L = 25$. The predicted variation in

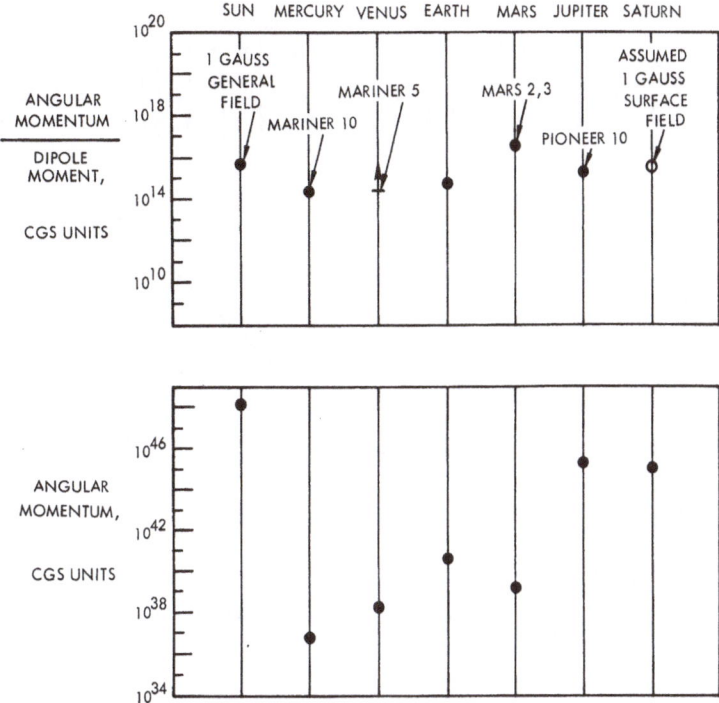

Fig. 3. A re-examination of the 'magnetic Bode's law' that assumes a constant ratio of angular momentum to planetary dipole moment.

field-strength with L-value for Saturn is shown in Figure 4, and the positions of the inner satellites are also marked here. Since the wind density has also decreased by a factor of four, in comparison with the 5-AU value, it turns out that for this model the Saturn magnetosphere is almost as large as the Jovian one. Plasma physics effects are relatively more important for Saturn because of the higher β-values, and all characteristic wave frequencies are lower by factors of 2 to 4, in comparison with corresponding Jupiter values, but to the extent that we can neglect satellite and ring perturbations, the magnetospheres of Jupiter and Saturn may be very similar.

4.1. COLD PLASMA AND SPACECRAFT CHARGING

The thermal plasma distribution, which directly affects the stable trapping limit and the distortion of the magnetic field, can also control the spacecraft potential with respect to the plasma. In order to investigate the consistency of assuming that *only* the Brice-Mendis-Axford distributions are present at Jupiter, we have to be concerned with spacecraft sheath problems and with current balance to the spacecraft. The basic problem is that a very energetic plasma (the radiation belt particles) is surely present. If there is insufficient cold plasma, in principal the spacecraft could charge to a very high level $[e|\phi| \simeq \kappa T_e \simeq$ hundreds of kilovolts] with respect to the remaining plasma; this charging can certainly lead to misinterpretation of scientific data, and it can concievably lead to engineering problems.

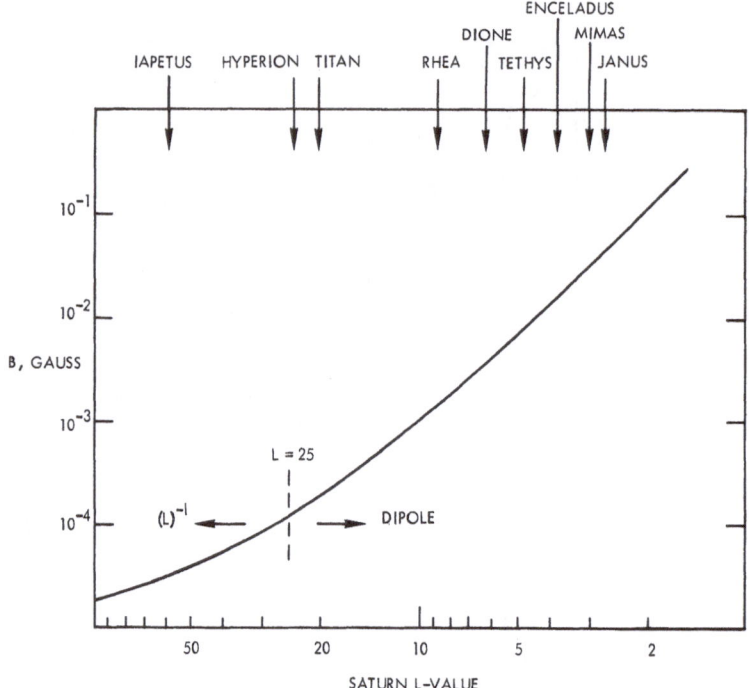

Fig. 4.   Predicted field variation at Saturn, based on a simple generalization of one Jupiter model.

The basic phenomenon has been discussed by DeForest (1972) using ATS-5 data taken in Earth orbit (at 6.6 R_E) during substorms. When $\kappa T_e \simeq 10$–20 kilovolts during electron injection events, when the cold plasma density is low, and when photo-emission becomes ineffective for a given surface element, this surface develops a negative charge,

$$e|\phi| \simeq \kappa T_e \log(j_-^{(S)}/j_+),\tag{4}$$

where $j_-^{(S)}$ is the saturation electron current and j_+ is the actual positive current to the surface. In practice, j_+ has significant contributions from secondary electron emission, and it turns out that during these charging events, $e|\phi|$ is of order κT_e; DeForest has reported charging up to 10 kilovolts (negative) during substorms.

Figure 5 illustrates one ATS-5 charging event that developed when the photo-emission current became negligible because the spacecraft entered eclipse. It can be seen that the 'zero' energy protons detected in sunlight (dots) appeared at an energy near 5 keV (crosses) when the spacecraft became charged. The lowest energy electrons appeared to lose several kilovolts at the same time, and clearly here the thermal protons were accelerated by the sheath fields while the electrons were retarded. The moderately energetic electron flux (say $E > 20$–30 keV) was only slightly reduced by charging to several kilovolts, but one can see large fluctuations in the more energetic electron fluxes. These variations characteristically develop dring charging events (C. McIlwain, private communication), and we shall comment again on this phenomenon.

ATS-5 (de Forest, JGR 77, 651, 1972)

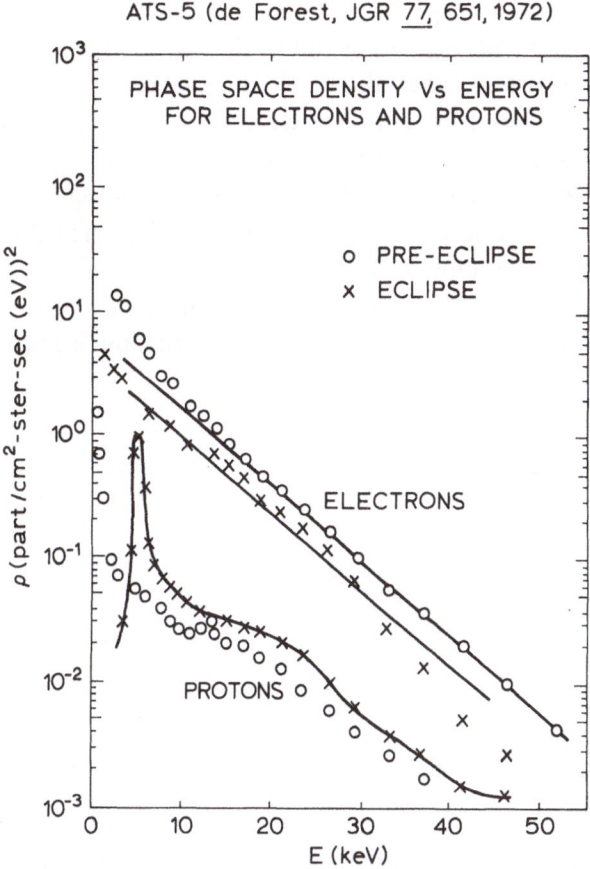

Fig. 5. An ATS-5 example of spacecraft charging to keV-levels when the photoemission current becomes ineffective during eclipse. Thermal protons are accelerated to keV energies by the sheath fields, low energy electrons are retarded and the energetic electron fluxes show large variations.

DeForest also analyzed the response from another ATS-5 plasma probe mounted away from a conducting band, and he presented direct evidence that nonconducting surfaces produced local charge variations with associated electric fields of several hundred volts per meter, when κT_e was several kilovolts. More recently, Fredricks and Scarf (1973) used engineering data from other synchronous spacecraft with nonconducting outer surfaces to infer the presence of local sheath electric fields exceeding one kilovolt/meter during noneclipse substorm injection events. Flight data and laboratory simulations indicated that portions of surfaces of a spacecraft charge to many kilovolts negative and that they also suffer discharges (arcs or coronas). The large amplitude electromagnetic pulses with high frequency spectra irradiate cabling, and cause anomalous changes of state of electronics subsystems, degradation of aluminized mylar super insulating material, degradation of optical systems, etc.

The relevance of this effect for Jupiter and Saturn has been discussed by Scarf (1973) and by Axford and Mendis (1974). It is easy to see that much higher charging levels can be expected at the outer planets. If particle energization comes from diffusion and convection with conservation of μ, then the 10 keV electrons detected at 6.6 R_E would have $E > 100$ keV at a corresponding Jupiter L-shell simply because the planetary field is twelve times stronger than the Earth's surface field. Since the UCSD particle detector on Pioneer 10 measured fluxes of electrons with $E > 160$ keV, we examine the spacecraft charging problem at Jupiter with the assumption that $j_e(E > 160$ keV) represents the Jupiter analog of the electron flux shown in Figure 5. We must also be concerned with scaling of the photoemission flux at Jupiter and with a detailed current balance to the exposed conducting surfaces on Pioneer 10.

The top view of Figure 6 shows that only six percent of the outer conducting surfaces of Pioneer 10 were in sunlight, and since j_+ (photoemission) is approximately 2×10^8

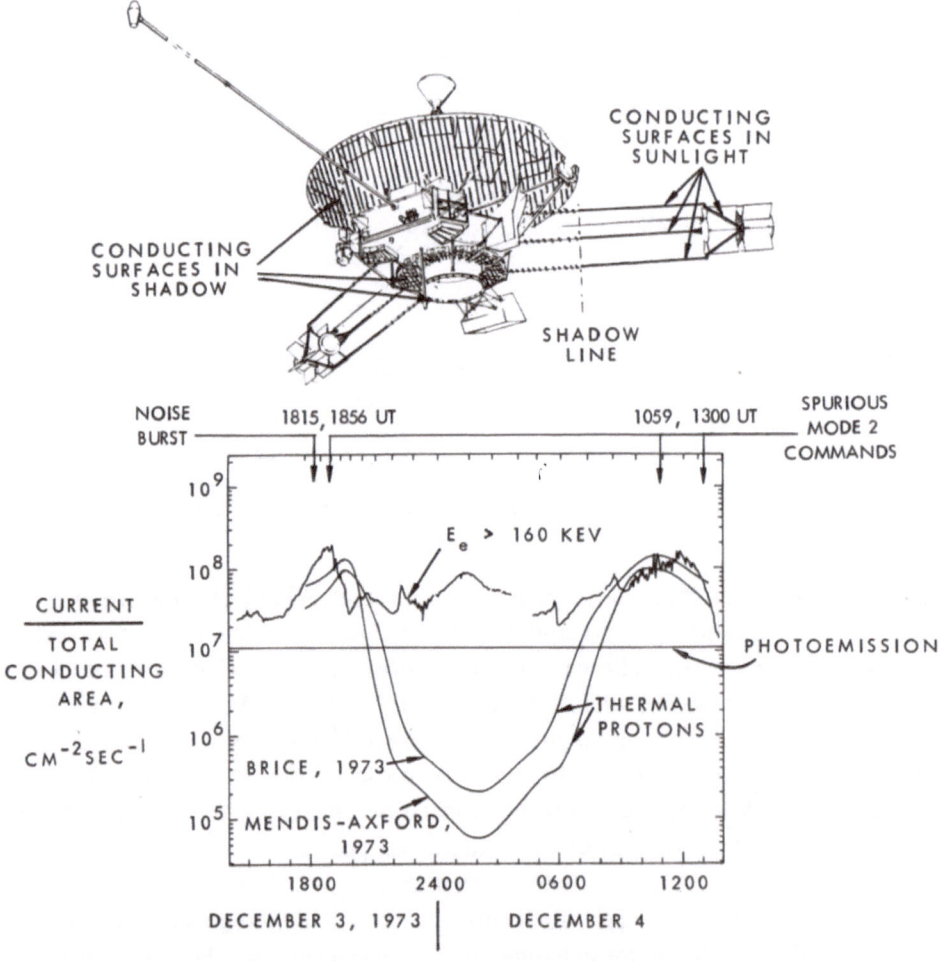

Fig. 6. *Top:* View of Pioneer 10 showing that only 6-percent of the conducting outer surface was in sunlight. *Bottom:* Current balance assuming $I_-/A = j_-(E > 160$ keV) and $I_+/A = [j_+$ (thermal protons) + $+ 0.06\,j_+$ (photoemission)]. With this model, the spacecraft could charge to negative keV-values near 1800–1900 UT, December 3, and near 1000–1200 UT December 4. Note the spurious commands and the large flux ripples detected in these regions.

$cm^{-2} s^{-1}$ at 5.2 AU, the net positive current associated with photoemission on Pioneer turns out to be so small that we can treat the charging problem as if the spacecraft had been in eclipse at all times. The bottom drawing then shows the current balance to Pioneer 10, and since j_+ (thermal)$\simeq j(E > 160$ keV) near $L = 9-11$, this is consistent with charging to several hundreds of volts or kilovolts netative near these L-shells. It can be seen that flux ripples characteristic of charging and a number of spacecraft anomalies were indeed detected in these regions. In fact, a number of other spacecraft parameters (RTG #4 current, traveling wave tube cathode current, shunt current, etc.) showed measurable changes as early as 1800–1900 UT on December 3, and L. A. Frank has presented data from the plasma analyzer showing fairly large fluxes of several hundred volt to kilovolt protons near $L = 9-11$. These protons could be real, but they can simply be cold protons accelerated to the spacecraft by sheath fields, as in Figure 5. It can be seen that the charging interpretation predicts rapidly-rising sheath fields (to perhaps 10's of kilovolts as j (thermal) decreases near 2000–2100 UT, December 3 and 0600 to 0800 UT, December 4, unless another source of thermal plasma was present.

In fact, it is now known that very large charging levels *can* be sustained and can generally be undetected by a spacecraft. In some ground tests, conducting surfaces were raised to 20 kilovolts above spacecraft ground and arcs were readily generated. However, in a series of these tests, it was found that only rarely were spacecraft anomalies induced by the arcs. These tests indicate that insulators do break down and become conducting, so that one might try to explain the charge state of Pioneer 10 near periapsis (where j_+ thermal is small) by assuming that the effective illuminating conducting area was greater than six percent in the presence of breakdown induced by large sheath fields. In this case, photoemission currents would rise and be able to balance the $j(E_e > 160$ keV) current. However this explanation appears unlikely because no shadow transients were detected. We conclude that it is more likely that *another source of low energy plasma kept the spacecraft charge down within $L \simeq 6-7$.*

TABLE I

Possible origin of non-thermal low energy plasma at Jupiter or Saturn

- Secondary electron emission from satellite atmospheres
- Secondary electron emission from planetary atmospheres
- Escape and ionization (photo, charge exchange, impact) of satellite atmospheres
- Sputtering on the satellite or saturn ring surfaces
- Acceleration of polar wind plasma
- Local acceleration associated with satellite $V \times B$ electric fields
- Local acceleration associated with satellite differential charging (Sun-shadow asymmetries, satellite wake effects)
- Resistive acceleration by current-driven waves interacting with particles in the presence of parallel electric fields (auroral analogy, anomalous resistivity)
- Cyclotron acceleration by resonant wave-particle interactions
- Heating by Landau damping of waves (SAR arc analogy)
- Betatron acceleration
- Particle heating in field-merging regions
- Charge exchange with the interstellar medium

Some very plausible sources for non-thermal low energy plasma at Jupiter and/or Saturn are listed in Table I. The questions are: (a) how can one know *whether or not* these particles are present to keep the spacecraft charge down? (b) even with an on-board plasma probe, how can one distinguish between thermal plasma locally accelerated by sheath fields and actual ambient low energy plasma? (c) how can one *compare* Saturn and Jupiter in an *unambiguous* way without knowing whether or not serious charging effects have entered at one or both planets?

5. Discussion

The basic problem that is faced in trying to discuss the Jupiter and Saturn magnetospheres concerns the lack of knowledge of the plasma distribution. We do know that at Jupiter the plasma is certainly dense enough and has sufficient streaming and thermal energy to cause a huge field distortion and to contribute strongly to the pressure balance at the distant magnetosphere boundary. We have some indirect evidence that the thermal plasma locally determines the limits of the energetic particle flux that can be stably trapped in the inner belt of Jupiter. We can also conjecture that thermal or low energy plasma serves to keep the spacecraft potential within reasonable bounds in the inner magnetosphere (say $L_J \leqslant 11$). Finally, it appears likely that some sources of plasma provide important additions to the baseline Brice-Mendis-Axford plasma distribution within $L_J \simeq 7$–8. However, real knowledge of the plasma characteristics can only be obtained if an *in situ* sheath-independent measurement program is carried out.

Similar ambiguities were *not* encountered in the early studies of the Earth's plasma environment because the necessary sheath-independent measurements of plasma density were available even before the days of rockets and satellites. These sheath-independent density measurements involve use of long-wavelength plasma waves to determine density profiles, and ground-based wave techniques have given very good density estimates for the ionosphere (ionosondes, incoherent scatter) and the magnetosphere (whistler analysis, micropulsation analysis). Absolute and accurate density profiles are still being obtained by flying wave experiments on earth-orbiting spacecraft. The Alouette-Isis-ISEE topside sounding experiments provide local and non-local density information, and the measurement of the lower hybrid resonance also gives very good electron densities.

Figure 7 illustrates another excellent technique that has recently been used to evaluate N_e. This method, which is probably the best for Jupiter and Saturn, is based on measurement of the propagation cutoff of electromagnetic waves at the local electron plasma frequency. Electromagnetic radiation is emitted above the plasma frequency in the Earth's magnetosphere, and it is believed that this is synchrotron radiation from energetic electrons. Of particular interest is the very sharp low frequency cutoff of the radiation. This cutoff is at the local electron plasma frequency, f_p^-, and it occurs because the free space electromagnetic modes cannot propagate at

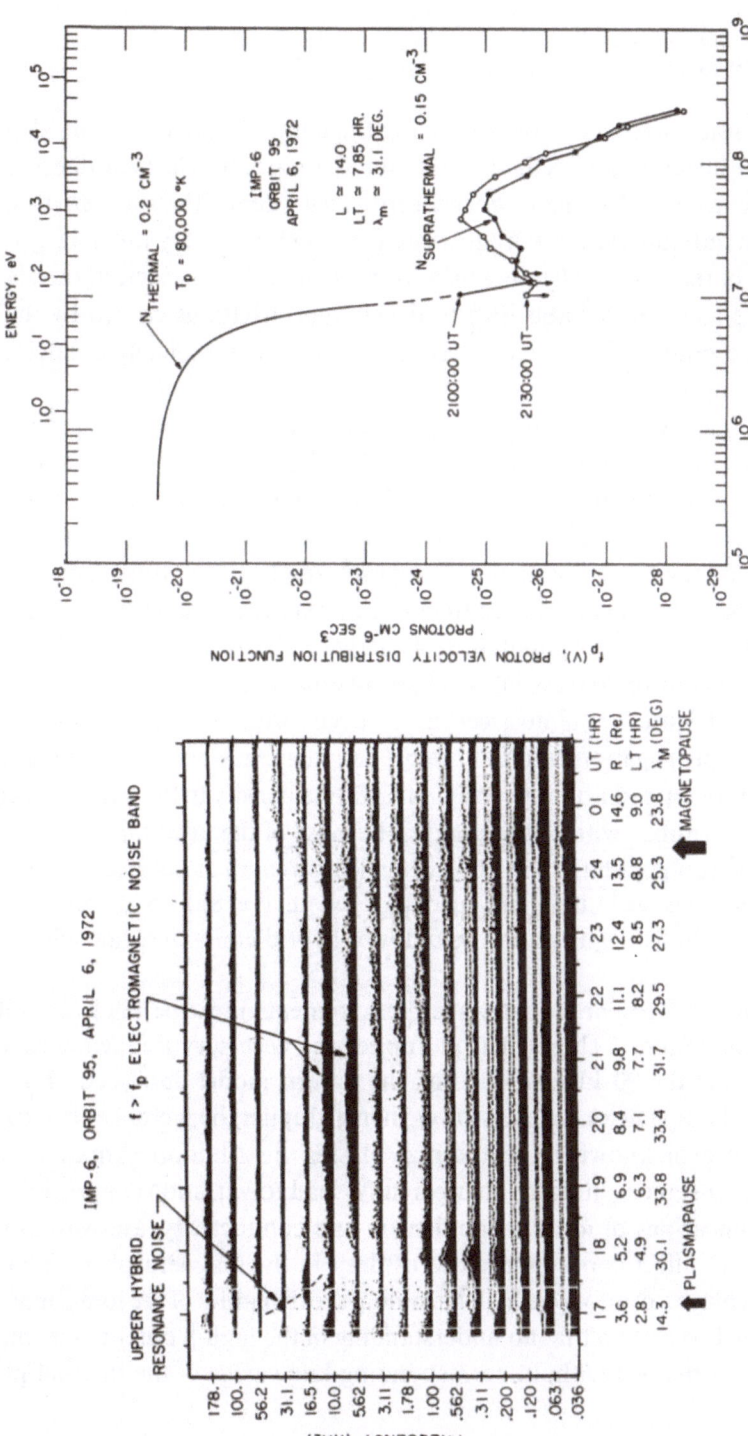

Fig. 7. *Left-hand panel*: Detection of the f_p^- noise band with the IMP-6 plasma wave spectrum analyzer. *Right-hand panel*: Use of the IMP-6 plasma wave spectrum analyzer and the IMP-6 LEPEDEA to determine total plasma density (independent of sheath effects) and density of suprathermal plasma.

frequencies less than f_p^- where N is related to f_p^- via

$$N_e = \frac{\pi m^-}{e^2} (f_p^-)^2.$$ (5)

The electron density determined by measuring the cutoff frequency is considered to be an extremely accurate and reliable value for the absolute electron density. The electron density measured is the *total* electron density. Since the wavelength of the electromagnetic radiation at these frequencies ($\lambda \simeq 100$ km) is very much larger than the dimensions of the plasma sheath which forms around the spacecraft, this density determination is completely insensitive to density perturbations caused by the presence of the spacecraft. This density determination is also completely unaffected by spacecraft charging.

The lower frequency cutoff of the f_p^- noise band can readily be measured with a suitable spectrum analyzer, and the left-hand side of Figure 7 shows one example from IMP 6 that was used to determine the total electron density. The plasma wave and LEPEDEA observations on the right side of Figure 7 show that there were two low energy particle populations (thermal and suprathermal) at this time. The absolute total density measurement can be used to derive any spacecraft charging corrections.

Because of the much greater population of energetic electrons in the Jovian magnetosphere, compared to the terrestrial magnetosphere, synchrotron radiation similar to that shown in Figure 7 is almost certain to occur with even greater intensity at Jupiter. Since the propagation cutoff at the plasma frequency must also be present in the Jovian synchrotron radiation spectrum, the plasma density in the Jovian magnetosphere can be determined with this technique. Because of the uncertainty concerning the energetic electron population and magnetic field strength at Saturn, we cannot be as confident that this technique can also be used in the Saturn magnetosphere; however, other methods, such as analysis of lower hybrid emissions, can still provide densities.

The only real requirement is that wave measurements must be made at suitable wave frequencies. Figure 1 shows that at Jupiter the necessary frequency coverage extends well below the 20 kHz range. For the Saturn model described above, the appropriate local frequencies are even lower than at Jupiter, by factors of two to four.

The importance of knowing the thermal plasma distribution cannot be overemphasized. For instance, on Earth changes in N_e lead to variations in precipitation, associated modifications of ionospheric density and conductivity, and variations in line-tying that can affect convection patterns deep in the magnetosphere. Thus, the thermal plasma plays a central role in determining the dynamics of the low-β magnetosphere on Earth. It is certain that no understanding of the high-β outer planet magnetospheres will be obtained without corresponding knowledge of the thermal plasma distributions.

Acknowledgements

I thank Drs W. Fillius, E. J. Smith, J. H. Wolfe, L. A. Frank, and D. A. Gurnett for very helpful discussions of their observations, and I am particularly grateful to Dr Fillius for furnishing me with the electron data shown in Figure 6. I have also benefited from discussions with Drs N. Brice, C. F. Kennel, F. V. Coroniti, W. I. Axford, S. DeForest, and N. L. Sanders. This work was performed under the auspices of the TRW Independent Research and Development Program.

References

Axford, W. I. and Mendis, D. A.: 1974, *Ann. Rev. Earth Planetary Sci.* **2**, 419.

Brice, N. M. and Ioannidis, G. A.: 1970, *Icarus* **13**, 173.

Brice, N. M.: 1972, *Proc. Jupiter Radiat. Belt Workshop*, NASA-JPL Tech. Memo 33-543, p. 283.

Brice, N. M. and McDonough, T. R.: 1973, *Icarus* **18**, 206.

Brown, L. W.: 1974, *EOS* **55**, 413.

Carr, T. and Gulkis, S.: 1969, *Ann. Rev. Astron. Astrophys.* **7**, 577.

Coroniti, F. V.: 1974, *Astrophys. J. Suppl.* **27**, 261.

Coroniti, F. V.: 1975, this volume, p. 391.

DeForest, S. E.: 1972, *J. Geophys. Res.* **77**, 651.

Dolginov, Sh. Sh., Yeroshenko, Ye. G., and Zhuzgov, L. N.: 1973, *J. Geophys. Res.* **78**, 4779.

Fredericks, R. W. and Scarf, F. L.: 1973, in R. J. L. Grard (ed.), *Photon and Particle Interactions with Surfaces in Space*, D. Reidel Publ. Co., Dordrecht, p. 277.

Gleeson, L. J. and Axford, W. I.: 1974, *EOS* **55**, 404.

Hill, T. W., Dessler, A. J., and Michell, F. C.: 1974, *Geophys. Res. Letters* **1**, 3.

Ioannidis, G. A. and Brice, N. M.: 1971, *Icarus* **14**, 360.

Kennel, C. F.: 1972, *Proc. Jupiter Radiat. Belt Workshop*, NASA-JPL Tech. Memo 33-543, p. 347.

Ness, N. *et al.*: 1974, *Science* **185**, 151.

Scarf, F. L.: 1974, AIAA Paper 73-566, July 1973.

Scarf, F. L.: 1973, *Cosmic Electrodyn.* **3**, 437.

Smith, E. J., Davis, L., Jr., Jones, D. E., Colburn, D. S., Coleman, P. J., Jr., Dyal, P., and Sonett, C. P.: 1974, *Science* **183**, 305.

Thorne, R. and Coroniti, F. V.: 1972, *Proc. Jupiter Radiat. Belt Workshop*, NASA-JPL Tech Memo 33-543, 363.

Wolfe, J. H., Collard, H. R., Mihalov, J. D., and Intriligator, D. S.: 1974, *Science* **183**, 302.

Wolfe, J. H., Mihalov, J. D., Collard, H. R., McKibbin, D. D., Frank, L. A., and Intriligator, D. S.: 1974, *J. Geophys. Res.* **79**, 3489.

IS JUPITER'S MAGNETOSPHERE LIKE A PULSAR'S OR EARTH'S?

C. F. KENNEL

Dept. of Physics, UCLA and Institute of Geophysics and Planetary Physics, UCLA

and

F. V. CORONITI

Dept. of Physics, UCLA and Dept. of Planetary and Space Science, UCLA

1. Introduction

Can Jupiter teach us about pulsars? The many prima facie analogies between Jupiter and pulsars – both are oblique magnetic rotators generating and containing healthy fluxes of relativistic particles, both are sources of cosmic rays and radio emissions, they even have comparable magnetic moments – make the above question an interesting one. At a deeper level, the recent Pioneer 10 encounter revealed a magnetic structure in Jupiter's outer magnetosphere reminiscent of hydromagnetic outflow solutions postulated for pulsars (Michel, 1969, 1971) and also suggested for Jupiter (Piddington, 1969; Ioannidis and Brice, 1971; Hill *et al.*, 1974; Michel and Sturrock, 1974).

One way to approach the above question is to turn it around: Can pulsar physics teach us about Jupiter? In this paper, we treat Jupiter's magnetosphere as an astrophysicist would when a new exotic object has just been discovered. We will impose upon the data – and upon our conceptualizations! – a variety of oversimplified theoretical models whose function is to illuminate broad areas of consistency or conflict between theory and experiment. With such a procedure, we must expect that what the models fail to explain may be fully as interesting as any experimental numbers they happen to fit.

We compare two possible models of Jupiter's magnetosphere – a pulsar-like radial outflow model and an Earth-like convection model. In Section 2, we ask what kind of super-Alfvénic radial outflow model does the available Pioneer 10 data seem to require. We concentrate on estimating the total particle and energy fluxes which must be provided by Jupiter – or its magnetosphere within the Alfvén radius – to power the outflow. In Section 3, we consider a convection model, concentrating upon weakening the objections previously held by theoreticians against a dominant role for convection in Jupiter's magnetosphere. In Section 4, we report our preliminary and incomplete consideration of one fundamental assumption underlying all outflow models and nearly all convection models. We ask to what extent can Jupiter actually enforce corotation on its magnetosphere. Since much of our paper is a compilation of the simple order of magnitude estimates derivable from the various models we have posed, the reader may wish to turn first to Section 5 where the point of view which

V. Formisano (ed.), The Magnetospheres of the Earth and Jupiter, 451–477. All Rights Reserved

emerges from this compilation is summarized. At present, there appears to be sufficient difficulty with the outflow model that convection ought to be taken seriously.

2. A Radial Outflow Model for Jupiter's Magnetosphere

2.1. INTRODUCTION

In this chapter we ask to what extent Jupiter behaves like a spinar (Morrison, 1969);
to what extent does the hydromagnetic interaction of Jupiter's spin with the solar
wind determine the structure, energy, and evolution of its magnetosphere and
possibly its spin. A basic requirement for a spinar-type solution is that Jupiter
possesses sources of both particles and energy within the magnetosphere which
exceed any external solar wind particle and energy source. To aid the imagination we
will assume that Jupiter has a radial outflow solution similar to that constructed by
Mestel (1968) for the magnetic deceleration of rotating stars. We then use Pioneer 10
measurements to infer the number and energy source strengths required to drive the
postulated radial outflow (Kennel and Coroniti, 1974).

Following Mestel (1968) we assume for simplicity that Jupiter's rotational and
magnetic dipole axes are aligned, and that Jupiter possesses a *centered* dipole moment
with equatorial field strength of 4 G. Near Jupiter the magnetic field is assumed to be
that of a rotating dipole field. This condition persists out to a certain critical equatorial
radius, where the flow and magnetic stresses become equal. Beyond this radius we
assume that a two-dimensional radial outflow solution of the type discussed by

ALIGNED ROTATOR

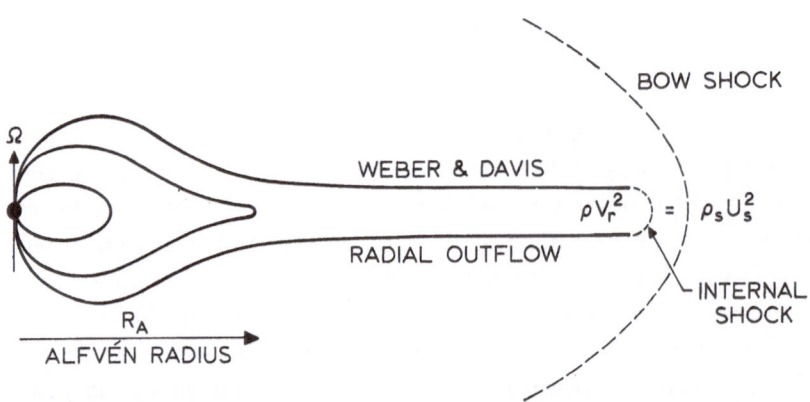

Fig. 1.   Radial Outflow Model. Jupiter is assumed to have a centered dipole magnetic field, with dipole
and spin axis aligned. The approximately dipolar field is assumed to corotate within the Alfvén radius.
Beyond the Alfvén radius, a two-dimensional radial outflow solution is assumed. The Alfvén radius is
fixed by assuming that the dipole and radial field components are comparable at the Alfvén point of the
radial outflow. The radial outflow terminates in a fast shock at the magnetopause. Approximate dynamic
pressure balance prevails across the magnetopause and bow shock. Pioneer 10 measured the upstream
solar wind flow parameters, the location of the magnetopause, and the
magnetic field near the magnetopause.

Weber and Davis (1967) prevails near the spin-magnetic equatorial plane. The magnetic field has only radial and azimuthal components $B_r B_\varphi$; the absence of the field component B_z, normal to the outflow disk, is a shortcoming of this solution. The observations indicate that if a radial outflow solution exists, it exists in a thin outflow disk of half-height h above the spin-magnetic equatorial plane. We do not discuss here what might happen to field lines above the radial outflow disk, although reconnection of Jovian and solar wind field lines might produce a magnetopause similar to Earth's there. The critical radius is defined to be the Alfvén point of the Weber-Davis solution, where $4\pi\varrho U^2/B_r^2 = 1$, and ϱU^2 is the dynamic pressure of the radial outflow. The Jovicentric distance r_a to the Alfvén point is estimated by assuming that $B_r(r_a)$ is roughly equal to the vacuum dipole field $B_D(r_a)$ at distance r_a. Our model is sketched in Figure 1. While Pioneer 10 did not measure a synoptic set of hydromagnetic flow parameters, it did provide us with the mass density ϱ_s, flow speed U_s, and dynamic pressure $\varrho_s U_s^2$ of the solar wind upstream of Jupiter's bow shock prior to magnetopause encounter, the Jovicentric distance to the magnetopause r_m, and, from the 10-h 'flapping' of the disk, a rough estimate of its height h. In addition, Pioneer 10 measured the magnetic field strength B at the magnetopause, the sign of the azimuthal component B_φ, and rough average values of $|B_\varphi/B_r|$ in the distant magnetosphere. We will restrict our present discussion to published data from Pioneer 10's first magnetopause crossing, but in principle it is an easy matter to perform the same analysis for other magnetopause crossings. The data are sufficient to permit us to estimate certain basic parameters of the assumed outflow solution.

2.2. REQUIRED PARTICLE AND ENERGY OUTFLOWS

If, as postulated, the radial outflow is super-Alfvénic, it must be terminated by a fast shock near the magnetopause which decelerates the flow as it enters the magnetosheath. The magnetopause fast shock differs from Jupiter's bow shock, which decelerates the solar wind. The flows behind the bow and magnetopause shocks would be separated by a tangential discontinuity if strict magnetohydrodynamics were applicable. Near the edge of the disk, $z \approx h$, the flow is presumably Alfvénic, and so above the disk a fast shock is not a necessary part of the magnetopause structure. In steady state, pressure equality should apply across this system of shocks. Near the subsolar point, where Pioneer 10 first encountered the magnetopause, this implies the rough equality

$$N_m M_+ U_m^2 \simeq N_s M_H U_s^2, \qquad (2.1)$$

where N_m and U_m are the number density and radial flow velocity of the internal flow at the magnetopause, and M_s and U_s are the corresponding values in the solar wind ahead of the bow shock. M_H is the mass of a hydrogen ion, and M_+ the mass of the ions flowing out from Jupiter.

We estimate U_m as follows: In the Weber-Davis solution, assuming $r_m \gg r_a$, we have the relation

$$B_\varphi/B_r = -r\Omega/U, \qquad (2.2)$$

where r is the Jovicentric distance and Ω is the angular velocity of the field lines at $r = r_a$. Pioneer 10 magnetic observations (Smith *et al.*, 1974; E. J. Smith, private communication) indicate that while B_φ/B_r is highly variable within the disk, the time-averaged B_φ/B_r is negative – consistent with the 'garden hose' field expected with radial outflow – and its magnitude $|B_\varphi/B_r| < 1$. While it is not yet entirely clear to us whether this information applies to the center of the disk as well as to its edges, we will explore its consequences. Inserting into the relation $U_m = r_m\Omega_J|B_r/B_\varphi|$ the observed $r_m \simeq 100\,R_J$, $\Omega = \Omega_J = 1.74 \times 10^{-4}$ rad s^{-1} – Jupiter's spin frequency, we find that $U_m \simeq 10^8|B_r/B_\varphi|$ cm s^{-1}. In other words, the required flow energy is $5(M_+/M_H)$ $(B_r/B_\varphi)^2$ keV. From (2.1) we may now estimate N at the magnetopause

$$N_m = N_s(M_H/M_+)\,(U_s/r_m\Omega)^2\,(B_\varphi/B_r)^2 \tag{2.3}$$

and the particle number flux

$$N_mU_m = (N_sU_s^2/r_m\Omega_J)\,(M_H/M_+)|B_\varphi/B_r|. \tag{2.4}$$

The total particle outflux $\dot{\eta} = N_mU_mA_m$, where A_m is the *dayside* frontal area of the outflow disk, follows immediately by estimating $A_m \simeq 2\pi r_m h$, whereupon

$$\dot{\eta} = N_sU_s^2(M_H/M_+)\,(2\pi/\Omega_J)\,|hB_\varphi/B_r|. \tag{2.5}$$

The energy outflow $\dot{W} = \tfrac{1}{2}N_mM_+U_m^3A_m$ is similarly estimated

$$\dot{W} \simeq \pi\varrho_sU_s^2\Omega_Jr_m^2|hB_r/B_\varphi|. \tag{2.6}$$

We may normalize $\dot{\eta}$ and $\dot{W}$ to $\dot{\eta}_s$ and $\dot{W}_s$, the solar wind number flux and flow energy flux crossing the area πr_m^2

$$\dot{\eta}/\dot{\eta}_s = 2(M_H/M_+)\,(U_s/U_m)\,(h/r_m)$$
$$\dot{W}/\dot{W}_s \simeq (2\Omega_Jh/U_s)\,|B_\varphi/B_r|\,(M_+/M_H) = 2(U_m/U_s)\,(M_+/M_H)\,(h/r_m). \tag{2.7}$$

According to Wolfe *et al.* (1974), before the first shock encounter, $N_s \simeq 3 \times 10^{-2}$ cm^{-3} and $U_s \simeq 420$ km s^{-1}. Thus, expressing h in units of $R_J = 7 \times 10^9$ cm

$$N_m \approx 3 \times 10^{-3}|B_\varphi/B_r|^2 \text{ cm}^{-3}$$
$$\dot{\eta} \approx 10^{28}(M_H/M_+)\,|hB_\varphi/B_r| \text{ par s}^{-1}$$
$$\dot{\eta}/\eta_s \approx 10^{-2}(M_H/M_+)\,|hB_\varphi/B_r|$$
$$\dot{W} \approx 10^{20}|hB_r/B_\varphi|\,(M_+/M_H) \text{ ergs s}^{-1}$$
$$\dot{W}/W_s \simeq 5 \times 10^{-2}|hB_r/B_\varphi|. \tag{2.8}$$

It seems that typical values for h and $|B_\varphi/B_r|$ are a few R_J and $\tfrac{1}{3}$ respectively, so that $|hB_\varphi/B_r|$ is $0(1)$ and $|hB_r/B_\varphi|$ is $0(10)$.

Equation (2.8) does not support the notion that particle and energy input from the solar wind could be neglected, even if Jupiter had a radial outflow solution. For example, the earth's magnetosphere captures $10^{-3} \sim 10^{-2}$ of the particles crossing πr_m^2 at the Earth; there the number of particles circulating through its magnetospheric convection pattern is $\approx 10^{26-27}$ s^{-1} and $\eta_s \approx 10^{29}$ s^{-1}. Similarly, the energy dissipated by the solar wind into the Earth's magnetosphere is $\approx 10^{18}$ ergs s^{-1} and $\dot{W}_s \simeq 10^{20}$ ergs s^{-1}. Thus, Jupiter, strictly speaking, probably cannot be a pure spinar. Moreover, it is difficult to see how Jupiter generates particle fluxes $\approx 10^{28}$ s^{-1} (assuming $M_+ = M_H$) and energy fluxes $\approx 10^{21}$ ergs s^{-1} within its Alfvén radius. For example, if all the ions produced by solar UV ionization in Jupiter's dayside ionosphere were sucked into the radial outflow before recombining, a gross upper limit, only 10^{28} par s^{-1} would flow out (Hill et al., 1974). Similarly, Io's atmosphere produces a torus of neutral gas near the orbit of Io. Some 10^{27} neutrals s^{-1} are required to maintain this torus (R. Carlson, private communication). Even if all the neutrals were lost by ionization and no charge exchange occurred, Io's ring could not provide the required plasma number flux. Frank (1974) has reported generation of a few hundred eV ions near Europa, but no source strength has been given. In any case, even if the number flux could be accounted for it is difficult to see how the plasma generates some tens of keV mean energy at the Alfvén point, since the corotation energy at the Alfvén point (to be computed in Section 4) is only a keV or so.

2.3. SUMMARY

Several serious questions bedevil the simple radial outflow model for Jupiter's magnetosphere posed in this section. Particle and energy sources of 10^{28} par s^{-1} and 10^{21} ergs s^{-1} respectively must be found within the Alfvén radius. It is unlikely that photoionization in the Jovian ionosphere can produce the requisite particle source. Even if the requisite internal number and energy sources could be found, our estimates do not make a compelling case that particle and energy input from the solar wind can be safely neglected.

3. A Convection Model of Jupiter's Magnetosphere

3.1. INTRODUCTION

It is now abundantly clear that magnetic field line reconnection occurs regularly and is, in fact, responsible for convection in the Earth's magnetosphere. The Dungey (1961) model of the Earth's magnetosphere is essentially correct. Two tests indicate that reconnection at the nose of the Earth's magnetosphere occurs. First, the field lines in the Earth's polar caps are definitely open, permitting rapid access of solar cosmic ray electrons (Lin and Anderson, 1966). Second, the intensity of the magnetospheric convective circulation pattern is largest when the solar wind field is southward, the theoretically optimum configuration for reconnection at the nose of the Earth's magnetosphere (see Arnoldy, 1971 and references therein). The phenomenological studies between various measurables within the magnetosphere and conditions in the solar wind seem now to be providing answers to two questions concerning field line reconnection on which laboratory experimentation and theory shed at best a dim light. These are 'how does the reconnection rate depend upon the relative orientations of the magnetic field directions on either side of the neutral sheet' and 'how fast can the reconnection rate be'. The answers seem to be that except possibly for the special case where the magnetic fields are parallel on both sides of the neutral sheet, some reconnection will occur (Mozer *et al.*, 1974). Moreover, the response of the magnetosphere to changes in the solar wind field direction (Burch, 1974) suggests that the nose reconnection rate follows small changes in the solar wind field 'sweetly and docilely'. Satellite observations of several hundred kV potentials across the Earth's polar caps (Gurnett and Frank, 1973) – a significant fraction of the total solar wind $U_s \times B$ potential across the width of the Earth's magnetosphere – indicate that at times, reconnection can be very rapid. In fact, the auroral and magnetospheric substorm may well be a consequence of changes in the reconnection rate. The simplest picture of a substorm – still controversial outside UCLA – holds that it has two phases, a 'growth' and a 'breakup' phase (McPherron. 1970; Coroniti and Kennel, 1973). Nose reconnection starts the growth phase in this picture. Following an increase in the nose reconnection rate, the convective flow increases in intensity; magnetic flux is added to the geomagnetic tail, so that the polar caps increase in area; and the entire magnetosphere goes through an indentifiable sequence of configurational changes (Coroniti and Kennel, 1972). When this has proceeded long enough, explosive reconnection occurs in the Earth's plasma sheet 15–30 R_E from the Earth (Nishida and Nagayama, 1973), thereby initiating the 'breakup' phase of rapid injection of plasma into the dipolar region of the geomagnetic field and great intensification and poleward motions of the auroral arcs bounding the equatorward edge of the polar cap. Recently Siscoe and Crooker (1974) have found a theoretical relation between nose reconnection rate and the rate of energy injection into the inner magnetosphere which places the tail reconnection region at 15–30 R_E, in agreement with observation.

All in all, sound advice for those constructing models of other magnetospheres seems to be that one neglects reconnection at his peril. If the magnetic field configuration allows for the possibility of reconnection, it is much better to assume that it does occur then to assume that it does not (Kennel, 1974). There is, therefore, no doubt that reconnection will occur in Jupiter's magnetosphere. The real question is whether it will have significant effects. We might ask, for example whether the solar wind can dissipate as much energy into Jupiter's magnetosphere as it seems Jupiter must provide to power the postulated radial outflow discussed in Section 2. We can compute an upper limit to the reconnection energy dissipation rate as follows. The solar wind $U_s \times B_s$ emf Φ across the width $3 r_m$ of Jupiter's magnetosphere is given by

$$\Phi = \frac{3 U_s B_s}{e} r_m \simeq 10 \text{ MV}, \tag{3.1}$$

where $U_s = 400 \text{ km s}^{-1}$, $B_s \simeq 1 \gamma$, and r_m 100 R_J. The energy dissipation rate $\dot{W}$ is given by computing the total current in the reconnection   of the dayside magnetopause and multiplying by Φ, assuming all the solar wind flux crossing Jupiter's magnetosphere is reconnected. The current per unit length along the magnetosphere is $c\Delta B/4\pi$ where ΔB is the jump in magnetic field strength at the magnetopause. The total current I is then approximately $(c\Delta B/4\pi) l_{\text{eff}}$ where l_{eff} is the effective length, normal to the ecliptic plane, of the reconnection region

$$\dot{W} = (c\Delta B/4\pi)\, l_{\text{eff}}\Phi \simeq 2 \times 10^{21} \text{ ergs s}^{-1}, \tag{3.2}$$

where we choose $\Delta B \approx 4\gamma$, the measured field at the magnetopause (Smith et al., 1974) and $l_{\text{eff}} \approx r_{\text{m}} \simeq 100\ R_J$ above. For the Earth, (3.2) yields 10^{19} ergs s^{-1}.

The reconnection upper limit energy dissipation rate, (3.2), and the energy outflow required by the radial outflow model, (2.8), are comparable in magnitude. This suggests that even if Jupiter did possess a strong radial outflow, it would be unwise to neglect reconnection and convection driven by the solar wind. Moreover, the reconnection dissipation rate is sufficiently large to make reasonable the consideration of a pure convection model where all the particles and energy come from the solar wind rather than from Jupiter's inner magnetosphere. This we shall do in the remainder of this chapter. In view of our discussion in Section 2, a reconnection model possesses several attractive features. Since the radially extended magnetic field observed in Jupiter's outer magnetosphere reveals the presence of significant hydromagnetic stresses, it is likely that the hydromagnetic outflow theory discussed in Section 2 may indicate at least the order of magnitude of the grossest features of any hydromagnetic flow solution. If so, a convection model may not have any particular difficulties supplying the requisite number and energy fluxes circulating through Jupiter's magnetosphere. Moreover, since convection in the Earth's magnetosphere easily creates plasma temperatures in the ring current of some tens of keV when the solar wind emf is of order 100 kV, supplying high temperature plasma if it is needed for Jupiter seems to be no problem either.

3.2. LENGTH OF JUPITER'S MAGNETIC TAIL AND CONVECTION FLOW TIME

The length of Jupiter's magnetic tail may be computed, following Dungey (1965) if Φ and the radius of the polar cap r_{pc} are known. The electric field in Jupiter's ionosphere is of order $\Phi/2r_{\text{pc}}$ in magnitude; the convection speed is therefore $\sim c\Phi/2r_{\text{pc}}B_I$ where B_I is the ionospheric magnetic field. The foot of a field line in the ionosphere crosses the polar cap in a time $\tau = 4r_{\text{pc}}^2 B_I/c\Phi$, and the length of the tail $L_T \simeq U_s \tau$.

We define $r_{\text{pc}} = R_J\sqrt{L_{\text{pc}}}$ where L_{pc} is the L-shell of the last closed field line, and estimate Φ as before by $3r_{\text{m}}(U_s B_s/c)\beta$ where $\beta < 1$ parameterizes the efficiency of reconnection, whereupon

$$\tau \simeq \frac{R_J}{r_{\text{m}}} \frac{B_J}{B_s} \frac{R_J}{U_s L_{\text{pc}}\beta} \approx \frac{14}{\beta} \text{ hours} \tag{3.3a}$$

$$\frac{L_T}{R_J} = \frac{R_J}{r_{\text{m}}} \frac{B_I}{B_s} \frac{1}{L_{\text{pc}}\beta} \approx \frac{300}{\beta}. \tag{3.3b}$$

In (3.3) we chose $R_J/r_{\text{m}} \simeq 10^{-2}$, $B_I = 8$ G, $B_s \simeq 1\ \gamma$. L_{pc} is not known. We chose $L_{\text{pc}} \simeq 30$ corresponding to the radial Jovicentric distance where distortions from a dipolar field begin to become small (Smith et al., 1974). Choosing $\beta \simeq 0.1$, we arrive at $L_T \simeq 3000$ R_J, roughly the estimate Kennel (1973) arrived at by a completely different means. We note that Jupiter rotates once in 14 h minimum convection time and if $\beta \ll 1$ may rotate many times in a convection time.

3.3. LOCATING JUPITER'S PLASMAPAUSE AND MAGNETOPAUSE

Given Jupiter's magnetic moment $\mathbf{M}$ and its spin $\mathbf{\Omega}$ and one simple fact – that reconnection imposes a more or less uniform electric field E_c of order $\beta U_s B_s/c$ across the magnetosphere – can one locate Jupiter's plasmapause and magnetopause? We begin by reconstructing Brice and Ioannidis' (1970) model for Jupiter's plasmapause. We assume that Jupiter is an aligned rotator, $\mathbf{M} \parallel \mathbf{\Omega}$, and that corotation is imposed at the foot of all field lines with the angular velocity Ω_J. Then in the spin-magnetic equatorial plane the convection potential φ_c is given by

$$\varphi_c = E_c r \sin\theta, \tag{3.4}$$

where r is the Jovicentric distance, and θ is measured clockwise from the midnight meridian. The corotation electric field E_{CR}, for a dipole magnetic field, is

$$E_{CR} = \frac{\Omega r B}{c} = \frac{\Omega R_J B_0}{c}\left(\frac{R_J}{r}\right)^2 \tag{3.5}$$

and points radially outwards. B_0 is the equatorial surface field strength. At local dusk, $\theta = \pi/2$, there is a stagnation point in the flow where the corotation and convection speeds just cancel. The Jovicentric distance r_p to the dusk plasmapause may be found by equating the magnitudes of E_c and E_{CR}

$$\left(\frac{r_p}{R_J}\right) = \left[\frac{\Omega_J R_J B_0}{cE_c}\right]^{1/2} = \left[\frac{\Omega_J R_J B_0}{\beta U_s B_s}\right]^{1/2}. \tag{3.6}$$

The corotation potential is given by

$$\varphi_{CR} - \varphi_{CR}(r_p) = \frac{\Omega_J R_J^2 B_0}{c}\left[\frac{R_J}{r} - \frac{R_J}{r_p}\right] \tag{3.7}$$

and the total corotation potential $\varphi_T = \varphi_{CR} + \varphi_c$ is

$$\varphi_T = \frac{\Omega_J R_J^2 B_0}{c}\left(\frac{R_J}{r} - \frac{R_J}{r_p}\right) + E_c R_J\left[\frac{r}{R_J}\sin\theta - \frac{r_p}{R_J}\right], \tag{3.8}$$

where we have defined $\varphi_T(r_0) = 0$.

The plasmapause is the curve in the spin-magnetic equator on which $\varphi = 0$. At local dawn, $\theta = -\pi/2$, the radius r_p^* of the plasmapause is defined by the condition $\varphi_T(r_p^*, -\pi/2) = 0$. This leads to the equation $y^2 + 2y - 1$, where $y = r_p^*/r_p$, which has the nontrivial solution $r_p^*/r_p = -1 + \sqrt{2} \simeq 0.4$. Thus the plasmasphere has minimum radius 0.4 r_p and maximum radius r_p. Using $\Omega_J = 1.75 \times 10^{-4}$ rad s^{-1}, $B_0 = 4$ G, $U_s = 4 \times 10^7$ cm s^{-1}, and $B_s = 1$ γ, we find $r_p \simeq 100/\sqrt{\beta}$ R_J.

In computing their plasmapause position Brice and Ioannidis (1970) had assumed that the energy density of the convecting plasma was very nearly zero. There were two reasons for this: first, they assumed an undistorted dipole magnetic field everywhere; and second, they neglected all gradient drifts in arguing that all convecting particles would follow equipotentials and would therefore avoid the plasmapause defined by $\varphi_T = 0$. Therefore, the only consistent way of locating the magnetopause was to assume that it formed where solar wind dynamic pressure was balanced by twice the magnetic pressure. The nose radius r_m^0 of this dipolar magnetopause is located by the standard relation

$$\frac{r_m^0}{R_J} \simeq \left[\frac{B_0^2}{2\pi\varrho_s U_s^2}\right]^{1/6} \approx 55\, R_J \tag{3.9}$$

using $B_0 = 4\,\text{G}$, and $\varrho_s U_s^2 \simeq 8 \times 10^{-11}$ ergs cm^{-3}, corresponding to upstream solar wind parameters prior to the first Pioneer 10 magnetopause crossing (Wolfe *et al.*, 1974).

Let us compare the mean radius of the plasmapause $0.7\, r_p = \tilde{r}_p$ with the dipolar nose radius r_m^0

$$\frac{\tilde{r}_p}{r_m^0} = 0.7 \left(\frac{\Omega_J R_J}{\beta B_s}\right)^{1/2} \left(\frac{2\pi B_0 \varrho_s}{U_s}\right)^{1/6} \approx \frac{1.25}{\sqrt{\beta}}. \tag{3.10}$$

Thus Brice and Ioannidis found that the plasmapause extended beyond the magnetopause. This led them and their followers (Kennel, 1973) to suppose that convection could never be important in Jupiter's magnetosphere, since convection could never penetrate close to the planet. In fact, with these numbers it was difficult to see how the flux carried by convection (if it occurred) could ever penetrate to the frontside of the magnetosphere.

There was, however, something Brice overlooked: it is very likely that the flow speed near the plasmapause would be super-Alfvénic. This means that the convective flow energy density would exceed the magnetic energy density near the plasmapause and that the dipolar magnetic field would be strongly distorted. We will establish the plausibility of this point by a *reductio ad absurdem*: we will assume a dipolar field and then compute the ratio of the E/B convection speed to the Alfvén speed C_A at the mean radius of the plasmapause, $0.7\, r_p$, where r_p is also computed assuming a dipolar field. Let us call this ratio $\mathscr{R}$. Substituting $E_c \simeq (U_s B_s/c)$, and $r_p/R_J = [\Omega_J R_J B_0/\beta U_s U_s]^{1/2}$ and reducing, we find

$$\mathscr{R} = \frac{\sqrt{N}}{\beta^2} (0.7)^6 \frac{B_0}{(U_s B_s)^2} (\Omega_J R_J)^3 \sqrt{4\pi M_H}, \tag{3.11}$$

where N is the number density at the plasmapause and M_H is the proton mass. Substituting $U_s = 4 \times 10^7$ cm s^{-1}, $B_s = 1\,\gamma$, $\Omega_J R_J = 1.2 \times 10^6$ cm s^{-1}, $B_0 = 4\,\text{G}$, we find for Jupiter

$$\mathscr{R}_J = \frac{20\sqrt{N}}{\beta^2}, \tag{3.12}$$

whereas substituting $B_0 = \frac{1}{3}$ G, $U_s = 4 \times 10^7$, $B_s = 5\ \gamma$, $\Omega_E R_E = 4 \times 10^4$ cm s^{-1}, we find for a much less restrictive condition for Earth

$$\mathcal{R}_E = \frac{3 \times 10^{-6}\ \sqrt{N}}{\beta^2}. \tag{3.13}$$

Thus, for the flow to be sub-Alfvénic at Jupiter's plasmapause the plasma density must satisfy $N < 4 \times 10^{-3} \beta^4$ cm^{-3}, whereas at Earth it must satisfy $N < 10^{11} \beta^4$ cm^{-3}. On this basis we may conclude that it is very likely that the convection flow is sub-Alfvénic at the Earth's plasmapause and super-Alfvénic at Jupiter's.

What does this all mean? It means first of all that Brice and Ioannidis' (1970) computation of the plasmapause location was incorrect since for all but the most unrealistically low densities the super-Alfvénic flow stresses will strongly distort the dipolar magnetic field. It also means that the standard computation of the magneto-pause nose radius is incorrect, since the flow energy density beyond the plasmapause exceeds the magnetic energy density near the magnetic equator. This suggests that convection will push out the magnetopause in the magnetic equatorial plane. A full hydromagnetic theory of this kind of flow is very difficult, and we are far from even complete conceptual understanding of it, much less an analytic theory. Nonetheless, the above arguments suggest that convection in Jupiter's outer magnetosphere would mimic what one expects from radial outflow solutions. Both would have relatively thin disks of super-Alfvénic flow and radially extended magnetic fields. Figure 2 summarizes our arguments.

At this point it is useful to note that the Earth may have Jupiter-like magnetopauses during strong convection events – magnetic storms and substorms. The magneto-meter experiment on OGO-1 (Heppner *et al.*, 1967) found a large region of constant magnetic field strength near the morning magnetopause, which was pushed out farther than the calculated magnetopause position based upon a dipole field during sub-storms. Figure 3 shows one of their events in which the magnetic field strength *increased* as the spacecraft crossed the magnetopause into the magnetosheath. Heppner *et al.* (1967) argued that this was due to the presence of high pressure plasma near the magnetopause. The above arguments suggest, however, that, following sub-storm breakup and during strong convection events in general, the flow in the Earth's outer magnetosphere may be super-Alfvénic near the dipole equator, just as it is super-Alfvénic in the plasmasheet at these times. In view of the absence of a good theory of super-Alfvénic convection in Jupiter's outer magnetosphere, the possibility that Earth may have such solutions is in our opinion a very strong reason for taking a convection model of Jupiter's magnetosphere seriously. If the analogy is a true one, then the observed great variability of Jupiter's magnetopause location might have a simple explanation: it is due to substorms.

The above arguments indicate that further studies of the Earth's dayside magneto-pause might be extremely illuminating, both in and for themselves, and as a possible analog for the behavior of Jupiter's magnetopause. For example, it is thought that the Earth's magnetopause moves inward prior to substorm breakup (Aubry *et al.*, 1970;

HIGH β SUPER-ALFVÉNIC CONVECTION

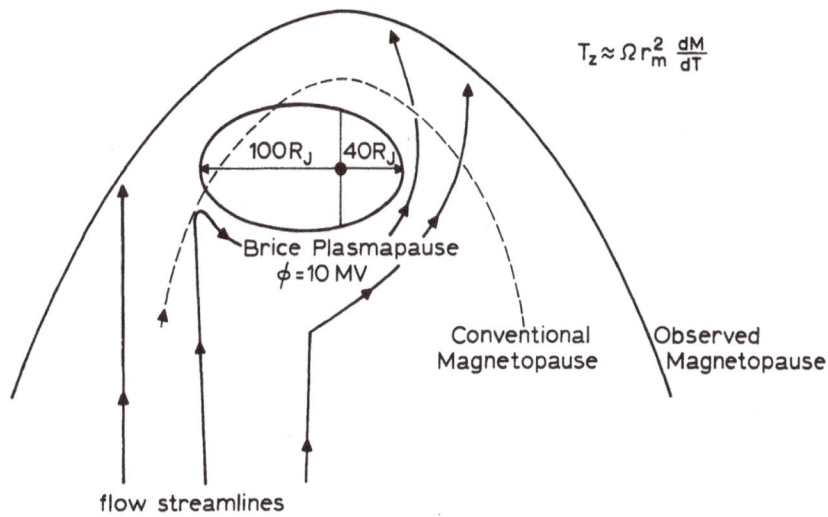

flow streamlines

Fig. 2. High β Super-Alfvénic Convection. The Brice plasmapause, computed assuming an undistorted dipole field, and therefore zero-β sub-Alfvénic convection, would lie at 40 R_J at dawn, 100 R_J at local evening. The conventional magnetopause, computed by balancing the upstream solar wind dynamic pressure prior to Pioneer 10's first magnetopause encounter with the magnetic pressure of the undistorted dipole, would lie at 55 R_J. Since the conventional plasmapause intersected the conventional magnetopause, it was thought that convection would be unimportant compared to corotation, particularly in view of the fact that there would be little room for the convective return of flux to the dayside magnetopause. However, convection beyond the plasmapause is likely to be super-Alfvénic. This might create a radially extended magnetic field and push the magnetopause beyond its conventional location near the magnetic equator.

Coroniti and Kennel, 1973); does it move outward following breakup? Does it begin to move outward in less than the Alfvén travel time between the tail neutral line and the dayside magnetopause? Is the field near the magnetopause radially extended? Does the magnetopause bulge near the equatorial plane following breakup?

3.4. SUMMARY

(1) Reconnection at the nose of Jupiter's magnetosphere can dissipate as much energy as the radial outflow solution discussed in Section 2. Even if an internally driven outflow exists, it would therefore be unwise to neglect convection.

(2) Brice and Ioannidis' original plasmapause solution (1970) overlooked the likelihood that convection would be super-Alfvénic in Jupiter's outer magnetosphere. This implies that the standard computation of Jupiter's magnetopause nose radius is incorrect. Just as in the radial outflow solution, convection would push out Jupiter's magnetopause. Since either solution has a super-Alfvénic flow near the magnetopause, shocks are a necessary part of the magnetopause solution.

(3) OGO-1 may have observed Jupiter-like Earth magnetopauses in the local morning sector during substorms. If so, these observations are a good reason for taking a convection of Jupiter's magnetosphere seriously.

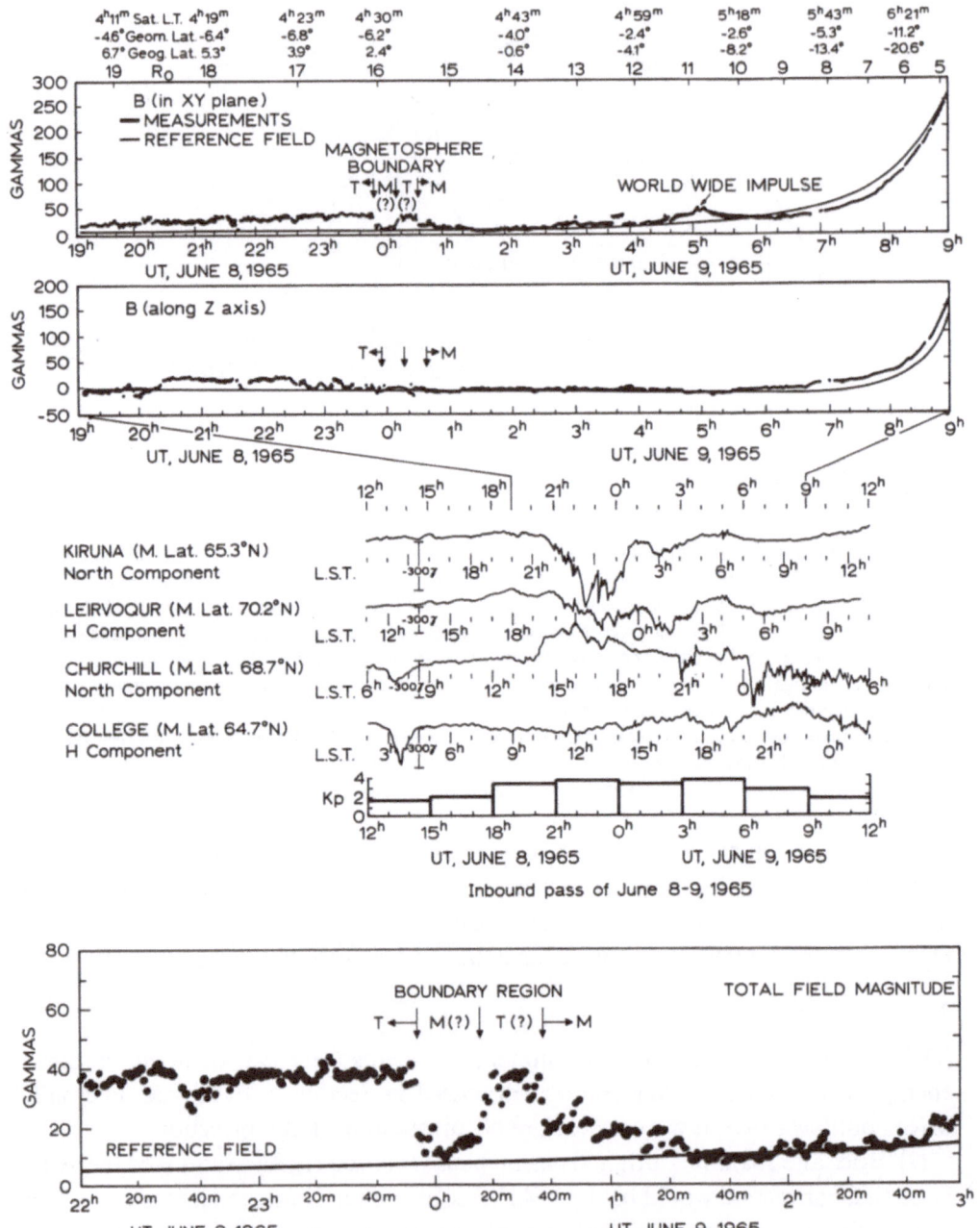

Total field intensity during the period of magnetopause crossing on June 8-9, 1965.

Fig. 3.   A Jupiter-Like Magnetopause at Earth. Reproduced here is a figure from Heppner *et al.* (1967). The top inset shows magnetic measurements from a considerable portion of the pass of June 8–9, 1965; the middle inset indicates that a substorm was in progress; and the bottom inset shows that the magnetic field was larger in the magnetosheath than in the magnetosphere. Could super-Alfvénic convection have caused this magnetopause?

(4) Pioneer 10 encountered magnetopauses from 100 to 240 R_J on its outbound pass through Jupiter's dawn sector. Since with super-Alfvénic convection the magneto-pause position depends not only on variations in the solar wind dynamic pressure but also on variations in the convection dynamic pressure, the observed variability of Jupiter's magnetopause location could be due to substorms if, as at Earth, convection is time-variable.

4. Coupling of Jupiter's Ionosphere and Atmosphere to either Radial Outflow or Convection Magnetospheres

4.1. INTRODUCTION

We now turn to one feature which distinguishes planetary from pulsar magneto-spheres – the existence of a neutral nonconducting atmosphere separating the highly conducting planet from the conducting ionosphere – where convection first inter-acts with neutral material gravitationally bound to the planet. This means that hydromagnetic stresses cannot be communicated directly between the planet and ionosphere the way such stresses are known to be communicated between the Earth's ionosphere and magnetosphere – by a circuit involving field-aligned currents between ionosphere and magnetosphere which exert stresses as they close by currents flowing perpendicular to the magnetic field in the ionosphere and magnetosphere. The effect permits significant convection in the Earth's ionosphere, since a convection electric field can exist in the ionosphere yet be very small in the Earth's crust. In effect, the ionospheric field lines can 'slide over' the field lines below the ionosphere which are held in place by the high conductivity of the Earth.

For aligned rotators, rotation does not induce an electric field, in the non-rotating frame, between the conducting planet and conducting ionosphere. Since the con-dustivity law in the ionosphere has the form $\mathbf{j} = \sigma(\mathbf{E} + \mathbf{V}_n \times \mathbf{B}/c)$ where σ is the con-ductivity tensor and $\mathbf{V}_n$ the neutral velocity, there can be an equivalent corotation electric field $(\mathbf{\Omega} \times r) \times \mathbf{B}/c$ induced by the rotation of the neutral atmosphere with angular velocity $\mathbf{\Omega}$. This makes it clear, however, that corotation can only be en-forced on the magnetosphere through upward diffusion of atmospheric angular momentum, which then couples to the ionosphere through ion-neutral collisions. In steady state, the angular momentum acquired by the magnetosphere must balance that provided by the atmosphere (Hines, 1974).

In this section we investigate the validity for an aligned rotator of one of the key assumptions underlying the discussions of both Sections 2 and 3. In Section 2 we assumed that the solid body angular frequency Ω_J was imposed on all the flux tubes in-volved in the radial outflow. The model of the plasmapause discussed in Section 3 tacitly assumed that planetary corotation could be imposed upon convecting field lines. While significant differences exist between aligned and oblique rotators, nonetheless, our discussion of the aligned rotator case raises the question of how and to what extent corotation can be imposed on Jupiter's ionosphere, magneto-sphere, and upper atmosphere.

4.2. Coupling of atmospheric torque to radial outflow

In this section we first compute the spindown torque T_z implied by the radial outflow solution of Section 2. This torque is exerted on Jupiter's ionosphere and atmosphere by a system of field aligned currents threading Jupiter's polar cap, which we sketch in Figure 4. We then estimate the torque exerted upon the magnetic field lines of ionospheric levels by the upward viscous diffusion of angular momentum.

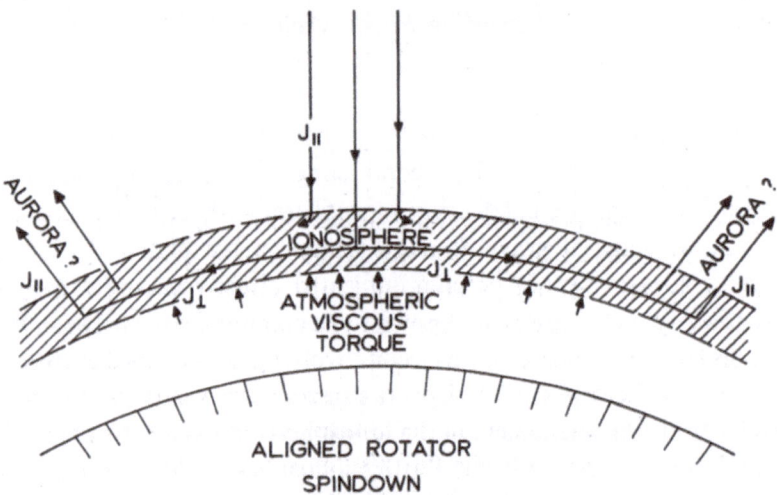

Fig. 4. Alligned Rotator Coupling of Torque. The hydromagnetic spin-down torque is communicated to the ionosphere and atmosphere by a system of currents in at the dipole and spin axis, across the ionosphere, and out at the boundary of the polar cap. Where the current flows out, an aurora boreolis may be found, if Earth-like physics prevails. The $\mathbf{J}_\perp \times \mathbf{B}$ torque is exerted first on the atmosphere at ionospheric levels. This should be balanced by the diffusion of angular momentum upward from the ionospheric layers below the ionosphere.

Knowing the mass outflux $\dot{M} = \eta \dot{M}_+$, we could compute the spindown torque T_z if we knew the Alfvén radius, since according to Weber and Davis (1967), $T_z = \Omega r_a^2 \dot{M}$. We estimate r_a as follows: Assuming that U depends weakly on r, then $N_a/N_m = (r_m/r_a)^2$. We then compute the location where $4\pi N_a M_+/B_D(r_a) = 1$, where B_D is the vacuum dipole field. The result is given by

$$r_a/R_J = [B_D^2/4\pi \varrho_s U_s^2]^{1/4} (R_J/r_m)^{1/2}. \tag{4.1}$$

For the parameters leading to (2.8), $r_a \simeq 35\ R_J$. This result does not contradict observation, since the measurable B_φ was encountered beyond 35 R_J (E. J. Smith, private communication), but it is not clear that these observations support this theory.

The torque T_z then becomes

$$T_z = \Omega_J r_a^2 \dot{M} \simeq 2 \times 10^{23}\ |hB_\varphi/B_r|\ \text{dyn-cm}. \tag{4.2}$$

And the rotational energy invested in the flow, $T_z \Omega_J$ is

$$T_z \Omega \simeq 3.5 \times 10^{19}\ |hB_\varphi/B_r|\ \text{ergs s}^{-1}. \tag{4.3}$$

We note that $T_z\Omega_J/\dot{W}<1$, again posing the question of where the energy in a radial outflow would come from. Assuming Jupiter's moment of inertia $I\simeq10^{49}$ gm cm^{-2}, we may estimate Jupiter's spindown time τ from the above torque to be 3×10^{14} yr, so the torque is cosmogonically insignificant.

In his model of pulsars Sturrock (1971) pointed out that a hydromagnetic torque is communicated to the neutron star by a system of currents which for an aligned dipole, flow in along field lines over the poles; across the field in the neutron star crust, where a $\mathbf{J}\times\mathbf{B}$ force opposing rotation is exerted; and out along the magnetic field line connecting to the Alfvén point. For Jupiter, the cross-field current flows in the ionosphere, not in the planet.

It is worth noting that the field-aligned current flowing out of the equatorward edge of Jupiter's polar cap should connect to field lines in the outflow disk. Using $B\approx4\times10^{-5}$ G and $h\approx$ a few R_J to estimate the magnetic flux in the disk, we find that field-aligned current should leave the ionosphere in an annular ring of a few hundred km thickness poleward of the field lines connecting to the Alfvén point. The field-aligned current density then turns out to be $\approx10^9$ el cm^{-2} s^{-1}, probably large enough to be unstable in Jupiter's topside ionosphere. If such upward field-aligned currents behave as they do at Earth, we would expect them to be carried by beams of energetic electrons and to produce an aurora. This current configuration is sketched in Figure 4.

The angular momentum radiating outward is taken first from the atmospheric neutral at ionospheric levels, which in an aligned rotator can only be replaced by viscous angular momentum diffusion upward from below the ionosphere. Furthermore, the current configuration of Figure 4 makes it clear that only the polar cap atmosphere exerts a torque on the radial outflow. Hines (1974) has estimated the viscous diffusion of angular momentum as follows: The atmospheric angular momentum density is $\varrho r^2\Omega$, where ϱ is the atmospheric mass density, r the distance from the spin axis, and Ω the spin frequency. The angular momentum flux is $D(\mathrm{d}/\mathrm{d}z)(\varrho r^2\Omega)$, where D is a kinematic diffusivity and z denotes altitude. Then, treating it as a thin disk, the torque exterted by the entire polar cap is

$$T_z=2\pi\int_0^{r_{pc}} r\,\mathrm{d}rD(\mathrm{d}/\mathrm{d}z)(\varrho r^2\Omega)\simeq(\pi/2)\,r_{pc}^4D(\mathrm{d}/\mathrm{d}z)(\varrho\Omega), \tag{4.4}$$

where r_{pc} is the radius of the polar cap. We estimate an upper limit to the torque by assuming that at one atmospheric scale height H below the ionosphere $\Omega=\Omega_J$. Then

$$T_z\simeq(\pi/2)\,(\varrho r_{pc}^4D\Omega_J/H). \tag{4.5}$$

We estimate r_{pc} by $(R_J^3/r_a)^{1/2}$.

According to Atreya et al. (1974) sunlight forms an ionosphere with a peak ion density at the level where $\varrho\approx10^{10}$ H atoms cm^{-3}. H is the order of 10 km. While D is highly uncertain, they chose $D\approx10^6$ cm^{-2} s^{-1} as an illustrative value. With these values Equation (4.5) becomes

$$T_z \approx 10^{19} \varrho_{10} D_6 / H \text{ dyn cm}^{-1}, \tag{4.6}$$

where ϱ_{10} is in units of 10^{10} H atoms cm$^{-3} = 1.6 \times 10^{-14}$ g cm^{-3}, D_6 is measured in units of 10^6 cm^2 s^{-1}, and H is normalized to 10 km.

The atmospheric torque, Equation (4.6), is some 4 orders of magnitude smaller than the hydromagnetic torque inferred by assuming a radial outflow solution. The atmospheric torque could exceed the estimate (4.6) if for example, D_6 is large. According to Atreya *et al.* (1974), D is highly uncertain. Alternately, the peak conductivity region of Jupiter's polar cap ionosphere could be formed at a denser layer of Jupiter's atmosphere, where ϱ_{10} is large. If the photochemistry of Atreya *et al.* (1974) prevails, this seems unlikely, since we have already applied their midlatitude model to the polar cap. On the other hand, there could conceivably be energetic electron precipitation to the polar cap. All in all, the four orders of magnitude difference between the hydromagnetic and atmospheric torques makes it an interesting question whether at least the aligned rotator can support corotation.

Another way to perceive the above question, if not the answer to it, is to estimate the moments of inertia of the polar cap ionospheres of Jupiter and the Earth. The moment of inertia of a thin uniform disk is $I = MR^2/4$, a sufficient approximation for our purposes. We estimate M by $\pi R^2 H \varrho$ where H and ϱ are the neutral scale heights and density of the atmosphere where the strongest hydromagnetic coupling occurs. We estimate R by $R_p / \sqrt{L_{pc}}$ where R_p is either R_E or R_J. Thus, over all

$$I = \pi/2 \left(R_p^4 / L_{pc}^2 \right) H \varrho \tag{4.7}$$

and the ratio I_J / I_E is

$$I_J / I_E \simeq (R_J / R_E)^4 \, (H_J / H_E) \, (\varrho_J / \varrho_E) \, (L_{pc}, \hat{E} / L_{pc}, \hat{J})^2. \tag{4.8}$$

Magnetosphere-ionosphere coupling at Earth occurs in the E-region, where there are 10^{13} NO$^+$ atoms cm^{-3}, whereas according to Atreya *et al.* (1974) the coupling region for Jupiter is 10^{10} H atoms cm^{-3}. $H_J \simeq 2 H_E \simeq 10$ km, $(R_J/R_E)^4 \approx 10^4$, and $L_{pc, E} \simeq 10$, $L_{pc, J} \simeq 35$, so that overall $I_J \simeq I_E$. Since the greater scale of Jupiter's magnetosphere suggests that much larger hydromagnetic stresses will be exterted upon its ionosphere and atmosphere than on the Earth's, the equality of the moments of inertia of their polar cap atmospheres leads one to wonder whether Jupiter's atmosphere, acting as a flywheel, can spin-up its magnetosphere for long.

4.3. Coupling of convection of Jupiter's ionosphere and atmosphere

4.3.1. *Convection in the Polar Cap Ionosphere*

Coroniti *et al.* (1973) and Coroniti (1974a) first pointed out the significant effects of the low inertia of Jupiter's atmosphere at ionospheric levels on Jupiter's magnetosphere. Fleshing out an idea originally proposed by Brice and McDonough (1973), they argued that intermittent convection events – in effect, substorms – would couple to planetary scale atmospheric neutral wind modes. These in turn would couple at low Jovian latitudes to fluctuating dynamo electric fields to drive radial diffusion of radiation belt particles with the $D = D_0 L^3$ diffusion coefficient required by the observed profiles of synchrotron radiation. They based their arguments on Jupiter's small ionosphere atmosphere coupling time $\tau = (M_N N_N / M_I N_I)\, v_{in}^{-1}$ (Fedder and Bansks, 1972), where M_N and M_I are the neutral and ion masses and N_N and N_I their number densities. They estimated $\tau_J \approx 40$ min, much shorter than any conceivable convection time. In effect, like the Earth's F-region atmosphere, Jupiter's atmosphere at ionospheric levels would tend to follow the ionospheric convection pattern.

The above argument has led us to ask whether an aligned rotator with Jupiter's ionospheric parameters could have a corotating polar cap atmosphere at ionospheric altitudes when reconnection drives convection through the polar cap. We shall make our point with another *reductio ad absurdem*. Suppose, as sketched in Figure 5, that corotation is rigidly enforced throughout the polar cap. We note that a magnetosheath field line which reconnects at the nose of Jupiter's magnetosphere

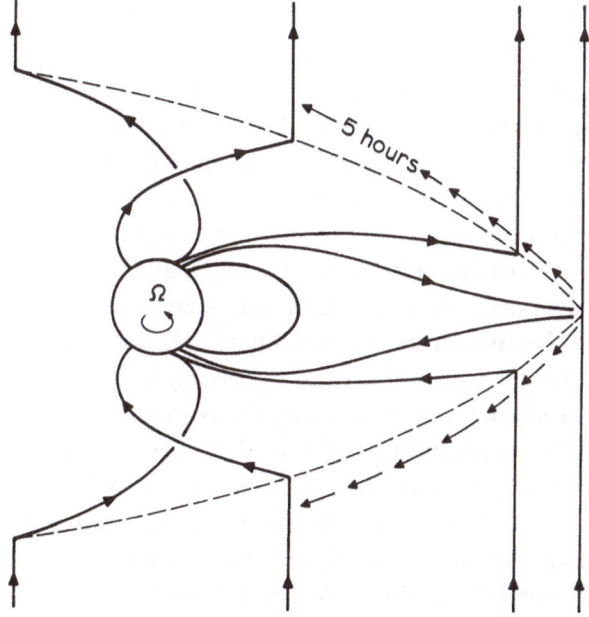

Fig. 5.   Coupling of Solar Wind Torque to Jupiter's Polar Cap Ionosphere and Atmosphere. Suppose the ionospheric feet of all field lines corotated. Then 5 hours after a field-line reconnected at the nose, it would be over the polar cap. It's foot would have rotated to local midnight. Five hours later, the field line would still not have reached the tail reconnection point, yet its foot would be back at local noon. The twist in the field lines corresponds to an Alfvén wave which carries a field-aligned current into the ionosphere with the same sense as in Figure 4. It therefore communicates angular momentum between the ionosphere and solar wind. If Jupiter's polar cap atmosphere has a small moment of inertia and viscous coupling to the lower atmosphere is weak, Jupiter's polar cap atmosphere will not corotate.

will take 5 hours to travel 100 R_J at a speed of 400 km s^{-1}. At this point, the magneto-sheath end of the field line would be over the polar cap while its ionospheric end would have corotated to local midnight. Five hours later, the magnetosheath end would still not have reached the tail reconnection point, while the ionospheric end would have corotated back to local noon. If such a picture of convection were correct, the foot of the field line would trace out a cycloidal pattern as it progresses across the polar cap. But then, the ionospheric electric field would have to be double-valued. This dilemma is resolved by the twist in the field lines above the polar cap – an Alfvén wave carrying field-aligned current whose function is to communicate angular momentum between the ionosphere and magnetosheath. This torque spins down the ionosphere and atmosphere and spins up the magnetosheath flow. In view of Jupiter's low atmospheric inertia, we suspect that its polar cap atmosphere should spin about once in a characteristic convection time.

4.3.2. *Enforcement of Corotation*

A basic assumption underlying Brice's models of the plasmapauses of Jupiter and the Earth is that corotation is enforced at the feet of the field lines. Assuming that the convective flow in the tail is symmetric around the local midnight meridian in the distant geomagnetic tail, convection at asymptotically long distances carries no net angular momentum towards the Earth. However, in the Brice solutions more mass and angular momentum flows past the dawn meridian plane beyond the plasma-pause than flows past the evening meridian. The planet's ionosphere must have exerted a torque on the flow at this point. If the dayside magnetopause were very far away the difference in angular momentum flowing past dawn and evening would even-tually be restored to the ionosphere as the flow again approaches symmetry about the local noon meridian plane. In this case, convection would exert no net torque on the ionosphere. However, the flux tubes cross the magnetopause into the magnetosheath before complete symmetry is re-attained, and since a significant fraction of the difference between the angular momentum fluxes at dawn and evening is transferred to the magnetosheath, angular momentum is probably lost from this system, although some of the angular momentum acquired by the magnetosheath could be restored to the Earth again as the magnetosheath plasma flows over the polar cap.

We will estimate the Earth's angular momentum loss to be the difference between that flowing past dawn and evening in the Brice plasmapause model. We again assume the magnetic field to be an aligned undistorted dipole and the convection electric field to be spatially uniform. A dipole field line has the equation $r = LR_p \cos^2 \theta$, where L is McIlwain's (1961) L-shell parameter, R_p is either planet's radius, and θ is the magnetic latitude. The distance from the spin axis $r_\perp$ at any point on a dipolar field line is $r_\perp = r \cos \theta = LR_p \cos \theta^3$. The azimuthal velocity v_φ of any point on the field line is related to its equatorial value V_φ by $v_\varphi = V_\varphi(L) \cos^3 \theta$. The angular momentum flux of any element of mass on a given flux tube is given by $\varrho LR_p V_\varphi^2 \cos^9 \theta$ where ϱ is the mass density. The element of meridional plane area of a flux tube dA is given by d$A = L$ d$R_p^2 \cos^4 \theta$ dθ, so that the angular momentum flux d$\dot{J}$ carried by a flux tube between L and $L+dL$ is

$$d\dot{J}/dL = \varrho L^2 R_p^3 V_\varphi^2 \int_0^{\theta^*} \cos^{13}\theta \, d\theta, \tag{4.9}$$

where $\theta^* \simeq \cos^{-1}(1/\sqrt{L})$ denotes the latitude where the flux tube hits the atmosphere.

By entirely similar reasoning the mass flux $\dot{M}$ flowing between L and $L+dL$ is given by

$$d\dot{M}/dL = \varrho(L) \, v_\varphi(L) \, R_p^2 L \int_0^{\theta^*} \cos^7\theta \, d\theta. \tag{4.10}$$

In both (4.9) and (4.10) we assume that the particles are isotropic in pitch angle so that the mass density $\varrho(L)$ is independent of magnetic latitude. The total angular momentum and mass fluxes flowing between any two L_1 and L_2 are

$$\dot{J} = R_p^3 \int_{L_{13}}^{L_2} \varrho(L) \, V_\varphi^2 L^2 \, dL_\varphi \int_0^{\theta^*} \cos^{13}\theta \, d\theta \tag{4.11}$$

$$\dot{M} = R_p^2 \int_{L_1}^{L_2} \varrho(L) \, v_\varphi L \, dL \int_0^{\theta^*} \cos^7\theta \, d\theta. \tag{4.12}$$

For the Brice model of convection, the azimuthal equatorial speed in the dawn $(+)$ and evening $(-)$ meridians is

$$v_\varphi = c/B\left[E_c \pm \frac{\Omega_p B_0 R_p^3}{cr^2}\right] = (c/B_0) \, E_c L^3 \pm \Omega R_p L. \tag{4.13}$$

The net torque is

$$\dot{T}_z = R_p^3 \int_0^{\theta^*} \cos^{13}\theta^* \left\{ \left[\int_{0.4L_0}^{L_m} \varrho(L) \left[\frac{cE_c}{B_0} L^3 + \Omega R_p L \right]^2 L^2 \, dL \right.\right.$$

$$\left.\left. - \int_{L_0}^{L_m} \varrho(L) \left[\frac{cE_c}{B_0} L^3 - \Omega R_p L \right]^2 L^2 \, dL \right] \right\}, \tag{4.14}$$

where we have neglected a weak dependence of θ^* on L. L_0 is the L-shell of the evening plasmapause and L_m is that of the magnetopause in the dawn-evening meridian plane.

For the Earth, $L_m \gg L_0$, and the convection speed greatly exceeds the corotation speed at the magnetopause. Making these approximations we find

$$T_z \simeq R_E^3 \int\limits_0^{\theta\theta} \cos^{13}\theta \left\{ \int\limits_0^{L_m} 4\frac{cE_c}{B_0} \Omega R_E L^6 \varrho(L)\, dL \right\}, \qquad (4.15)$$

where we have noted that the integrals in (4.14) depend weakly on their lower limit.

We model $\varrho(L)$ by assuming that in the distant tail all the convecting flux tubes have the same volume and density and therefore total mass. If, further, no mass is lost as they convect towards the Earth, the density in the dawn or evening meridians will vary inversely as the volume of the flux tube, i.e., $\varrho = \varrho_0 L^{-4}$ where ϱ_0 is a constant. With this assumption

$$T_z \simeq \frac{4\varrho_0}{3} \frac{cE_c}{B_0} R_E^4 L_m^3 \Omega \int\limits_0^{\theta*} \cos^{13}\theta\, d\theta. \qquad (4.16)$$

With the same approximations that led to (4.16), the total mass flux carried by convection is approximately

$$\dot{M} \simeq 2\varrho_0 R_E^2 \frac{cE_c}{B_0} L_m. \qquad (4.17)$$

So that, finally

$$T_z \simeq (2\alpha/3)\, \Omega r_m^2 \dot{M}, \qquad (4.18)$$

where

$$\alpha = \int\limits_0^{\theta*} \cos^{13}\theta\, d\theta \Big/ \int\limits_0^{\theta*} \cos^7\theta\, d\theta \qquad (4.19)$$

and $r_m \equiv L_m R_E$. Because of the many approximations made it is proper to take only the basic scaling $T_z \sim \Omega r_m^2 \dot{M}$, and not the numerical factor $2\alpha/3$ seriously

For the Earth, $\dot{M} = 200$ g s^{-1}, corresponding to $\dot{N} = 10^{26}$ protons s^{-1}. Then, using $\Omega_E = 7 \times 10^{-5}$ rad s^{-1} and $r_m \simeq 10^{10}$ cm, $T_z = 1.5 \times 10^{18}$ dyn-cm, corresponding to an energy dissipation rate $T_z \Omega_E \simeq 10^{14}$ ergs s^{-1}. Our estimate (4.13) does not conflict with either Coleman's (1971) or Hirshberg's (1972), both of whom estimated upper limits to the torque of the order 10^{20-21} dyn-cm. Hines (1974) has estimated the torque produced by viscous diffusion in the Earth's atmosphere to be of order $10^{20}\, \Delta\Omega/\Omega$ dyn-cm, where $\Delta\Omega$ is the difference in angular velocity between the Earth and the neutrals at E-region levels. Combining Hines' estimate with (4.19), we conclude that $\Delta\Omega/\Omega \leqslant -1\%$, so that the Earth's atmosphere can enforce corotation on the convective return flow in Brice's Earth plasmapause calculation.

Can a plasmapause be formed at Jupiter? A precise evaluation of this question cannot yet be formulated. Equation (4.13) cannot be correct, since two of its basic assumptions – undistorted dipole field and sub-Alfvénic flow everywhere – are violated. Nonetheless, if a plasmapause is formed, the torque should be expressed in the form $T_z = \Omega_J r^{*2} \dot{M}$ where $R_J < r^* < r_m$. Let us estimate the convection mass flux $\dot{M}$ by $\delta \dot{M}_s$, where $\dot{M}_s = M_H N_s U_s \pi r_m^2 \simeq 2 \times 10^{30}$ protons s^{-1}, the solar wind mass flux across the cross-section of Jupiter's magnetosphere. We compute the lower estimate to the torque

$$T_z > \Omega_J R_J^2 \delta \dot{M}_s = 3 \times 10^{22} \delta \text{ dyn cm}^{-1}. \tag{4.20}$$

At Earth, the trapping factor $\delta \simeq 10^{-3}$–10^{-2}. Thus, unless $\delta < 10^{-3}$, even the lower estimate torque exceeds the atmospheric torque (4.6).

We do not feel that the question of whether or not Jupiter's atmosphere can enforce corotation can be resolved at present. In his accompanying review article, Coroniti (1974b) suggests that energetic electron precipitation could significantly modify the structure of the ionosphere, and, therefore, the basis on which the atmospheric torque (4.6) was estimated. Until model ionospheres including electron precipitation are computed we have no solid foundation on which to base an estimate of atmospheric torque. A particularly urgent need is to evaluate the effects of soft electron precipitation from the solar wind on Jupiter's polar cap ionosphere. Such computations have yet to be carried out even for Earth, except for the polar cusp (Kennel and Rees, 1972).

We *can* ask, however, what would happen if no corotation were imposed. Then the convection streamlines from the tail would be straight lines towards the Sun until that point where, according to Bernouilli's Law, the magnetic pressure of the dipole deflects the flow. While the flow streamlines reach the dayside magnetopause, the energetic electrons from Jupiter's plasmasheet might have a sharp inner boundary, as they do at Earth. We may estimate the Jovicentric distance of the midnight meridian inner boundary following Kennel (1969). Should microscopic plasma turbulence keep the electron pitch angle isotropic, the electron precipitation lifetime will approach the electron minimum lifetime, T_M. According to Kennel and Petschek (1966), $T_M = = 2T_B L^3$ where T_B is the electron quarterbounce time, $2L^3$ is the 'mirror ratio' for dipolar field lines, and L is the McIlwain (1961) L-parameter. The inner boundary of the plasma sheet is formed at that point where T_M first equals the flow characteristic time T_F. Beyond this point, electrons are lost from the flow before the tubes of force can cross a scale length.

We shall estimate T_B by $LR_J/v_\parallel$ where $v_\parallel$ is the electron velocity component perpendicular to the magnetic field. We estimate T_F to be the dipolar magnetic field gradient length $LR_J/3$ divided by the convection speed cE_c/B. Assuming that at the inner boundary the plasma pressure does not distort the dipolar field significantly, we may estimate the L-shell of the inner boundary L_B:

$$L_B = \left[\frac{v_\parallel B_0}{6\beta U_s B_s} \right]^{1/6} = 19 \left(\frac{v_\parallel}{\beta c} \right)^{1/6} R_J. \tag{4.21}$$

Coroniti's radiation belt model (1974a) assumed that radial diffusion carried electrons from a distant plasmapause to the synchrotron radiation region near the planet. An inner boundary was formed at $L \approx 20$ where the minimum lifetime matched the radial diffusion scale time. However, (4.21) indicates that if corotation were not a factor, convection could transport energetic electrons to $L = 20$ directly. Of course, the estimate (4.21) was based on the assumption that electrons precipitate at their maximum rate. As argued previously, such precipitation could affect the coupling of atmospheric rotation to the magnetosphere.

4.4. SUMMARY

An aligned rotator with an ionosphere similar to that of Atreya *et al.* (1974) would have difficulty enforcing corotation, due to the very low inertia of the atmosphere at ionospheric altitudes. Thus, one basic assumption underlying both radial outflow models and most convection models may be questionable.

5. Discussion

We have treated Jupiter's magnetosphere as an astrophysicist would when faced with incomplete observations and correspondingly undeveloped theoretical conceptualizations. We have imposed various simplified models on the data in order to identify broad areas of consistency or conflict between theory and experiment. Our major goal has been to try to decide between a radial outflow pulsar-like model of Jupiter's magnetosphere and an Earthlike convection model.

In Section 2, we imposed on the data a simple super-Alfvénic radial outflow model for an aligned rotating dipole similar to those constructed by Mestel (1968) for non-relativistic flows around rotating magnetic stars and those by Michel (1969) for relativistic flows away from rapidly rotating magnetized neutron stars. Our aim was to estimate the internal particle and energy sources required to drive a super-Alfvénic outflow consistent with the Pioneer 10 measurements reported in *Science*. These came out to be stringent: 10^{28} particles s^{-1} and at least 10^{21} ergs s^{-1}. It is not obvious that photoelectrons from Jupiter's ionosphere can be an adequate source of particles. On the other hand, secondary electrons from the precipitation of energetic electrons to Jupiter's atmosphere could conceivably be a more potent source – a source we did not estimate. However, this does beg the question of what magnetospheric processes energize the precipitating electrons, which would be important for a final self-consistent treatment. Jupiter's satellites could also be a significant particle source, though Io's neutral particle ring may be inadequate. None of the above arguments address the question of where the outflowing particles get their energy, which we estimated to be of order 50 keV. It must be remembered that our estimates depend completely upon interpreting the measured B_φ/B_r as the garden-hose angle of a super-Alfvénic outflow. At present, the information on B_φ/B_r near the magnetopause is sparse. However, if $|B_\varphi/B_r| < 1$ and if the model applies at all, the outflowing particles would have high energy and low density.

Even if Jupiter had particle and energy sources strong enough to drive a radial flow, our estimates indicated that they would at best be comparable with the solar wind particle and energy fluxes across the frontal area of Jupiter's magnetopause. This suggested that the solar wind could be a significant source of particles and energy. This conclusion was bolstered by our estimate of the energy input due to reconnection at Jupiter's magnetopause – again roughly 10^{21} ergs s^{-1}. This raised a class of questions which we did not address specifically. A whole spectrum of hydromagnetic models exists between the two extremes of a pure radial outflow, with all energy and mass fluxes provided by Jupiter itself or its inner magnetosphere, and a pure convection model, where all the energy and particles come from the solar wind. Such mixed model magnetospheres offer many challenging theoretical problems which never have been addressed. For example, if reconnection as well as radial outflow is important, radial outflow solutions with magnetic field components normal to the flow direction are needed. Furthermore, the flows could well be super-Alfvénic, so that one might expect shocks and other discontinuities within the magnetosphere where the convection and radial flows clash. If we could understand such flows, we might learn something about X-ray sources in binary orbit, like Hercules X-1. It is commonly thought that these may be spinning magnetized neutron stars which gain energy by accretion from the atmosphere or stellar wind of its stellar companion (e.g., Davidson and Ostriker, 1973). Since pulsars not in binary orbit are known to be energy and particle emitters, the possibility exists that a mixed model might apply to the X-ray sources.

We did consider qualitatively the other extreme case of a pure convection magnetosphere, concentrating primarily on weakening the objections previously held by the theoretical community against an important role for convection in Jupiter's magnetosphere. These objections were:

(1) The plasmapause, according to Brice and Ioannidis (1970), was so distant that convection could not transport plasma anywhere near the planet. It could not, therefore, significantly energize particles. Furthermore, the computed plasmapause was near or beyond the position of the magnetopause expected prior to Pioneer 10 encounter. There was little or no room within the magnetosphere to return magnetic flux to the dayside magnetopause as is required by a reconnection model.

(2) After Pioneer 10 encounter, it appeared that the observed radially extended magnetic field observed on Jupiter's outer magnetosphere simply didn't look like the Earth's magnetic field. Besides the magnetopause was too far out, 100 R_J and more, whereas Brice and Ioannidis (1970) and Kennel (1973) had estimated no more than 50 R_J. Jupiter's outer magnetosphere, at first glance, does resemble what one expects from a radial outflow model.

However, in Section 3, we compared the plasma densities required to make the convection flows sub-Alfvénic at the plasmapauses of Earth and Jupiter, as is implicitly required by conventional plasmapause models. We concluded that it is likely that Jupiter's convection flow will be super-Alfvénic in its outer magnetosphere. In this case, the dynamic pressure of the convection flow would exceed that of the magnetic field at least near the dipole equator, and one might well expect a radially extended magnetic field in the outer magnetosphere. In addition, the magnetopause would be pushed out by convection. We then noted that OGO-1 may have observed Jupiter-like Earth magnetopauses near local dawn during substorms. If the analogy is a proper one, then the observed variability in Jupiter's magnetopause location might be due in part to substorms or other variable convection events.

In Section 4, we discussed one basic presumption underlying nearly all theories of Jupiter's magnetosphere: that corotation be enforced. Again, we investigated first the simplest possible model – an aligned rotator. In addition, we used published models of Jupiter's ionosphere and magnetosphere to estimate the coupling of angular momentum between atmosphere, ionosphere, and magnetosphere. The conclusion of this straightforward, if oversimplified, procedure seems clear: The angular momentum flux which can diffuse upward through Jupiter's polar cap atmosphere seems insufficient to impose corotation upon a radial outflow with parameters similar to those in Section 2, or upon the convective return flow of Section 3 to form a plasmapause. Goertz *et al.* (1974) argue that the observed system III longitudes of the appearances of the peak electron flux regions in the magnetodisk can be explained by a slippage of the ionospheric feet of magnetospheric field lines with respect to Jupiter, consistent with a weak coupling between planet and magnetosphere.

If there were no corotation imposed at all – another extreme limit – we estimated that convection could carry plasmasheet electrons to about $L \simeq 20$ where they might form a precipitation inner boundary similar to that of Earth (Vasyliunas, 1968; Kennel, 1969). In Coroniti's (1974a) radiation belt model, a similar inner boundary was formed at $L = 20$ where radial diffusion and electron precipitation have similar scale times. The similarity in results is no accident, because in Coroniti's (1974a) model radial diffusion is driven by sporadic convection events, and he specifically presumed that convection would carry the upper atmosphere around with it, so that the electric field would penetrate to low Jovian L-shells.

Section 4 should not be allowed to blind us to the salient fact that the Pioneer 10 energetic particle experiments observed a 'flapping' of the high intensity flux region of the magnetodisk with roughly a ten hour, and not 20, 30, or 100 h period. Thus, some aspect of Jupiter's rotation is enforced on the magnetosphere. At our present primitive level of theoretical understanding, we do not know why. Perhaps the current ionospheric models, which do not take energetic electron precipitation into account, underestimate the viscous coupling of angular momentum to the atmosphere. On the other hand, it is also possible that the assumption of an aligned rotator is at fault. In pulsar theory, there is a fundamental physical difference between aligned and oblique rotators. An aligned rotator can only lose angular momentum hydromagnetically by emitting particles (Goldreich and Julian, 1969; Mestel, 1969). On the other hand, an oblique rotator can lose angular momentum even in vacuum (Pacini, 1967). Since rotation now induces a time varying magnetic field, it generates magnetic dipole radiation. Similarly, Jupiter's oblique dipole may produce an electric field in its ionosphere which drives field-aligned currents causing the magnetodisk to flap. Angular momentum would then be carried off by Alfvén waves propagating through the outer magnetosphere. Thus, Section 4 raised interesting questions without settling them. Nonetheless, we believe that Section 4 indicates that there is a commonality of interest between Jupiter's atmospheric and ionospheric communities, on the one hand, and its fields and particles community on the other. The dynamics of Jupiter's high latitude upper atmosphere may be controlled hydromagnetically and the structure of its ionosphere may be significantly affected by electron precipitation. Conversely, we may not be able to understand the hydromagnetics of Jupiter's magnetosphere until the nature of the boundary conditions at the ionosphere are elucidated. Should Saturn or the other outer planets also have magnetospheres, energetic electron precipitation might also be important for their ionospheres, since the solar photon flux is even smaller than at Jupiter.

How are we to decide between the convection and the radial outflow models? Pioneers 10 and 11 will be unable to make a simple yet decisive test. For example, a directional plasma detector, sensitive in the energy range between 1 keV – the co-rotation energy at the inner edge of the magnetodisk – and 50 keV – the corotation energy at the magnetopause – could determine whether the flow is antisolar or from the planet at the dawn meridian. As it is, our best information may come from much more elaborate versions of what was done in order of magnitude fashion in this paper: comparison of magnetic field measurements with hydromagnetic models. Several simple signatures of convection ought to appear in the magnetic field data. For example, B_φ/B_r could have a sign opposite to that of the conventional garden-hose field. Here, there are two cases. The convection speed could exceed the local corotation speed but be less the Alfvén speed. An onset of rapid convection would then bend the field line towards the Sun. The bend would then propagate as an Alfvén wave down the field-line to the ionosphere where it would exert a stress on the atmosphere. We would expect that the anomalous B_φ/B_r would persist for times comparable with the effective ionosphere-atmosphere coupling time – a few hours at most. On the other hand, where the convection speed exceeds both the Alfvén and corotation speed, a sunward B_φ component might persist for the duration of the convection event. In a reconnection magnetosphere, we expect that tangential magnetopause stresses will sweep the field lines back so that in the local midnight-noon sector, B_φ/B_r will have the same sign as the garden-hose field sufficiently near the magnetopause. Goertz et al. (1974) have reported that B_φ/B_r is often though not predominantly anti-gardenhose, however, they are still somewhat uncertain about their data reduction procedure for these cases. In any case, study of sporadic B_φ/B_r anomalies might illuminate the problem of convection. In addition, one might try to establish that magnetopause motions occur with no change in solar wind dynamic pressure and/or are correlated following a suitable delay with northward switches of the solar wind magnetic field. Similarly, convection might cause anomalies in the time of magnetodisk crossings.

The absence of plasma wave detectors on Pioneers 10 and 11 means that we really do not know whether precipitation of electrons to the atmosphere occurs. Observation of one whistler emission would have settled that. With suitable spectral information, and the cold plasma density inferred by detection of the plasma frequency we could estimate the fluxes and precipitation rates of electrons in the 1–100 keV range currently not measured. A radial profile of whistler amplitude would enable us to infer electron precipitation fluxes with sufficient accuracy to permit believable new ionospheric models to be computed.

Despite some shortcomings, Pioneer 10, and the excitement induced by the preparations for, and the fact of, our first encounter with Jupiter have provided us a good start on a research program which will certainly be vigorous until the end of the century. For the stakes are high. Not only is Jupiter's magnetosphere intrinsically interesting, but it may have already shed a little light on the Earth's magnetosphere. Because of Jupiter's rotation and large magnetic moment, we are convinced its magnetosphere will be a useful astrophysical analog. It's just that, right now – at the beginning – we don't yet know what kind of anlog it will turn out to be.

Acknowledgements

We are pleased to acknowledge interesting discussions with F. L. Scarf, P. Morrison, R. Carlson, G. Siscoe, J. Maggs, A. Cameron, P. Coleman, M. McElroy, F. Busse, and L. Knopoff. This work was supported by NASA grant NGL 05-007-190.

References

Arnoldy, R. L.: 1971, *J. Geophys. Res.* **76**, 5189.
Atreya, S. K., Donohye, T. M., and McElroy, M. B.: 1974, *Science* **184**, 154.
Aubry, M. P., Russell, C. T., and Kivelson, M. G.: 1970, *J. Geophys. Res.* **75**, 7018.
Brice, N. M. and Ioannidis, G. A.: 1970, *Icarus* **13**, 173.
Brice, N. and McDonough, T. R.: 1973, *Icarus* **18**, 206.
Burck, J. L.: 1974, NASA GFSC preprint X-646-73-390.
Coleman, P. J. Jr.: 1971, *J. Geophys. Res.* **76**, 3800.
Coroniti, F. V.: 1974a, *Astrophys. J. Suppl.* **27**, 261.
Coroniti, F. V.: 1975, this volume, p. 391.
Coroniti, F. V. and Kennel, C. F.: 1972, *J. Geophys. Res.* **77**, 3361.
Coroniti, F. V. and Kennel, C. F.: 1973, *J. Geophys. Res.* **78**, 2837.
Coroniti, F. V., Kennel, C. F., and Thorne, R. M.: 1973, *Trans. Am. Geophys. Union* **54**, 446.
Davidson, K. and Ostriker, J. P.: 1973, *Astrophys. J.* **179**, 585.
Dungey, J. W.: 1961, *Phys. Rev. Letters* **6**, 47.
Dungey, J. W.: 1965, *J. Geophys. Res.* **70**, 1753.
Fedder, J. A. and Banks, P. M.: 1972, *J. Geophys. Res.* **77**, 2328.
Frank, L. A.: 1954, *55th Ann. Mtq., Am. Geophys. Union.*
Goertz, C. K., Northrup, T. G., and Thomsen, M. F.: 1974, Dept. Phys. Astron., Univ. Iowa, preprint.
Goldreich, P. and Julian, W. H.: 1969, *Astrophys. J.* **157**, 869.
Gurnett, D. A. and Frank, L. A.: 1973, *J. Geophys. Res.* **78**, 145.
Heppner, J. P., Sugiura, M., Skillman, T. L., Ledley, B. G., and Campbell, M.: 1967, *J. Geophys. Res.* **72**, 5417.
Hill, T. W., Dessler, A. J., and Michel, F. C.: 1974, *Geophys. Res. Letters* **1**, 1.
Hines, C. P.: 1974, *J. Geophys. Res.* **79**, 1543.
Hirshberg, J.: 1972, *J. Geophys. Res.* **77**, 4855.
Ioannidis, G. and Brice, N. M.: 1971, *Icarus* **14**, 360.
Kennel, C. F.: 1969, *Rev. Geophys. Space Phys.* **7**, 379.
Kennel, C. F.: 1973, *Space Sci. Rev.* **14**, 511.

Kennel, C. F.: 1974, *Comments Astrophys. Space Phys.*, in press.
Kennel, C. F. and Coroniti, F. V.: 1974, *Astrophys. J. Letters*, submitted.
Kennel, C. F. and Petscheck, H. E.: 1966, *J. Geophys. Res.* **71**, 1.
Kennel, C. F. and Rees, M. H.: 1972, *J. Geophys. Res.* **77**, 2294.
Lin, R. P. and Anderson, K. A.: 1966, *J. Geophys. Res.* **71**, 4213.
McIlwain, C. D.: 1961, *J. Geophys. Res.* **66**, 3681.
Mestel, L.: 1968, *Monthly Notices Roy. Astron. Soc.* **138**, 359.
McPherron, R. L.: 1970, *J. Geophys. Res.* **75**, 5592.
Michel, F. C.: 1969, *Phys. Rev. Letters* **23**, 247.
Michel, F. C.: 1971, *Comments Astrophys. Space Phys.* **3**, 227.
Michel, F. C. and Sturrock, P. A.: 1974, *Planetary Space Sci.*, submitted.
Morrison, P.: 1969, *Astrophys. J.* **157**, L73.
Mozer, F. S., Gonzales, W. D., Bogott, F., Kelley, M. C., and Schutz, S.: 1974, *J. Geophys. Res.* **79**, 56.
Nishida, A. and Nagayama, N.: 1973, *J. Geophys. Res.* **78**, 3782.
Pacini, F.: 1967, *Nature* **216**, 567.
Piddington, J. H.: 1969, *Cosmic Electrodynamics*, Wiley-Interscience.
Siscoe, G. L. and Crooker, N. U.: 1974, MIT Preprint CSR-P-74-115.
Smith, E. J., Davis, L., Jr., Jones, D. E., Colburn, D. S., Coleman, P. J., Dyal, P., and Sonett, C. P.: 1974, *Science* **183**, 305.
Sturrock, P. A.: 1971, *Astrophys. J.* **164**, 529.
Vasyliunas, V. M.: 1968, *J. Geophys. Res.* **73**, 2839.
Weber, E. J. and Davis, L., Jr.: 1967, *Astrophys. J.* **148**, 217.
Wolfe, J. H., Collard, H. R., Mihalov, J. D., and Intriligator, D. S.: 1974, *Science* **183**, 303.

INDEX OF NAMES

ASTROPHYSICS AND SPACE SCIENCE LIBRARY

Edited by

J. E. Blamont, R. L. F. Boyd, L. Goldberg, C. de Jager, Z. Kopal, G. H. Ludwig, R. Lüst,
B. M. McCormac, H. E. Newell, L. I. Sedov, Z. Švestka, and W. de Graaff

on the Symposium on the Magellanic Clouds, held in Santiago de Chile, March 1969, on the Occasion of the Dedication of the European Southern Observatory. 1971, XII + 189 pp.

24. B. M. McCormac (ed.), *The Radiating Atmosphere. Proceedings of a Symposium Organized by the Summer Advanced Study Institute, held at Queen's University, Kingston, Ontario, August 3–14, 1970.* 1971, XI + 455 pp.

25. G. Fiocco (ed.), *Mesospheric Models and Related Experiments. Proceedings of the 4th ESRIN-ESLAB Symposium, held at Frascati, Italy, July 6–10, 1970.* 1971, VIII + 298 pp.

26. I. Atanasijević, *Selected Exercises in Galactic Astronomy.* 1971, XII + 144 pp.

27. C. J. Macris (ed.), *Physics of the Solar Corona. Proceedings of the NATO Advanced Study Institute on Physics of the Solar Corona, held at Cavouri-Vouliagmeni, Athens, Greece, 6–17 September 1970.* 1971, XII + 345 pp.

28. F. Delobeau, *The Environment of the Earth.* 1971, IX + 113 pp.

29. E. R. Dyer (general ed.), *Solar-Terrestrial Physics/1970. Proceedings of the International Symposium on Solar-Terrestrial Physics, held in Leningrad, U.S.S.R., 12–19 May 1970.* 1972, VIII + 938 pp.

30. V. Manno and J. Ring (eds.), *Infrared Detection Techniques for Space Research, Proceedings of the 5th ESLAB-ESRIN Symposium, held in Noordwijk, The Netherlands, June 8–11, 1971.* 1972, XII + 344 pp.

31. M. Lecar (ed.), *Gravitational N-Body Problem. Proceedings of IAU Colloquium No. 10, held in Cambridge, England, August 12–15, 1970.* 1972, XI + 441 pp.

32. B. M. McCormac (ed.), *Earth's Magnetospheric Processes. Proceedings of a Symposium Organized by the Summer Advanced Study Institute and Ninth ESRO Summer School, held in Cortina, Italy, August 30–September 10, 1971.* 1972, VIII + 417 pp.

33. Antonin Rükl, *Maps of Lunar Hemispheres.* 1972, V + 24 pp.

34. V. Kourganoff, *Introduction to the Physics of Stellar Interiors.* 1973, XI + 115 pp.

35. B. M. McCormac (ed.), *Physics and Chemistry of Upper Atmospheres. Proceedings of a Symposium Organized by the Summer Advanced Study Institute, held at the University of Orléans, France, July 31–August 11, 1972.* 1973, VIII + 389 pp.

36. J. D. Fernie (ed.), *Variable Stars in Globular Clusters and in Related Systems. Proceedings of the IAU Colloquium No. 21, held at the University of Toronto, Toronto, Canada, August 29–31, 1972.* 1973, IX + 234 pp.

37. R. J. L. Grard (ed.), *Photon and Particle Interaction with Surfaces in Space. Proceedings of the 6th ESLAB Symposium, held at Noordwijk, The Netherlands, 26–29 September, 1972.* 1973, XV + 577 pp.

38. Werner Israel (ed.), *Relativity, Astrophysics and Cosmology. Proceedings of the Summer School, held 14–26 August, 1972, at the BANFF Centre, BANFF, Alberta, Canada.* 1973, IX + 323 pp.

39. B. D. Tapley and V. Szebehely (eds.), *Recent Advances in Dynamical Astronomy, Proceedings of the NATO Advanced Study Institute in Dynamical Astronomy, held in Cortina d'Ampezzo, Italy, August 9–12, 1972.* 1973, XIII + 468 pp.

40. A. G. W. Cameron (ed.), *Cosmochemistry. Proceedings of the Symposium on Cosmochemistry, held at the Smithsonian Astrophysical Observatory, Cambridge, Mass., August 14–16, 1972.* 1973, X + 173 pp.

41. M. Golay, *Introduction to Astronomical Photometry.* 1974, IX + 364 pp.

42. D. E. Page (ed.), *Correlated Interplanetary and Magnetospheric Observations. Proceedings of the 7th ESLAB Symposium, held at Saulgau, W. Germany, 22–25 May, 1973.* 1974, XIV + 662 pp.

43. Riccardo Giacconi and Herbert Gursky (eds.), *X-Ray Astronomy.* 1974, X + 450 pp.

44. B. M. McCormac (ed.), *Magnetospheric Physics. Proceedings of the Advanced Summer Institute, held in Sheffield, U.K., August 1973.* 1974, VII + 399 pp.

45. C. B. Cosmovici (ed.), *Supernovae and Supernova Remnants. Proceedings of the International Conference on Supernovae, held in Lecce, Italy, May 7–11, 1973.* 1974, XVII + 387 pp.

46. A. P. Mitra, *Ionospheric Effects of Solar Flares.* 1974, XI + 294 pp.

50. Zdeněk Kopal and Robert W. Carder, *Mapping of the Moon.* 1974, VIII + 237 pp.